Third Edition

Introductory Algebra

Third Edition

Introductory Algebra

D. Franklin Wright
Cerritos College

Bill D. New
Cerritos College

Wm. C. Brown Publishers

Book Team

Editor *Earl McPeek*
Developmental Editor *Theresa Grutz*
Designer *K. Wayne Harms*
Production Editor *Eugenia M. Collins*
Art Editor *Gayle A. Salow*
Visuals Processor *Jodi Wagner*

WCB

Wm. C. Brown Publishers

President *G. Franklin Lewis*
Vice President, Publisher *George Wm. Bergquist*
Vice President, Publisher *Thomas E. Doran*
Vice President, Operations and Production *Beverly Kolz*
National Sales Manager *Virginia S. Moffat*
Advertising Manager *Ann M. Knepper*
Marketing Manager *David F. Horwitz*
Editor in Chief *Edward G. Jaffe*
Manager of Visuals and Design *Faye M. Schilling*
Production Editorial Manager *Colleen A. Yonda*
Production Editorial Manager *Julie A. Kennedy*
Publishing Services Manager *Karen J. Slaght*

A study guide and solutions manual are available in your college
bookstore. The titles are: *Student's Study Guide to accompany
Introductory Algebra* 3e and *Student's Solutions Manual to accompany
Introductory Algebra* 3e. They have been written to help you study,
review, and master the course material. Ask the bookstore manager to
order a copy for you if it is not in stock.

Cover photo by Mike Mitchell, Cedar Rapids, Iowa

Library of Congress Catalog Card Number: 89–85951

ISBN 0–697–05945–6

Printed in the United States of America by Wm. C. Brown Publishers,
2460 Kerper Boulevard, Dubuque, IA 52001

10 9 8 7 6 5 4 3 2 1

CONTENTS

The Purpose

Introductory Algebra (third edition) provides a smooth transition from arithmetic to the more abstract skills and reasoning abilities developed in a beginning algebra course. With feedback from users, insightful comments from many reviewers, and an active editorial staff, we have confidence that students and instructors alike will find that this text is indeed a superior teaching and learning tool.

We have assumed only that the students have basic arithmetic skills and no previous knowledge of algebra. Chapter 1 begins with a review of whole numbers, fractions, decimals, and percents while integrating the algebraic concepts of variables and solutions to equations. Geometric figures and related formulas are introduced to provide practice with algebraic concepts and as reference for word problems throughout the text. The similarities between arithmetic and algebra are stressed whenever possible to give the students confidence with familiar ideas and understanding of the nature of algebra.

The Style

The style of the text is informal and nontechnical while maintaining mathematical accuracy. Each topic is developed in a straightforward step-by-step manner. Each section contains many carefully developed and worked out examples to lead the students successfully through the exercises and prepare them for examinations. Whenever appropriate, information is presented in list form for organized learning and easy reference. Common Errors are highlighted and explained so that students can avoid such pitfalls and better understand the correct corresponding techniques. Practice Problems with answers are provided in almost every section to allow the students to "warm up" and to provide the instructor with immediate classroom feedback.

New to the Third Edition

Chapter Introductions

Each chapter now begins with an introduction previewing the chapter coverage and preparing students for the new topics.

Section Objectives

Each section begins with a listing of objectives, clearly identifying the skills to be learned.

Practice Problems

Practice Problems and Calculator Problems can now be found in almost every section.

Chapter Summaries

Each chapter summary has been revised and reorganized. The summaries are now subdivided into Key Terms and Formulas, Properties and Rules, and Procedures.

Cumulative Reviews

Beginning with Chapter 2, each chapter contains a Cumulative Review of topics chosen from previous chapters to continually refresh the students' memories, maintain skill levels, and help the students understand the bulding-block nature of algebra.

Reorganization

Most sections have been rewritten and reorganized so that the explanations are more complete and easier to understand.

Additional Examples

More examples are included with a careful analysis of each step in the development.

Emphasis on Word Problems

Solving equations and solving word problems have been treated with special emphasis, and these topics appear in every chapter except Chapter 4. In Chapter 4, the skills related to exponents, scientific notation, and operations with polynomials are developed.

Applications

In addition to integrated applications throughout the text, several sections entitled "Additional Applications" illustrate the use of algebra and algebraic concepts in real-life situations. The students see and manipulate formulas from such fields of study as physics, nursing, chemistry, construction, and agriculture that they might otherwise see only in specialized programs.

The Content

Chapter 1, Whole Numbers and Fractions, is a review of arithmetic and an introduction to algebra. Variables and exponents are introduced and are used in the review of fractions, decimals, percents, least common multiple, and order of operations. Also included are geometric formulas and testing the solutions to equations.

    Chapter 2, Signed Numbers, develops the algebraic concept of integers and the basic skills of operating with signed numbers. The discussions and the exercises include decimal numbers and fractions as signed numbers. This allows for more varied and interesting exercises and word problems and a better understanding of positive and negative numbers.

    Chapter 3, Solving First-Degree Equations, develops the techniques for solving first-degree equations in a step-by-step manner over two sections. Emphasis is placed on understanding the processes with a section on solving formulas for specified terms.

    Chapter 4, Exponents and Polynomials, presents the skills with exponents and polynomials necessary for success in the remaining chapters.

Chapter 5, Factoring Polynomials and Solving Equations, has been re-organized to include solving quadratic equations by factoring as well as the basic skills needed for factoring polynomials. Thus, any instructor short of time can be assured that the students have been exposed to the topic of qua-dratic equations and have seen the value of factoring in solving equations.

Chapter 6, Rational Expressions, provides still more practice and appli-cations with factoring and solving equations.

Chapter 7, Applications, Including Inequalities, begins with a review of solving equations then introduces inequalities and applications with inequal-ities. Chapters 6 and 7 have been reorganized so that inequality applications, previously in a separate section, are now in Section 6.2 where first-degree in-equalities are introduced. Also, to alleviate the previous concentration of ap-plications, one section of applications has been moved to Chapter 6.

Chapter 8, Linear Equations and Systems of Equations, now contains graphing, the various forms of the equations of linear equations, and tech-niques for solving systems of linear equations. Word problems involving sys-tems of equations are also included.

Chapter 9, Real Numbers and Radicals, discusses the real numbers in detail with emphasis on square roots, cube roots, and calculator skills.

Chapter 10, Quadratic Equations, develops the quadratic formula over three sections, and a variety of applications are provided including the Py-thagorean Theorem, work (time to complete a job), geometry, and distance-rate-time problems.

Chapter 11, Additional Topics, provides the instructor and students with some flexibility at the end of the course. Fractional exponents, the distance between two points and the midpoint formula, linear inequalities, functions and function notation, and graphing parabolas are included. Any of these topics can be considered a bonus for a beginning algebra course and can be covered in any order of interest.

We recommend that the topics be covered in the order presented throughout the first ten chapters, since most sections assume knowledge of the material in previous sections. Since Chapter 1 contains review material from arithmetic, some instructors may choose to start the course with Chapter 2 and have the students study Chapter 1 on an independent basis.

Special Tools for Learning Provided in Every Chapter

"Did You Know?"   Historical features of interest

Introduction   Comments on the nature of and reasons for the topics to be covered

Highlighted Definitions   Provide concise meanings in an easily identifiable format

Procedures   Summarize step-by-step techniques for easy reference and review

Section Exercises   Paired and graded problems reinforce section concepts and skills

Chapter Summary   Coded for reference to corresponding sections and formatted according to definitions, lists of rules, and procedures

Chapter Review   Problems for extra practice

Chapter Test   Practice under test conditions

Cumulative Review   Provides for continual reinforcement

Each section begins with a list of Objectives.

Most sections contain Practice Problems with Solutions.

Additional Features

1. Answers to the odd-numbered exercises and to all of the Chapter Review, Chapter Test, and Cumulative Review questions are in the back of the book.
2. There are over 490 numbered examples completely worked out with detailed analysis. There are also over 200 Practice Problems. Practice Problems can be found in almost every section.
3. Key ideas, procedures, and rules are in list form and boldface print for easy identification and reference.
4. The new wide format gives an "open" look to the text, and the new design allows for easy identification of the various parts of the text: explanations, examples, definitions, practice problems, rules, procedures, common errors, and exercises.
5. Additional Applications sections show real-life applications designed to show the students that algebra does have value and does appear in everyday situations in a variety of jobs.

Calculators   Calculator exercises are provided in many sections throughout the text. We believe that the calculator is now part of every student's life and it is important that students understand both the capabilities and the limitations of calculators. Having learned the necessary skills in the mathematics classroom, they will be able to apply this knowledge in their daily lives as well as in other classes.

The Exercises   More than 3200 carefully selected and graded section exercises are provided. They proceed from relatively easy problems to more difficult ones.

Ancillary Package

For the Instructor

The *Instructor's Manual* has been expanded to include chapter overviews, objectives, Points for Emphasis, chapter summaries, chapter quizzes, challenge problems, and all solutions for the quizzes and challenge problems, and concise answer key for all text exercises.

The *Instructor's Solutions Manual* provides complete solutions for all problems in the section exercises, chapter reviews, chapter tests, and cumulative reviews.

Wm. C. Brown Publishers Math TestPak is a free, computerized testing service with two convenient options. You may use your own Apple® IIe, IIc, or IBM PC to produce your test. It is menu-driven with on-line help screens to guide you through the test-making process. You may select items from the bank, edit the existing items, add new items of your own, or have the system randomly select items for you by chapter, section, or objective.

Wm. C. Brown Publishers Call-in Testing Service offers customized student test masters and answer keys based on your selections from the *Test Item File*. Within two working days of your request, the master and answer key will be mailed to you.

The printed *Test Item File* in an 8½″ × 11″ format contains all of the questions on the **Wm. C. Brown Publishers Math TestPak**. It will serve as a ready-reference if you use your own computer to generate tests or use the Wm. C. Brown Publishers Call-in Testing Service.

Wm. C. Brown Publishers Computerized Gradebook Software is available for the IBM PC and Apple II family of computers.

For the Student

The *Student Study Guide* provides alternative explanations and examples, additional problems, self-testing materials, and built-in answer keys. It is available for student purchase.

The *Student's Solutions Manual* contains helpful hints, summaries, and detailed solutions to every other odd-numbered exercise problem. It is available for student purchase.

Algebra Problem Solver, Interactive Student Tutorial Software, by Michael Hoban and Kathirgama Nathan, is available to help your students master introductory algebra topics. Contact your Wm. C. Brown Publishers sales representative for details.

Math Lab Software helps students master algebraic concepts in an interactive format. This program generates problems at random on a given topic, provides immediate feedback, and channels the student into a correct problem-solving procedure.

Apple is a registered trademark of Apple Computer, Inc.

On the **Videotapes,** the instructor introduces a concept, provides detailed explanations of example problems that illustrate the concept, including applications, and concludes with a summary. The tapes are available in ½" VHS format and are free to qualified adopters.

The **Audiotapes** start with a brief synopsis of the section and worked-out sample problems, explaining each step and offering helpful hints and useful warnings. Important concepts and procedures are stressed. These tapes are free to qualified adopters.

Acknowledgments

The following reviewers have been helpful with their constructive and critical comments:

Jeffery C. Barnett
Fort Hays State University

George L. Holloway
Los Angeles Valley College

Walker E. Hunt
San Antonio College

William H. Keils
San Antonio College

Charlotte K. Lewis
University of New Orleans

Robert C. Limburg
St. Louis Community College–Florissant Valley

J. Larry Martin
Missouri Southern State College

Marcel Maupin
Oklahoma State University Technical Branch

Harold Oxsen
Diablo Valley College

Dorothy Schwellenbach
Hartnell College

Richard C. Spangler
Tacoma Community College

Nancy Spears
Everett Community College

Ann E. Steen
Santa Fe Community College

Gerry C. Vidrine
Louisiana State University

J. Terry Wilson
San Jacinto College

Thomas H. Wilson
Sinclair Community College

We would particularly like to thank the staff at Wm. C. Brown Publishers—Earl McPeek (editor), Theresa Grutz (developmental editor), and Gene Collins (production editor)—for their continued enthusiasm and hard work above and beyond. Pat Wright has done another outstanding job of manuscript preparation which has made everyone's job just that much easier. Thanks also to our students and colleagues at Cerritos College for their interest and comments over the years.

Thank You All

Frank Wright
Bill New

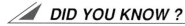

DID YOU KNOW ?

Almost all of mathematics uses the language of sets to simplify notation and to help in understanding concepts. The theory of sets is one of the few branches of mathematics that was initially developed almost completely by one person, Georg Cantor.

Georg Cantor was born in St. Petersburg, Russia, but spent his adult life in Germany, first as a student at the University of Berlin and later as a professor at the University of Halle. Cantor's research and development of set ideas was met by ridicule and public attacks on his character and work. In 1885, Cantor suffered the first of a series of mental breakdowns probably caused by the attacks on his work by other mathematicians, most notably a former teacher Leopold Kronecker. It is suggested that Kronecker kept Cantor from becoming a professor at the prestigious University of Berlin.

We are not used to seeing intolerance toward new ideas among the scientific community, and, as you study mathematics, you will probably have difficulty imagining how anyone could have felt hostile toward or threatened by Cantor's ideas. However, his idea of a set with an **infinite** number of elements was thought of as revolutionary by the mathematical establishment.

The problem is that infinite sets have the following curious property: a part may be numerically equal to a whole. For example,

$$N = \{1, 2, 3, 4, \quad 5, \ldots, \quad n, \ldots\}$$
$$\Updownarrow \Updownarrow \Updownarrow \Updownarrow \quad \Updownarrow \qquad \Updownarrow$$
$$E = \{2, 4, 6, 8, 10, \ldots, 2n, \ldots\}$$

Sets N and E are numerically equal because set E can be put in one-to-one correspondence (matched or counted) with set N. So there are as many even numbers as there are natural numbers! This is contrary to common sense and, in Cantor's time, contrary to usual mathematical assumptions.

The essence of mathematics lies in its freedom.

Georg Cantor (1845–1918)

◢ CHAPTER OUTLINE

*A*ny Olympic athlete will tell you that practice is the key to success. Chapter 1 allows you to practice the basic arithmetic skills you already have with whole numbers, fractions, decimals, and percents. At the same time, you will find an integrated introduction to elementary algebraic concepts using variables and formulas.

   Your success in this course may depend on how well you start and how much you "practice" the basic skills represented in this first chapter. Good luck and try your best. Keep a positive attitude.

◢ 1.1 Whole Numbers, Exponents, and Order of Operations

OBJECTIVES

In this section, you will be learning to:

1. Operate with whole numbers.
2. Complete statements by using the properties of whole numbers.
3. Name the whole number properties that justify given statements.
4. Find the averages of collections of whole numbers.
5. Evaluate powers of whole numbers.
6. Write products of whole numbers in exponential form.
7. Find the prime factorizations of whole numbers.
8. Evaluate whole number expressions by using the rules for order of operations.

Much of algebra is a general form of arithmetic using letters and other symbols. Thus, we begin our study of algebra by reviewing and solidifying basic skills and terminology from arithmetic.

   In this section, we will review the four basic operations with whole numbers—addition, subtraction, multiplication, and division—and discuss the use of exponents and the rules of order of operations in simplifying expressions.

Addition

The study of algebra requires that we know a variety of types of numbers and their names. The numbers

$$1, 2, 3, 4, 5, 6, 7, 8, 9, 10, 11, \ldots$$

are called the **counting numbers** or **natural numbers.** The three dots indicate that the pattern is to continue without end. Putting 0 with the set of natural numbers gives the **whole numbers**

$$0, 1, 2, 3, 4, 5, 6, \ldots.$$

Thus, 0 is a whole number but not a natural number.

   Addition with whole numbers can be indicated by writing the numbers either horizontally with a plus (+) sign between the numbers or in column form with directions to add. For example,

$$25 + 8 + 12 \qquad \text{or} \qquad \text{Add: } \begin{array}{r} 25 \\ 8 \\ \underline{12} \end{array}$$

The result of addition is called the **sum.**

$$25 + 8 + 12 = 45 \qquad \text{or} \qquad \begin{array}{r} 25 \\ 8 \\ \underline{12} \\ 45 \end{array} \leftarrow \text{sum}$$

The statement $7 + 6 = 6 + 7$ is called an **equation.** The equal sign ($=$) means that the number represented on the left is the same as the number represented on the right.

We illustrate two fundamental **properties of addition** using **variables,** that is, symbols or letters that can represent more than one number.

Commutative Property of Addition	If a and b are whole numbers, then $$a + b = b + a$$

Associative Property of Addition	If a, b, and c are whole numbers, then $$a + b + c = (a + b) + c = a + (b + c)$$

EXAMPLES   The commutative property states that two numbers can be added in either order to get the same sum.

  1. $5 + 2 = 2 + 5$

  2. $8 + 10 = 10 + 8$

  3. $17 + x = x + 17$ for any value of the **variable** x.
     If $x = 3$, then $17 + 3 = 3 + 17$.
     If $x = 5$, then $17 + 5 = 5 + 17$. ∎

EXAMPLES   The associative property allows us to group (associate) numbers together differently but still get the same sum.

  4. $(3 + 7) + 4 = 3 + (7 + 4)$
     We have $(3 + 7) + 4 = (10) + 4 = 14$
     and $3 + (7 + 4) = 3 + (11) = 14.$

  5. $(5 + 9) + y = 5 + (9 + y)$ for any value of the **variable** y.
     If $y = 18$, then $(5 + 9) + 18 = (14) + 18 = 32$
     and $5 + (9 + 18) = 5 + (27) = 32.$
     Thus, $(5 + 9) + 18 = 5 + (9 + 18).$ ∎

Subtraction

What number would you add to 15 to get 18? The answer is 3, of course. This number is called the **difference** between 18 and 15 and can be found by using a process called **subtraction.** Subtraction is indicated with a minus ($-$) sign, and

$$18 - 15 = 3 \text{ because } 18 = 15 + 3.$$

To subtract large numbers, we must be aware of the place value of each digit in writing a number and of the technique called **borrowing.** For example, $473 - 195$ is difficult to do in your head if you don't readily know what to add to 195 to get 473. We can write

$$473 = 400 + 70 + 3 = 400 + 60 + 13 = 300 + 160 + 13$$
$$\underline{-195 = 100 + 90 + 5 = 100 + 90 + 5 = 100 + 90 + 5}$$
$$200 + 70 + 8 = 278$$

We "borrowed" 10 from 70, then borrowed 100 from 400. In familiar shorthand,

$$\begin{array}{ccc} 3 & 16 & 1 \\ \cancel{4} & \cancel{7} & 3 \\ -1 & 9 & 5 \\ \hline 2 & 7 & 8 \end{array}$$

The commutative and associative properties are **not** true for subtraction. For example,

$$8 - 5 \neq 5 - 8 \quad \text{and} \quad 17 - (6 - 3) \neq (17 - 6) - 3$$

("$\neq$" is read "is not equal to.")

Multiplication

Multiplication is shorthand for repeated addition. For example,

$$613 + 613 + 613 + 613 = 4 \cdot 613 = 2452$$

The repeated number (613) and the number of times it is being used (4) are both called **factors** of the result (2452), which is now called the **product.**

Multiplication can be indicated in any of the following ways:

a. $4 \cdot 613$ **b.** $4(613)$ **c.** $(4)613$

d. $(4)(613)$ **e.** 4×613 **f.** $\begin{array}{r} 613 \\ \underline{\times 4} \end{array}$

Generally we will avoid types (e) and (f) because the times sign ($\times$) can be confused with the letter x. Also, multiplication can be indicated by writing a number next to a variable or two or more variables next to each other. For example, $7y$ means $7 \cdot y$ and abc means $a \cdot b \cdot c$. Thus, $3a$, ab, and $5xyz$ all indicate multiplication.

Multiplication is both commutative and associative. That is, $3 \cdot 5 = 5 \cdot 3$ and $6 \cdot (4 \cdot 7) = (6 \cdot 4) \cdot 7$.

Commutative Property of Multiplication

If a and b are whole numbers, then

$$a \cdot b = b \cdot a$$

Associative Property of Multiplication	If a, b, and c are whole numbers, then $$a \cdot b \cdot c = (a \cdot b) \cdot c = a \cdot (b \cdot c)$$

EXAMPLES   For the commutative property of multiplication:

6. $4 \cdot 20 = 20 \cdot 4$ Notice that the dots are necessary between numbers because 420 does not equal 204. The dots are not necessary between variables and numbers.

7. $3w = w3$ Note that $w3$ means the product of w and 3. For clarity we usually write the number to the left of the variable, such as $3y$ or $7x$ or $18ab$.   ∎

EXAMPLES   For the associative property of multiplication:

8. $2 \cdot (12 \cdot 5) = (2 \cdot 12) \cdot 5$
since $2 \cdot (12 \cdot 5) = 2(60) = 120$
and $(2 \cdot 12) \cdot 5 = (24) \cdot 5 = 120$.

9. $8 \cdot x \cdot y = (8x)y = 8(xy)$
For example, if x is 2 and y is 3, we have
$(8 \cdot 2) \cdot 3 = (16) \cdot 3 = 48$
and $8 \cdot (2 \cdot 3) = 8(6) = 48$.
So, $(8 \cdot 2) \cdot 3 = 8(2 \cdot 3)$.   ∎

One property of numbers that we find extremely useful in algebra combines both addition and multiplication. We can illustrate this property, called **the distributive property of multiplication over addition,** in the following way. If you were to multiply $8(10 + 1)$, you would probably add 10 and 1 first and get $8(10 + 1) = 8(11) = 88$. We can get the same result as follows:

$$8(\overset{\frown}{10 + 1}) = 8 \cdot 10 + 8 \cdot 1 = 80 + 8 = 88$$

The fact that $8(10 + 1) = 8 \cdot 10 + 8 \cdot 1$ illustrates the distributive property.

Distributive Property of Multiplication over Addition	If a, b, and c are whole numbers, then $$a \cdot (b + c) = a \cdot b + a \cdot c$$

EXAMPLES   For the distributive property:

10. $3(5 + 7) = 3 \cdot 5 + 3 \cdot 7 = 15 + 21 = 36$
11. $4(x + 8) = 4 \cdot x + 4 \cdot 8 = 4x + 32$ ∎

Division

What number would you multiply by 2 to get 14? Think $2 \cdot \boxed{?} = 14$. The answer is 7. This number is called the **quotient** of 14 and 2 and can be found by using a reverse multiplication called **division**. Division can be indicated using a divide ($\div$) sign, or a bar ($-$) called fraction form, or a $\overline{)}$ sign. Thus, we can write

$$14 \div 2 = 7, \qquad \frac{14}{2} = 7, \qquad 2\overline{)14}^{\,7}$$

and $14 \div 2 = 7$ because $14 = 2 \cdot 7$.

In general, for nonzero b, $\dfrac{a}{b} = x$ if and only if $a = b \cdot x$.

Division by 0 Is Undefined

(a) Consider $\dfrac{6}{0} = \square$. Whatever $\square$ is, $6 = 0 \cdot \square$. But $0 \cdot \square = 0$ whatever $\square$ represents, so that $6 = 0 \cdot \square = 0$. This is impossible.

(b) Next consider $\dfrac{0}{0} = \square$. Then $0 = 0 \cdot \square$, which is true regardless of what $\square$ represents. This means $\dfrac{0}{0}$ could be any number, an unacceptable situation.

Thus, $\dfrac{a}{0}$ **is undefined for any whole number** a.

Does $14 \div 2 = 2 \div 14$? Does $36 \div (12 \div 3) = (36 \div 12) \div 3$? The answer to both questions is No. In other words, division, like subtraction, is neither commutative nor associative.

EXAMPLES

12. Add:

$$\begin{array}{r} 25 \\ 46 \\ 193 \\ +\ 50 \\ \hline 314 \end{array} \quad \text{sum}$$

13. Subtract:

$$\begin{array}{r} 742 \\ -361 \\ \hline 381 \end{array} \quad \text{difference}$$

14. Multiply:

$$\begin{array}{r} 82 \\ \times 34 \\ \hline 328 \\ 246 \\ \hline 2788 \end{array} \quad \text{product}$$

15. Divide:

$$\begin{array}{r} 46 \\ 7\overline{)322} \\ 28 \\ \hline 42 \\ 42 \\ \hline 0 \end{array} \quad \text{quotient}$$

Average

Another topic related to operations with whole numbers is **average,** such as a batter's average in baseball or the average value of stocks on the stock market.

To Find the Average of a Collection of Numbers	1. Add all the numbers in the collection.
	2. Divide this sum by the number of numbers in the collection.
	This quotient is the **average.**

EXAMPLE

16. Find the average of the numbers 82, 91, 63, 51, and 48.

Solution

$$
\begin{array}{r}
82 \\
91 \\
63 \\
51 \\
48 \\
\hline
335
\end{array}
\qquad
\begin{array}{r}
67 \\
5\overline{)335} \\
\underline{30} \\
35
\end{array}
\text{ average}
$$

Since 5 numbers are added, the sum is divided by 5 to find the average.

Exponents

If, when writing the product of several numbers, one or more of the **factors** is repeated (such as $7 \cdot 7 = 49$ or $3 \cdot 3 \cdot 3 \cdot 3 = 81$), there is a shorthand notation using **exponents** to indicate the repetition. An **exponent** is a number that tells how many times a factor occurs in a product. The exponent is written to the right and slightly above the factor, and the factor is called the **base** of the exponent. The product is called a **power** of the factor. Thus, we can write

$$
7 \cdot 7 = 7^2 = 49
$$

exponent 2 / base 7 / power

7^2 is read "seven squared" or "seven to the second power." We say that 49 is the second power of 7.

$$
2 \cdot 2 \cdot 2 = 2^3 = 8
$$

2^3 is read "two cubed" or "two to the third power."

$$
10 \cdot 10 \cdot 10 \cdot 10 = 10^4 = 10,000
$$

10^4 is read "ten to the fourth power."

In general,

$$
a^n = \underbrace{a \cdot a \cdot a \ldots a}_{n \text{ factors}}
$$

Prime Factorization

In working with fractions and simplifying algebraic expressions, we sometimes want to see factors in their simplest form. This can involve **prime numbers**. A **prime number is a whole number that has exactly two different factors.** For example,

13 is prime   13 has exactly two different factors, 13 and 1.

24 is not prime   24 has 1, 2, 3, 4, 6, 8, 12, and 24 as factors.

Composite numbers are whole numbers, other than 0 and 1, that are not prime. The two numbers 0 and 1 are neither prime nor composite. The prime numbers less than 50 are

2, 3, 5, 7, 11, 13, 17, 19, 23, 29, 31, 37, 41, 43, 47.

Finding the **prime factorization** of a composite number is particularly useful in working with fractions and serves here to illustrate the use of exponents.

To Find the Prime Factorization of a Composite Number	1. Find any two factors of the number.
	2. Continue to factor each of the factors until all factors are prime numbers.
	3. Write the product of all these prime factors. (Use exponents for repeated factors.)

EXAMPLES   Find the prime factorizations of the composite numbers.

17. 72

Solution $72 = 8 \cdot 9 = 2 \cdot 4 \cdot 3 \cdot 3 = 2 \cdot 2 \cdot 2 \cdot 3 \cdot 3 = 2^3 \cdot 3^2$

Or you might write
$72 = 12 \cdot 6 = 3 \cdot 4 \cdot 2 \cdot 3 = 3 \cdot 2 \cdot 2 \cdot 2 \cdot 3 = 2^3 \cdot 3^2$
No matter how you start, the prime factorization will be the same.

18. 60

Solution $60 = 6 \cdot 10 = 2 \cdot 3 \cdot 2 \cdot 5 = 2^2 \cdot 3 \cdot 5$

Or you might write
$60 = 2 \cdot 30 = 2 \cdot 15 \cdot 2 = 2 \cdot 3 \cdot 5 \cdot 2 = 2^2 \cdot 3 \cdot 5$ ∎

Order of Operations

We know that division is not associative. So there is some question as to how to evaluate an expression such as $36 \div 12 \div 3$.

Does $36 \div 12 \div 3 = (36 \div 12) \div 3 = 3 \div 3 = 1?$

Or does $36 \div 12 \div 3 = 36 \div (12 \div 3) = 36 \div 4 = 9?$

We must agree on one answer.

What is the value of $14 \div 2 + 3 \cdot 2$?

Does $14 \div 2 + 3 \cdot 2 = 7 + 3 \cdot 2 = 10 \cdot 2 = 20$?

Or does $14 \div 2 + 3 \cdot 2 = 7 + 6 = 13$?

Again, we must agree on one answer.

General agreement has been reached by mathematicians on the following order of operations for evaluating numerical expressions.

Rules for Order of Operations	1. Work within symbols of inclusion (parentheses, brackets, or braces), beginning with the innermost pair.
	2. Find any powers indicated by exponents.
	3. Perform any multiplications or divisions in the order they appear **from left to right**.
	4. Perform any additions or subtractions in the order they appear **from left to right**.

EXAMPLES Simplify each expression using the rules for order of operations.

19. $36 \div 12 \div 3$

Solution $36 \div 12 \div 3 = 3 \div 3$ Divide from left to right.
$= 1$

20. $14 \div 2 + 3 \cdot 2$

Solution $14 \div 2 + 3 \cdot 2 = 7 + 6$ Divide and multiply from left to
$= 13$ right before adding.

21. $3^2 - 8 \div 4$

Solution $3^2 - 8 \div 4 = 9 - 8 \div 4$ Exponents
$= 9 - 2$ Division
$= 7$ Subtraction

22. $[3(4 + 6) + 14] \div 4 + 7$

Solution

$$[3(4 + 6) + 14] \div 4 + 7 = [3(10) + 14] \div 4 + 7 \qquad \text{Parentheses}$$
$$= [30 + 14] \div 4 + 7 \qquad \text{Multiplication in brackets}$$
$$= [44] \div 4 + 7 \qquad \text{Addition in brackets}$$
$$= 11 + 7 \qquad \text{Division}$$
$$= 18 \qquad \text{Addition} \quad \blacksquare$$

Practice Problems   Find the value of each expression.

1. $2^5 - 8 \cdot 3 =$ 8
2. $(16 \div 2^2 + 6) \div 5 + 5 =$ 7
3. $14 \div 2 + 2 \cdot 6 + 30 \div 3$ 29
4. $3[8 + 2(1 + 3)] - 15 \div 3$ 43

EXERCISES 1.1

Complete the expressions in Exercises 1–10 using the given property.

1. $7 + 3 =$ $\underline{3 + 7 = 10}$ Commutative property of addition
2. $(6 \cdot 9) \cdot 3 =$ $\underline{9 \cdot 3 \cdot 6 = 162}$ Associative property of multiplication
3. $19 \cdot 4 =$ $\underline{4 \cdot 19 = 76}$ Commutative property of multiplication
4. $18 + 5 =$ $\underline{5 + 18 = 23}$ Commutative property of addition
5. $6(5 + 8) =$ $\underline{(6 \cdot 5) + (6 \cdot 8) = 78}$ Distributive property
6. $16 + (9 + 11) =$ $\underline{(16 + 9) + 11 = 36}$ Associative property of addition
7. $2 \cdot (3x) =$ $\underline{(2 \cdot 3)x}$ Associative property of multiplication
8. $3(x + 5) =$ $\underline{5 \cdot 3 + x \cdot 3}$ Distributive property
9. $3 + (x + 7)$ $\underline{(3 + x) + 7}$ Associative property of addition
10. $9(x + 5) =$ $\underline{9 \cdot x + 9 \cdot 5}$ Distributive property

Name the properties illustrated in Exercises 11–20 and show that each equation is true if, for example, $x = 4$.

11. $6 \cdot x = x \cdot 6$ Comm
12. $19 + x = x + 19$ Assoc.
13. $8 + (4 + x) = (8 + 4) + x$ Assoc.
14. $(2 \cdot 7) \cdot x = 2 \cdot (7 \cdot x)$ Assoc.
15. $5(x + 18) = 5x + 90$ Distr.
16. $(x + 14) + 3 = x + (14 + 3)$ Assoc.
17. $(6 \cdot x) \cdot 9 = 6 \cdot (x \cdot 9)$ Assoc.
18. $11 \cdot x = x \cdot 11$ Comm.
19. $x + 34 = 34 + x$ Comm.
20. $3(x + 15) = 3x + 45$ Distributive

Use the distributive property to evaluate the expressions in Exercises 21–24.

21. $6(3 + 8) = 66$
$6 \cdot 3 + 8 \cdot 6 = 66$
22. $7(8 + 5) = 91$
$7 \cdot 8 + 7 \cdot 5 = 91$
23. $10(9 + 2) = 110$
$10 \cdot 9 + 2 \cdot 10 = 110$
24. $13(5 + 3) = 104$
$13 \cdot 5 + 13 \cdot 3$

Add in Exercises 25 and 26.

25. 97
   132
    61
     5
  295

26. 153
   201
    89
    62
  505

Subtract in Exercises 27 and 28.

27. 365
    79
  286

28. 482
   288
  194

Multiply in Exercises 29–32.

29. 45
   27
 315
 90
1215

30. (34)(28)
   34
   28
  272
  68
  952

31. $46 \cdot 50$
  46
  50
2300

32. $72 \cdot 24$
  72
  24
 288
 144
1728

Answers to Practice Problems 1. 8 2. 7 3. 29 4. 43

Divide in Exercises 33–36.

33. $238 \div 17 = 14$ **34.** $\dfrac{336}{15} = 224$ **35.** $\dfrac{0}{0} = UNDEFINED$ **36.** $69 \div 0 = UNDEFINED$

Find the average of each set of numbers in Exercises 37–40.

37. $21, 14, 16 = 17$ **38.** $47, 39, 55 = 47$ **39.** $27, 18, 25, 30 = 25$ **40.** $37, 45, 52, 58 = 48$

Evaluate the exponential expressions in Exercises 41–46.

41. $7^2 = 49$ **42.** $8^2 = 64$ **43.** $5^3 = 5 \cdot 5 \cdot 5 = 125$ **44.** $4^3 = 64$ **45.** $3^4 = 81$ **46.** $2^4 = 16$

Write each expression in exponential form in Exercises 47–52.

47. $2 \cdot 2 \cdot 2 \cdot 2 \cdot 2 = 2^5$

48. $7 \cdot 7 \cdot 7 \cdot 7 = 7^4$

49. $2 \cdot 2 \cdot 5 \cdot 5 \cdot 5 \cdot a \cdot a = 2^2 \cdot 5^3 \cdot a^2$

50. $3 \cdot a \cdot a \cdot a \cdot b \cdot b = 3a^3 b^2$

51. $7 \cdot 7 \cdot a \cdot a \cdot a \cdot a \cdot b \cdot b \cdot b = 7^2 A^4 \cdot b^3$

52. $5 \cdot 2 \cdot 2 \cdot a \cdot a \cdot b \cdot b \cdot b \cdot b = 5 \cdot 2^2 a^2 b^4$

Find the prime factorization for each number in Exercises 53–58.

53. $54 = 2 \cdot 3^3$ **54.** $96 = 2^5 \cdot 3$ **55.** $168 = 2^3 \cdot 7 \cdot 3$ **56.** $153 = 3^2 \cdot 17$ **57.** $196 = 2^2 \cdot 7^2$ **58.** 214

Find the value of each expression in Exercises 59–74 using the rules for order of operations.

59. $3 \cdot 5 - 2 = 13$ **60.** $4 + 8 \div 2 = 8$ **61.** $2 \cdot 4^2 + 1 = 33$

62. $5 \cdot 3^2 - 6 = 39$ **63.** $18 \div 3 \cdot 6 - 3 = 33$ **64.** $3 \cdot 5 + 60 \div 3 \cdot 2 = 50$

65. $6 \cdot 3 \div 2 + 4 - 2 = 11$ **66.** $(4^2 + 6) \div 11 - 1 \cdot 2 = 0$ **67.** $7(4 - 2) \div 7 + 3 = 5$

68. $2^3 + 4 \div 2 + 2 = 8$

69. $3^3 \div 9 + (6 + 4^2) \div 2 = 14$

70. $(4^2 - 3^2) \div 7 + [2 + 2(3^2)] \div (2 \cdot 5) = 3$ **71.** $14 + [11 \cdot 4 - (6 \cdot 3 + 1)] = 39$

72. $6 + 3[4 - 2(3 - 1)]$ **73.** $7 - [4 \cdot 3 - (4 + 3 \cdot 2)]$ **74.** $2[6 + 4(1 + 7)] \div 4$

$6 + 3[4 - 2 \cdot 2] = 6$ $7 - [4 \cdot 3 - 10]$ $2[6 + 4(8)] \div 4$

$7 - [12 - 10] =$ $2[38] \div 4$

$7 - 2 = 5$ $76 \div 4 = 19$

1.2 Fractions

A **fraction** is a number that can be written in the form $\dfrac{a}{b}$ which means $a \div b$.

$$\dfrac{a \leftarrow \text{numerator}}{b \leftarrow \text{denominator}}$$

$b \neq 0$ since division by 0 is undefined.

In this chapter, the fractions discussed will have only whole numbers or whole-number variables for numerator and denominator. This discussion will form the basis for our work with more complicated algebraic fractions in Chapter 6.

Multiplication and Division

To multiply two fractions, multiply the numerators and multiply the denominators.

$$\frac{a}{b} \cdot \frac{c}{d} = \frac{a \cdot c}{b \cdot d}$$

For example,

$$\frac{2}{3} \cdot \frac{7}{5} = \frac{2 \cdot 7}{3 \cdot 5} = \frac{14}{15}$$

The number 1 is called the **multiplicative identity** since the product of 1 with any number is that number. That is,

$$\frac{a}{b} \cdot 1 = \frac{a}{b}$$

Thus, if $k \neq 0$, we have

$$\frac{a}{b} = \frac{a}{b} \cdot 1 = \frac{a}{b} \cdot \frac{k}{k} = \frac{a \cdot k}{b \cdot k}$$

The Fundamental Principle of Fractions

$$\frac{a}{b} = \frac{a \cdot k}{b \cdot k}, \quad \text{where } k \neq 0$$

We can use the Fundamental Principle to build a fraction to **higher terms** (find an equal fraction with a larger denominator) or reduce to **lower terms** (find an equal fraction with a smaller denominator).

To reduce a fraction, factor both the numerator and denominator, then use the Fundamental Principle to "divide out" any common factors. If the numerator and the denominator have no common prime factors, the fraction has been **reduced to lowest terms. Finding the prime factorizations of the numerator and denominator before reducing, while not necessary, will help guarantee a fraction is in lowest terms.**

EXAMPLES

1. Raise $\frac{3}{7}$ to higher terms with a denominator of 28.

   Solution   Use $k = 4$ since $7 \cdot \mathbf{4} = 28$.

   $$\frac{3}{7} = \frac{3 \cdot \mathbf{4}}{7 \cdot \mathbf{4}} = \frac{12}{28}$$

2. Raise $\frac{5}{8}$ to higher terms with a denominator of $16a$.

   Solution   Use $k = 2a$ since $8 \cdot \mathbf{2a} = 16a$.

   $$\frac{5}{8} = \frac{5 \cdot \mathbf{2a}}{8 \cdot \mathbf{2a}} = \frac{10a}{16a}$$

3. Reduce $\frac{12}{20}$ to lowest terms.

   Solution $\dfrac{12}{20} = \dfrac{2 \cdot 2 \cdot 3}{2 \cdot 2 \cdot 5} = \dfrac{2}{2} \cdot \dfrac{2}{2} \cdot \dfrac{3}{5} = 1 \cdot 1 \cdot \dfrac{3}{5} = \dfrac{3}{5}$

$$\text{or} \quad \frac{12}{20} = \frac{4 \cdot 3}{4 \cdot 5} = \frac{4}{4} \cdot \frac{3}{5} = 1 \cdot \frac{3}{5} = \frac{3}{5}$$

$$\text{or} \quad \frac{12}{20} = \frac{\cancel{2} \cdot \cancel{2} \cdot 3}{\cancel{2} \cdot \cancel{2} \cdot 5} = \frac{3}{5}$$

4. Factor and reduce $\dfrac{15}{28} \cdot \dfrac{4}{9}$ before finding the product.

Solution $\quad \dfrac{15}{28} \cdot \dfrac{4}{9} = \dfrac{\cancel{3} \cdot 5 \cdot \cancel{4}}{\cancel{4} \cdot 7 \cdot 3 \cdot \cancel{3}} = \dfrac{5}{21}$

5. Find the product $\dfrac{4a}{12b} \cdot \dfrac{3b}{7a}$ in lowest terms. Note that the number 1 is implied to be a factor even if it is not written.

Solution $\quad \dfrac{4a}{12b} \cdot \dfrac{3b}{7a} = \dfrac{\cancel{4} \cdot \cancel{a} \cdot \cancel{3} \cdot \cancel{b} \cdot 1}{\cancel{4} \cdot \cancel{3} \cdot \cancel{b} \cdot 7 \cdot \cancel{a}}$ Here we write the factor 1 because all other factors have been "divided out."

$$= \frac{1}{7}$$

We could write

$$\frac{4a}{12b} \cdot \frac{3b}{7a} = \frac{4 \cdot a \cdot 3 \cdot b}{4 \cdot 3 \cdot b \cdot 7 \cdot a}$$

$$= \frac{4}{4} \cdot \frac{3}{3} \cdot \frac{a}{a} \cdot \frac{b}{b} \cdot \frac{1}{7}$$

$$= 1 \cdot 1 \cdot 1 \cdot 1 \cdot \frac{1}{7} = \frac{1}{7}$$ ∎

If $a \neq 0$ and $b \neq 0$, the **reciprocal** of $\dfrac{a}{b}$ is $\dfrac{b}{a}$, and

$$\frac{a}{b} \cdot \frac{b}{a} = 1$$

Now consider the division problem $\dfrac{2}{3} \div \dfrac{5}{6}$. Using the definition of division in terms of multiplication, if

$$\frac{2}{3} \div \frac{5}{6} = \square, \text{ then } \frac{2}{3} = \square \cdot \frac{5}{6}$$

Since $\qquad \dfrac{2}{3} = \dfrac{2}{3} \cdot 1 = \dfrac{2}{3} \cdot \left(\dfrac{6}{5} \cdot \dfrac{5}{6} \right) = \dfrac{2}{3} \cdot \dfrac{6}{5} \cdot \dfrac{5}{6}$

we have $\qquad \dfrac{2}{3} \div \dfrac{5}{6} = \dfrac{2}{3} \cdot \dfrac{6}{5}$

Another approach to understanding division with fractions is to write the indicated division as a fraction.

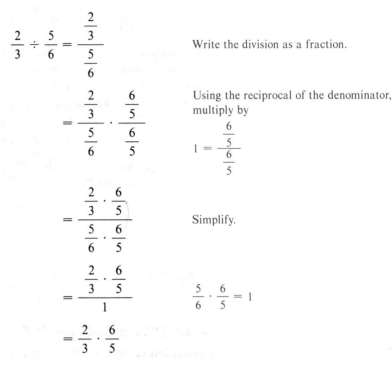

$$\frac{2}{3} \div \frac{5}{6} = \frac{\dfrac{2}{3}}{\dfrac{5}{6}}$$ Write the division as a fraction.

$$= \frac{\dfrac{2}{3} \cdot \dfrac{6}{5}}{\dfrac{5}{6} \cdot \dfrac{6}{5}}$$ Using the reciprocal of the denominator, multiply by

$$1 = \frac{\dfrac{6}{5}}{\dfrac{6}{5}}$$

$$= \frac{\dfrac{2}{3} \cdot \dfrac{6}{5}}{\dfrac{5}{6} \cdot \dfrac{6}{5}}$$ Simplify.

$$= \frac{\dfrac{2}{3} \cdot \dfrac{6}{5}}{1}$$ $\dfrac{5}{6} \cdot \dfrac{6}{5} = 1$

$$= \frac{2}{3} \cdot \frac{6}{5}$$

In either case, we have

$$\frac{2}{3} \div \frac{5}{6} = \frac{2}{3} \cdot \frac{6}{5} = \frac{2 \cdot 2 \cdot \cancel{3}}{\cancel{3} \cdot 5} = \frac{4}{5}$$

That is, **to divide by a nonzero fraction, multiply by its reciprocal.**

$$\frac{a}{b} \div \frac{c}{d} = \frac{a}{b} \cdot \frac{d}{c}$$

EXAMPLES   Divide and reduce all answers.

6. $\dfrac{3}{4} \div \dfrac{2}{5}$

Solution $\dfrac{3}{4} \div \dfrac{2}{5} = \dfrac{3}{4} \cdot \dfrac{5}{2} = \dfrac{15}{8}$

In algebra, $\dfrac{15}{8}$, an improper fraction, is preferred to the mixed number $1\dfrac{7}{8}$. Improper fractions are perfectly acceptable as long as they are reduced, that is, as long as the numerator and denominator have no common prime factors.

7. $\dfrac{26}{35} \div \dfrac{39}{20}$

Solution $\dfrac{26}{35} \div \dfrac{39}{20} = \dfrac{26}{35} \cdot \dfrac{20}{39} = \dfrac{2 \cdot \cancel{13} \cdot 2 \cdot 2 \cdot \cancel{5}}{\cancel{5} \cdot 7 \cdot 3 \cdot \cancel{13}} = \dfrac{8}{21}$

Note carefully that we factored and reduced without actually multiplying the numerators and denominators. It would not be wise to multiply first because we would then just have to factor two large numbers. For example,

$$\frac{26}{35} \div \frac{39}{20} = \frac{26}{35} \cdot \frac{20}{39} = \frac{520}{1365}$$

and now we have to factor 520 and 1365. But these were already factored for us since $26 \cdot 20 = 520$ and $35 \cdot 39 = 1365$.

8. $\dfrac{21a}{5b} \div 3a$

Solution

$$\frac{21a}{5b} \div 3a = \frac{21a}{5b} \cdot \frac{1}{3a} = \frac{\cancel{3} \cdot 7 \cdot \cancel{a}}{5 \cdot \cancel{3} \cdot \cancel{a} \cdot b} = \frac{7}{5b}$$ **Note:** The reciprocal of $3a$ is $\dfrac{1}{3a}$ since $3a = \dfrac{3a}{1}$. ∎

Addition and Subtraction

Finding the sum of two or more fractions with the same denominator is similar to adding whole numbers of some particular item. For example, the sum of 5 apples and 6 apples is 11 apples. Similarly, the sum of 5 seventeenths and 6 seventeenths is 11 seventeenths, or $\dfrac{5}{17} + \dfrac{6}{17} = \dfrac{11}{17}$.

To add two fractions $\dfrac{a}{b}$ and $\dfrac{c}{b}$ with common denominator b, add the numerators a and c and use the common denominator.

$$\frac{a}{b} + \frac{c}{b} = \frac{a + c}{b}$$

A formal proof of this relationship involves the fact that $\dfrac{a}{b} = \dfrac{1}{b} \cdot a$ and the distributive property.

$$\frac{a}{b} + \frac{c}{b} = \frac{1}{b} \cdot a + \frac{1}{b} \cdot c$$

$$= \frac{1}{b}(a + c) \qquad \text{Distributive property}$$

$$= \frac{a + c}{b}$$

EXAMPLES

9. $\dfrac{3}{8} + \dfrac{4}{8}$

   Solution $\dfrac{3}{8} + \dfrac{4}{8} = \dfrac{3+4}{8} = \dfrac{7}{8}$

10. $\dfrac{9}{10} + \dfrac{3}{10}$

   Solution $\dfrac{9}{10} + \dfrac{3}{10} = \dfrac{9+3}{10} = \dfrac{12}{10} = \dfrac{\cancel{2} \cdot 6}{\cancel{2} \cdot 5} = \dfrac{6}{5}$

11. $\dfrac{2}{15} + \dfrac{3}{15} + \dfrac{1}{15} + \dfrac{6}{15}$

   Solution $\dfrac{2}{15} + \dfrac{3}{15} + \dfrac{1}{15} + \dfrac{6}{15} = \dfrac{2+3+1+6}{15}$

$$= \dfrac{12}{15} = \dfrac{\cancel{3} \cdot 4}{\cancel{3} \cdot 5} = \dfrac{4}{5} \qquad ■$$

To find the sum of fractions with different denominators, we need the concepts of **multiples** and **least common multiple. Multiples** of a number are the products of that number with the counting numbers.

Counting numbers: 1, 2, 3, 4, 5, 6, 7, 8, 9, . . .
Multiples of 6:   6, 12, 18, 24, 30, 36, 42, 48, 54, . . .
Multiples of 8:   8, 16, 24, 32, 40, 48, 56, 64, 72, . . .

For the multiples of 6 and 8, the common multiples are

$$24, 48, 72, 96, 120, . . .$$

The smallest of these, called the **least common multiple (LCM),** is 24. Note that the LCM is **not** $6 \cdot 8 = 48$. In this case the LCM is 24, and it is smaller than the product of 6 and 8.

We can use prime factorizations to find the LCM of two or more numbers using the following steps.

To Find the LCM

1. Find the prime factorization of each number.

2. Form the product of all prime factors that appear by using each prime factor the greatest number of times that it appears in any one prime factorization.

EXAMPLES

12. Find the LCM for the numbers 27, 15, and 60.

   Solution $27 = 3 \cdot 3 \cdot 3 = 3^3$
   $\left. \begin{array}{l} 15 = 3 \cdot 5 \\ 60 = 2 \cdot 2 \cdot 3 \cdot 5 = 2^2 \cdot 3 \cdot 5 \end{array} \right\}$ LCM $= 2^2 \cdot 3^3 \cdot 5 = 540$

13. Find the LCM for $4x$, x^2y, $6x^2$, and $18y^3$. (**Hint:** Use each variable as a factor.)

Solution

$$\left.\begin{array}{l} 4x = 2^2 \cdot x \\ x^2y = x^2 \cdot y \\ 6x^2 = 2 \cdot 3 \cdot x^2 \\ 18y^3 = 2 \cdot 3^2 \cdot y^3 \end{array}\right\} \quad \text{LCM} = 2^2 \cdot 3^2 \cdot x^2 \cdot y^3 = 36x^2y^3$$

∎

Now, to add fractions with different denominators:

1. Find the LCM of the denominators.
2. Change each fraction to an equal fraction with the LCM as denominator.
3. Add the new fractions.

EXAMPLES

14. $\dfrac{1}{4} + \dfrac{3}{8} + \dfrac{3}{10}$

Solution

$$\left.\begin{array}{l} 4 = 2^2 \\ 8 = 2^3 \\ 10 = 2 \cdot 5 \end{array}\right\} \quad \text{LCM} = 2^3 \cdot 5 = 40$$

To get a common denominator, we see

$$40 = 4 \cdot 10 = 8 \cdot 5 = 10 \cdot 4.$$

$$\dfrac{1}{4} + \dfrac{3}{8} + \dfrac{3}{10} = \dfrac{1 \cdot \mathbf{10}}{4 \cdot \mathbf{10}} + \dfrac{3 \cdot \mathbf{5}}{8 \cdot \mathbf{5}} + \dfrac{3 \cdot \mathbf{4}}{10 \cdot \mathbf{4}} \qquad \text{Multiply each fraction by 1.}$$

$$= \dfrac{10}{40} + \dfrac{15}{40} + \dfrac{12}{40} \qquad \text{Each fraction has the same denominator.}$$

$$= \dfrac{37}{40} \qquad \text{Add the fractions.}$$

15. $\dfrac{5}{21a} + \dfrac{5}{28a}$

Solution

$$\left.\begin{array}{l} 21a = 3 \cdot 7 \cdot a \\ 28a = 2^2 \cdot 7 \cdot a \end{array}\right\} \quad \begin{array}{l} \text{LCM} = 2^2 \cdot 3 \cdot 7 \cdot a = 84a \\ \qquad\qquad = 21a \cdot \mathbf{4} \\ \qquad\qquad = 28a \cdot \mathbf{3} \end{array}$$

$$\dfrac{5}{21a} + \dfrac{5}{28a} = \dfrac{5 \cdot \mathbf{4}}{21a \cdot \mathbf{4}} + \dfrac{5 \cdot \mathbf{3}}{28a \cdot \mathbf{3}} \qquad \begin{array}{l}\text{In each fraction, the numerator} \\ \text{and denominator are multiplied} \\ \text{by the same number to get } 84a \text{ as} \\ \text{denominator.}\end{array}$$

$$= \dfrac{20}{84a} + \dfrac{15}{84a} = \dfrac{35}{84a}$$

$$= \dfrac{\cancel{7} \cdot 5}{\cancel{7} \cdot 12a} = \dfrac{5}{12a}$$

∎

The difference of two fractions with a common denominator is found by subtracting the numerators and using the common denominator.

$$\frac{a}{b} - \frac{c}{b} = \frac{a - c}{b}$$

If the two fractions do not have the same denominator, find equal fractions with the least common denominator, just as with addition.

EXAMPLES

16. $\dfrac{5}{8} - \dfrac{1}{8}$

    Solution $\dfrac{5}{8} - \dfrac{1}{8} = \dfrac{5 - 1}{8} = \dfrac{4}{8} = \dfrac{\not 4 \cdot 1}{\not 4 \cdot 2} = \dfrac{1}{2}$

17. $\dfrac{1}{45} - \dfrac{1}{72}$

    Solution $\dfrac{1}{45} - \dfrac{1}{72}$

$$\left.\begin{array}{l} 45 = 3^2 \cdot 5 \\ 72 = 2^3 \cdot 3^2 \end{array}\right\} \quad \begin{array}{l} \text{LCM} = 2^3 \cdot 3^2 \cdot 5 = 360 \\ \phantom{\text{LCM}} = 45 \cdot 8 \\ \phantom{\text{LCM}} = 72 \cdot 5 \end{array}$$

$$\frac{1}{45} - \frac{1}{72} = \frac{1 \cdot 8}{45 \cdot 8} - \frac{1 \cdot 5}{72 \cdot 5} = \frac{8}{360} - \frac{5}{360} = \frac{3}{360} = \frac{\not 3 \cdot 1}{\not 3 \cdot 120} = \frac{1}{120} \quad \blacksquare$$

To evaluate an expression such as $\dfrac{1}{2} + \dfrac{3}{8} \div \dfrac{3}{4}$ that involves more than one operation, we use the same rules for order of operations discussed in Section 1.1.

EXAMPLES

18. $\dfrac{1}{2} + \dfrac{3}{8} \div \dfrac{3}{4}$

    Solution $\dfrac{1}{2} + \dfrac{3}{8} \div \dfrac{3}{4} = \dfrac{1}{2} + \dfrac{3}{8} \cdot \dfrac{4}{3}$ Divide.

$$= \frac{1}{2} + \frac{\not 3 \cdot \not 4 \cdot 1}{2 \cdot \not 4 \cdot \not 3} \qquad \text{Reduce.}$$

$$= \frac{1}{2} + \frac{1}{2} \qquad\qquad\quad \text{Add.}$$

$$= \frac{2}{2} = 1$$

19. $\dfrac{2}{7} \cdot \dfrac{14}{3} + \dfrac{5}{8} \cdot \dfrac{2}{5}$

Solution

$$\dfrac{2}{7} \cdot \dfrac{14}{3} + \dfrac{5}{8} \cdot \dfrac{2}{5} = \dfrac{2 \cdot 14}{7 \cdot 3} + \dfrac{5 \cdot 2}{8 \cdot 5} \qquad \text{Multiply.}$$

$$= \dfrac{2 \cdot \cancel{7} \cdot 2}{\cancel{7} \cdot 3} + \dfrac{\cancel{5} \cdot \cancel{2} \cdot 1}{\cancel{2} \cdot 4 \cdot \cancel{5}} \qquad \text{Reduce.}$$

$$= \dfrac{4}{3} + \dfrac{1}{4}$$

$$= \dfrac{4 \cdot 4}{3 \cdot 4} + \dfrac{1 \cdot 3}{4 \cdot 3} \qquad \begin{array}{l}\text{Common denominator}\\ \text{is 12.}\end{array}$$

$$= \dfrac{16}{12} + \dfrac{3}{12} = \dfrac{19}{12} \qquad \text{Add.}$$

20. Divide the sum of $\dfrac{5}{8}$ and $\dfrac{3}{4}$ by $\dfrac{3}{2}$.

Solution First find the sum of $\dfrac{5}{8}$ and $\dfrac{3}{4}$.

$$\dfrac{5}{8} + \dfrac{3}{4} = \dfrac{5}{8} + \dfrac{3 \cdot 2}{4 \cdot 2} = \dfrac{5}{8} + \dfrac{6}{8} = \dfrac{11}{8}$$

Now divide the sum by $\dfrac{3}{2}$.

$$\dfrac{11}{8} \div \dfrac{3}{2} = \dfrac{11}{8} \cdot \dfrac{2}{3} = \dfrac{11 \cdot \cancel{2}}{\cancel{2} \cdot 4 \cdot 3} = \dfrac{11}{12}$$

The answer is $\dfrac{11}{12}$.

21. Suppose that a True–False test contains 60 questions. If $\dfrac{3}{4}$ of the questions are True, how many are False?

Solution Since $\dfrac{3}{4}$ are True, then $\dfrac{1}{4}$ are False. We find $\dfrac{1}{4}$ of 60. A fraction of a number means to multiply.

$$\dfrac{1}{4} \cdot 60 = \dfrac{1}{\cancel{4}} \cdot \dfrac{\overset{15}{\cancel{60}}}{1} = 15$$

Fifteen questions are False. ∎

Practice Problems

Perform the indicated operations and reduce all answers to lowest terms.

(handwritten work at left:)
$\frac{5}{1} \cdot \frac{9}{100} \quad \frac{9}{20}$

20

$\frac{7}{8} \cdot \frac{5}{4} \cdot \frac{3}{20} \qquad \frac{7}{16}$

1. $5 \cdot \left(\dfrac{3}{10}\right)^2$

2. $\dfrac{3}{4} \div \dfrac{4}{3}$ *(handwritten:)* $\frac{3}{4} \cdot \frac{3}{4} = \frac{9}{16}$

3. $\dfrac{7}{3} \div \dfrac{4}{5} \cdot \dfrac{3}{20}$

4. $\dfrac{2}{2a} - \dfrac{1}{2a}$ *(handwritten:)* $\frac{1}{2a}$

5. $\left(\dfrac{5}{24} + \dfrac{7}{36}\right) \cdot 3$

6. $\dfrac{1}{3} \div \dfrac{1}{2} + \dfrac{1}{5} \cdot \dfrac{5}{3}$

EXERCISES 1.2

(handwritten:) $\frac{5}{2} \quad \frac{15+14}{72} \cdot \frac{3}{72} \cdot \frac{3}{1} \quad \frac{29 \cdot 8}{72} \quad \frac{29}{24} \quad \frac{1}{3} \cdot \frac{2}{1} + \frac{1}{5} \cdot \frac{5}{3}$
$\frac{24}{}$
$\frac{2}{3} + \frac{1}{3} = 1$

Find the missing numerators in Exercises 1–4 by changing each fraction to an equal fraction with the indicated denominator.

1. $\dfrac{5 \cdot 8}{6 \cdot 8} = \dfrac{40}{48}$

2. $\dfrac{3 \cdot 4}{13 \cdot 4} = \dfrac{12}{52}$

3. $\dfrac{0 \cdot 7}{9 \cdot 7} = \dfrac{0}{63b}$

4. $\dfrac{7 \cdot 6}{24 \cdot 6} = \dfrac{42}{144x}$

Reduce each fraction to lowest terms in Exercises 5–8.

5. $\dfrac{18}{45} = \dfrac{2 \cdot 3 \cdot 3}{3 \cdot 3 \cdot 5} = \dfrac{2}{5}$

6. $\dfrac{35}{63} = \dfrac{5 \cdot 7}{3 \cdot 3 \cdot 7} = \dfrac{5}{9}$

7. $\dfrac{150}{350} = \dfrac{3 \cdot 5}{7 \cdot 5}$

8. $\dfrac{72}{108} = \dfrac{8 \cdot 9 = 2 \cdot 2 \cdot 2 \cdot 3 \cdot 3}{2 \cdot 54 = 2 \cdot 2 \cdot 3 \cdot 3 \cdot 3} = \dfrac{2}{3}$

Find (a) the factors and (b) the first six multiples for each of the numbers in Exercises 9–12.

9. 6 *(handwritten:)* $1, 2, 3, 6$; $6, 12, 18, 24, 30, 36$

10. 12 *(handwritten:)* $2 \cdot 6$; $12, 24, 36, 48, 60$; 66

11. 15 *(handwritten:)* $1, 3, 5, 15$; $15, 30, 45, 60, 75, 90$

12. 18 *(handwritten:)* $1, 2, 3, 6, 9, 18$; $18, 36, 54, 72, 90, 108$

Find the LCM for each set of numbers in Exercises 13–18.

13. 24, 15, 10 *(handwritten:)* $24 = 24, 48, 72, 96, 120$; $15 = 15, 30, 45, 60, 75, 90$; $10 = 120$; 120

14. 49, 25, 35

15. $8x$, $10y$, $20xy$

16. $20xz$, $24xy$, $32yz$

17. $14x^2$, $21xy$, $35x^2y^2$

18. $60x$, $105x^2y$, $120xy$

Perform the indicated operations in Exercises 19–53. Reduce each answer to lowest terms.

19. $\dfrac{3}{8} \cdot \dfrac{4}{9} = \dfrac{1}{6}$

20. $\dfrac{4}{5} \cdot \dfrac{3}{7} = \dfrac{12}{35}$

21. $\dfrac{3}{8} \div \dfrac{3}{4} = \dfrac{1}{2}$

22. $\dfrac{4}{5} \div \dfrac{8}{7} = \dfrac{4}{5} \cdot \dfrac{7}{8} = \dfrac{7}{10}$

23. $\dfrac{2}{9} + \dfrac{5}{9} = \dfrac{7}{9}$

24. $\dfrac{2}{7} + \dfrac{8}{7} + \dfrac{4}{7} = \dfrac{14}{7} = 2$

25. $\dfrac{14}{23} - \dfrac{9}{23} = \dfrac{5}{23}$

26. $\dfrac{7}{15} - \dfrac{2}{15} = \dfrac{5}{15} = \dfrac{1}{3}$

27. $\dfrac{11}{12} \cdot \dfrac{13}{12} = \dfrac{143}{144}$

28. $\dfrac{5}{16} \cdot \dfrac{9}{16} = \dfrac{45}{16}$

29. $\dfrac{5}{9} + \dfrac{7}{9} = \dfrac{12}{9} = \dfrac{4}{3}$

30. $\dfrac{11}{15a} - \dfrac{7}{15a} + \dfrac{2}{15a} = \dfrac{2}{15}$

31. $\dfrac{19}{25x} - \dfrac{7}{25x} - \dfrac{2}{25x} = \dfrac{2}{5x}$

32. $\dfrac{16x}{7} \cdot \dfrac{49}{64x} = \dfrac{7}{4}$

33. $\dfrac{26b}{51a} \cdot \dfrac{4a}{39} = \dfrac{8Nb}{153A}$

34. $\dfrac{15}{4} \cdot \dfrac{5}{6} \cdot \dfrac{16}{9} = \dfrac{100}{18} = 2.50$; $2 \cdot 3 \cdot 3$

35. $\dfrac{9a}{15} \cdot \dfrac{10}{3b} \cdot \dfrac{1}{5a^2}$

36. $\dfrac{5}{4a} \div \dfrac{1}{5a} = \dfrac{5}{4a} \cdot \dfrac{5a}{1}$

37. $\dfrac{4}{13} \div 0$ *(handwritten:)* UNDEFINED

38. $0 \div \dfrac{7}{3x}$ *(handwritten:)* UNDEFINED

Answers to Practice Problems 1. $\dfrac{9}{20}$ 2. $\dfrac{9}{16}$ 3. $\dfrac{7}{16}$ 4. $\dfrac{1}{2a}$ 5. $\dfrac{29}{24}$ 6. 1

39. $\dfrac{10}{3} \div 6x$ $\dfrac{10}{3} \cdot \dfrac{1}{6x} = \dfrac{10}{18x} = \dfrac{5}{9x}$

40. $\dfrac{13}{6x} \div \dfrac{2}{5} \cdot \dfrac{3x}{25}$

41. $\dfrac{7a}{10x} \div \dfrac{4}{5} \cdot \dfrac{3x^2}{28}$

42. $\dfrac{5}{6} + \dfrac{1}{8}$

43. $\dfrac{11}{12} + \dfrac{7}{15}$

44. $\dfrac{27}{40} - \dfrac{5}{8}$

45. $\dfrac{2}{7} - \dfrac{1}{6}$

46. $\dfrac{5}{24} + \dfrac{7}{18} - \dfrac{3}{8}$

47. $\dfrac{8}{35} + \dfrac{6}{10} - \dfrac{5}{14}$

48. $\dfrac{8}{9} - \dfrac{5}{12} - \dfrac{1}{4}$

49. $\dfrac{2}{3x} + \dfrac{1}{4x} - \dfrac{1}{6x}$

50. $\dfrac{7}{10y} - \dfrac{1}{5y} + \dfrac{1}{3y}$

51. $\dfrac{11}{24x} - \dfrac{7}{18x} + \dfrac{5}{36x}$

52. $\dfrac{7}{6b} - \dfrac{9}{10b} + \dfrac{4}{15b}$

53. $\dfrac{1}{28a} + \dfrac{8}{21a} - \dfrac{5}{12a}$

Simplify the expressions in Exercises 54–63 using the rules for order of operations.

54. $\dfrac{3}{8} \cdot \dfrac{4}{5} + \dfrac{1}{15}$

55. $\dfrac{1}{3} \div \dfrac{1}{2} - \dfrac{5}{6} \cdot \dfrac{3}{4}$

56. $\left(\dfrac{5}{6}\right)^2 \div \dfrac{5}{12} - \dfrac{3}{8}$

57. $\left(\dfrac{2}{5}\right)^2 \cdot \dfrac{3}{8} + \dfrac{1}{5} \cdot \dfrac{3}{4}$

58. $\dfrac{3}{4} \div \dfrac{3}{16} - \dfrac{2}{3} \cdot \dfrac{3}{4}$

59. $\dfrac{3}{4} + \dfrac{5}{6} \div \dfrac{5}{8} - \dfrac{1}{3}$

60. $\dfrac{5}{7} + \dfrac{1}{2} \cdot \dfrac{2}{3} - \dfrac{1}{12}$

61. $\left(\dfrac{7}{8} - \dfrac{7}{16}\right) \cdot \dfrac{6}{7} - \dfrac{1}{4}$

62. $\left(\dfrac{7}{10} + \dfrac{3}{5}\right) \div \left(\dfrac{1}{2} + \dfrac{3}{7}\right)$

63. $\left(\dfrac{2}{3} - \dfrac{1}{5}\right) \div \left(\dfrac{5}{7} - \dfrac{1}{3}\right)$

64. Multiply the quotient of $\dfrac{19}{2}$ and $\dfrac{13}{4}$ by $\dfrac{13}{12}$.

65. Find the product of $\dfrac{11}{4}$ with the quotient of $\dfrac{8}{5}$ and $\dfrac{11}{6}$.

66. Divide the product of $\dfrac{5}{8}$ and $\dfrac{9}{10}$ by the product of $\dfrac{9}{10}$ and $\dfrac{4}{5}$.

67. The product of $\dfrac{9}{16}$ and $\dfrac{13}{7}$ is added to the quotient of $\dfrac{10}{3}$ and $\dfrac{7}{4}$. What is the sum?

68. Find the quotient of the sum of $\dfrac{4}{5}$ and $\dfrac{11}{15}$ divided by the difference between $\dfrac{7}{12}$ and $\dfrac{4}{15}$.

69. Seventy-two boys reported to football practice. If $\dfrac{2}{3}$ of the boys made the team, how many were on the team? $\dfrac{2}{3} \cdot \cancel{72}^{24} = \dfrac{24 \cdot 2}{1} = 48$

70. In scale for a blueprint of a building, 1 in. represents 48 ft. If on the drawing the building is $\dfrac{7}{8}$ in. by $\dfrac{5}{4}$ in., what will be the actual size of the building?

71. The seventeenth hole at the local golf course is a par 4 hole. Ralph drove his ball 258 yards. If this distance was $\dfrac{3}{4}$ the length of the hole, how long is the seventeenth hole?

72. How much space on a shelf will Betty need to hold 3 books $\dfrac{3}{4}$-in. thick and 4 books $\dfrac{7}{8}$-in. thick?

$\dfrac{3}{1} \quad \dfrac{3}{4} = \dfrac{9}{4} \qquad \dfrac{9+28}{4} =$

$\dfrac{4}{1} \quad \dfrac{7}{8} \quad \dfrac{28}{8}$

$\dfrac{23}{4} \qquad \dfrac{9 \times 2 = 18 + 28}{4 \times 2 = 8 \quad 8}$

$\dfrac{18 + 28}{8} = \dfrac{46}{8} = 23 \cdot \dfrac{1}{4 \cdot \dfrac{1}{1}}$

1.3 Decimals, Percents, and Word Problems

OBJECTIVES

In this section, you will be learning to:

1. Change decimals to percents.
2. Change percents to decimals.
3. Change fractions to decimals.
4. Operate with decimals.
5. Find percents of numbers.
6. Find fractional parts of numbers.
7. Solve word problems involving whole numbers, decimals, fractions, and percents.

Our daily lives are bombarded with decimal numbers and percents: stock market reports, batting averages, won-lost records, salary raises, measures of pollution, checkbook balancing, daily vitamin dosages, interest on savings, and on and on. Since decimal numbers and percents play such a prominent role in our society, we need to understand how to operate with them and apply them correctly in a variety of practical situations.

Decimals

A **decimal number** (or simply a **decimal**) is a fraction that has a power of ten in its denominator. Thus,

$$\frac{3}{10}, \frac{51}{100}, \frac{24}{10}, \frac{17}{1}, \text{ and } \frac{3}{1000}$$

are all decimal numbers. Decimal numbers can be written with a decimal point and a place-value system that indicates whole numbers to the left of the decimal point and fractions less than 1 to the right of the decimal point. The values of each place are indicated in Figure 1.1.

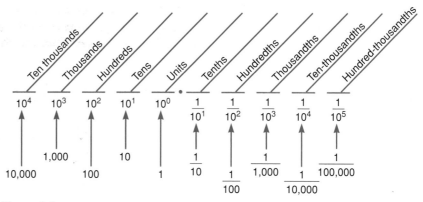

Figure 1.1

To add or subtract decimal numbers, line up the decimal points, one under the other, then add or subtract as with whole numbers. Place the decimal point in the answer in line with the other decimal points. This technique guarantees that digits having the same place value will be added together or subtracted from each other.

EXAMPLES

1. Add: $37.498 + 5.63 + 42.781$

   Solution

$$
\begin{array}{r}
37.498 \\
5.630 \\
+42.781 \\
\hline
85.909
\end{array}
$$

2. Subtract: $26.872 - 13.99$

Solution

$$\begin{array}{r} 26.872 \\ -13.990 \\ \hline 12.882 \end{array}$$ ∎

To find the product of two decimal numbers, multiply as with whole numbers, then place the decimal point so that the number of digits to its right is equal to the sum of the number of digits to the right of the decimal points in the numbers being multiplied.

EXAMPLES   Find the products.

3. $\begin{array}{r} 4.78 \\ \times\ 0.3 \\ \hline \end{array}$

Solution $\begin{array}{r} 4.78 \\ \times\ 0.3 \\ \hline 1.434 \end{array}$

4. $\begin{array}{r} 16.4 \\ \times 0.517 \\ \hline \end{array}$

Solution $\begin{array}{r} 16.4 \\ \times 0.517 \\ \hline 1148 \\ 164 \\ 820 \\ \hline 8.4788 \end{array}$ ∎

To find the quotient of two decimal numbers, move the decimal point in the divisor to the right to get a whole number. Move the decimal point in the dividend the same number of places. Divide as with whole numbers.

EXAMPLE

5. Find the quotient: $3.2\overline{)51.52}$

Solution $3.2.\overline{)51.5.2}$ $32.\overline{)515.2}$ ← decimal point in quotient

one   one
place   place

$$\begin{array}{r} 16.1 \\ 32\overline{)515.2} \\ \underline{32} \\ 195 \\ \underline{192} \\ 3\ 2 \\ \underline{3\ 2} \end{array}$$ ∎

Percents

Decimals and percents (%) are closely related because "percent" means hundredths. Thus, 70% and 0.70 have the same meaning. Similarly,

$$65\% = 0.65 \quad \text{and} \quad 125\% = 1.25.$$

To change a decimal to a percent, move the decimal point two places to the right and write the % sign. (This is the same as multiplying by 100, then adding the % sign.)

To change a percent to a decimal, move the decimal point two places to the left and drop the % sign. (This is the same as multiplying by 0.01, then dropping the % sign.)

EXAMPLES

6. Change the decimals to percents.
   a. 0.3   **b.** 0.73   **c.** 1.4   **d.** 0.356
   Solution

   a. 0.3 = 30%   **b.** 0.73 = 73%
   c. 1.4 = 140%   **d.** 0.356 = 35.6%

7. Change the percents to decimals.
   a. 28%   **b.** 4.3%   **c.** 5%   **d.** 50%
   Solution

   a. 28% = 0.28   **b.** 4.3% = 0.043
   c. 5% = 0.05   **d.** 50% = 0.50   ∎

In many practical problems, such as sales tax, discounts, and budgeting, we are to find a percent of a number. **To find a percent of a number, change the percent to a decimal, then multiply.**

EXAMPLES

8. Find 40% of 15.
   Solution Multiply to find a percent of a number.

   $$\begin{array}{r} 15 \\ \times 0.40 \\ \hline 6.00 \end{array}$$

   40% of 15 is 6.

9. Find 160% of 35.
   Solution
   $$\begin{array}{r} 35 \\ \times 1.60 \\ \hline 21\ 00 \\ 35 \\ \hline 56.00 \end{array}$$

   160% of 35 is 56.

   Note that more than 100% of a number is larger than the number.   ∎

Finding a fractional part of a number is similar to finding a percent of a number. For example, find $\frac{3}{4}$ of 12 or $\frac{7}{8}$ of 1000. As with percents, we multiply. There are two procedures that are frequently used.

1. Multiply using the techniques for fractions.

$$\frac{3}{4} \cdot 12 = \frac{3}{\cancel{4}} \cdot \frac{\cancel{12}^{3}}{1} = 9$$

Thus, 9 is $\frac{3}{4}$ of 12.

or **2.** Change the fraction to a decimal (by dividing the numerator by the denominator), then multiply using the decimal.

$$\frac{3}{4} = .75 \quad \text{and}$$

$$
\begin{array}{r}
12 \\
\times .75 \\
\hline
60 \\
8\ 4 \\
\hline
9.00
\end{array}
$$

This second procedure is particularly useful when calculators are used. When using a calculator to find percents or fractional parts of a number, remember to change the percents and fractions to decimals first, then multiply.

CALCULATOR EXAMPLES   Use your calculator to verify the following results.

10. $8.6321 + 7.5476 + 2.143 + 17.8293 = 36.152$

11. $(14.763)(0.47)(321.6) = 2231.457$

12. 18.3% of 210.55 is 38.53065.

13. $\dfrac{7}{8}$ of 1000 is 875.   ■

Word Problems

Now we are interested in applying what we have learned about whole numbers, fractions, decimals, and percents to practical situations (word problems). The student must:

1. Read each problem thoroughly (maybe several times).
2. On the basis of personal experience and understanding, decide what operations to perform and on what numbers.
3. Check to see that the answer is reasonable.

EXAMPLES

14. Six monthly payments to purchase a used car were $150, $175, $230, $200, $180, and $259. What was the total of the payments? What was the average monthly payment?

Solution   To find the total, add all six numbers. To find the average monthly payment, divide the sum by 6.

$$
\begin{array}{r}
\$\ 150 \\
175 \\
230 \\
200 \\
180 \\
259 \\
\hline
\$1194
\end{array}
\quad \text{total}
$$

$$
\begin{array}{r}
\$199 \qquad \text{average monthly payment} \\
6\overline{)1194} \\
\underline{6} \\
59 \\
\underline{54} \\
54 \\
\underline{54}
\end{array}
$$

15. A bicycle was purchased for $\frac{3}{4}$ of the asking price of $160. If 6% sales tax was added to the purchase price, what was paid for the bicycle?

Solution Multiply to find a fraction of a number.

$$\frac{3}{4} \cdot 160 = \frac{3}{\cancel{4}} \cdot \frac{\cancel{160}^{40}}{1} = \$120 \qquad \text{the purchase price}$$

$$
\begin{array}{ll}
\$120 & \$120.00 \\
\underline{.06} & \underline{7.20} \\
\$7.20 \quad \text{sales tax} & \$127.20 \quad \text{total price including sales tax}
\end{array}
$$

Practice Problems

1. Change 0.3 to a percent. 30%
2. Change 6.4% to a decimal. 0.064
3. Find the sum: 73.6 + 18.15 + 2.98 94.73
4. Multiply: (2.57)(3.1)
5. Find 12% of 200.
6. What is $\frac{3}{4}$ of 92? 69

EXERCISES 1.3

Change the decimals to percents in Exercises 1–4.

1. 0.91 **2.** 0.625 **3.** 1.37 **4.** 2.125

Change the percents to decimals in Exercises 5–8.

5. 69% **6.** 7.5% **7.** 11.3% **8.** 162%

Change the fractions to decimals in Exercises 9–12.

9. $\frac{3}{8}$ **10.** $\frac{4}{5}$ **11.** $\frac{1}{20}$ **12.** $\frac{3}{50}$

Add or subtract as indicated in Exercises 13–20.

13. 243.7 + 65.22 + 8.31 **14.** 29.51 + 17.2 + 72.4 **15.** 420.43 − 156.92
16. 87.0 − 66.31 **17.** 65.13 + 44.81 − 75.9 **18.** 84.1 + 17.63 − 12.98
19. 147.0 + 79.6 − 85.43 **20.** 6.49 + 103.81 − 59.62

Find the indicated products in Exercises 21–24.

21. 60.4 $\times$ 1.8 **22.** 21.6 $\times$ 0.83 **23.** (1.23)(6.2) **24.** (5.83)(0.24)

Find the indicated quotients in Exercises 25 and 26.

25. $5.1\overline{)77.01}$ **26.** $0.023\overline{)0.4922}$

Answers to Practice Problems **1.** 30% **2.** 0.064 **3.** 94.73 **4.** 7.967 **5.** 24 **6.** 69

27. What is 11% of 93?

28. What is 6% of 480?

29. What is 38% of 147?

30. 81% of 76 is equal to _____ .

31. 102% of 87 is equal to _____ .

32. 112% of 620 is equal to _____ .

33. What is 108% of 735?

34. What is 125% of 350?

35. 9.5% of 570 is equal to _____ .

36. 10.3% of 986 is equal to _____ .

37. Find 24.3% of 85.

38. Find 118.6% of 68.

39. Find $\frac{1}{2}$ of 76.

40. Find $\frac{4}{5}$ of 29.

41. Find $\frac{1}{4}$ of 63.

42. Find $\frac{5}{8}$ of 37.

43. Find $\frac{7}{25}$ of 30.

44. Find $\frac{9}{16}$ of 60.

45. During the first four months of the year, a company showed profits of $7483, $9157, $10,544, and $9280. What was the total profit? What was the average profit per month?

46. A family's electricity bills were $68.45, $56.83, $49.78, $46.90, and $42.39 for the months January through May. Find the average monthly cost of electricity for these months.

47. Sound travels approximately 1080 feet per second. If the sound of thunder is heard 9 seconds after the lightning is seen, how far away was the lightning?

48. The odometer on a car changed from 36,849 to 37,485 in 12 hours of driving. Find the average miles traveled per hour.

49. How tall is a stack of 14 boards if each board is 0.75 inch thick?

50. Five boards, each exactly 2 feet long, are cut from a board 12 feet long. If the waste for each cut is 0.012 foot, what is the length of the remaining piece?

51. How many pieces of wood $\frac{2}{3}$ foot long can be cut from a board 6 feet long?

52. Twelve people in a math class made above 90% on a test. If this was $\frac{3}{8}$ of the entire class, how many people were in the class?

53. Georgia must pay 11% interest on a loan of $6000. How much interest will she pay in one year?

54. A salesperson works on a commission of 9%. What would be the commission on sales of $8420?

55. A salesman's salary is determined as follows: $800 per month plus a commission of 7% based on the amount of sales exceeding $5000. Find the amount of last month's check if he sold $11,470 in merchandise.

56. To buy a used car, you can pay $3250 cash or you can put down $800 and make 24 monthly payments of $131.50. How much can you save by paying cash?

57. A ski shop is selling out its new skis and its used rental skis. If the new skis are priced at $160.00 per pair and the used ones at $69.50 per pair, what will be the total received if 8 pairs of new skis and 13 pairs of used ones are sold?

58. Two couples went out to dinner. The dinner prices were $15.95, $17.25, $27.95, and $17.95. If a sales tax of 6% was added to the bill, what would be the total bill? If they left a tip of 15% of the total bill, what was the amount of the tip? (Round off to the nearest cent.)

Calculator Problems

Use a calculator to solve Exercises 59–68.

59. $27.081 + 13.1152 - 20.539$

60. $(0.6534)(142.5) + 6.831$

61. $(11.36)(19.83) - (2.64)^2$

62. Find 16.67% of 24.3.

63. Find 1.05% of 5.3.

64. Five bars of iron weigh 5.75, 6.375, 4.5625, 5.125, and 7.5 pounds. Find the total weight and the average weight.

65. A piece of wood $7\frac{11}{16}$ (or 7.6875) inches long is cut from a board 18 inches long. If the cut itself wastes $\frac{1}{8}$ (or 0.125) inch, how long is the remaining piece?

66. Four pieces of ribbon, each measuring $3\frac{5}{8}$ (or 3.625) inches, are cut from a piece of ribbon 20 inches long. How long is the remaining piece?

67. How many pieces of wire, each $2\frac{3}{8}$ (or 2.375) inches long, can be cut from a piece of wire 22 inches long?

68. Natalie works part time. She worked 3.25, 4.5, and 3.75 hours for one three-day period. She is paid $5.65 per hour. How much did she earn? If 12% of her wages is held out for taxes, how much did she pay in taxes?

1.4 Variables and Solutions to Equations

A **variable** is a symbol or letter used to represent numbers. A variable in algebra is similar to a pronoun such as "he" or "she" in an English sentence. Consider the two sentences:

He can drive a car. He lives in Canada.

The pronoun "he" can be replaced in both sentences by the name of any male driver living in Canada.

Just as pronouns can represent unknown persons in English sentences, variables are used to represent unknown numbers in **algebraic expressions.** An algebraic expression is an expression that indicates the sum, product, difference, or quotient of constants and variables. For example,

$$x + y, \; 3a - 7, \; 2(x + 3), \text{ and } 8a^2 + 3b^2 - 5$$

are all algebraic expressions. The symbols we use as variables are generally letters such as $a, b, x, y,$ and z or letters from other alphabets, such as the Greek letters α (alpha), β (beta), and γ (gamma).

If an expression contains variables, then the value of that expression depends on the numbers substituted for the variables. **Any value substituted for a variable must be substituted for every occurrence of that variable in an expression.**

For example, using the rules for order of operations, evaluate the following expression:

$$2x + 3 + 5x \qquad \text{for } x = 4 \text{ and again for } x = 3.$$

a. If $x = 4$, then

$$2x + 3 + 5x = 2 \cdot 4 + 3 + 5 \cdot 4$$
$$= 8 + 3 + 20 = 31$$

b. If $x = 3$, then

$$2x + 3 + 5x = 2 \cdot 3 + 3 + 5 \cdot 3$$
$$= 6 + 3 + 15 = 24$$

EXAMPLES

1. Evaluate the expression
   $3a + 5 - 2a$ for $a = 2$
   Solution $3a + 5 - 2a = 3 \cdot 2 + 5 - 2 \cdot 2$
   $$= 6 + 5 - 4$$
   $$= 11 - 4 = 7$$

2. Evaluate the expression
   $3a^2 - 7a$ for $a = \dfrac{5}{2}$

   Solution $3a^2 - 7a = 3 \cdot \left(\dfrac{5}{2}\right)^2 - 7 \cdot \dfrac{5}{2}$

   $$= 3 \cdot \dfrac{25}{4} - \dfrac{35}{2}$$

   $$= \dfrac{75}{4} - \dfrac{70}{4} = \dfrac{5}{4}$$ ∎

An **equation** states that two algebraic expressions are equal. That is, both expressions represent the same number. All of the following are examples of equations:

$$6 + 7 = 10 + 3, \qquad x + 3 = 11, \qquad \text{and} \qquad 2y - 5 = y + 7.$$

We will discuss techniques for solving equations throughout the text. In this section, you will be asked only to verify that a certain number is a **solution** to an equation.

Any value for a variable that gives a true statement when that value is substituted for the variable is a **solution to the equation.** For example,

   a. $x = 5$ **is** a solution to $3x + 4 = 19$
   since $3 \cdot 5 + 4 = 19$ is true.

   b. $x = 3$ **is not** a solution to $3x + 4 = 19$
   since $3 \cdot 3 + 4 = 9 + 4 = 13$ and $13 \neq 19$.

EXAMPLES Determine whether or not the given number is a solution to the given equation by substituting it into the equation.

3. $5x = 20$; $x = 4$
   Solution Substitute 4 for x:

   $$5x = 20$$
   $$5 \cdot 4 = 20 \qquad \text{true}$$

   4 is a solution of the equation.

4. $3q - 8 = 13$; $q = 7$

Solution Substitute 7 for q:

$$3q - 8 = 13$$
$$3 \cdot 7 - 8 = 13$$
$$21 - 8 = 13$$
$$13 = 13 \quad \text{true}$$

7 is a solution of the equation.

5. $5a + 4 = 6 + 3a$; $a = 2$

Solution Substitute 2 for a:

$$5a + 4 = 6 + 3a$$
$$5 \cdot 2 + 4 = 6 + 3 \cdot 2$$
$$10 + 4 = 6 + 6$$
$$14 = 12 \quad \text{false } (14 \neq 12)$$

2 is **not** a solution of the equation.

6. $5a + 4 = 6 + 3a$; $a = 1$

Solution Substitute 1 for a:

$$5a + 4 = 6 + 3a$$
$$5 \cdot 1 + 4 = 6 + 3 \cdot 1$$
$$5 + 4 = 6 + 3$$
$$9 = 9 \quad \text{true}$$

1 is a solution of the equation.

CALCULATOR EXAMPLE

7. Use a calculator to verify that $y = 3.75$ is a solution to $4.1y - 1.6 = 13.775$.

Solution $4.1(3.75) - 1.6 = 15.375 - 1.6 = 13.775$

Practice Problems

Evaluate for $a = 3$.

1. $2a - 6$

2. $3a^2 - a + 1$

3. Determine whether $x = 5$ is a solution to the equation $4x - 1 = 19$.

◢ EXERCISES 1.4

Evaluate each expression in Exercises 1–20 for the given value of the variable.

1. $x + 7$; $x = 2$ **2.** $x + 11$; $x = 0$ **3.** $8 - x$; $x = 4$

4. $17 - x$; $x = 6$ **5.** $2x - 1$; $x = 1$ **6.** $3y + 5$; $y = 5$

Answers to Practice Problems **1.** 0 **2.** 25 **3.** $x = 5$ is a solution.

7. $5y + 1$; $y = 0$ **8.** $8x + 3$; $x = 3$ **9.** $11 - 2x$; $x = 5$
10. $14 - 5x$; $x = 2$ **11.** $3y + 4 - y$; $y = 3$ **12.** $x + 5 + 2x$; $x = 1$
13. $4x + x + 3$; $x = 6$ **14.** $8 - 3y + y$; $y = 4$ **15.** $x^2 + 3$; $x = 6$
16. $3x^2 + x$; $x = 0$ **17.** $5y^2 - 3y + 1$; $y = 2$ **18.** $2x^2 + 4x + 4$; $x = 3$
19. $4x^2 + x + 7$; $x = 0$ **20.** $x^2 + 9x + 4$; $x = 5$

Evaluate the expressions in Exercises 21–26 if (a) $x = \dfrac{1}{2}$ and (b) $x = \dfrac{1}{3}$.

21. $x + 4$ **22.** $x + 2$ **23.** $2x + 5$ **24.** $3x + 6$ **25.** $3x - \dfrac{1}{4}$ **26.** $4x - \dfrac{1}{6}$

Evaluate the expressions in Exercises 27–32 if (a) $x = 1.4$ and (b) $x = 0.25$.

27. $2x + 3$ **28.** $4x + 2$ **29.** $5x + 8$ **30.** $7x + 4$ **31.** $8x - 1.3$ **32.** $9x - 2.03$

Determine whether or not the given number is a solution to the given equation in Exercises 33–52.

33. $3x = 18$; $x = 6$ **34.** $8x = 32$; $x = 4$ **35.** $5x + 1 = 17$; $x = 3$
36. $4x - 3 = 11$; $x = 4$ **37.** $4y - 1 = 17$; $y = 4$ **38.** $8 - 3y = 2$; $y = 2$
39. $15 - 2x = 5$; $x = 5$ **40.** $23 - 6y = 5$; $y = 3$ **41.** $7y + 3 = 2y + 8$; $y = 1$
42. $5x + 2 = 4x - 3$; $x = 5$ **43.** $6 - x = 4x - 9$; $x = 3$ **44.** $11 - 3x = 2x + 1$; $x = 2$
45. $2x + 3 = 8x$; $x = \dfrac{1}{2}$ **46.** $6 - 3x = 6x + 3$; $x = \dfrac{1}{3}$ **47.** $9y = y + 2$; $y = \dfrac{1}{4}$

48. $6y + 4 = 5y + 4$; $y = \dfrac{1}{5}$ **49.** $4x + 6 = 6.8$; $x = 0.2$ **50.** $5y + 3 = 2.8$; $y = 0.5$

51. $3x + 4 = 7x$; $x = 1.4$ **52.** $2x + 3 = 4x - 2$; $x = 2.5$

Calculator Problems

Use a calculator to determine whether or not the given number is a solution to the given equation in Exercises 53–58.

53. $6.2x + 1.57 = 15.52$; $x = 2.25$ **54.** $5.3y + 3.261 = y + 9.496$; $y = 1.45$
55. $4.2y - 1.573 = 2.3y + 2.9$; $y = 2.36$ **56.** $1.3x + 4.614 = 2.4x + 0.6$; $x = 3.62$
57. $0.3x + 8.537 = 3.1x + 2.713$; $x = 2.08$ **58.** $0.5x + 3.172 = 2x - 2.012$; $x = 3.456$

1.5 *Formulas in Geometry*

OBJECTIVE
In this section, you will be learning to recognize and use appropriate geometric formulas for computations.

A **formula** is an equation that represents a general relationship between two or more quantities or measurements. Several variables may appear in a formula, and, if values are known for all but one of these variables, the remaining value can be found. Formulas are useful in mathematics, economics, chemistry, medicine, physics, and many other fields of study. In this section, we will discuss some formulas related to simple geometric figures such as rectangles, circles, and triangles.

Perimeter

The **perimeter** of a geometric figure is the total distance around the figure. Perimeters are measured in units of length such as inches, feet, yards, miles, centimeters, and meters. The formulas for the perimeters of various geometric figures are shown in Figure 1.2.

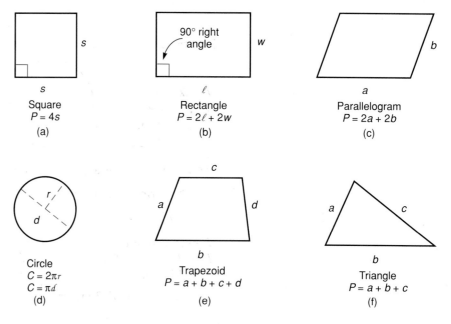

Square	Rectangle	Parallelogram
$P = 4s$	$P = 2\ell + 2w$	$P = 2a + 2b$
(a)	(b)	(c)

Circle	Trapezoid	Triangle
$C = 2\pi r$	$P = a + b + c + d$	$P = a + b + c$
$C = \pi d$	(e)	(f)
(d)		

Figure 1.2

Some special comments about circles:

1. The perimeter of a circle is called its **circumference, (***C***)**.
2. The distance from the center to a point on the circle is called its **radius, (***r***)**.
3. The distance from one point on a circle to another point on the circle measured through its center is called its **diameter, (***d***)**. The diameter is twice the radius. (*d* is the same as 2*r*.)
4. Pi (*π*) is the symbol used for the constant 3.1415926535. . . . For our purposes, we will use *π* = 3.14, but you should understand that this is only an approximation. By definition, the value of *π* can be found by dividing the circumference of any circle by its diameter.

EXAMPLES

1. Find the perimeter of a rectangle with a length of 10 feet and a width of 8 feet.

Solution

a. Sketch the figure.

b. $P = 2\ell + 2w$
$P = 2 \cdot 10 + 2 \cdot 8$
$= 20 + 16 = 36$ ft

The perimeter is 36 feet.

ℓ = 10 ft

w = 8 ft

2. Find the circumference of a circle with a diameter of 3 centimeters.

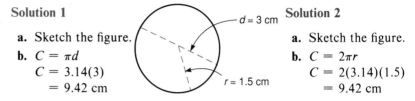

Solution 1

a. Sketch the figure.

b. $C = \pi d$
$C = 3.14(3)$
$= 9.42$ cm

d = 3 cm

r = 1.5 cm

Solution 2

a. Sketch the figure.

b. $C = 2\pi r$
$C = 2(3.14)(1.5)$
$= 9.42$ cm

The circumference is approximately 9.42 centimeters.

3. Find the perimeter of the triangle with sides labeled as in the figure.

Solution

$P = a + b + c$
$P = 3 + 6.2 + 8.1$
$= 17.3$ in.

The perimeter is 17.3 inches.

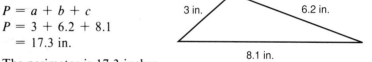

3 in.   6.2 in.

8.1 in.

■

Area

Area is a measure of the interior, or enclosure, of a surface. Area is measured in square units such as square feet, square inches, square meters, and square miles. For example, the area enclosed by a square of 1 inch on each side is 1 sq in. [Figure 1.3(a)], while the area enclosed by a square of 1 centimeter on each side is 1 sq cm [Figure 1.3(b)].

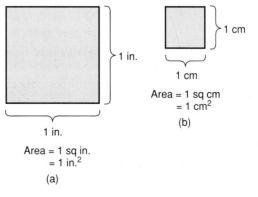

1 in.

1 in.

Area = 1 sq in.
= 1 in.²

(a)

1 cm

1 cm

Area = 1 sq cm
= 1 cm²

(b)

Figure 1.3

The formulas for finding the areas of several geometric figures are shown in Figure 1.4.

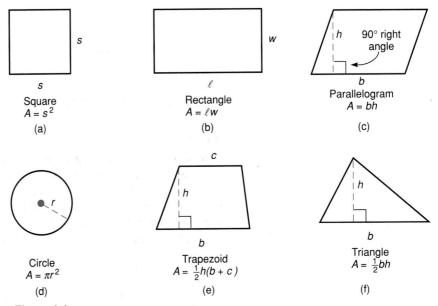

Square
$A = s^2$
(a)

Rectangle
$A = \ell w$
(b)

Parallelogram
$A = bh$
(c)

Circle
$A = \pi r^2$
(d)

Trapezoid
$A = \frac{1}{2}h(b + c)$
(e)

Triangle
$A = \frac{1}{2}bh$
(f)

Figure 1.4

EXAMPLES

4. Find the area of a triangle with a height of 3 centimeters and a base of 4 centimeters.

Solution

a. Sketch the figure.

b. $A = \dfrac{1}{2} \cdot b \cdot h$

$A = \dfrac{1}{2} \cdot 3 \cdot 4$

$= \dfrac{1}{2} \cdot 12 = 6$ sq cm or 6 cm²

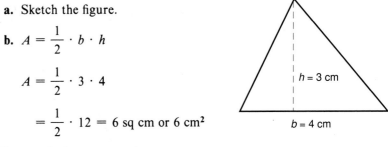

$h = 3$ cm

$b = 4$ cm

The area is 6 square centimeters.

5. Find the area of a circle with a radius of 6 inches.

Solution

a. Sketch the figure.

b. $A = \pi r^2$
$A = 3.14(6^2)$
$= 3.14(36)$
$= 113.04$ sq in. or 113.04 in.^{2}

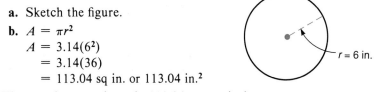

$r = 6$ in.

The area is approximately 113.04 square inches.   ∎

Volume

Volume is a measure of the space enclosed by a three-dimensional figure. Volume is measured in cubic units, such as cubic inches, cubic centimeters, and cubic feet. For example, the volume enclosed by a cube of 1 inch on each edge is 1 cu in. (or 1 in.^{3}) [Figure 1.5(a)], while the volume enclosed by a cube of 1 centimeter on each edge is 1 cu cm (or 1 cm^3) [Figure 1.5(b)].

The formulas for finding the volumes of some geometric solids are given in Figure 1.6.

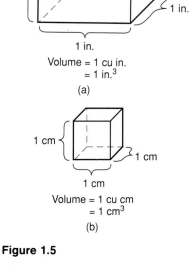

1 in.

1 in.

1 in.

Volume = 1 cu in.
= 1 in.^{3}

(a)

1 cm

1 cm

1 cm

Volume = 1 cu cm
= 1 cm^3

(b)

Figure 1.5

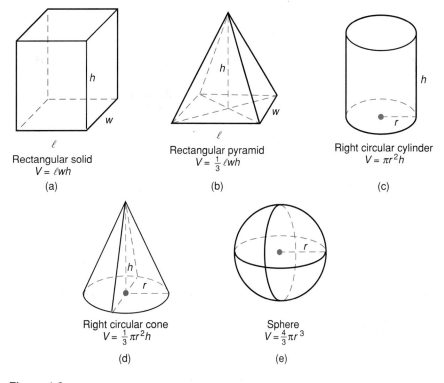

h

w

ℓ

Rectangular solid
$V = \ell wh$

(a)

h

w

ℓ

Rectangular pyramid
$V = \frac{1}{3}\ell wh$

(b)

h

r

Right circular cylinder
$V = \pi r^2 h$

(c)

h

r

Right circular cone
$V = \frac{1}{3}\pi r^2 h$

(d)

r

Sphere
$V = \frac{4}{3}\pi r^3$

(e)

Figure 1.6

EXAMPLES

6. Find the volume of a right circular cylinder with a radius of 2 centimeters and a height of 5 centimeters.

Solution

a. Sketch the figure.

b. $V = \pi r^2 h$

$V = 3.14(2^2) \cdot 5$

$= 3.14(4) \cdot 5$

$= 3.14(20)$

$= 62.80 \text{ cm}^3$

The volume is approximately 62.80 cubic centimeters.

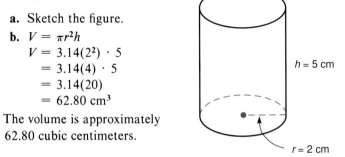

$h = 5 \text{ cm}$

$r = 2 \text{ cm}$

7. Find the volume of a sphere with a radius of 4 feet.

Solution

a. Sketch the figure.

b. $V = \dfrac{4}{3} \pi r^3$

$V = \dfrac{4}{3}(3.14) \cdot 4^3$

$= \dfrac{4(3.14) \cdot 64}{3}$

$= \dfrac{803.84}{3} \text{ft}^3$ or about 267.95 ft^3

The volume is approximately 267.95 cubic feet.

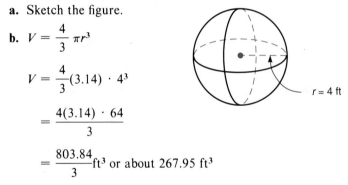

$r = 4 \text{ ft}$

Practice Problems

1. Find the area of a square with sides 4 cm long.

2. Find the area of a circle with diameter 6 in.

3. Find the perimeter of a rectangle 3.5 meters long and 1.6 meters wide.

◢ EXERCISES 1.5

In Exercises 1–15, select the answer from the right-hand column that correctly matches the statement.

1. The formula for the perimeter of a square _____

2. The formula for the circumference of a circle _____

3. The formula for the perimeter of a triangle _____

4. The formula for the area of a rectangle _____

5. The formula for the area of a square _____

6. The formula for the area of a trapezoid _____

7. The formula for the area of a triangle _____

8. The formula for the area of a parallelogram _____

9. The formula for the perimeter of a rectangle _____

10. The formula for the volume of a rectangular pyramid _____

11. The formula for the volume of a rectangular solid _____

12. The formula for the area of a circle _____

13. The formula for the volume of a right circular cylinder _____

14. The formula for the volume of a sphere _____

15. The formula for the volume of a right circular cone _____

a. $V = \dfrac{1}{3}\ell wh$

b. $A = \dfrac{1}{2}h(b + c)$

c. $A = s^2$

d. $A = \dfrac{1}{2}bh$

e. $P = 4s$

f. $A = \ell w$

g. $P = 2\ell + 2w$

h. $V = \dfrac{1}{3}\pi r^2 h$

i. $V = \dfrac{4}{3}\pi r^3$

j. $C = 2\pi r$

k. $P = a + b + c$

l. $V = \ell wh$

m. $V = \pi r^2 h$

n. $A = \pi r^2$

o. $A = bh$

Find (a) the perimeter and (b) the area of each figure in Exercises 16–21.

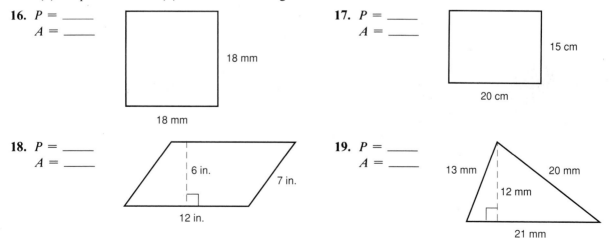

16. $P =$ _____
    $A =$ _____

18 mm
18 mm

17. $P =$ _____
    $A =$ _____

15 cm
20 cm

18. $P =$ _____
    $A =$ _____

6 in.
7 in.
12 in.

19. $P =$ _____
    $A =$ _____

13 mm
20 mm
12 mm
21 mm

20. $P =$ _____
    $A =$ _____

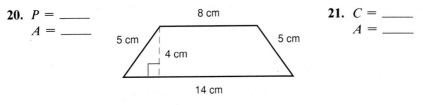

21. $C =$ _____
    $A =$ _____

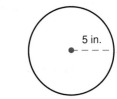

Find the volume of each figure in Exercises 22–26.

22. $V =$ _____

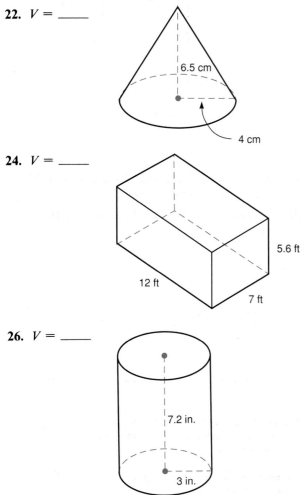

23. $V =$ _____

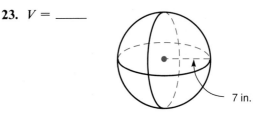

24. $V =$ _____

25. $V =$ _____

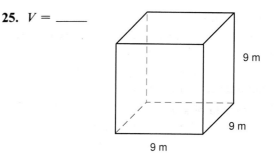

26. $V =$ _____

Solve the problems in Exercises 27–38.

27. What is the perimeter of a rectangle whose length is 17 in. and width is 11 in.?

28. Find the area of a square whose sides are 9 ft long.

29. The base of a triangle is 14 cm and the height is 9 cm. Find the area.

30. Find the circumference of a circle with a radius of 8 m (use $\pi = 3.14$).

31. The radius of the base of a cylindrical tank is 14 ft. If the tank is 10 ft high, find the volume.

32. The sides of a triangle are 6.2 m, 8.6 m, and 9.4 m. Find the perimeter.

33. What is the volume of a cube whose edge is 5 ft?

34. A rectangular garden plot is 60 ft long and 42 ft wide. Find the area.

35. A parallelogram has a base of 20 cm and a height of 13.6 cm. Find the area.

36. A rectangular box is 18 in. long, 10.3 in. wide, and 8 in. high. Find the volume.

37. The diameter of the base of a right circular cone is 15 in. If the height of the cone is 9 in., find the volume.

38. Find the volume of a sphere whose radius is 10 cm.

Find the perimeter and area for each figure in Exercises 39 and 40.

39. $P =$ _____
$A =$ _____

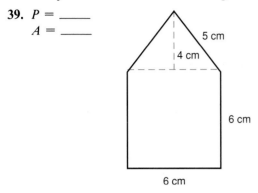

40. $P =$ _____
$A =$ _____

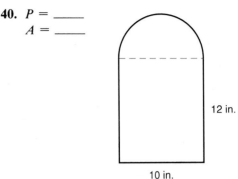

Calculator Problems

41. Find the area of a rectangle that is 16.54 meters long and 12.82 meters wide.

42. The radius of a circle is 8.32 inches. Find the circumference. (Use $\pi = 3.14$.)

43. Find the area of a trapezoid with bases of 22.36 inches and 17.48 inches and a height of 6.53 inches.

44. Find the area of a triangle with a base of 63.52 cm and a height of 41.78 cm.

45. The radius of a sphere is 114.8 cm. Find the volume.

CHAPTER 1 SUMMARY

Key Terms and Formulas

The **counting numbers** (or natural numbers) are 1, 2, 3, 4, 5, ... [1.1]

The **whole numbers** are 0, 1, 2, 3, 4, 5, ... [1.1]

An **exponent** is a number that tells how many times a factor occurs in a product. The factor is the **base** of the exponent and the product is the **power**. [1.1]

The **reciprocal** of $\frac{a}{b}$ is $\frac{b}{a}$; and $\frac{a}{b} \cdot \frac{b}{a} = 1$. [1.2]

A **decimal number** (or simply a **decimal**) is a fraction that has a power of ten in its denominator. [1.3]

A **variable** is a symbol or letter that can represent more than one number. [1.4]

An **equation** states that two algebraic expressions are equal. [1.4]

A **formula** is an equation that represents a general relationship between two or more quantities or measurements. [1.5]

The **perimeter** of a geometric figure is the total distance around the figure. [1.5]

Area is a measure of the interior, or enclosure, of a surface. [1.5]

Volume is a measure of the space enclosed by a three-dimensional figure. [1.5]

Formulas in Geometry

Perimeter

$P = 4s$ (square)

$P = 2\ell + 2w$ (rectangle)

$P = 2a + 2b$ (parallelogram)

$P = a + b + c$ (triangle)

$C = 2\pi r$ (circle)

$C = \pi d$ (circle)

$P = a + b + c + d$ (trapezoid)

Area

$A = s^2$ (square)

$A = \ell w$ (rectangle)

$A = bh$ (parallelogram)

$A = \dfrac{1}{2}bh$ (triangle)

$A = \pi r^2$ (circle)

$A = \dfrac{1}{2}h(b + c)$ (trapezoid)

Volume

$V = \ell wh$ (rectangular solid)

$V = \dfrac{1}{3}\ell wh$ (rectangular pyramid)

$V = \pi r^2 h$ (right circular cylinder)

$V = \dfrac{1}{3}\pi r^2 h$ (right circular cone)

$V = \dfrac{4}{3}\pi r^3$ (sphere)   [1.5]

Properties and Rules

Rules for Order of Operations   [1.1]

1. Work within symbols of inclusion (parentheses, brackets, or braces), beginning with the innermost pair.
2. Find any powers indicated by exponents.
3. Perform any multiplications or divisions in the order they appear **from left to right.**
4. Perform any additions or subtractions in the order they appear **from left to right.**

Division by 0 Is Undefined   [1.1]

$\dfrac{a}{0}$ is undefined for any whole number a.

Commutative Property of Addition   [1.1]
$$a + b = b + a$$

Associative Property of Addition   [1.1]
$$(a + b) + c = a + (b + c)$$

Commutative Property of Multiplication   [1.1]
$$a \cdot b = b \cdot a$$

Associative Property of Multiplication   [1.1]
$$(a \cdot b) \cdot c = a \cdot (b \cdot c)$$

Distributive Property of Multiplication over Addition   [1.1]
$$a \cdot (b + c) = a \cdot b + a \cdot c$$

Properties of Fractions   [1.2]

The following properties of fractions are true if no denominator is 0:

$$\frac{a}{b} \cdot \frac{c}{d} = \frac{a \cdot c}{b \cdot d}$$

$$\frac{a}{b} = \frac{a \cdot k}{b \cdot k}, \qquad \text{where } k \neq 0$$

$$\frac{a}{b} \div \frac{c}{d} = \frac{a}{b} \cdot \frac{d}{c}$$

$$\frac{a}{b} + \frac{c}{b} = \frac{a + c}{b}$$

$$\frac{a}{b} - \frac{c}{b} = \frac{a - c}{b}$$

Procedures

To Find the Average of a Collection of Numbers [1.1]
1. Add all the numbers in the collection.
2. Divide this sum by the number of numbers in the collection.

This quotient is the **average.**

To Find the Prime Factorization of a Composite Number [1.1]
1. Find any two factors of the number.
2. Continue to factor each of the factors until all factors are prime numbers.
3. Write the product of all these prime factors. (Use exponents for repeated factors.)

To Find the LCM [1.2]
1. Find the prime factorization of each number.
2. Form the product of all prime factors that appear by using each prime factor the greatest number of times that it appears in any one prime factorization.

To Change a Decimal to a Percent [1.3]
Move the decimal point two places to the right and write the % sign.

To Change a Percent to a Decimal [1.3]
Move the decimal point two places to the left and drop the % sign.

To Find a Percent of a Number [1.3]
Change the percent to a decimal, then multiply.

CHAPTER 1 REVIEW

Perform the indicated operations in Exercises 1–12. [1.1]

1. $18 + 7 + 8 + 3$ **2.** $9 + 16 + 11 + 4$ **3.** $26 - 17$ **4.** $33 - 18$

5. $(14)(9)$ **6.** $(27)(17)$ **7.** $273 \div 7$ **8.** $744 \div 6$

9. Add: 75
         34
      608
      123

10. Subtract: 738
             542

11. Multiply: 42
           79

12. Divide: $17\overline{)408}$

Use the distributive property to evaluate Exercises 13 and 14. [1.1]

13. $7(3 + 6)$ **14.** $8(9 + 7)$

Find the average of each set of numbers in Exercises 15 and 16. [1.1]

15. 17, 14, 20   **16.** 18, 22, 27, 37

Find the indicated products in Exercises 17 and 18. [1.1]

17. 3^4 **18.** 11^2

Write Exercises 19 and 20 in exponential form. [1.1]

19. $2 \cdot 2 \cdot 2 \cdot a \cdot a \cdot b$ **20.** $3 \cdot 3 \cdot 5 \cdot 5 \cdot a \cdot b \cdot b \cdot b$

Find the prime factorization for Exercises 21 and 22. [1.1]

21. 60   **22.** 153

Find the LCM for each set of numbers in Exercises 23 and 24. [1.2]

23. 18, 27, 36   **24.** $12x, 9xy, 24xy^2$

Find the value of each expression in Exercises 25 and 26 using the rules for order of operations.   [1.1]

25. $(13 \cdot 5 - 5) \div 2 \cdot 3$

26. $4[6 - (4 \cdot 5 - 2 \cdot 9) - 1]$

Reduce each fraction to lowest terms in Exercises 27 and 28.   [1.2]

27. $\dfrac{20}{44}$

28. $\dfrac{36}{54}$

Find the missing numerator by changing each fraction in Exercises 29 and 30 to an equal fraction with the indicated denominator.   [1.2]

29. $\dfrac{5}{12} = \dfrac{}{72}$

30. $\dfrac{4}{5} = \dfrac{}{65}$

Name each property illustrated in Exercises 31–33.   [1.1]

31. $x + 17 = 17 + x$

32. $(y + 11) + 4 = y + (11 + 4)$

33. $6(x + 7) = 6x + 42$

Perform the indicated operations in Exercises 34–41. Reduce each answer to lowest terms.   [1.2]

34. $\dfrac{3}{8} + \dfrac{7}{10}$

35. $\dfrac{4}{9} - \dfrac{5}{12}$

36. $\dfrac{4}{5y} + \dfrac{2}{3y}$

37. $\dfrac{7}{4a} - \dfrac{9}{6a}$

38. $\dfrac{20}{6} \cdot \dfrac{9}{4}$

39. $\dfrac{11}{9} \div \dfrac{22}{15}$

40. $\dfrac{7b}{15} \div \dfrac{2b}{6}$

41. $\dfrac{8x}{21} \cdot \dfrac{7}{12x}$

Simplify Exercises 42–45 using the rules for order of operations.   [1.2]

42. $\dfrac{7}{18} + \dfrac{5}{24} - \left(\dfrac{3}{8}\right)^2$

43. $\left(\dfrac{3}{5}\right)^2 \div \dfrac{8}{5} \cdot \dfrac{8}{3}$

44. $\dfrac{1}{3} + \dfrac{2}{5} \cdot \dfrac{5}{8} \div \dfrac{3}{2} - \dfrac{1}{2}$

45. $\left(\dfrac{3}{4} - \dfrac{1}{3}\right) \div \left(\dfrac{2}{3} + \dfrac{2}{7}\right)$

Change the decimals to percents in Exercises 46 and 47.   [1.3]

46. 0.73

47. 0.065

Change the percents to decimals in Exercises 48 and 49.   [1.3]

48. 7%

49. 12.5%

Perform the indicated operations in Exercises 50–57.   [1.3]

50. $15.8 + 9.1 + 7.63$

51. $19.31 - 14.62$

52. $24.3 + 6.81 - 16.51$

53. $(25.3)(2.3)$

54. $(14.7)(0.36)$

55. Find 18% of 60.

56. What is 106% of 55?

57. 11.5% of 380 is equal to _____ .

Evaluate the expressions in Exercises 58 and 59 if $x = 5$.   [1.4]

58. $3x + 2 - x$

59. $x^2 + 2x$

Evaluate the expressions in Exercises 60 and 61 if $x = \dfrac{1}{4}$.   [1.4]

60. $2x + 3$

61. $3x - \dfrac{1}{2}$

Evaluate the expressions in Exercises 62 and 63 if $x = 1.2$.   [1.4]

62. $4x - 3.1$

63. $5x - 2.34$

Determine whether or not the given number is a solution to the given equation in Exercises 64–67. [1.4]

64. $5x + 3 = 10$; $x = 2$

65. $7x - 8 = 13$; $x = 3$

66. $4x + 1 = 2x + 2$; $x = \dfrac{1}{2}$

67. $3x + 7 = 11.5$; $x = 1.5$

68. A rectangle is 23 ft long and 15 ft wide. Find the perimeter. [1.5]

69. Find the area of a circle with a radius of 9 meters. [1.5]

70. Find the volume of a rectangular box that is 11 in. long, 9 in. wide, and 7 in. high. [1.5]

71. Lucia is making punch for a party. She is making 5 gallons. The recipe calls for $\dfrac{2}{3}$ cup of sugar per gallon of punch. How many cups of sugar does she need? [1.2]

72. Ozzie is building some bookshelves. He needs 3 pieces 25 in. long, 1 piece 19 in. long, 2 pieces 64 in. long, and 3 pieces 38 in. long. The wood costs $0.75 per 12 in. How much will the wood cost? [1.3]

73. Julie grew corn for a school experiment. She planted 6 seeds. At the end of 7 weeks, she measured each cornstalk. The heights were 18.6, 20.4, 23.5, 19.7, 22.1, and 20.5 in. What was the average height of Julie's plants? [1.1]

74. A rancher is selling 72 head of cattle. It cost him $17,280 to raise and feed these cattle. He wants to make a profit of $125 per head. What price per head must he receive for his cattle? [1.3]

75. Matt's Little League team is selling magazine subscriptions to earn extra money. If they receive 15% of sales, how much will they earn if they sell $453.40 worth of subscriptions? [1.3]

CHAPTER 1 TEST

1. Add: $32 + 688 + 1013 + 61$

2. Subtract: $453 - 87$

3. $437 \div 19$

4. Identify the property illustrated: $8x + 20 = 4(2x + 5)$

5. Identify the property illustrated: $(6x)y = 6(xy)$

6. Identify the property illustrated: $x + (5 + y) = x + (y + 5)$

7. Find the average of the numbers: 34, 61, 72, 45

8. Write in exponential form: $2 \cdot 3 \cdot 3 \cdot x \cdot y \cdot y \cdot y$

9. Find the prime factorization of 315.

10. Find the LCM of $10xy^2$, $18x^2y$, and $15xy$.

11. Find the value of $4[5^2 - (6 + 5 \cdot 3) \div 7]$.

12. Reduce to lowest terms: $\dfrac{72}{156}$

13. Find the missing numerator: $\dfrac{5}{9} = \dfrac{}{108}$

Perform the indicated operations and reduce to lowest terms.

14. $\dfrac{7}{12} \cdot \dfrac{9}{28} \div \dfrac{5}{16}$

15. $\dfrac{4}{15} + \dfrac{1}{3} - \dfrac{3}{10}$

16. $\dfrac{7}{8} - \dfrac{1}{3} \div \dfrac{5}{6} + \dfrac{1}{4}$

17. $28.63 + 7.9 - 15.47$

18. Find the product: $(31.6)(0.43)$

19. Find 12.5% of 340.

20. Evaluate the expression $4x + 6 - x^2$ if $x = 3$.

21. Determine whether or not $x = 4$ is a solution to the equation $2x + 1 = 4x - 7$.

22. Find the circumference of a circle with a radius of 14 centimeters.

23. Find the area of a triangle if the base is 11 inches and the height is 6 inches.

24. Find the area of a rectangle 12 ft long and $8\dfrac{1}{2}$ ft wide.

25. Jaime bought a pair of skis priced at $240. He paid 20% down in cash. The balance was paid off in 6 equal payments. How much was each payment?

2

SIGNED NUMBERS

◢ DID YOU KNOW ?

Arithmetic operations defined on the set of positive integers, negative integers, and zero are studied in this chapter. The integer zero will be shown to have interesting properties under the operations of addition, subtraction, multiplication, and division.

Curiously, zero was not recognized as a number by early Greek mathematicians. When Hindu scientists developed the place-value numeration system we currently use, the zero symbol was initially a place holder but not a number. The spread of Islam transmitted the Hindu number system to Europe where it became known as the Hindu-Arabic system and replaced Roman numerals. The word **zero** comes from the Hindu word meaning "void," which was translated into Arabic as "sifr" and later into Latin as "zephirum," hence the derivation of our English words "zero" and "cipher."

Almost all of the operational properties of zero were known to the Hindus. However, the Hindu mathematician Bhaskara the Learned (1114–85?)

asserted that a number divided by zero was zero, or possibly infinite. Bhaskara did not seem to understand the role of zero as a divisor, since division by zero is undefined and hence an impossible operation in mathematics.

Albert Einstein, in his development of a proof that the universe was stable and unchangeable in time, divided both sides of one of his intermediate equations by a complicated expression that under certain circumstances could become zero. When the expression became zero, Einstein's proof did not hold and the possibility of a pulsating, expanding, or contracting universe had to be considered. This error was pointed out to Einstein and he was forced to withdraw his proof that the universe was stable. The moral of this story is that although zero seems like a "harmless" number, its operational properties are different from those of the positive or negative integers.

How can it be that mathematics, being after all a product of human thought independent of experience, is so admirably adapted to the objects of reality.

Albert Einstein (1879–1955)

CHAPTER OUTLINE

2.1 Number Lines

2.2 Absolute Value

2.3 Addition with Signed Numbers

2.4 Subtraction with Signed Numbers

2.5 Multiplication and Division with Signed Numbers

2.6 Word Problems

*T*his chapter is the springboard from arithmetic to higher-level algebraic concepts and solving equations. You will learn about positive and negative numbers and how to operate with these numbers. Believe it or not, the key number is 0.

Pay particularly close attention to the idea of the magnitude of a number, called its absolute value, and the terminology used to represent different types of numbers.

2.1 Number Lines

OBJECTIVES

In this section, you will be learning to:

1. Determine if given numbers are greater than, less than, or equal to other given numbers.
2. Determine if number statements are true or false.
3. Graph sets of numbers on the number line.

The closest thing to a picture of a set of numbers is its graph on a number line. We generally will use horizontal and vertical lines for number lines. For example, choose some point on a horizontal line and label it with the number 0 (Figure 2.1).

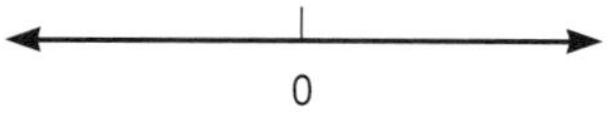

Figure 2.1

Now choose another point on the line to the right of 0 and label it with the number 1 (Figure 2.2).

Figure 2.2

We now have a number line. Points corresponding to all the whole numbers are determined. The point corresponding to 2 is the same distance from 1 as 1 is from 0, 3 from 2, 4 from 3, and so on (Figure 2.3).

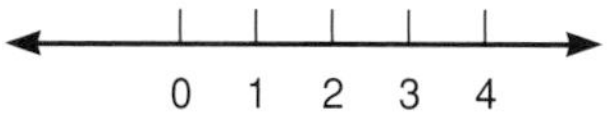

Figure 2.3

The **graph** of a number is the point that corresponds to the number, and the number is called the **coordinate** of the point. We will follow the convention of using the terms "number" and "point" interchangeably. Thus, a point can be called "seven" or "two." The graph of 7 is indicated by marking the point corresponding to 7 with a large dot (Figure 2.4).

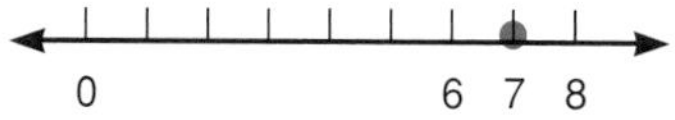

Figure 2.4

The graph of the set $A = \{2, 4, 6\}$ is shown in Figure 2.5.

Figure 2.5

On a horizontal number line, the point one unit to the left of 0 is the **opposite of 1.** It is called **negative 1** and is symbolized -1. Similarly, the point two units to the left of 0 is the opposite of 2, called negative 2, and symbolized -2, and so on (Figure 2.6).

The opposite of 1 is -1;	The opposite of -1 is $-(-1) = +1$;
the opposite of 2 is -2;	the opposite of -2 is $-(-2) = +2$;
the opposite of 3 is -3;	the opposite of -3 is $-(-3) = +3$;
and so on.	and so on.

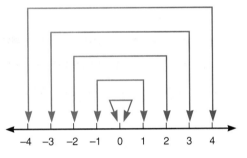

Figure 2.6

The set of numbers consisting of the whole numbers and their opposites is called the set of **integers.** The natural numbers are called **positive integers.** Their opposites are called **negative integers. Zero is its own opposite and is neither positive nor negative** (Figure 2.7). Note that the opposite of a positive integer is a negative integer, and the opposite of a negative integer is a positive integer.

Integers:	$\ldots, -3, -2, -1, 0, 1, 2, 3, \ldots$
Positive integers:	$1, 2, 3, 4, 5, \ldots$
Negative integers:	$\ldots, -4, -3, -2, -1$

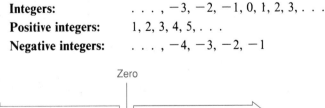
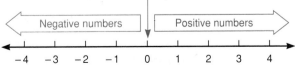

Figure 2.7

EXAMPLES

1. Find the opposite of 7.
   Solution -7

2. Find the opposite of -3.
   Solution $-(-3)$ or $+3$

3. Graph the set $\{-3, -1, 1, 3\}$.
   Solution

   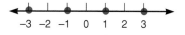

4. Graph the set $\{\ldots, -5, -4, -3\}$.
   Solution

   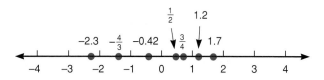

   The three dots above the number line indicate that the pattern in the graph continues without end.   ■

The integers are not the only numbers that can be represented on a number line. Fractions and decimal numbers such as $\dfrac{1}{2}$, $\dfrac{3}{4}$, $-\dfrac{4}{3}$, and 1.2 can also be represented (Figure 2.8).

Figure 2.8

Numbers that can be written as fractions with integers as numerator and denominator are called **rational numbers.** They include positive and negative decimal numbers and the integers themselves. For example, the following are rational numbers:

$$1.3 = \frac{13}{10} \quad \text{and} \quad 5 = \frac{5}{1} \quad \text{and} \quad -4 = \frac{-4}{1}.$$

All rational numbers have corresponding points on a number line.

Rational numbers $\begin{cases} \text{numbers that can be written as } \dfrac{a}{b}, \\ \text{where } a \text{ and } b \text{ are integers, } b \neq 0 \end{cases}$

Other numbers on a number line such as $\sqrt{2}$, $\sqrt{3}$, π, and $\sqrt[3]{5}$ are called **irrational numbers.** These numbers will be discussed in some detail in Chapter 10 and in later courses in mathematics. All the rational and irrational numbers, positive and negative, can be referred to as **signed numbers.**

On a horizontal number line, **smaller numbers are always to the left of larger numbers.** Each number is smaller than any number to its right and larger than any number to its left. Two symbols used to indicate order are

$$<, \quad \text{read "is less than"}$$
$$\text{and } >, \quad \text{read "is greater than."}$$

Using the number line in Figure 2.9, you can see the following relationships:

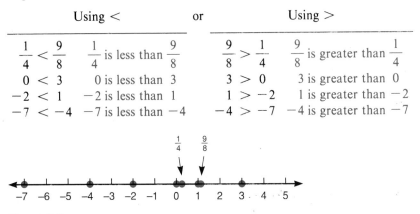

Using $<$		or	Using $>$	
$\dfrac{1}{4} < \dfrac{9}{8}$	$\dfrac{1}{4}$ is less than $\dfrac{9}{8}$		$\dfrac{9}{8} > \dfrac{1}{4}$	$\dfrac{9}{8}$ is greater than $\dfrac{1}{4}$
$0 < 3$	0 is less than 3		$3 > 0$	3 is greater than 0
$-2 < 1$	-2 is less than 1		$1 > -2$	1 is greater than -2
$-7 < -4$	-7 is less than -4		$-4 > -7$	-4 is greater than -7

Figure 2.9

Two other symbols commonly used are

$$\leq, \quad \text{read "is less than or equal to"}$$
$$\text{and } \geq, \quad \text{read "is greater than or equal to."}$$

For example, $5 \geq -10$ is true since 5 is greater than -10. Also, $5 \geq 5$ is true since 5 does equal 5.

Table of Symbols	$=$	is equal to	$\neq$	is not equal to
	$<$	is less than	$>$	is greater than
	$\leq$	is less than or equal to	$\geq$	is greater than or equal to

EXAMPLES

5. Determine whether each of the following statements is true or false.
   a. $7 < 15$ True, since 7 is less than 15.
   b. $3 > -1$ True, since 3 is greater than -1.
   c. $4 \geq -4$ True, since 4 is greater than -4.
   d. $2.7 \leq 2.7$ True, since 2.7 is equal to 2.7.
   e. $-5 < -6$ False, since -5 is greater than -6.

6. Graph the set of numbers $\left\{ -\dfrac{3}{4}, 0, 1, 1.5, 3 \right\}$.

Solution

$-\dfrac{3}{4}$ 1.5

-1 0 1 2 3 4 5

7. Graph all natural numbers less than or equal to 3.

Solution

-1 0 1 2 3 4 5

Note: Remember that the natural numbers are 1, 2, 3, 4, ...

8. Graph all integers less than 1.

Solution

· · ·

-3 -2 -1 0 1

Practice Problems

Fill in the blank with the appropriate symbol: $<$, $>$, or $=$.

1. $-2 \underline{\ \leq\ } 1$

2. $1\dfrac{6}{10} \underline{\ \geq\ } 1.6$

3. $-(-4.1) \underline{\ \geq\ } -7.2$

4. Graph the set of all negative integers on a number line.

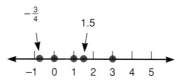 **EXERCISES 2.1**

Fill in the blank in Exercises 1–15 with the appropriate symbol: $<$, $>$, or $=$.

1. $6 \underline{\ >\ } 4$ **2.** $-3 \underline{\ \leq\ } 1$ **3.** $-2 \underline{\ >\ } -4$ **4.** $5 \underline{\ =\ } -(-5)$

5. $\dfrac{1}{3} \underline{\ <\ } \dfrac{1}{2}$ **6.** $-\dfrac{2}{3} \underline{\ \leq\ } \dfrac{1}{8}$ **7.** $-\dfrac{2}{8} \underline{\ =\ } -\dfrac{1}{4}$ **8.** $-8 \underline{\ \leq\ } 0$

9. $1.6 \underline{\ \leq\ } 2.3$ **10.** $-\dfrac{3}{4} \underline{\ >\ } -1$ **11.** $\dfrac{9}{16} \underline{\ <\ } \dfrac{3}{4}$ **12.** $-\dfrac{1}{2} \underline{\ \geq\ } -\dfrac{1}{3}$

13. $-2.3 \underline{\ =\ } -2\dfrac{3}{10}$ **14.** $5.6 \underline{\ \leq\ } -(-8.7)$ **15.** $-\dfrac{4}{3} \underline{\ \leq\ } -\left(-\dfrac{1}{3}\right)$

· · ·

-4 -3 -2 -1 0

Answers to Practice Problems   1. $<$ **2.** $=$ **3.** $>$ **4.**

Determine whether each statement in Exercises 16–35 is true or false.

16. $0 = -0$ **17.** $-22 < -16$ **18.** $-9 > -8.5$ **19.** $11 = -(-11)$

20. $-17 \leq -17$ **21.** $-6 < -8$ **22.** $4.7 \geq 3.5$ **23.** $-\dfrac{1}{3} \leq 0$

24. $-8.1 < -8.1$ **25.** $-7.3 \leq -8.6$ **26.** $\dfrac{3}{5} > \dfrac{1}{4}$ **27.** $-2.3 < 1$

28. $-9 > -7.69$ **29.** $4.6 > 4.1$ **30.** $0 > -25$ **31.** $4 + 3 < 2 + 5$

32. $4 - 2 < 5 - 1$ **33.** $14.3 > 8.1 + 5.9$

34. $13.6 - 7.8 > 2.3 + 1.5$ **35.** $6.3 + 5.2 \geq 12.0 - 0.5$

In Exercises 36–55 graph each set of numbers on a number line.

36. $\{1, 2, 5, 6\}$ **37.** $\{-3, -2, 0, 1\}$ **38.** $\{2, -3, 1, 0, -1\}$ **39.** $\{-2, -1, 4, -3\}$

40. $\left\{0, -1, \dfrac{5}{4}, 3, 1\right\}$ **41.** $\left\{-2, -1, -\dfrac{1}{3}, 2\right\}$ **42.** $\left\{-\dfrac{3}{4}, 0, 2, 3.6\right\}$ **43.** $\left\{-3.4, -2, 0.5, 1, \dfrac{5}{3}\right\}$

44. $\left\{-\dfrac{7}{2}, -1.5, 1, \dfrac{4}{3}, 2\right\}$ **45.** $\left\{-4, -\dfrac{7}{3}, -1, 0.2, \dfrac{5}{2}\right\}$

46. All natural numbers greater than 4

47. All natural numbers less than 7

48. All positive integers less than or equal to 3

49. All negative integers greater than or equal to -3

50. All integers less than -6

51. All integers greater than or equal to -1

52. All whole numbers less than 8

53. All whole numbers less than or equal to 4.3

54. All negative integers greater than or equal to -5.7

55. All integers greater than or equal to 1.6

◢ 2.2 Absolute Value

In working with the number line in Section 2.1, you may have noticed that any integer and its opposite lie the same number of units from 0 on the number line. For example, both $+7$ and -7 are seven units from 0 (Figure 2.10). The $+$ and $-$ signs indicate direction and the 7 indicates distance.

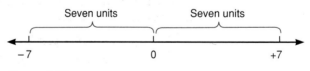

Figure 2.10

    The distance a number is from 0 on a number line is called its **absolute value** and is symbolized by two vertical bars, $|\quad|$. Thus, $|+7| = 7$ and $|-7| = 7$. Similarly,

$$|3| = 3$$

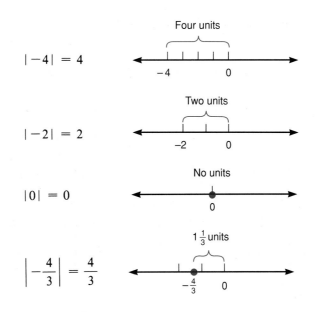

$\|-4\| = 4$	
$\|-2\| = 2$	
$\|0\| = 0$	
$\left\|-\dfrac{4}{3}\right\| = \dfrac{4}{3}$	

Absolute Value

The **absolute value** of a number is its distance from 0. The absolute value of a number is never negative.

Symbolically,

$$\begin{cases} |a| = a & \text{if } a \text{ is a positive number or } 0 \\ |a| = -a & \text{if } a \text{ is a negative number.} \end{cases}$$

EXAMPLES

1. $|6.3| = 6.3$

   The number 6.3 is 6.3 units from 0. Also, 6.3 is positive so its absolute value is the same as the number itself.

2. $|-5.1| = -(-5.1) = 5.1$

   The number -5.1 is 5.1 units from 0. Also, -5.1 is negative so its absolute value is its opposite.

3. If $|x| = 7$, what are the possible values for x?

   Solution $x = 7$ or $x = -7$ since $|7| = 7$ and $|-7| = 7$.

4. If $|x| = 1.35$, what are the possible values for x?

   Solution $x = 1.35$ or $x = -1.35$ since $|1.35| = 1.35$ and $|-1.35| = 1.35$.

5. True or False: $|-4| \leq 4$

   Solution True, since $|-4| = 4$ and $4 \leq 4$.

6. True or False: $\left| -5\frac{1}{2} \right| > 5\frac{1}{2}$

Solution False, since $\left| -5\frac{1}{2} \right| = 5\frac{1}{2}$ and $5\frac{1}{2} \not> 5\frac{1}{2}$ ($\not>$ is read "is not greater than").

7. If $|x| = -3$, what are the possible values for x?

Solution There are no values of x for which $|x| = -3$. The absolute value can never be negative.

8. If $|x| < 3$, what are the possible integer values for x? Graph these numbers on a number line.

Solution The integers are within 3 units of 0: $-2, -1, 0, 1, 2$.

9. If $|x| \geq 4$, what are the possible integer values for x? Graph these numbers on a number line.

Solution The integers must be 4 or more units from 0:
$\ldots, -7, -6, -5, -4, 4, 5, 6, 7, \ldots$

Practice Problems

1. True or false: $3.6 \leq |-3.6|$

2. List the numbers that satisfy the equation $|x| = 8$.

3. List the numbers that satisfy the equation $|x| = -6$.

4. Graph the integers that satisfy the inequality $|x| < 4$.

◹ EXERCISES 2.2

In Exercises 1–10, graph the sets on a number line.

1. $\{-5, 3, 2, |-1|, |4|\}$

2. $\{-7, |-3|, |-2|, 0, 1\}$

3. $\{0, -1, |-5|, |2|, -3\}$

4. $\{-2, |-6|, 5, |-2|, 7\}$

5. $\{|-8|, 0, 8, |6|, -8\}$

6. $\{-3, 1, |-3|, |3|, -2\}$

7. $\left\{ -2, \left| -\frac{3}{2} \right|, -1, 2, |2.7| \right\}$

8. $\left\{ \left| -\frac{2}{3} \right|, -1, |-2|, |-1.5| \right\}$

9. $\left\{ 0, \left| -\frac{4}{3} \right|, -2.5, |2|, -2.1 \right\}$

10. $\left\{ -1.6, |-2.5|, \frac{3}{2}, |2.5|, 3 \right\}$

In Exercises 11–20, determine whether the statements are true or false.

11. $|-5| = |5|$ **12.** $|-8| \geq 4$ **13.** $|-6| \geq 6$ **14.** $|-7| < |7|$

15. $|-1.9| < 2$ **16.** $|-1.6| < |-2.1|$ **17.** $\left|-\dfrac{5}{2}\right| < 2$ **18.** $\dfrac{2}{3} < |-1|$

19. $3 > \left|-\dfrac{4}{3}\right|$ **20.** $|-3.4| < 0$

List the numbers, then graph the numbers on a number line that satisfy the equations in Exercises 21–40.

21. $|x| = 4$ **22.** $|y| = 6$ **23.** $|x| = 9$ **24.** $13 = |x|$ **25.** $0 = |y|$

26. $-2 = |x|$ **27.** $|x| = -3$ **28.** $|x| = 3.5$ **29.** $|x| = 4.7$ **30.** $|y| = |-8|$

31. $|-12| = |y|$ **32.** $|x| = \dfrac{5}{2}$ **33.** $|x| = \dfrac{4}{7}$ **34.** $|x| = \dfrac{4}{3}$ **35.** $|x| = \left|-\dfrac{5}{4}\right|$

36. $|y| = -\dfrac{5}{8}$ **37.** $|y| = -2.7$ **38.** $|x| = 4.16$ **39.** $|x| = -\dfrac{3}{7}$ **40.** $|x| = 11.624$

On a number line, graph the integers that satisfy the conditions stated in Exercises 41–48.

41. $|x| \leq 4$ **42.** $|x| < 6$ **43.** $|x| > 5$ **44.** $|x| > 2$

45. $|x| < 7$ **46.** $|x| \leq 2$ **47.** $|x| = x$ **48.** $|x| = -x$

Choose the response that correctly completes each sentence in Exercises 49–54.

49. $|x|$ is (never, sometimes, always) equal to x.

50. $|x|$ is (never, sometimes, always) a negative number.

51. $|x|$ is (never, sometimes, always) greater than or equal to x.

52. $|x|$ is (never, sometimes, always) equal to $-x$.

53. $|x|$ is (never, sometimes, always) equal to 0.

54. $|x|$ is (never, sometimes, always) equal to a positive number.

2.3 Addition with Signed Numbers

OBJECTIVES

In this section, you will be learning to:

1. Add signed numbers.
2. Determine if given signed numbers are solutions for specified equations.
3. Complete statements about signed numbers.

Picture a straight line in an open field and numbers marked on a number line. An archer stands at 0 and shoots an arrow to $+3$, then stands at 3 and shoots the arrow 5 more units in the positive direction (to the right). Where will the arrow land? (Figure 2.11.)

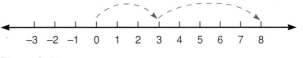

Figure 2.11

Naturally, you have figured that the answer is $+8$. What you have done is add the two positive numbers, $+3$ and $+5$.

$$(+3) + (+5) = +8 \qquad \text{or} \qquad 3 + 5 = 8$$

Suppose another archer shoots arrows in the same manner as the first but in the opposite direction. Where would his second arrow light? The arrow lights at −8. You have just added −3 and −5 (Figure 2.12).

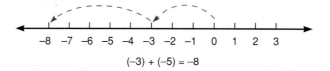

$$(-3) + (-5) = -8$$

Figure 2.12

If an arrow is shot to +3 and then the archer goes to +3 and turns around and shoots an arrow 5 units in the opposite direction, where will the arrow stick? Would you believe at −2? (Figure 2.13.)

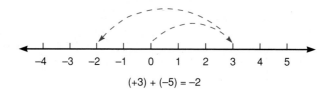

$$(+3) + (-5) = -2$$

Figure 2.13

For our final archer, the first shot is to −3. Then, after going to −3, he turns around and shoots 5 units in the opposite direction. Where is the arrow? It is at +2 (Figure 2.14).

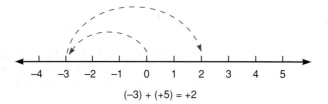

$$(-3) + (+5) = +2$$

Figure 2.14

In summary:

1. The sum of two positive numbers is positive.

$$(+3) \;+\; (+5) \;=\; +8$$

$\uparrow \qquad \uparrow \quad \uparrow \qquad\qquad \uparrow$

positive plus positive is positive

2. The sum of two negative numbers is negative.

$$(-3) \;+\; (-5) \;=\; -8$$

$\uparrow \qquad \uparrow \quad \uparrow \qquad\qquad \uparrow$

negative plus negative is negative

3. The sum of a positive number and a negative number may be negative or positive (or zero) depending on which number is further from 0.

$$(+3) \ + \ (-5) \ = \ -2 \qquad (+5) \ + \ (-3) \ = \ +2$$
$$\uparrow \quad \uparrow \quad \uparrow \qquad \uparrow \qquad\qquad \uparrow \quad \uparrow \quad \uparrow \qquad \uparrow$$

positive plus negative is **negative**   positive plus negative is **positive**

Practice Problems

Find each sum.

1. $(-14) + (-6) =$ ~-20~

2. $(+16) + (-10) =$ ~6~

3. $(-12) + (+8) =$ ~-4~

4. $11 + 7 =$ ~18~

5. $(-13) + (+13) =$ ~0~

6. $(-11) + (-8) =$ ~-19~

7. $(+6) + (-7) =$ ~-1~

8. $(+100) + (-100) =$ ~0~

9. $(-5.2) + (+16.3) =$ ~11.1~

10. $9.7 + 4.1 =$ ~13.8~

11. $\left(-4\frac{1}{2}\right) + \left(-3\frac{1}{4}\right) =$ ~7¾~

12. $7.8 + (-7.8) =$ ~0~

You probably did quite well and understand how to add numbers. The rules can be written out in the following rather formal manner.

Rules for Adding Signed Numbers

1. To add two numbers with like signs, add their absolute values and use the common sign:

$$(+7) + (+3) = +(|+7| + |+3|) = +(7 + 3) = +10$$
$$(-7) + (-3) = -(|-7| + |-3|) = -(7 + 3) = -10$$

2. To add two numbers with unlike signs, subtract their absolute values (the smaller from the larger) and use the sign of the number with the larger absolute value:

$$(-12) + (+10) = -(|-12| - |+10|) = -(12 - 10) = -2$$
$$(+12) + (-10) = +(|+12| - |-10|) = +(12 - 10) = +2$$
$$(-15) + (+15) = (|-15| - |+15|) = (15 - 15) = 0$$

Since equations in algebra are almost always written horizontally, you should become used to working with sums written horizontally. However, there are situations (as in long division) where sums (and differences) are written vertically with one number directly under another.

Answers to Practice Problems   **1.** -20 **2.** $+6$ **3.** -4 **4.** 18   **5.** 0   **6.** -19
7. -1 **8.** 0   **9.** 11.1   **10.** 13.8   **11.** $-7\frac{3}{4}$ **12.** 0

EXAMPLES Find each sum.

| 1. $\begin{array}{r} -10 \\ \underline{7} \\ -3 \end{array}$ | 2. $\begin{array}{r} -4 \\ 6 \\ \underline{-15} \\ -13 \end{array}$ | 3. $\begin{array}{r} -5 \\ -8 \\ \underline{-9} \\ -22 \end{array}$ | 4. $\begin{array}{r} -10.5 \\ +3.2 \\ \underline{+6.8} \\ -0.5 \end{array}$ |

Now that we know how to add positive and negative numbers, we can determine whether or not a particular signed number satisfies an equation.

EXAMPLES Determine whether or not the given number is a solution to the given equation by substituting and adding.

5. $x + 5 = -2$; $x = -7$
   Solution $(-7) + 5 = -2$ is true, so -7 is a solution.

6. $y + (-4) = -6.3$; $y = -2.3$
   Solution $(-2.3) + (-4) = -6.3$ is true, so -2.3 is a solution.

7. $14 + z = -3$; $z = -11$
   Solution $14 + (-11) = -3$ is false since $14 + (-11) = +3$. So, -11 is **not** a solution. ■

EXERCISES 2.3

Find the sum in Exercises 1–46.

1. $4 + 9 = 13$
2. $8 + (-3)$
3. $(-9) + 5$
4. $(-7) + (-3)$
5. $(-9) + 9 = 0$
6. $2 + (-8)$
7. $11 + (-6)$
8. $(-12) + 3$
9. $-18 + 5 = -13$
10. $26 + (-26)$
11. $-5 + |-3|$
12. $11 + |-2|$
13. $(-2) + (-8)$
14. $10 + (-3)$
15. $17 + (-17)$
16. $(-7) + 20$
17. $21 + (-4)$
18. $(-5) + (-3)$
19. $(-12) + (-7)$
20. $26 + (-26)$
21. $-4\frac{2}{3} + \left(-5\frac{1}{6}\right)$
22. $\left(-6\frac{3}{5}\right) + \left(-8\frac{7}{10}\right)$
23. $9\frac{5}{16} + \left(-12\frac{3}{4}\right)$
24. $-12\frac{3}{4} + 9\frac{5}{8}$
25. $38.5 + (-16.48)$
26. $(-20.3) + (-11.81)$
27. $(-33.62) + (-21.9)$
28. $(-21.6) + 18.5$
29. $-3 + 4 + (-8)$
30. $(-9) + (-6) + 5$
31. $(-9) + (-2) + (-5)$
32. $(-21) + 6 + 15$
33. $-13 + (-1) + (-12)$
34. $-19 + (-2) + (-4)$
35. $27 + (-14) + (-13)$
36. $-33 + 29 + 2$
37. $-43 + (-16) + 27$
38. $-68 + (-3) + 42$
39. $-38 + 49 + (-6)$
40. $102 + (-93) + (-6)$
41. $\begin{array}{r} -21 \\ \underline{-62} \end{array}$
42. $\begin{array}{r} -12 \\ \underline{17} \end{array}$

43. $\begin{array}{r} -15 \\ 8 \\ \underline{19} \end{array}$
44. $\begin{array}{r} -7 \\ 23 \\ \underline{-9} \end{array}$
45. $\begin{array}{r} -163 \\ 204 \\ \underline{-73} \end{array}$
46. $\begin{array}{r} -93 \\ -87 \\ \underline{147} \end{array}$

In Exercises 47–55, determine whether or not the given number is a solution to the given equation by substituting and then evaluating.

47. $x + 4 = 2$; $x = -2$
48. $x + (-7) = 10$; $x = -3$
49. $-10 + x = -14$; $x = -4$

50. $2x + 9 = 7;\quad x = 1$ **51.** $17 + x = 10;\quad x = -7$ **52.** $2x + (-1) = 9;\quad x = 5$

53. $3x + (-12) = 6;\quad x = 2$ **54.** $x + 3.5 = 2.8;\quad x = -1.7$ **55.** $x + \dfrac{3}{4} = -\dfrac{1}{4};\quad x = -1$

Choose the response that correctly completes the sentence in Exercises 56–60.

56. If x is a positive number and y is a negative number, then $x + y$ is (never, sometimes, always) a negative number.

57. If x and y are signed numbers, then $x + y$ is (never, sometimes, always) equal to 0.

58. If x and y are positive numbers, then $x + y$ is (never, sometimes, always) equal to 0.

59. If y is a signed number, then $y + (-y)$ is (never, sometimes, always) equal to 0.

60. If x and y are negative numbers, then $x + y$ is (never, sometimes, always) a positive number.

Calculator Problems

Use a calculator to evaluate Exercises 61–65.

61. $47.832 + (-29.572) + 66.919$

62. $56.473 + (-41.031) + (-28.638)$

63. $(-16.945) + (-27.302) + (-53.467)$

64. $29.832 + 47.951 + (-38.176)$

65. $(-0.8154) + 2.9147 + (-1.836)$

2.4 Subtraction with Signed Numbers

OBJECTIVES

In this section, you will be learning to:

1. Find the additive inverse of a signed number.
2. Subtract signed numbers.
3. Evaluate signed number expressions.
4. Determine if given signed numbers are solutions for specified equations.

In basic arithmetic, subtraction is defined in terms of addition. For example, we know that the difference $32 - 25$ is equal to 7 because $25 + 7 = 32$. A student in arithmetic does not know how to find a difference such as $15 - 20$, where a larger number is subtracted from a smaller number, because negative numbers are not yet defined and there is no way to add a positive number to 20 and get 15. Now, with our new knowledge of negative numbers, we will define subtraction in such a way that larger numbers may be subtracted from smaller numbers. We will still define subtraction in terms of addition, but we will apply our new rules of addition with positive and negative numbers.

Before we proceed to develop the techniques for subtraction, we will state and illustrate an important relationship between a number and its opposite.

Additive Inverse	The **opposite** of a signed number is called its **additive inverse.** The sum of a number and its additive inverse is 0. Symbolically, for any number a, $$a + (-a) = 0$$

EXAMPLES   Find the additive inverse (opposite) of each number.

1. 3

    Solution   The additive inverse of 3 is -3.

$$3 + (-3) = 0$$

2. -7

    Solution   The additive inverse of -7 is $-(-7) = +7$.

$$(-7) + (+7) = 0$$

3. -4.8

    Solution   The additive inverse of -4.8 is $-(-4.8) = +4.8$.

$$(-4.8) + (+4.8) = 0$$

4. 0

    Solution   The additive inverse of 0 is $-0 = 0$.

$$(0) + (0) = 0$$ ■

In Section 2.3, we added numbers such as $5 + (-2) = 3$ and $26 + (-9) = 17$. Note that in each case, subtraction will give the same answers; that is, $5 - 2 = 3$ and $26 - 9 = 17$. Thus, it seems that subtraction and addition are closely related—and, in fact, they are.

    From arithmetic, $5 - 2$ is asking, "What number **added** to 2 gives 5?" That is,

$$5 - 2 = 3 \quad \text{because} \quad 5 = 2 + 3$$
$$26 - 9 = 17 \quad \text{because} \quad 26 = 9 + 17$$

    What do we mean by $4 - (-1)$? Do we mean, "What number added to -1 gives 4?" Precisely. Thus,

$$4 - (-1) = 5 \quad \text{because} \quad 4 = (-1) + 5$$
$$4 - (-2) = 6 \quad \text{because} \quad 4 = (-2) + 6$$
$$4 - (-3) = 7 \quad \text{because} \quad 4 = (-3) + 7$$

But note the following results:

$$4 + (+1) = 5$$
$$4 + (+2) = 6$$
$$4 + (+3) = 7$$

    The following relationship between subtraction and addition becomes the basis for subtraction with all numbers:

$$4 - (-1) = 4 + (+1) = 5 \qquad (-4) - (-1) = (-4) + (+1) = -3$$
$$4 - (-2) = 4 + (+2) = 6 \qquad (-4) - (-2) = (-4) + (+2) = -2$$
$$4 - (-3) = 4 + (+3) = 7 \qquad (-4) - (-3) = (-4) + (+3) = -1$$

EXAMPLES

5. $(-1) - (-4) = (-1) + (+4) = +3$
6. $(-1) - (-5) = (-1) + (+5) = +4$
7. $(-1) - (-6) = (-1) + (+6) = +5$
8. $(-10) - (-2) = (-10) + (+2) = -8$
9. $(+10) - (-5.7) = (-10) + (+5.7) = -4.3$ ■

Practice Problems

1. $(-4) - (-4) = (-4) + (+4) = 0$
2. $(-3) - (-8) = (-3) + (+8) = +5$
3. $(-3) - (+8) = (-3) + (-8) = -11$
4. $(25.4) - (46.7) = 25.4 + (-46.7) = 21.3$
5. $\left(13\frac{3}{4}\right) - \left(8\frac{1}{2}\right) = \left(13\frac{3}{4}\right) + \left(-8\frac{1}{2}\right) =$
6. $14 - (-5) = 14 + (5) = 19$
7. $14 - (5) = 14 + (-5) = 9$

(handwritten work in margin:)

$\left(13\frac{3}{4}\right) + \left(-8\frac{1}{2}\right)$

$\frac{55 \times 1}{4 \times 1} - \frac{17 \times 2}{2 \times 2}$

$\frac{55}{4} - \frac{34}{4} = 21 \; 4/\frac{21}{20} = 5\frac{1}{4}$

You may have noticed that in subtraction, the **opposite** of the number being subtracted is **added**. For example, we could write

$$(-1) - (-4) = (-1) + [-(-4)] = (-1) + (+4) = +3$$

or

$$(-10) - (-3) = (-10) + [-(-3)] = (-10) + (+3) = -7$$

Subtraction

For any signed numbers *a* and *b*,

$$a - b = a + (-b)$$

This definition translates as, "To subtract *b* from *a*, **add** the **opposite** of *b* to *a*." In practice, the notation $a - b$ is thought of as addition of signed numbers. That is, since $a - b = a + (-b)$, we think of the plus sign, $+$, as being present in $a - b$. **In fact, an expression such as 4 − 19 can be thought of as "four plus negative nineteen."** We have

$$4 - 19 = 4 + (-19) = -15$$
$$-25 - 30 = -25 + (-30) = -55$$
$$-3 - (-17) = -3 + (+17) = 14$$
$$24 - 11 = 24 + (-11) = 13$$

Answers to Practice Problems   1. 0   2. 5   3. −11   4. −21.3   5. $5\frac{1}{4}$ 6. 19

7. 9

Generally, the second form is omitted and we go directly to the answer by computing the sum mentally.

$$4 - 19 = -15$$
$$-25 - 30 = -55$$
$$-3 - (-17) = 14$$
$$24 - 11 = 13$$

The numbers may also be written vertically, that is, one underneath the other. In this case, the sign of the number being subtracted (the bottom number) is changed and addition is performed. This format is used in long division, as discussed in Chapter 6.

EXAMPLES

10. Subtract (**Add**) | **11. Subtract** (**Add**)

$$\begin{array}{r} 43 \\ -25 \\ \hline 68 \end{array} \qquad \begin{array}{r} 43 \\ +25 \\ \hline 68 \end{array} \qquad\qquad \begin{array}{r} -38 \\ +11 \\ \hline 49 \end{array} \qquad \begin{array}{r} -38 \\ -11 \\ \hline -49 \end{array}$$

12. Subtract (**Add**) | **13. Subtract** (**Add**)

$$\begin{array}{r} -73 \\ -32 \\ \hline 41 \end{array} \qquad \begin{array}{r} -73 \\ +32 \\ \hline -41 \end{array} \qquad\qquad \begin{array}{r} 17.6 \\ 69.3 \\ \hline 51.7 \end{array} \qquad \begin{array}{r} 17.6 \\ -69.3 \\ \hline -51.7 \end{array}$$ ∎

Now, using subtraction as well as addition, we can determine whether or not a number is a solution to an equation of a slightly more complex nature.

EXAMPLES Determine whether or not the given number is a solution to the given equation by substituting and then evaluating.

14. $x - (-5) = 6$; $x = 1$
Solution $1 - (-5) = 1 + (+5) = 6$,   so 1 is a solution.

15. $5 - y = 7$; $y = -2$
Solution $5 - (-2) = 5 + (+2) = 7$,   so -2 is a solution.

16. $z - 14 = -3$; $z = 10$
Solution $10 - 14 = -4$ and $-4 \neq -3$,   so 10 is **not** a solution.

17. $6 - 8 = x - (-3)$; $x = -5$
Solution $6 - 8 = -5 - (-3)$

$$6 - 8 = -2 \quad \text{and} \quad -5 - (-3) = -5 + (+3) = -2$$

so -5 is a solution. ∎

Practice Problems

1. What is the additive inverse of 8.5? -8.5
2. Find the difference: $-6 - (-5)$ -1
3. Simplify: $-6 - 4 - (-2)$
4. True or false: $-5 + (-3) < -5 - (-3)$ TRUE
5. Is $x = 15$ a solution to the equation $2x - 1 = -29$? NOT A SOLUTION

$2 \cdot 15 - 1 = 29$

EXERCISES 2.4

Find the additive inverse for the numbers given in Exercises 1–10.

1. 11 -11   2. 17 -17   3. -6 6   4. -23 23   5. 4.7 -4.7

6. -3.4 +3.4   7. 0 0   8. $-\dfrac{2}{3}$ $\frac{2}{3}$ 9. $-\dfrac{5}{16}$ $\frac{5}{16}$ 10. -2.57 2.57

Simplify the expressions in Exercises 11–22.

11. $8 - 3$ 5   12. $5 - 7$ 2   13. $-4 + 6$ -10   14. $3 - (-4)$ 7

15. $5 - (-7)$ 12   16. $-18 + 17$ -35   17. $-8 + (+11)$ +3   18. $0 - (-12)$ 12

19. $-4\dfrac{5}{8} - 2\dfrac{1}{4}$ 20. $8\dfrac{2}{3} - 7\dfrac{1}{4}$ $\frac{26}{3}$ $\frac{29}{4}$ $\frac{149}{12} - \frac{87}{12} = \frac{57}{12}$ 21. $16.34 - (-8.4)$ 16.34 $+ 8.4$ 22. $(14.71) - 23.8$

Subtract the bottom number from the top number in Exercises 23–34.

23. 27
    $+42$
    $\overline{15}$

24. 19
    $\dfrac{26}{-7}$

25. 23
    $\dfrac{+7}{30}$

26. 41
    $\dfrac{-8}{49}$

27. -21
    $\dfrac{36}{-57}$

28. -47
    $\dfrac{13}{-60}$

29. -17
    $\dfrac{-17}{\emptyset}$

30. $14\dfrac{3}{5}$
    $17\dfrac{9}{10}$

31. $-7\dfrac{3}{4}$ -31/4
    $18\dfrac{1}{8}$ -257/8 145/8

32. $-11\dfrac{3}{8}$
    $+5\dfrac{5}{6}$ 6

33. -41.62
    $\dfrac{-10.58}{-31.04}$

34. 16.4
    $\dfrac{-5.83}{10.57}$

Perform the indicated operations in Exercises 35–46.

35. $-6 + (-4) - 5$ -15   36. $-7 - (-2) + 6$ 37. $6 + (-3) + (-4)$ -1 +3 -4 +3 + 3 -1

38. $-3 + (-7) + 2$ -10 +2 -8   39. $-5 - 2 - (-4)$ -5 -7 -7 +4 -3   40. $-8 - 5 - (-3)$

41. $-2 - 2 + 11$ -2 11 -4 +7   42. $-3 - (-3) + (-6)$ 0+ -6   43. $-\dfrac{2}{3} + \dfrac{4}{5} - \left(-\dfrac{1}{2}\right)$

44. $\dfrac{4}{3} - \dfrac{1}{4} - \dfrac{5}{6}$ 45. $9.37 - 16.42 - (8.21)$ 46. $-11.63 + 5.83 - 7.29$

Fill in the blank in Exercises 47–60 with the proper symbol: $<$, $>$, or $=$.

47. $-6 + (-2)$ $\leq$ $3 + (+8)$ -8 -5   48. $-4 - (-3)$ $=$ $-4 + (-3)$ -7 -7

49. $5 - 8$ _____ $8 - 5$ -5   50. $7 - (-3)$ _____ $-3 - 7$

51. $11 + (-3)$ _____ $11 - 3$ 52. $0 - 6$ _____ $0 - (-6)$

Answers to Practice Problems   1. -8.5 2. -1 3. -8 4. True
5. Not a solution

53. $-8 - (-8)$ _____ $-14 - 13$

54. $-7 - (-3)$ _____ $4 - 9$

55. $0 - \left(-\dfrac{1}{8}\right)$ _____ $0 - \dfrac{1}{8}$

56. $-\dfrac{4}{5} - \left(-\dfrac{1}{2}\right)$ _____ $\dfrac{3}{5} - \dfrac{3}{10}$

57. $-\left(-\dfrac{7}{8}\right)$ _____ $\dfrac{1}{2} - \left(-\dfrac{5}{8}\right)$

58. $6.5 - 4.3$ _____ $9.9 - (-1.3)$

59. $-15.71 - 8.46$ _____ $-(10.07 + 14.1)$

60. $7.25 - 21.62$ _____ $-13.31 - 2.53$

In Exercises 61–72, determine whether or not the given number is a solution to the given equation by substituting and then evaluating.

61. $x + 5 = -3$; $x = -8$

62. $x - 6 = -9$; $x = -3$

63. $15 - y = 17$; $y = -2$

64. $11 - x = 8$; $x = 3$

65. $2x - 3 = -5$; $x = 4$

66. $3y - 2 = 10$; $y = 4$

67. $-9 - 3x = 6$; $x = 5$

68. $4x + 13 = 3$; $x = 2$

69. $5x - 2 = 3$; $x = 1$

70. $-18 - 4y = -30$; $y = 3$

71. $2x + 5 = 7.2$; $x = 1.1$

72. $3x - 3.1 = 0.2$; $x = 1.2$

Calculator Problems

Use a calculator to find the value of each expression in Exercises 73–75.

73. $14.685 - 22.753 + 8.33$

74. $-21.5832 + 15.614 - 9.591$

75. $27.681 - 14.117 - (-6.841)$

Fill in the blank with the proper symbol: $<$, $>$, or $=$.

76. $64.851 - (-39.26)$ _____ $124.82 - 16.513$

77. $-19.824 - 23.417$ _____ $12.793 - (-14.387)$

78. $-43.931 - (-28.677)$ _____ $-(13.665 + 21.425)$

2.5 Multiplication and Division with Signed Numbers

OBJECTIVES

In this section, you will be learning to:

1. Multiply signed numbers.
2. Divide signed numbers.
3. Complete statements about the products and quotients of signed numbers.
4. Determine if signed number equations are true or false.
5. Determine if given signed numbers are solutions for specified equations.

Multiplication with whole numbers is shorthand for repeated addition. That is,

$$7 + 7 + 7 + 7 + 7 = 5 \cdot 7 = 35$$

and

$$\underbrace{7 + 7 + 7 + \cdots + 7}_{105 \ 7s} = 105 \cdot 7 = 735$$

Similarly, multiplication with signed numbers can be considered shorthand for repeated addition. For example,

$$(-6) + (-6) + (-6) = 3(-6) = -18$$
$$(-2) + (-2) + (-2) + (-2) + (-2) = 5(-2) = -10$$

Repeated addition with a negative number results in a product of a positive number and a negative number. Since the sum of negative numbers is negative, the product of a positive number and a negative number will be negative. In fact, the product of any positive number with a negative number will be negative.

EXAMPLES

1. $5(-3) = (-3) + (-3) + (-3) + (-3) + (-3) = -15$
2. $7(-10) = -70$
3. $42(-1) = -42$
4. $3.1(-5) = -15.5$
5. $+\dfrac{1}{4}\left(-\dfrac{1}{2}\right) = -\dfrac{1}{8}$ ∎

The product of two negative numbers does not relate to repeated addition. The following discussion is based on intuition, and no formal proof of the results will be given here. Notice the pattern of the results and see if you can supply the missing products.

$$4(-7) = -28 \qquad 0(-7) = 0$$
$$3(-7) = -21 \qquad -1(-7) = \,?$$
$$2(-7) = -14 \qquad -2(-7) = \,?$$
$$1(-7) = -7 \qquad -3(-7) = \,?$$

Did you get the following answers (noting that each product is 7 more than the previous product)?

$$-1(-7) = +7$$
$$-2(-7) = +14$$
$$-3(-7) = +21$$

You were correct if you did. Although one example does not prove that a procedure is correct, our intuition is good this time. The product of two negative numbers is positive. Again, the rule can be extended to include any signed numbers.

EXAMPLES

6. $(-4)(-9) = +36$
7. $-7(-5) = +35$
8. $-2.1(-6) = +12.6$
9. $\left(-\dfrac{1}{3}\right)\left(-\dfrac{5}{8}\right) = +\dfrac{5}{24}$ ∎

What happens if a number is multiplied by 0? For example, $3(0) = 0 + 0 + 0 = 0$. In fact, multiplication by 0 always gives a product of 0.

EXAMPLES

10. $6 \cdot 0 = 0$
11. $(3.7) \cdot 0 = 0$
12. $0 \cdot 8\dfrac{1}{2} = 0$ ∎

The rules for multiplication can be summarized as follows:

Rules for Multiplying Signed Numbers	1. The product of two positive numbers is positive. 2. The product of two negative numbers is positive. 3. The product of a positive number and a negative number is negative. 4. The product of 0 with any signed number is 0.

The rules can be stated using variables.

If a and b are positive numbers,

1. $a \cdot b = ab$
2. $a(-b) = -ab$
3. $(-a)(-b) = ab$
4. $a \cdot 0 = 0$ if a is any signed number.

Practice Problems

Find the following products.

1. $5(-3) =$ ~~-15~~
2. $-6(-4) =$ 24
3. $-8(4) =$ -32
4. $-13(0) =$ 0
5. $-9.1(-2) =$ 18.2
6. $3(-20.6) =$ 61.8
7. $\left(+\dfrac{3}{4}\right)\left(-\dfrac{2}{15}\right) = -\dfrac{1}{10}$

The rules for multiplication lead directly to the rules for division since division is defined in terms of multiplication.

For any numbers a and b where $b \neq 0$,

$$\frac{a}{b} = x \quad \text{means} \quad a = b \cdot x$$

If a is any number, then $\dfrac{a}{0}$ is **undefined** (Section 1.1).

EXAMPLES

13. $\dfrac{36}{9} = 4$ because $36 = 9 \cdot 4$.

14. $\dfrac{-36}{9} = -4$ because $-36 = 9(-4)$.

Answers to Practice Problems   1. -15 2. 24   3. -32 4. 0   5. 18.2
6. -61.8 7. $-\dfrac{1}{10}$

15. $\dfrac{36}{-9} = -4$ because $36 = -9(-4)$.

16. $\dfrac{-36}{-9} = +4$ because $^{•}-36 = -9(4)$. ∎

The rules for division can be stated as follows:

Rules for Dividing Signed Numbers	**1.** The quotient of two positive numbers is positive.
	2. The quotient of two negative numbers is positive.
	3. The quotient of a positive number and a negative number is negative.

The rules can be stated using variables.

If *a* and *b* are positive numbers,

1. $\dfrac{a}{b} = \dfrac{a}{b}$

2. $\dfrac{-a}{-b} = \dfrac{a}{b}$

3. $\dfrac{-a}{b} = -\dfrac{a}{b}$ and $\dfrac{a}{-b} = -\dfrac{a}{b}$

You might find the following common rules about multiplication and division with two nonzero numbers helpful in remembering the signs of answers.

1. If the numbers have the same sign, both the product and quotient will be positive.
2. If the numbers have different signs, both the product and quotient will be negative.

The quotient of two integers may not always be another integer. Just as with whole numbers, the quotient may be a fraction. For example, $\dfrac{-3}{6} = -\dfrac{1}{2}$ and $\dfrac{2}{-8} = -\dfrac{1}{4}$. We will discuss these ideas more thoroughly in Chapter 6.

Practice Problems	Find the quotients.
	1. $\dfrac{-30}{10}$ **2.** $\dfrac{40}{-10}$ **3.** $\dfrac{-20}{-10}$
	4. $\dfrac{-7}{0}$ **5.** $\dfrac{0}{13}$

Answers to Practice Problems   1. -3 **2.** -4 **3.** 2 **4.** Undefined   **5.** 0

Now that we have all the rules for addition, subtraction, multiplication, and division with signed numbers, we can discuss the solutions to equations that involve any or all of these operations.

EXAMPLES Determine whether or not the given number is a solution to the given equation by substituting and then evaluating.

17. $7x = -21$; $x = -3$

Solution $7(-3) = -21$, so -3 is a solution.

18. $-8y = 56$; $y = -7$

Solution $-8(-7) = +56$, so -7 is a solution.

19. $\dfrac{y}{-4} = -10$; $y = -40$

Solution $\dfrac{-40}{-4} = +10$ and $+10 \neq -10$, so -40 is **not** a solution.

20. $-5x + 7 = -3$; $x = 2$

Solution $-5(2) + 7 = -10 + 7 = -3$, so 2 is a solution. ∎

EXERCISES 2.5

Find the product in Exercises 1–20.

1. $4 \cdot (-3)$ **2.** $(-5) \cdot 6$ **3.** $(-8)(-7)$ **4.** $12 \cdot 4$

5. $19 \cdot 3$ **6.** $(-11)(-2)$ **7.** $(-14)(-4)^2$ **8.** $(-3.2)^2(7)$
                                                    16

9. $(5.4)(-6)$ **10.** $(-1.1)(-0.6)$ **11.** $(-1.3)(-2.1)$ **12.** $\left(\dfrac{1}{2}\right)\left(-\dfrac{1}{3}\right)$

13. $\left(-\dfrac{2}{5}\right)\left(\dfrac{3}{2}\right)$ **14.** $\left(-\dfrac{1}{8}\right)\left(-\dfrac{4}{3}\right)$ **15.** $\left(-\dfrac{7}{6}\right)\left(-\dfrac{9}{14}\right)$ **16.** $(-6)(-3)(-9)$

17. $-8 \cdot 4 \cdot 9$ **18.** $-3 \cdot 2 \cdot (-3)$ **19.** $(-7)(-16) \cdot 0$ **20.** $(-9) \cdot 11 \cdot 4$

Find the quotient in Exercises 21–35.

21. $\dfrac{-8}{-2}$ **22.** $\dfrac{-20}{10}$ **23.** $\dfrac{-30}{5}$ **24.** $\dfrac{-26}{-13}$ **25.** $\dfrac{39}{-13}$

26. $\dfrac{-51}{3}$ **27.** $\dfrac{-91}{-7}$ **28.** $\dfrac{0}{6}$ **29.** $\dfrac{0}{-7}$ **30.** $\dfrac{-3}{0}$

31. $\dfrac{16}{0}$ **32.** $\dfrac{-3.4}{2}$ **33.** $\dfrac{4.24}{-4}$ **34.** $\left(-\dfrac{2}{3}\right) \div \dfrac{3}{4}$ **35.** $\left(-\dfrac{1}{2}\right) \div \left(-\dfrac{7}{8}\right)$

Correctly complete the sentences in Exercises 36–45 with *positive, negative, 0,* or *undefined.*

36. If x is a positive number, then $x(-x)$ is a _____ number.

37. If x is a negative number, then $x(-x)$ is a _____ number.

38. If x is a negative number and y is a negative number, then xy is a _____ number.

39. If x is a positive number and y is a negative number, then xy is a _____ number.

40. If x and y are positive numbers, then $\dfrac{x}{y}$ is a _____ number.

41. If x is a negative number and y is a natural number, then $\dfrac{x}{y}$ is a _____ number.

42. If x and y are negative numbers, then $\dfrac{x}{y}$ is a _____ number.

43. If x is a signed number, then $x \cdot 0$ is _____ .

44. If x is a nonzero number, then $\dfrac{0}{x}$ is _____ .

45. If x is a signed number, then $\dfrac{x}{0}$ is _____ .

Determine whether the equations in Exercises 46–55 are true or false.

46. $(-4) \cdot (6) = 3 \cdot 8$

47. $(-7) \cdot (-9) = 3 \cdot 21$

48. $(-12)(6) = 9(-8)$

49. $(-6)(9) = (18)(-3)$

50. $6(-3) = (-14) + (-4)$

51. $7 + 8 = (-10) + (-5)$

52. $-7 + 0 = (-7) \cdot (0)$

53. $17 + (-3) = (16) + (-4)$

54. $-4(6 + 3) = (-24) + (-12)$

55. $14 + 6 = -2[(-7) + (-3)]$

In Exercises 56–65, determine whether or not the given number is a solution to the given equation by substituting and then evaluating.

56. $11x = -55; \quad x = -5$

57. $-7x = 84; \quad x = -12$

58. $\dfrac{x}{7} = -6; \quad x = 42$

59. $\dfrac{y}{-6} = 12; \quad y = -72$

60. $\dfrac{y}{10} = -9; \quad y = -90$

61. $-9x = -72; \quad x = -8$

62. $5x + 3 = 18; \quad x = 3$

63. $-3x + 7 = -8; \quad x = 5$

64. $4x - 3 = -23; \quad x = -5$

65. $7x - 6 = -34; \quad x = -4$

Calculator Problems

Use a calculator to find the value of each expression in Exercises 66–70.

66. $(2.73)(-0.241)(-1.8)$

67. $(-4.613)(-0.45)(-1.66)$

68. $(5.314)(-1.7)(24)$

69. $(-77.459) \div 29$

70. $(-62.234) \div (-37)$

2.6 Word Problems

OBJECTIVE

In this section, you will be learning to solve word problems that require the use of signed numbers.

The word problems in this section involve simple applications of the sums and differences of integers. Subtraction is used, for example, to find the change in values between two readings on a thermometer or the change between two distances or two altitudes. To calculate the change, including direction (positive or negative), **subtract the beginning value from the end value.**

EXAMPLES

1. On a winter day, the temperature dropped from $35°$ F at noon to $6°$ F below zero ($-6°$ F) at 7 P.M. What was the change in temperature?

Solution end value $-$ beginning value

$$-6° \quad - \quad (35)° \quad = -6° + (-35°) = -41°$$

2. A jet pilot flew her plane from an altitude of 30,000 ft to an altitude of 12,000 ft. What was the change in altitude?

    Solution $\underbrace{\text{end value}}$ $-$ $\underbrace{\text{beginning value}}$

    $$12,000 \;-\; 30,000 \;=\; -18,000 \text{ ft}$$

3. Sue weighed 130 lb when she started to diet. The first week she lost 7 lb, the second week she gained 2 lb, and the third week she lost 5 lb. What was her weight after 3 weeks of dieting?

    Solution $130 + (-7) + (+2) + (-5) = 123 + (+2) + (-5)$
    $$= 125 + (-5)$$
    $$= 120 \text{ lb} \qquad \blacksquare$$

Practice Problems

1. What should be added to -8 to get a sum of 20?

2. At noon, the temperature was 40° F. At 3:00 P.M., the temperature was 32° F. What was the change in temperature?

3. Subtract $5\dfrac{1}{4}$ from the sum of $3\dfrac{1}{2}$ and $1\dfrac{1}{2}$.

Handwritten work:
$\begin{array}{r} 4\,0 \\ +3\,2 \\ \hline 8 \end{array}$

$\begin{array}{r} 2\,0 \\ +\,18 \\ \hline 2\text{-}8 \end{array}$

$\approx -\dfrac{1}{4}$

EXERCISES 2.6

Handwritten work:
$\dfrac{21}{4} \qquad \dfrac{7}{2}\;\dfrac{3}{2} = \dfrac{10}{2} \qquad \dfrac{21}{4} - \dfrac{10}{2} \qquad \dfrac{21}{4} - \dfrac{20}{4} = \dfrac{1}{4}$

Write each problem as a sum or difference, then simplify.

1. 23 lb lost, 13 lb gained, 6 lb lost
    Handwritten: $\begin{array}{r} -29 \\ +\,13 \\ \hline -16 \end{array}$

2. $24 earned, $17 spent, $2 earned

3. 4° rise, 3° rise, 9° drop

4. $10 withdrawal, $25 deposit, $18 deposit, $9 withdrawal

5. $47 earned, $22 spent, $8 earned, $45 spent

6. $20 won, $42 lost, $58 won, $11 lost

7. 14° rise, 6° drop, 11° rise, 15° drop

8. $53 withdrawal, $8 withdrawal, $48 deposit, $17 withdrawal

9. $187 profit, $241 loss, $82 profit, $26 profit

10. Snap-O Mousetrap stock rose $3 Monday, rose $6 Tuesday, and dropped $7 Wednesday. What was the net change in price of the stock?

11. In the first quarter of a recent football game, Fumbles A. Lott carried the ball six times with the following results: a gain of 6 yd, a gain of 3 yd, a loss of 4 yd, a gain of 2 yd, no gain or loss, and a loss of 3 yd. What was his net yardage for the first quarter?

12. Beginning with 7° above zero, the temperature rose 4°, then dropped 2°, and then dropped 6°. Find the final temperature.

13. Starting at the third floor, an elevator went down 1 floor, up 3 floors, up 7 floors, and then down 4 floors. Find the final location of the elevator.

14. Bill lost 2 lb the first week of his diet, lost 6 lb the second week, gained 1 lb the third week, lost 4 lb the fourth week, and gained 3 lb the fifth week. What was the total loss or gain? If Bill weighed 223 lb at the time he began his diet, what was his weight after 5 weeks of dieting?

Answers to Practice Problems   1. 28   **2.** $-8°$ F   **3.** $-\dfrac{1}{4}$

15. Jeff works Friday, Saturday, and Sunday in a restaurant. His salary is $10 a night plus tips. Friday night he received $5 in tips but spent $2 for food. Saturday he bought a new shirt for $8, spent $3 for food, but received $12 in tips. Sunday he spent $3 for food and received $9 in tips. How much money did he have left after three days of work?

16. What should be added to -5 to get a sum of 17?

17. What should be added to -10 to get a sum of -23?

18. What should be added to 39 to get a sum of 16?

19. What should be added to 24.6 to get a sum of 13.8?

20. What should be added to -37.3 to get a sum of -54.7?

21. What should be added to $15\frac{7}{8}$ to get a sum of 18?

22. What should be added to $-12\frac{3}{10}$ to get a sum of $15\frac{1}{2}$?

23. From the sum of -4 and -17, subtract the sum of -12 and 6.

24. From the sum of 11 and -13, subtract the sum of 19 and -8.

25. Subtract $7\frac{3}{4}$ from the sum of $5\frac{1}{2}$ and $-2\frac{3}{16}$.

26. Add $4\frac{3}{5}$ to the difference between $2\frac{5}{6}$ and $7\frac{3}{10}$.

27. Mr. Adams received a bank statement indicating that he was overdrawn by $63. How much must he deposit to bring his balance to $157?

28. A campus sorority sold tickets to a pancake breakfast. The expenses totaled $87.50. If they realized a profit of $192.80, how much money did they receive from ticket sales?

29. At 2:00 P.M. the temperature was 77° F. At 8:00 P.M. the temperature was 58° F. What was the change in temperature?

30. The temperature at 5:30 A.M. was 37° F below zero; at noon, the temperature was 29° F above zero. What was the change in temperature?

31. Lotsa-Flavor Chewing Gum stock opened on Monday at $47 per share and closed Friday at $39 per share. Find the change in price of the stock.

32. The Greek mathematician and scientist Aristotle lived from 384 B.C. to 322 B.C. How long did he live?

33. If you travel from the top of Mt. Whitney, elevation 14,495 ft, to the floor of Death Valley, elevation 282 ft below sea level, what is the change in elevation?

34. A submarine submerged 280 ft below the surface of the sea fired a rocket that reached an altitude of 30,000 ft. What was the change in altitude?

35. A man who was born in 87 B.C. died in 35 B.C. How old was he when he died?

36. At the end of the first round of a golf tournament, Beth was 4 under par. At the end of the second round, she was 7 over par. How much over par did she shoot in the second round?

37. In a four-day golf tournament, Joe scored 2 over par the first day, 3 under par the second day, 3 under par the third day, and 1 under par the fourth day. How much over or under par was he for the tournament?

38. Widget Inc. stock closed Monday at $7\frac{3}{4}$. Tuesday it went up $\frac{5}{8}$ point, and Wednesday it went down $\frac{3}{8}$ point. What was the closing price?

39. On Monday, Alpha Corp. stock opened at $29\frac{3}{8}$ and closed at $31\frac{1}{4}$. What was the change in the price?

40. Claudia had a balance of $307.86 in her checking account. She wrote checks of $23.68, $42.50, and $17.43. She made a deposit of $35.80. What was the balance of her account?

CHAPTER 2 SUMMARY

Key Terms and Formulas

The **graph** of a number is the point that corresponds to the number, and the number is called the **coordinate** of the point.   [2.1]

The set of numbers consisting of the whole numbers and their opposites is called the set of **integers.**   [2.1]

Integers:
$\ldots, -3, -2, -1, 0, 1, 2, 3, \ldots$

Positive integers:
$1, 2, 3, 4, 5, \ldots$

Negative integers:
$\ldots, -4, -3, -2, -1$ [2.1]

Rational numbers

numbers that can be written as $\dfrac{a}{b}$,

where a and b are integers, $b \neq 0$ [2.1]

Symbols   [2.1]

$=$	is equal to
$<$	is less than
$\leq$	is less than or equal to
$\neq$	is not equal to
$>$	is greater than
$\geq$	is greater than or equal to

The **absolute value** of a number is its distance from 0. The absolute value of a number is never negative.   [2.2]

Symbolically,

$$\begin{cases} |a| = a & \text{if } a \text{ is a positive number or } 0 \\ |a| = -a & \text{if } a \text{ is a negative number.} \end{cases}$$

The **opposite** of a signed number is called its **additive inverse.** The sum of a number and its additive inverse is 0. Symbolically, for any number a,

$$a + (-a) = 0 \quad [2.4]$$

Properties and Rules

Zero is its own opposite and is neither positive nor negative.   [2.1]

On a horizontal number line, **smaller numbers are always to the left of larger numbers.**   [2.1]

Rules for Adding Signed Numbers   [2.3]
1. To add two numbers with like signs, add their absolute values and use the common sign.
2. To add two numbers with unlike signs, subtract their absolute values (the smaller from the larger) and use the sign of the number with the larger absolute value.

For any signed numbers a and b,

$$a - b = a + (-b) \quad [2.4]$$

Rules for Multiplying Signed Numbers   [2.5]
1. The product of two positive numbers is positive.
2. The product of two negative numbers is positive.

3. The product of a positive number and a negative number is negative.
4. The product of 0 with any signed number is 0.

Rules for Dividing Signed Numbers   [2.5]
1. The quotient of two positive numbers is positive.
2. The quotient of two negative numbers is positive.
3. The quotient of a positive number and a negative number is negative.

If a is any number, then $\dfrac{a}{0}$ is **undefined.**   [2.5]

For any numbers a and b where $b \neq 0$,

$$\frac{a}{b} = x \text{ means } a = b \cdot x \quad [2.5]$$

CHAPTER 2 REVIEW

Fill in the blanks in Exercises 1–6 with the proper symbol: $<$, $>$, or $=$. [2.1]

1. -3 _____ -2

2. -6 _____ $-(6)$

3. 2 _____ -2

4. $-(-5)$ _____ 1

5. $-\dfrac{2}{3}$ _____ $-\dfrac{3}{4}$

6. 1.8 _____ -0.3

Graph each of the sets of numbers in Exercises 7–16 on the number line. [2.1]

7. $\{-3, 2, 1, 5\}$

8. $\{-6, 0, 3, 1, -2\}$

9. $\{-4, 3, |-6|, 0, 1\}$

10. $\{|-3|, -3, 2, -1\}$

11. $\left\{-2, -\dfrac{2}{3}, 1, \left|-\dfrac{7}{3}\right|, 4\right\}$

12. $\{-2.1, -0.8, |-1.6|, 2.7\}$

13. All integers greater than -4

14. All negative integers greater than -8

15. All positive integers less than or equal to 6

16. All integers less than -8

List the numbers that satisfy the conditions stated in Exercises 17–22. [2.2]

17. $|x| = 6$

18. $|x| = 10$

19. $|y| = \dfrac{3}{5}$

20. $|y| = 2.9$

21. $|x| = -3.1$

22. $|y| = 1.724$

For Exercises 23–26 graph on a number line the integers that satisfy the stated condition. [2.2]

23. $|x| \geq 6$

24. $|x| \leq 5$

25. $|x| < 8$

26. $|x| > 10$

Perform the indicated operations in Exercises 27–44. [2.3–2.5]

27. $-13 + 12 + (-7)$

28. $-17 + (-3) - (-8)$

29. $-23 - (-8)$

30. $7.3 + (-5.5)$

31. $-\dfrac{4}{5} + \dfrac{2}{3}$

32. $-\dfrac{4}{3} + \dfrac{1}{4} - \dfrac{5}{6}$

33. $8.64 - 10.21 + 0.39$

34. $(-19)(6)$

35. $(-23)(-11)$

36. $(-8)(-5)4$

37. $\left(\dfrac{3}{8}\right)\left(-\dfrac{4}{5}\right)^2$

38. $(2.6)(-1.5)^2$

39. $-28 \div (-7)$

40. $11 \div 0$

41. $98 \div (-14)$

42. $0 \div \dfrac{5}{16}$

43. $-13.6 \div 2$

44. $-\dfrac{5}{7} \div \left(-\dfrac{3}{14}\right)$

Choose the response that correctly completes the sentences in Exercises 45–51. [2.3]

45. If x is a positive number and y is a negative number, then $x + y$ is (never, sometimes, always) a positive number.

46. If x is a positive number and y is a negative number, then $x \cdot y$ is (never, sometimes, always) a positive number.

47. If x is a signed number and y is a signed number, then $x - y$ is (never, sometimes, always) equal to $x + (-y)$.

48. If x is a signed number and y is a signed number, then $x + y$ is (never, sometimes, always) equal to 0.

49. If x is an integer, y is an integer, and $y \neq 0$, then $x \div y$ is (never, sometimes, always) an integer.

50. $|x|$ is (never, sometimes, always) less than x.

51. $|-x|$ is (never, sometimes, always) equal to $|x|$.

Determine whether or not the given number in Exercises 52–61 is a solution to the given equation.   [2.5]

52. $2x - 3 = 9$; $x = 6$

53. $5x + 7 = -3$; $x = -2$

54. $8 - x = 13$; $x = -5$

55. $7x - 1 = 20$; $x = -3$

56. $\dfrac{y}{8} + 1 = -1$; $y = -16$

57. $\dfrac{y}{4} + 8 = 12$; $y = 16$

58. $4x - 5 = -2$; $y = \dfrac{3}{4}$

59. $3x + 4 = 2$; $x = -\dfrac{2}{3}$

60. $2x + 1.7 = -3.5$; $x = -2.6$

61. $-3x - 4.1 = 8$; $x = 1.3$

Solve the word problems in Exercises 62–66.   [2.6]

62. Stock in the Go-Fly-A-Kite Company opened Monday at $11 and rose $2; Tuesday it dropped $4; Wednesday it rose $3; and Thursday it dropped $2. What was the closing price on Thursday?

63. Ralph started the month of January with $117. The first week he earned $21 and spent $9. The second week he earned $13 and spent $45. The third week he earned $19 and spent $16. The fourth week he earned $18 and spent $30. What is his balance after 4 weeks?

64. From the sum of -12 and -9, subtract the product of -3 and 8.

65. Subtract $8\dfrac{5}{8}$ from the sum of $6\dfrac{1}{4}$ and $3\dfrac{9}{16}$.

66. Subtract the product of 4.07 and -3 from 5.28.

CHAPTER 2 TEST

1. Fill in the blanks with the proper symbol: $<$, $>$, or $=$.
   a. -4 _____ -2 **b.** $-(-2)$ _____ 0

2. Graph the following set of numbers on the number line: $\left\{-2, -0.4, |-1|, \dfrac{7}{3}, 3.1\right\}$.

3. Graph the following set of numbers on the number line: all integers less than or equal to 2.

4. List the numbers that satisfy the condition $|y| = 7$.

5. List the numbers that satisfy the condition $|x| = \dfrac{7}{8}$.

List the integers that satisfy the conditions stated in Exercises 6 and 7, then graph the integers on the number line.

6. $|x| < 3$

7. $|x| \geq 9$

8. Add: $-15 + 45 + (-17)$

9. Find the sum: $17.04 + (-10.36)$

10. Find the sum: $-\dfrac{2}{3} + \dfrac{5}{8} + \dfrac{1}{6}$

11. Find the difference: $-28 - (-47)$

12. Perform the indicated operations:
   $-9 - (-6) + 14$

13. Find the product: $(-27)(-6)$

14. Divide: $64 \div (-16)$

15. Multiply: $(4.12)(-6.1)$

16. Multiply: $\left(-\dfrac{2}{3}\right)\left(-\dfrac{5}{6}\right)\left(-\dfrac{1}{4}\right)$

17. Divide: $-\dfrac{15}{8} \div \dfrac{5}{6}$

18. If x is a positive number and y is a negative number, then $x - y$ is (never, sometimes, always) a negative number.

19. If x is a positive number and y is a negative number, then $y \div x$ is (never, sometimes, always) a negative number.

20. Is $x = 5$ a solution to $3x - 9 = 6$?

21. Is $x = -1.5$ a solution to $-2x + 5 = 8$?

22. Is $x = \dfrac{4}{5}$ a solution to $5x - 2 = 6$?

23. From the sum of -12 and -18, subtract the product of -6 and -5.

24. All-American Inn stock opened on Monday at $\$6\dfrac{3}{8}$ per share and closed on Friday at $\$5\dfrac{3}{4}$ per share. Find the change in the price of the stock.

25. This morning Thuy had $27.63. During the day she earned $10.56, spent $14.29, and spent $8.43. How much money does she have tonight?

CUMULATIVE REVIEW (2)

Name each property illustrated in Exercises 1–3.

1. $(3x)y = 3(xy)$

2. $x(3 + 9) = (3 + 9)x$

3. $12x + 4 = 4(3x + 1)$

Simplify Exercises 4–8 using the rules for order of operations.

4. $6 + 4 \div 2$

5. $3 \cdot 2^3 - 5$

6. $(18 + 2 \cdot 3) \div 4 \cdot 3$

7. $\dfrac{1}{2} + \dfrac{3}{8} \div \dfrac{3}{4} - \dfrac{5}{6}$

8. $\left(\dfrac{3}{4} - \dfrac{5}{6}\right) \cdot \left(\dfrac{6}{5} + \dfrac{2}{3}\right)$

9. What is 15.2% of 65?

10. Evaluate the expression $3x - 2x^2$ if $x = 4$.

11. Determine whether or not $x = 1.3$ is a solution to $5x + 2.5 = 4$.

12. Determine whether or not $x = \dfrac{1}{2}$ is a solution to $2x + 3 = 6x + 1$.

13. A rectangle is 14 centimeters wide and $18\dfrac{1}{2}$ centimeters long. Find (**a**) the perimeter, and (**b**) the area.

14. Libby bought a T-shirt marked at $14.50. If 6.5% sales tax was added to the purchase price, how much did she pay for the shirt?

15. A monthly art collector's magazine costs $4.50 when purchased at the newsstand. The subscription for one year (12 issues) is advertised to be 60% of the newsstand price. What is the cost of a 1-year subscription?

16. Simplify the expression $[4 + (-7)] \cdot [-3 - (-5)]$.

17. List the numbers that satisfy the equation $|x| = 2.3$.

18. List the integers that satisfy the inequality $|x| \leq 5$.

19. Is $x = -8$ a solution to $\dfrac{x}{4} + 12 = 10$?

20. From the sum of -2 and -11, subtract the product of -4 and 2.

3 SOLVING FIRST-DEGREE EQUATIONS

C H A P T E R

◢ DID YOU KNOW ?

Traditionally, algebra has been defined to mean generalized arithmetic where letters represent numbers. For example, $3 + 3 + 3 + 3 = 4 \cdot 3$ is a special case of the more general algebraic statement that $x + x + x + x = 4 \cdot x$. The name *algebra* comes from an Arabic word, *al-jabr,* which means "to restore."

A famous Moslem ruler, Harun al-Rashid, the caliph made famous in *Tales of the Arabian Nights,* and his son Al-Mamun brought to their Baghdad court many famous Moslem scholars. One of these scholars was the mathematician Mohammed ibn-Musa al-Khowarizmi, who wrote a text (c. A.D. 800) entitled *ihm al-jabr w' al-muqabal.* The text included instructions for solving equations by adding terms to both sides of the equation, thus "restoring" equality. The abbreviated title of Al-Khowarizmi's text, *Al-jabr,* became our word for equation solving and operations on letters standing for numbers, **algebra.**

Al-Khowarizmi's algebra was brought to western Europe through Moorish Spain in a Latin translation done by Robert of Chester (c. A.D. 1140). Al-Khowarizmi's name may have sounded familiar to you. It eventually was translated as "algorithm," which came to mean any series of steps used to solve a problem. Thus we speak of the division algorithm used to divide one number by another. Al-Khowarizmi is known as the father of algebra, just as Euclid is known as the father of geometry.

One of the hallmarks of algebra is its use of specialized notation that enables complicated statements to be expressed using compact notation. Al-Khowarizmi's algebra used only a few symbols, with most statements written out in words. He did not even use symbols for numbers! **The development of algebraic symbols and notation occurred over the next 1000 years.**

Algebra is generous, she often gives more than is asked of her.

Jean le Rond d'Alembert (1717?–83)

CHAPTER OUTLINE

*N*ow you are ready to learn to solve equations. You will also find a variety of useful formulas and use your equation-solving skills in manipulating these formulas.

Of course, solving equations and working with formulas are useful skills for solving—yes, you guessed it—word problems. "Word problems" is not a dirty word. Remember that one of the major goals of mathematics is to develop the skills that will allow you, with reasoning, to solve a wide variety of problems you may see in your lifetime.

3.1 Simplifying and Evaluating Expressions

OBJECTIVES

In this section, you will be learning to:

1. Simplify algebraic expressions by combining like terms.
2. Evaluate expressions for given values of the variables.

An expression that involves only multiplication and/or division with constants and/or variables is called a **term. A single constant or variable is also a term.** Examples of terms are

$$3x, \qquad 16, \qquad -7y, \qquad a, \qquad 14x^2, \qquad \frac{-x}{y^3}$$

Expressions such as

$$3 + x, \qquad 5y - 7x, \qquad 3x + 4x, \qquad \text{and} \qquad 5x^2 - 3x^2 + 2x$$

are the algebraic sums of terms. **Like terms** (or **similar terms**) are terms that are constants or terms that contain the same variables that are of the same power.

Like Terms	Unlike Terms	
$-2x$ and $3x$	$5x$ and $4x^2$	(x is not of the same power in both terms.)
7 and -10		
$4a^2$ and $-9a^2$	-8 and $6y$	(The term -8 does not have the variable y in it.)
$5x^2y^3$ and $-2x^2y^3$		
$\frac{5}{8}x$ and $-\frac{2}{3}x$	$3c^2d$ and $7c^2$	(There is no d in the second term.)

The numerical part of a term is called the **coefficient** of the variable or variables in the term. For example,

a. in the term $8x$, 8 is the coefficient of x;

b. in the term $-5y^3z$, -5 is the coefficient of y^3z;

c. in the term $\frac{2x}{5}$ $\left(\text{or } \frac{2}{5}x\right)$, $\frac{2}{5}$ is the coefficient of x;

d. in the term $2.4y$, 2.4 is the coefficient of y.

If no coefficient is written next to a variable, the coefficient is understood to be 1. If a $-$ sign is next to a variable, the coefficient is understood to be -1. Thus,

$$x = 1 \cdot x, \qquad y^3 = 1 \cdot y^3, \qquad \text{and} \qquad -b = -1 \cdot b$$

By applying the distributive property (discussed in Section 1.1), we can **combine like terms.** The distributive property states that

$$a(b + c) = ab + ac$$

or

$$ab + ac = a(b + c)$$

or

$$ba + ca = (b + c)a$$

This last form is particularly useful when b and c are numerical coefficients. For example,

a. $3x + 5x = (3 + 5)x$ By the distributive property
          $= 8x$ Add the coefficients.

b. $3x^2 - 5x^2 = (3 - 5)x^2$ By the distributive property
           $= -2x^2$ Add the coefficients algebraically.

In (a) the like terms $3x$ and $5x$ are **combined** and the result is $8x$.
In (b) the like terms $3x^2$ and $-5x^2$ are **combined** and the result is $-2x^2$.

EXAMPLES   Combine like terms whenever possible.

1. $8x + 10x$
   Solution $8x + 10x = (8 + 10)x$ By the distributive property
                     $= 18x$

2. $6y - 2y$
   Solution $6y - 2y = (6 - 2)y$
                     $= 4y$

3. $4(x - 7) + 5(x + 1)$
   Solution

$$
\begin{aligned}
4(x - 7) + 5(x + 1) &= 4x - 28 + 5x + 5 \\
&= 4x + 5x - 28 + 5 \\
&= 9x - 23
\end{aligned}
$$
     Use the distributive property twice.
     Combine like terms.

4. $2x^2 + 3a + x^2 - a$
   Solution

$$
\begin{aligned}
2x^2 + 3a + x^2 - a &= 2x^2 + x^2 + 3a - a \\
&= (2 + 1)x^2 + (3 - 1)a \\
&= 3x^2 + 2a
\end{aligned}
$$
     Note: $+x^2 = +1x^2$
            $-a = -1a$

5. $\dfrac{x + 3x}{2} + 5x$

Solution

$$\dfrac{x + 3x}{2} + 5x = \dfrac{4x}{2} + 5x = \dfrac{4}{2} \cdot x + 5x$$
$$= 2x + 5x$$
$$= 7x$$

A fraction bar is a symbol of inclusion, like parentheses. So combine like terms in the numerator first.   ■

In most cases, if an expression is to be evaluated, like terms should be combined first and then the resulting expression evaluated. Before combining terms and evaluating expressions, we make special mention of one particular situation involving negative numbers. Is there a difference between $(-7)^2$ and -7^2, or are they the same?

In the expression $(-7)^2$, the base is -7, and -7 is to be squared:

$$(-7)^2 = (-7)(-7) = +49$$

However, for -7^2, the base is 7 and the rules for order of operations say to square 7 first:

$$-7^2 = -(7^2) = -49$$

or

$$-7^2 = -1 \cdot 7^2 = -1 \cdot 49 = -49$$

Suppose we want to evaluate $-x^2$ for $x = 3$ and for $x = -4$. We have

$$-x^2 = -3^2 = -1 \cdot 9 = -9 \qquad \text{for } x = 3$$

and

$$-x^2 = -(-4)^2 = -1(16) = -16 \qquad \text{for } x = -4$$

EXAMPLES   Simplify each expression by combining like terms; then evaluate the resulting expression using the given values for the variables. Refer to the Rules for Order of Operations in Section 1.2.

6. $2x + 5 + 7x;\ x = -3$

Solution $2x + 5 + 7x = 2x + 7x + 5$
$$= 9x + 5$$

$$9x + 5 = 9(-3) + 5$$
$$= -27 + 5$$
$$= -22$$

7. $3ab - 4ab + 6a - a;\ a = 2,\ b = -1$

Solution $3ab - 4ab + 6a - a = -ab + 5a$
$$-ab + 5a = -1(2)(-1) + 5(2) \qquad (\textbf{Note: } -ab = -1ab)$$
$$= 2 + 10$$
$$= 12$$

8. $\dfrac{5x + 3x}{4} + 2(x + 1); x = 5$

Solution $\dfrac{5x + 3x}{4} + 2(x + 1) = \dfrac{8x}{4} + 2x + 2$

$$= 2x + 2x + 2$$
$$= 4x + 2$$

$$4x + 2 = 4 \cdot 5 + 2$$
$$= 20 + 2$$
$$= 22 \qquad \blacksquare$$

Practice Problems

Simplify the following expressions by combining like terms.

1. $-2x - 5x$
2. $12y + 6 - y + 10$
3. $5(x - 1) + 4x$
4. $2b^2 - a + b^2 + a$

Simplify the expression, then evaluate the resulting expression if $x = 3$ and $y = -2$.

5. $2(x + 3y) + 4(x - y)$

EXERCISES 3.1

Combine like terms in each of the expressions in Exercises 1–35.

1. $8x + 7x$
2. $3y + 8y$
3. $5x - 2x$
4. $7x + (-3x)$
5. $6y^2 - y^2$
6. $16z^2 - 5z^2$
7. $23x^2 - 11x^2$
8. $18x^2 + 7x^2$
9. $4x + 2 + 3x$
10. $3x - 1 + x$
11. $2x - 3y - x$
12. $x + y + x - 2y$
13. $-2x^2 + 5y + 6x^2 - 2y$
14. $4a + 2a - 3b - a$
15. $3(x + 1) - x$
16. $2(x - 4) + x + 1$
17. $5(x - y) + 2x - 3y$
18. $4x - 3y + 2(x + 2y)$
19. $3(2x + y) + 2(x - y)$
20. $4(x + 5y) + 3(2x - 7y)$
21. $2x + 3x^2 - 3x - x^2$
22. $2y^2 + 4y - y^2 - 3y$
23. $2(x^2 + 3x) + 4(-x^2 + 2x)$
24. $3x^2 + 2x - 5 - x^2 + x - 4$
25. $3x^2 + 4xy - 5xy + y^2$
26. $2x^2 - 5xy + 11xy + 3$
27. $\dfrac{x + 5x}{6} + x$
28. $2y - \dfrac{2y + 3y}{5}$
29. $y - \dfrac{2y + 4y}{3}$
30. $z - \dfrac{3z + 5z}{4}$
31. $\dfrac{-3x + 5x}{-2} + x$
32. $\dfrac{-4x - 2x}{3} - 4x$
33. $\dfrac{4x - 2x}{2} + 3(x + 2x)$
34. $5(2x + 2) - \dfrac{3(2x + x)}{9}$
35. $\dfrac{3(5x - x)}{6} - \dfrac{4(2x - x)}{4}$

Evaluate the expressions in Exercises 36–45 if $x = 1$, $y = -3$, and $z = 2$.

36. $3xy^2$

37. x^2yz^2

38. $x^2(y + z)$

39. $xy - (x - y)$

40. $y(3x^2 + z) - 4z$

41. $x^2 + 3x - 4$

42. $\dfrac{4y^2 - z}{10z + y}$

43. $\dfrac{z - 4y}{z^2 - y}$

44. $\dfrac{2y^2 + 3y - 4}{1 - z - z^2}$

45. $\dfrac{3x^2 - 4x + 1}{y^2 + 3y - 4}$

Simplify the expressions in Exercises 46–60 by combining like terms. Then evaluate the resulting expression using the given value.

46. $3x + 4 - x$; $x = 6$

47. $5x - 7 + 8x$; $x = 4$

48. $3(x - 1) + 2(x + 2)$; $x = -5$

49. $4(y + 3) + 5(y - 2)$; $y = -3$

50. $-2x + 5y + 6x - 2y$; $x = 1, y = -2$

51. $4x + 2x - 3y - x$; $x = 2, y = -1$

52. $4x - 3y + 2(x + 2y)$; $x = -3, y = 1$

53. $5(x + y) + 2x - 3y$; $x = -2, y = 2$

54. $3xy + 5x - xy + x$; $x = 2, y = -1$

55. $7y - 2xy + 4xy - y$; $x = 4, y = 3$

56. $\dfrac{3x + 5x}{-2} + x$; $x = 5$

57. $\dfrac{-4x - 2x}{3} + 7x$; $x = -6$

58. $\dfrac{2(4x - x)}{3} - \dfrac{3(6x - x)}{3}$; $x = -4$

59. $\dfrac{8(5x + 2x)}{7} - \dfrac{2(3x + x)}{4}$; $x = 3$

60. $\dfrac{6(5x - x)}{8} + \dfrac{5(x + 5x)}{3}$; $x = 5$

49. $(4y + 12) + (5y - 10)$
$9y + 2 = -27$, $2 = -25$

Calculator Problems

Evaluate the expressions if $x = -0.83$ and $y = 2.53$.

61. $x(4x + y)$

62. $1.73x - 0.58y$

63. $x^2 + y^2$

64. $6.4x^2 + 0.42xy$

65. $0.8y^2 - 2.3y + 11.32$

◢ 3.2 Solving Equations [x + b = c and ax = c]

OBJECTIVES

In this section, you will be learning to:

1. Solve equations of the form $x + b = c$.

2. Solve equations that can be simplified to the form $x + b = c$.

3. Solve equations of the form $ax = c$.

4. Solve equations that can be simplified to the form $ax = c$.

Many practical applications involve setting up an equation (or several equations) involving an unknown quantity. The objective is to find the value of this unknown so that the question asked in the application is answered. This means that, before we can solve meaningful word problems, one of the basic skills needed is solving equations. In this section, we will discuss the basic techniques used for **solving first-degree equations.**

If an equation contains a variable, the value (or values) that gives a true statement when substituted for the variable is called the **solution** to the equation. The process of finding the solution is called **solving the equation.**

First-Degree Equation	If a, b, and c are constants and $a \neq 0$, then a **first-degree equation** is an equation that can be written in the form $$ax + b = c.$$

Note: The term **first-degree** is used because the variable x is understood to have an exponent of 1. That is, $x = x^1$. Also, variables other than x, such as y or z, may be used.

All the following equations are first-degree equations:

$$2x + 3 = 7, \qquad 5y + 6 = -13, \qquad z + 5 = 3z - 5.$$

In this section, we will discuss solving first-degree equations in the two forms

$$x + b = c$$

and $$ax = c.$$

Then, in Section 3.3, we will combine the ideas we have developed here and solve equations in the more general form

$$ax + b = c.$$

x + b = c

To begin, we need the addition property of equality.

Addition Property of Equality	If the same algebraic expression is added to both sides of an equation, the new equation has the same solutions as the original equation. Symbolically, if A, B, and C are algebraic expressions, then the equations $$A = B$$ and $$A + C = B + C$$ have the same solutions. We say that the equations are **equivalent.**

The objective of solving first-degree equations is to get the variable by itself on one side of the equation and any constants on the other side. The following procedure will accomplish this.

Procedure for Solving an Equation That Simplifies to the Form x + b = c	1. Combine any like terms on each side of the equation.
	2. Use the addition property of equality and add the opposite of one of the constant terms to both sides so that all constants are on one side.
	3. Use the addition property of equality and add the opposite of one of the variable terms so that all variables are on the other side.
	4. Check your answer by substituting it into the original equation.

EXAMPLES Solve each of the following equations.

1. $y - 8 = -2$

Solution $y - 8 + 8 = -2 + 8$ Add $(+8)$, the opposite of -8, to both sides.

$y = 6$ Simplify.

Check $y - 8 = -2$

$6 - 8 = -2$ Substitute $y = 6$.

$-2 = -2$ True statement

2. $x - \dfrac{2}{5} = \dfrac{3}{10}$

Solution $x - \dfrac{2}{5} + \dfrac{2}{5} = \dfrac{3}{10} + \dfrac{2}{5}$ Add $\left(\dfrac{2}{5}\right)$, the opposite of $-\dfrac{2}{5}$, to both sides.

$x = \dfrac{3}{10} + \dfrac{4}{10}$ Simplify. (The common denominator is 10.)

$x = \dfrac{7}{10}$ Simplify.

Check $x - \dfrac{2}{5} = \dfrac{3}{10}$

$\dfrac{7}{10} - \dfrac{2}{5} = \dfrac{3}{10}$ Substitute $x = \dfrac{7}{10}$.

$\dfrac{7}{10} - \dfrac{4}{10} = \dfrac{3}{10}$ Simplify.

$\dfrac{3}{10} = \dfrac{3}{10}$ True statement

3. $4x + 1 - x = -13 + 2x + 5$

Solution $3x + 1 = 2x - 8$ Combine like terms on each side.

$3x + 1 - \mathbf{1} = 2x - 8 - \mathbf{1}$ Add (-1), the opposite of $+1$, to both sides.

$3x = 2x - 9$ Simplify.

$3x + (\mathbf{-2x}) = 2x - 9 + (\mathbf{-2x})$ Add $(-2x)$, the opposite of $+2x$, to both sides.

$x = -9$ Simplify.

Check $4x + 1 - x = -13 + 2x + 5$
$4(-9) + 1 - (-9) = -13 + 2(-9) + 5$ Substitute $x = -9$.
$-36 + 1 + 9 = -13 - 18 + 5$ Simplify.
$-26 = -26$ True statement

4. $6y + 1.88 = 7y - 4.3$

Solution $6y + 1.88 + \mathbf{4.3} = 7y - 4.3 + \mathbf{4.3}$ Add (4.3), the opposite of -4.3, to both sides.

$6y + 6.18 = 7y$ Simplify.

$6y + 6.18 - \mathbf{6y} = 7y - \mathbf{6y}$ Add $(-6y)$, the opposite of $6y$, to both sides.

$6.18 = y$ Simplify.
Note: The variable can be on the right side as well as on the left side.

Check $6y + 1.88 = 7y - 4.3$
$6(6.18) + 1.88 = 7(6.18) - 4.3$ Substitute $y = 6.18$.
$37.08 + 1.88 = 43.26 - 4.3$ Simplify.
$38.96 = 38.96$ True statement   ■

Practice Problems   Solve the following equations.

1. $x + 5 = -16$ $= {}^-21$ 2. $5x - 2 = 4x - 2$ $= 0$
3. $7y - 1.5 = 6y + 3.2$ $= 4.7$

Answers to Practice Problems   1. $x = -21$ 2. $x = 0$ 3. $y = 4.7$

ax = c

To solve equations of the form $ax = c$, we can use the idea of the **reciprocal** of the coefficient a. For example, as we studied in Section 1.2, the reciprocal of $\frac{3}{4}$ is $\frac{4}{3}$ and $\frac{3}{4} \cdot \frac{4}{3} = 1$.

Also, we need the **multiplication property of equality.**

Multiplication Property of Equality	If both sides of an equation are multiplied by the same **nonzero** algebraic expression, the new equation has the same solutions as the original equation. Symbolically, if A, B, and C are algebraic expressions, then the equations $$A = B$$ and $$AC = BC \qquad \text{where } C \neq 0$$ have the same solutions. We say that the equations are **equivalent.**

Remember that the objective is to get the variable by itself on one side of the equation. That is, we want the variable to have 1 as its coefficient. The following procedure will accomplish this.

Procedure for Solving an Equation That Simplifies to the Form **ax = c**	1. Combine any like terms on each side of the equation. 2. Use the multiplication property of equality and multiply both sides of the equation by the reciprocal of the coefficient of the variable. (**Note:** This is the same as dividing both sides of the equation by the coefficient.) 3. Check your answer by substituting it into the original equation.

EXAMPLES   Solve each of the following equations.

5. $5x = 20$

Solution

$$\frac{1}{5} \cdot (5x) = \frac{1}{5} \cdot 20 \qquad \text{Multiply both sides by } \left(\frac{1}{5}\right), \text{ the reciprocal of 5.}$$

$$\left(\frac{1}{5} \cdot 5\right)x = \frac{1}{5} \cdot \frac{20}{1} \qquad \text{By the associative property of multiplication}$$

$$1x = 4 \qquad \qquad \frac{1}{5} \cdot 5 = 1$$

$$x = 4$$

Check $5x = 20$

$5 \cdot 4 = 20 \qquad \text{Substitute } x = 4.$

$20 = 20 \qquad \text{True statement}$

Multiplying by the reciprocal of the coefficient is the same as **dividing** by the coefficient itself. So, we can multiply both sides by $\frac{1}{5}$, as we did, or we can divide both sides by 5.

$$5x = 20$$
$$\frac{5x}{5} = \frac{20}{5} \qquad \text{Divide both sides by 5.}$$
$$x = 4 \qquad \text{Simplify.}$$

6. $-7y = 21$

   Solution $\dfrac{-7y}{-7} = \dfrac{21}{-7} \qquad$ Divide both sides by -7.

   $\qquad\qquad y = -3 \qquad$ Simplify.

   Check $\qquad -7y = 21$
   $\qquad\qquad -7(-3) = 21 \qquad$ Substitute $y = -3$.
   $\qquad\qquad\qquad 21 = 21 \qquad$ True statement

7. $5x + 4x = 3 - 21$

   Solution $5x + 4x = 3 - 21$
   $\qquad\qquad\qquad 9x = -18 \qquad$ Combine like terms.
   $\qquad\qquad\quad \dfrac{9x}{9} = \dfrac{-18}{9} \qquad$ Divide both sides by 9.
   $\qquad\qquad\qquad x = -2 \qquad$ Simplify.

   Check $\qquad\qquad 5x + 4x = 3 - 21$
   $\qquad 5(-2) + 4(-2) = 3 - 21 \qquad$ Substitute $x = -2$.
   $\qquad\qquad\quad -10 - 8 = -18 \qquad$ Simplify.
   $\qquad\qquad\qquad\quad -18 = -18 \qquad$ True statement

8. $1.3x = 9.1$

   Solution $\mathbf{10}(1.3x) = \mathbf{10}(9.1) \qquad$ Multiply both sides by 10 so that there will be no decimals.

   $\qquad\qquad\qquad 13x = 91 \qquad$ Simplify.

   $\qquad\quad \dfrac{1}{13} \cdot 13x = \dfrac{1}{13} \cdot 91 \qquad$ Multiply both sides by $\dfrac{1}{13}$.

   $\qquad\qquad\qquad\quad x = 7 \qquad$ Simplify.

   Check $\quad 1.3x = 9.1$
   $\qquad\quad 1.3(7) = 9.1 \qquad$ Substitute $x = 7$.
   $\qquad\qquad 9.1 = 9.1 \qquad$ True statement

When decimal coefficients or constants are involved, you might want to use a calculator. We could simply divide both sides by 1.3 and use a calculator:

$$1.3x = 9.1$$

$$\frac{\cancel{1.3}x}{\cancel{1.3}} = \frac{9.1}{1.3} = 7.0 \qquad \text{Use a calculator or just pencil and paper.}$$

$$x = 7.0$$

9. $\dfrac{4x}{5} = \dfrac{3}{10}$ This could be written $\dfrac{4}{5}x = \dfrac{3}{10}$ since

$$\dfrac{4}{5}x \text{ is the same as } \dfrac{4x}{5}.$$

Solution $\quad \dfrac{5}{4} \cdot \dfrac{4}{5}x = \dfrac{5}{4} \cdot \dfrac{3}{10}$ Multiply both sides by $\dfrac{5}{4}$.

$$1x = \frac{\overset{1}{\cancel{5}}}{4} \cdot \frac{3}{\underset{2}{\cancel{10}}} \qquad \text{Simplify.}$$

$$x = \frac{3}{8}$$

Check $\qquad \dfrac{4}{5}x = \dfrac{3}{10}$

$$\frac{4}{5} \cdot \frac{3}{8} = \frac{3}{10} \qquad \text{Substitute } x = \frac{3}{8}.$$

$$\frac{3}{10} = \frac{3}{10} \qquad \text{True statement}$$

10. $-x = 4$

Solution $\quad -x = 4$

$$-1x = 4 \qquad -1 \text{ is the coefficient of } x.$$

$$\frac{-1x}{-1} = \frac{4}{-1}$$

$$x = -4$$

Check $\qquad -x = 4$

$$-(-4) = 4 \qquad \text{Substitute } x = -4.$$

$$4 = 4 \qquad \text{True statement} \qquad \blacksquare$$

Practice Problems   Solve the following equations.

1. $4x = -20$ $= -5$

2. $\dfrac{3}{5}y = 33$ $= 55$

3. $1.7z + 2.4z = 8.2$ $= 2$

4. $3x = 7$ $= \dfrac{7}{3}$

5. $6x = -1.8$ $= 0.3$

6. $-x = -8$

EXERCISES 3.2

Use the addition property to solve Exercises 1–22.

1. $y - 5 = 1$
2. $x + 14 = 23$
3. $y + 5 = 5$
4. $x + 3 = -7$

5. $y - 9 = -4$
6. $x - \dfrac{2}{3} = \dfrac{7}{6}$
7. $y + \dfrac{3}{4} = \dfrac{1}{8}$
8. $x + 3.6 = 2.4$

9. $y - 14.6 = -16.3$
10. $5x = 4x - 1$
11. $x - 9 = 2x - 3$
12. $6x - 2 = 5x + 8$

13. $7x + 1 = 6x - 5$
14. $13x - 4 = 12x + 21$
15. $5x + \dfrac{2}{3} = 4x + \dfrac{1}{6}$

16. $10x - \dfrac{1}{2} = 9x - \dfrac{9}{10}$
17. $3x + 3.7 = 4x + 1.8$
18. $12x - 0.63 = 11x - 2.51$

19. $2x + 9 - x = 4x + 2 - 2x$
20. $6x - 2 + 3x = 4x + 8 + 4x$

21. $17 + x + 11x = 15x + 3 - 2x$
22. $-2x - 9 + 5x = 4x + 2$

Use the multiplication property to solve Exercises 23–44.

23. $2y = 10$
24. $-8x = 24$
25. $6x = -12$

26. $-9y = -54$
27. $x + 8x = 15 + 12$
28. $10x - 2x = 36 - 100$

29. $-\dfrac{1}{5}x = 5$
30. $\dfrac{3x}{4} = 15$
31. $\dfrac{2x}{5} = 4$ $\cdot \dfrac{5}{2} = 10$

32. $-\dfrac{5}{3}x = 10$
33. $\dfrac{5}{6}y = 17 + 13$
34. $-\dfrac{1}{8}x = 12 - 8$

35. $\dfrac{5}{3}x + \dfrac{2}{3}x = 21 + 7$
36. $\dfrac{5}{2}x + \dfrac{4}{2}x = 15 - 33$
37. $-3x = 7$

38. $4x = -3$
39. $\dfrac{3}{4}x = \dfrac{9}{10}$
40. $\dfrac{1}{6}x = \dfrac{3}{4}$

41. $3.2y = 6.4$
42. $0.2x = 1.6$
43. $6.4y - 2.0y = 1.2 - 10.0$

44. $1.5x - 6.5x = 4.6 - 30.1$

Answers to Practice Problems 1. $x = -5$ 2. $y = 55$ 3. $z = 2$ 4. $x = \dfrac{7}{3}$

5. $x = -0.3$ 6. $x = 8$

The profit, P, is equal to the revenue, R, minus the cost, C (P = R − C).

45. Find the revenue if the profit is $684.50 and the cost is $8329.00.

46. Find the cost if the profit is $93.25 and the revenue is $865.90.

The perimeter, P, of a triangle is equal to the sum of the sides, a, b, and c (P = a + b + c).

47. Two sides of a triangle measure 43 cm and 26 cm. Find the length of the third side if the perimeter is 98 cm.

48. One side of a triangle is $11\frac{1}{2}$ in. long. A second side is $8\frac{3}{4}$ in. long. If the perimeter is 24 in., find the length of the third side.

The selling price, S, of an item is equal to the sum of the cost, C, and the markup, M (S = C + M).

49. The selling price of an item that cost $9.80 is $13.50. Find the markup.

50. The selling price of a jacket is $39.90. If the markup is $16.70, find the cost.

The distance traveled, d, is equal to the product of the rate, r, and the time, t (d = rt).

51. How long will it take a truck to travel 350 mi if it travels at an average rate of 50 miles per hour?

52. How long will it take a train traveling at 40 miles per hour to go 140 mi?

The interest, I, is the product of the money invested (principal), P, the rate, r, and the time, t (I = Prt).

53. The interest on $3000 invested at 9% is $810. Find the time.

54. A savings account pays interest at a rate of 12%. How much must be invested to earn $900 interest in 5 years?

The volume, V, of a pyramid with a rectangular base is equal to one-third the product of the length, ℓ, the width, w, and the height, h $\left(V = \frac{1}{3}\ell wh \right)$.

55. Find the height of a pyramid with a rectangular base having a volume of 48 cu in., a length of 6 in., and a width of 4 in.

56. The volume of a pyramid with a rectangular base is 60 cu cm. The height is 9 cm and the length is 5 cm. Find the width.

Calculator Problems

Use a calculator to solve Exercises 57–62.

57. $y + 32.861 = -17.892$

58. $17.61x + 27.059 = 9.845 + 16.61x$

59. $14.38y - 8.65 + 9.73y = 17.437 + 23.11y$

60. $2.637x = 648.702$

61. $-0.3057y = 316.7052$

62. $0.5178y = -257.8644$

3.3 Solving Equations [ax + b = c]

OBJECTIVES	

In this section, you will be learning to solve equations of the form (or that can be simplified to the form) $ax + b = c$.

Now we are ready to apply all the properties and techniques that we have used in sections 3.1 and 3.2 to solving first-degree equations of the form $ax + b = c$. In most cases, the equation we are to solve will not be in exactly the form $ax + b = c$. The following general procedure allows for a wide variety of possible forms of the original equation.

To Solve a First-Degree Equation

1. Simplify each side of the equation by removing any grouping symbols and combining like terms.
2. Use the addition property of equality to add the opposites of constants and/or variables so that variables are on one side and constants on the other.
3. Use the multiplication property of equality to multiply both sides by the reciprocal of the coefficient of the variable (or divide both sides by the coefficient).
4. Check your answer by substituting it into the original equation.

Study each of the following examples carefully. Remember that **the objective is to get the variable on one side by itself.** Note that **the equations are written one under the other.** Do not write several equations on one line or set one equation equal to another equation.

EXAMPLES Solve each of the following equations.

1. $-5x - 1 = -11$

$$\text{Solution} \quad -5x - 1 + \mathbf{1} = -11 + \mathbf{1} \qquad \text{Add } +1 \text{ to both sides.}$$

$$-5x = -10 \qquad \text{Simplify.}$$

$$-\frac{1}{5}(-5x) = -\frac{1}{5}(-10) \qquad \text{Multiply both sides by } -\frac{1}{5}.$$

$$x = 2 \qquad \text{Simplify.}$$

$$\textit{Check} \quad -5x - 1 = -11$$

$$-5 \cdot 2 - 1 = -11 \qquad \text{Substitute } x = 2.$$

$$-10 - 1 = -11$$

$$-11 = -11 \qquad \text{True statement}$$

The solution is 2.

$2(y-7) = 4(y+1) - 26$

$2y - 14 = 4y + 4 - 26$

$2y - 14 = 4y - 22$

$-4y + 22 \quad -4y + 22$

$\dfrac{-2y}{2} = \dfrac{8}{2} = 4$

$2(y-7) = 4(y+1) - 26$

$2y - 14 = 4y + 4 - 26$

$2y - 14 = 4y - 22$

$+22 \qquad +22$

$2y + 8 = 4y$

$-2 \qquad -2$

$\dfrac{8}{2} = \dfrac{2}{2}$

$y = 4$

2. $2(y - 7) = 4(y + 1) - 26$
Solution

$2y - 14 = 4y + 4 - 26$	Use the distributive property.
$2y - 14 = 4y - 22$	Combine like terms.
$2y - 14 + \mathbf{22} = 4y - 22 + \mathbf{22}$	Add $+22$ to both sides. Here we will put the variables on the right side to get a positive coefficient of y.
$2y + 8 = 4y$	Simplify.
$\mathbf{-2y} + 2y + 8 = \mathbf{-2y} + 4y$	Add $-2y$ to both sides.
$8 = 2y$	Simplify. (Note that the coefficient of y is $+2$.)
$\dfrac{8}{2} = \dfrac{2y}{2}$	Divide both sides by 2.
$4 = y$	Simplify.

The number 4 does check, so the solution is 4.

3. $16.53 - 18.2z = 7.43$
Solution

$100(16.53 - 18.2z) = 100(7.43)$	Multiply both sides by 100 to clear the decimals. This will give integer constants and coefficients.
$100(16.53) - 100(18.2z) = 100(7.43)$	Use the distributive property.
$1653 - 1820z = 743$	Simplify.
$1653 - 1820z - \mathbf{1653} = 743 - \mathbf{1653}$	Add -1653 to both sides.
$-1820z = -910$	Simplify.
$\dfrac{-1820z}{\mathbf{-1820}} = \dfrac{-910}{\mathbf{-1820}}$	Divide both sides by -1820.
$z = 0.5$	
or $\qquad z = \dfrac{1}{2}$	

Check $16.53 - 18.2z = 7.43$	
$16.53 - 18.2(0.5) = 7.43$	Substitute $z = 0.5$.
$16.53 - 9.1 = 7.43$	Simplify.
$7.43 = 7.43$	True statement

The number 0.5 checks and it is the solution.

4. $\dfrac{2z}{5} + 2 = 6$

Solution $\dfrac{2z}{5} + 2 - 2 = 6 - 2$ Add -2 to both sides.

$$\dfrac{2z}{5} = 4$$ Simplify.

$$\dfrac{5}{2} \cdot \dfrac{2z}{5} = \dfrac{5}{2} \cdot 4$$ Multiply both sides by $\dfrac{5}{2}$.

$$z = 10$$ Simplify.

Check $\dfrac{2z}{5} + 2 = 6$

$$\dfrac{2 \cdot 10}{5} + 2 = 6$$ Substitute $z = 10$.

$$4 + 2 = 6$$ Simplify.

$$6 = 6$$ True statement

The number 10 checks and it is the solution.

5. $-2(5x + 13) - 2 = -6(3x - 2) - 41$

Solution

$-10x - 26 - 2 = -18x + 12 - 41$ Use the distributive property. Be careful with the signs.

$-10x - 28 = -18x - 29$ Simplify.
$-10x - 28 + \mathbf{18x} = -18x - 29 + \mathbf{18x}$ Add $+18x$ to both sides.

$8x - 28 = -29$ Simplify.
$8x - 28 + \mathbf{28} = -29 + \mathbf{28}$ Add $+28$ to both sides.

$8x = -1$ Simplify.
$\dfrac{8x}{8} = \dfrac{-1}{8}$ Divide both sides by 8.

$$x = -\dfrac{1}{8}$$ Simplify.

Checking will show that $-\dfrac{1}{8}$ is the solution.   ∎

About checking: Checking can be quite time-consuming and need not be done for every problem. This is particularly important on exams. You should check only if you have time after the entire exam is completed.

Practice Problems

Solve the following equations.

1. $3x + 4 = -2$

2. $x + 14 - 6x = 2x - 7$

3. $\dfrac{2y}{3} - 4 = 8$

4. $6.4z + 2.1 = 3.1z - 1.2$

5. $5 - (x - 3) = 14 - 4(x + 2)$

Example 3 illustrated how to multiply terms with decimals so that you get integer constants and coefficients. You may choose instead to use your calculator. Example 6 shows what you should write down while you have the calculator do the arithmetic calculations.

CALCULATOR EXAMPLE

6. Solve $16.53 - 18.2z = 7.43$.

    Solution $-18.2z = -9.1$ Add -16.53 to both sides.

$$\frac{-18.2z}{-18.2} = \frac{-9.1}{-18.2}$$ Divide both sides by -18.2.

$$z = 0.5$$ ■

△ **EXERCISES 3.3**

Solve the equations in Exercises 1–60.

1. $x + 8 = -3$ 2. $x + 14 = 12$ 3. $-6x = -12$ 4. $5x = 20$

5. $3x - 12 = 18$ 6. $5x - 7 = 18$ 7. $7x - 3 = -17$ 8. $3x + 7 = -2$

9. $5x + 6 = 26$ 10. $6x - 5 = -11$ 11. $-3x + 7 = 1$ 12. $9 - 4x = -7$

13. $5x + 6 = 7$ 14. $3x + 11 = 4$ 15. $-2y + 4 = -1$ 16. $4x + 3 = 8$

17. $3y + 1.6 = 7$ 18. $7x - 3.4 = 11.3$ 19. $9x + 4.7 = -3.4$ 20. $-5x + 2.9 = 9.9$

21. $\dfrac{x}{3} + 1 = 7$ 22. $\dfrac{x}{5} + 2 = -3$ 23. $\dfrac{3x}{5} - 1 = 2$ 24. $\dfrac{2x}{7} - 3 = 5$

25. $\dfrac{2x}{3} - 4 = 8$ 26. $\dfrac{5x}{4} + 1 = 11$ 27. $\dfrac{3}{4}x + 4 = 5$ 28. $\dfrac{4}{5}x - 2 = 1$

29. $-\dfrac{3}{2}x - 3 = 2$ 30. $-\dfrac{8}{3}x + 1 = 5$ 31. $3x + 18 = 7x - 6$ 32. $3x + 2 = x - 8$

Answers to Practice Problems **1.** $x = -2$ **2.** $x = 3$ **3.** $y = 18$ **4.** $z = -1$
5. $x = -\dfrac{2}{3}$

33. $5x - 3 = 2x + 6$

34. $x - 7 = 5x + 9$

35. $4x + 3 - x = x - 9$

36. $5x + 13 = x - 8 - 3x$

37. $4x + 3 = 2x + 4$

38. $5y - 2 = 2y + 5$

39. $8 - 3x = 4x + 10$

40. $14 - 2x = 8x - 1$

41. $\frac{2}{3}x + 1 = \frac{1}{3}x - 6$

42. $\frac{4}{5}x + 2 = \frac{2}{5}x - 4$

43. $\frac{3}{2}x + 1 = \frac{1}{2}x - 1$

44. $\frac{7}{3}x + 2 = \frac{1}{3}x + 6$

45. $0.23x + 0.18 = 0.08x - 0.27$

46. $0.9x + 3 = 0.4x + 1.5$

47. $1.1x - 4 = 0.5x + 2$

48. $0.67x + 3 = 2.7 + 0.63x$

49. $2(x - 3) = 5x + 9$

50. $3(x - 1) = 4x + 6$

51. $3(2x - 3) = 4x + 5$

52. $5(3x - 7) = 6x + 1$

53. $4(3 - 2x) = 2(x - 4)$

54. $-3(x + 5) = 6(x + 2)$

55. $2(x + 3) = 4(x + 5) - 7$

56. $7(2x - 1) = 5(x + 6) - 13$

57. $8(3x + 5) - 9 = 9(x - 2) + 14$

58. $5 - 3(2x + 1) = 4(x - 5) + 6$

59. $-2(x + 5) - 4 = 6(x - 2) + 2$

60. $8 + 4(2x - 3) = 5 - (x + 3)$

The perimeter, P, of a rectangle is the sum of twice the length, ℓ, plus twice the width, w $(P = 2\ell + 2w)$.

61. Eighty-four feet of fencing is needed to enclose a small garden plot. If the plot is 18 ft wide, find the length.

62. A rectangle is 36 cm long and has a perimeter of 85 cm. Find the width.

The area, A, of a trapezoid is the product of $\frac{1}{2}$ the height, h, times the sum of the lengths, a and b, of the two bases $\left[A = \frac{1}{2}h(a + b) \right]$.

63. The area of a trapezoid is 108 sq cm. The height is 12 cm and the length of one base is 8 cm. Find the length of the other base.

64. The lengths of the bases of a trapezoid are 10 in. and 15 in. Find the height if the area is 225 sq in.

When purchasing an item on the installment plan, you find the total cost, C, by multiplying the monthly payment, p, by the number of months, t, and adding the product to the down payment, d $(C = pt + d)$.

65. A refrigerator costs $857.60 if purchased on the installment plan. If the monthly payments are $42.50 and the down payment is $92.60, how long will it take to pay for the refrigerator?

66. A used automobile will cost $3,250 if purchased on an installment plan. If the monthly payments will be $115 for 24 months, what will be the down payment?

Calculator Problems

Use a calculator to solve Exercises 67–71. (Round off answers to the nearest hundredth.)

67. $0.17x - 23.014 = 1.35x + 36.234$

68. $48.512 - 1.63x = 2.58x + 87.635$

69. $327.93 + 22.62x = 17.76x + 463.52$

70. $0.32(x + 14.1) = 2.47x + 19.361$

71. $1.61(9.3 - 2x) = 0.24(3x - 12.9)$

3.4 *Writing Algebraic Expressions*

Algebra is a language of mathematicians, and to understand mathematics, you must understand the language. We want to be able to change English phrases into their "algebraic" equivalents, and vice versa. So, if a problem is stated in English, we can translate the phrases into algebraic symbols and proceed to solve the problem according to the rules developed for algebra.

The following examples illustrate how certain key words can be translated into algebraic symbols.

EXAMPLES

English Phrase	Algebraic Expression
1. 3 **multiplied by** the number represented by x the **product** of 3 and x 3 **times** x	$3x$
2. 3 **added to** a number the **sum** of z and 3 z **plus** 3	$z + 3$
3. 2 **times** the quantity found by **adding** a number to 1 **twice** the **sum** of x and 1 the **product** of 2 with the **sum** of x and 1	$2(x + 1)$
4. **twice** x **plus** 1 2 **times** x **increased by** 1 1 **more than** the **product** of 2 and a number	$2x + 1$
5. the **quotient** of two numbers x **divided by** y **divide** x by y	$\dfrac{x}{y}$
6. the **difference** between 5 **times** a number and 2 **times** the same number the **product** of 5 and a number **minus** the **product** of 2 and that number the **difference** between $5x$ and $2x$ $2x$ **subtracted from** $5x$	$5x - 2x$

Certain words, such as those in boldface in the previous examples, are the keys to the operations. Learn to look for these words and those from the following list.

Addition	Subtraction	Multiplication	Division
add	subtract (from)	multiply	divide
sum	difference	product	quotient
plus	minus	times	
more than	less than	twice	
increased by	decreased by	of (with fractions and percent)	

Practice Problems

Change the following phrases to algebraic expressions.

1. 7 less than a number
2. Twice the product of two unknown numbers
3. The quotient of y and 5
4. An unknown amount less than 10
5. 14 more than 3 times a number
6. The product of 5 with the difference of 2 and x
7. 4 less than the product of 2 with the quantity x minus 3
8. The sum of the product of 5 with a number and the product of 3 with that number

---

The words **quotient** and **difference,** as illustrated in Practice Problems 3 and 6, deserve special mention. The division and subtraction are done with the values in the same order that they are given in the problem. For example,

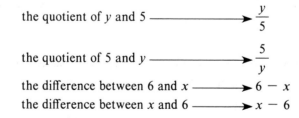

the quotient of y and 5 $\longrightarrow \dfrac{y}{5}$

the quotient of 5 and y $\longrightarrow \dfrac{5}{y}$

the difference between 6 and x $\longrightarrow 6 - x$

the difference between x and 6 $\longrightarrow x - 6$

 **EXERCISES 3.4**

Write in words what each of the expressions in Exercises 1–12 means.

1. $4x$

2. $x + 6$

3. $2x + 1$

4. $4x - 7$

5. $7x - 5.3$

6. $3.2(x + 2.5)$

7. $-2(x - 8)$

8. $10(x + 4)$

9. $5(2x + 3)$

10. $3(4x - 5)$

11. $6(x - 1) + \dfrac{2}{3}$

12. $9(x + 3) - \dfrac{8}{15}$

Write each pair of expressions in words in Exercises 13–16. Notice the differences between the expressions and between the corresponding English phrases.

13. $3x + 7$; $3(x + 7)$

14. $4x - 1$; $4(x - 1)$

15. $7x - 3$; $7(x - 3)$

16. $5(x + 6)$; $5x + 6$

Write the algebraic expression described by each of the word phrases in Exercises 17–35. Choose your own variable.

17. 6 added to a number

18. 7 more than a number

19. 4 less than a number

20. A number decreased by 13

21. 5 less than 3 times a number

22. The difference between twice a number and 10

23. The difference between x and 3, all divided by 7

24. 9 times the sum of a number and 2

25. 3 times the difference between a number and 8

26. 13 less than the product of 4 with the sum of a number and 1

27. 5 subtracted from three times a number

28. The sum of twice a number and four times the number

29. 8 minus twice a number all added to three times the number

30. The sum of a number and 9 more than the number

31. 4 more than the product of 8 with the difference between a number and 6

32. Twenty decreased by the sum of twice a number and 4.8

33. The difference between three times a number and 9, all decreased by five times the number

34. The sum of a number and three times itself, all increased by 8

35. The sum of a number and 5, all multiplied by 4, then decreased by twice the number

Write the algebraic expression described by each of the word phrases in Exercises 36–45.

36. The cost of x pounds of candy at \$4.95 a pound

37. The annual interest on x dollars if the rate is 11% per year

38. The number of days in t weeks and 3 days

39. The number of minutes in h hours and 20 minutes

40. The points scored by a football team on n touchdowns (6 points each) and 1 field goal (3 points)

41. The cost of renting a car for one day and driving m miles if the rate is $20 per day plus 20 cents per mile

42. The cost of purchasing a fishing rod and reel if the rod costs x dollars and the reel costs $8 more than twice the cost of the rod

43. A salesperson's weekly salary if he receives $250 as his base plus 9% of the weekly sales of x dollars

44. The selling price of an item that costs c dollars if the markup is 20% of the cost

45. The perimeter of a rectangle if the width is w centimeters and the length is 3 cm less than twice the width

◢ 3.5 Word Problems

In Section 3.4, we discussed translating English phrases into algebraic expressions. The phrase "8 added to twice a number" translates algebraically to $2x + 8$. How do you translate "4 more than a number"? If you said $x + 4$, you are correct. Now, the object is to translate an entire sentence into an equation and then to solve the equation. The two phrases above might be involved in a sentence such as the following:

"If **8 is added to twice a number,** the result is **4 more than the number.**"

Algebraically,

$$2x + 8 = x + 4 \qquad \text{"the result is" translates as } =$$

Solving, we have

$$2x + 8 = x + 4$$
$$2x + 8 - x = x + 4 - x$$
$$x + 8 = 4$$
$$x + 8 - 8 = 4 - 8$$
$$x = -4$$

Number Problems

In this section, the word problems will be simply exercises in translating sentences into equations and solving these equations. More sophisticated "application" problems will be discussed in later chapters. Such problems will involve geometric formulas, distance, interest, work, inequalities, and mixture.

EXAMPLES

1. Three times the sum of a number and 5 is equal to twice the number plus 5. Find the number.

    Solution   Let x = the unknown number.

3 times the sum of a number and 5	is equal to	twice the number plus 5
$3(x + 5)$	$=$	$2x + 5$

    $$3x + 15 = 2x + 5$$
    $$3x + 15 - 2x = 2x + 5 - 2x$$
    $$x + 15 = 5$$
    $$x + 15 - 15 = 5 - 15$$
    $$x = -10$$

    The number is -10.

2. If a number is decreased by 36 and the result is 76 less than twice the number, what is the number?

    Solution   Let n = the unknown number.

a number decreased by 36	the result is	76 less than twice the number
$n - 36$	$=$	$2n - 76$

    $$n - 36 - n = 2n - 76 - n$$
    $$-36 = n - 76$$
    $$-36 + 76 = n - 76 + 76$$
    $$40 = n$$

    The number is 40.

3. Joe pays \$300 per month to rent an apartment. If this is $\dfrac{2}{5}$ of his monthly income, what is his monthly income?

    Solution   Let x = Joe's monthly income.

$\dfrac{2}{5}$ of monthly income	is	\$300
$\dfrac{2}{5}x$	$=$	300

    $$\frac{5}{2} \cdot \frac{2}{5}x = \frac{5}{2} \cdot \frac{300}{1}$$
    $$x = 750$$

    Joe's monthly income is \$750.

4. A textbook is on sale for 25% off the original price. If you pay $22.50 for the book, what was the original price?

Solution   Let $p =$ the original price.

$$\underbrace{\text{original price}}_{p} \quad - \quad \underbrace{\text{discount}}_{0.25p} \quad \underbrace{\text{is}}_{=} \quad \underbrace{\text{what you pay}}_{22.50}$$

$$100(p - 0.25p) = 100(22.50) \qquad \text{Multiply both sides by 100.}$$
$$100p - 25p = 2250$$
$$75p = 2250$$
$$\frac{75p}{75} = \frac{2250}{75}$$
$$p = 30$$

The original price was $30.00.   ■

Consecutive Integers

Even integers are integers that are divisible by 2. The even integers are

$$E = \{. \ . \ . \ , -6, -4, -2, 0, 2, 4, 6, \ . \ . \ .\}$$

Odd integers are integers that are not even. The odd integers are

$$0 = \{. \ . \ . \ , -5, -3, -1, 1, 3, 5, \ . \ . \ .\}$$

Consecutive integers are two integers that differ by 1. That is, the second integer is 1 more than the first integer. For example, 21 and 22 are consecutive integers. -14 and -13 are consecutive integers. In general, if n is one integer, then $n + 1$ is the next consecutive integer.

An example of three consecutive integers is 51, 52, 53. Another example is $-9, -8, -7$. If n is one integer, then $n + 1$ is the next consecutive integer and $n + 2$ is the third consecutive integer.

Consecutive even integers are even integers that differ by 2; that is, the second integer is 2 more than the first. For example, 36 and 38 are two consecutive even integers. Also, $-12, -10,$ and -8 are three consecutive even integers. If n is an even integer, then $n + 2$ is the next consecutive even integer and $n + 4$ is the third consecutive even integer.

Consecutive odd integers are odd integers that differ by 2; again, the second integer is 2 more than the first. For example, -15 and -13 are two consecutive odd integers. Also, 17, 19, and 21 are three consecutive odd integers. If n is an odd integer, then $n + 2$ is the next consecutive odd integer and $n + 4$ is the third consecutive odd integer.

Notice that consecutive even integers and consecutive odd integers are represented in the same way: $n, n + 2, n + 4$, and so on.

EXAMPLES

5. Find three consecutive integers such that the sum of the first and third is 76 less than three times the second.

Solution   Let n = the first integer
$n + 1$ = the second integer
$n + 2$ = the third integer

$$n + (n + 2) = 3(n + 1) - 76$$
$$2n + 2 = 3n + 3 - 76$$
$$2n + 2 = 3n - 73$$
$$2n + 2 + 73 - 2n = 3n - 73 + 73 - 2n$$
$$75 = n$$
$$76 = n + 1$$
$$77 = n + 2$$

The three consecutive integers are 75, 76, and 77.

6. Three consecutive odd integers are such that their sum is -3. What are the integers?

Solution   Let n = the first odd integer
$n + 2$ = the second odd integer
$n + 4$ = the third odd integer

$$n + (n + 2) + (n + 4) = -3$$
$$3n + 6 = -3$$
$$3n = -9$$
$$n = -3$$
$$n + 2 = -1$$
$$n + 4 = +1$$

The three consecutive odd integers are -3, -1, and $+1$.   ∎

EXERCISES 3.5

Write an equation for each problem and then solve it.

1. If 7 is added to a number, the result is 43. Find the number.

2. A number decreased by 10 is equal to 23. What is the number?

3. A number subtracted from 12 is equal to twice the number. Find the number.

4. Five less than a number is equal to 13 decreased by the number. Find the number.

5. Three less than twice a number is equal to the number. What is the number?

6. Thirty-six is 4 more than twice a certain number. Find the number.

7. Fifteen decreased by twice a number is 27. Find the number.

8. Seven times a certain number is equal to the sum of twice the number and 35. What is the number?

9. The difference between twice a number and 5 is equal to the number increased by 8. Find the number.

10. Fourteen more than three times a number is equal to 6 decreased by the number. Find the number.

11. Two added to the quotient of a number and 7 is equal to negative three. What is the number?

12. The quotient of twice a number and 5 is equal to the number increased by 6. What is the number?

13. Three times the sum of a number and 4 is equal to -9. Find the number.

14. Four times the difference between a number and 5 is equal to the number increased by 4. Find the number.

15. When 17 is added to six times a number, the result is equal to 1 plus twice the number. What is the number?

16. If the sum of twice a number and 5 is divided by 11, the result is equal to the difference between 4 and the number. Find the number.

17. Twice a number increased by three times the number is equal to 4 times the sum of the number and 3. Find the number.

18. Find two consecutive integers whose sum is 53.

19. Find three consecutive integers whose sum is 69.

20. The sum of three consecutive integers is 207. What are the three integers?

21. The sum of two consecutive odd integers is 60. What are the integers?

22. Find two consecutive integers such that twice the first plus three times the second equals 83.

23. The sum of three consecutive even integers is -156. Find the integers.

24. Find three consecutive even integers such that the first plus twice the second is 54 less than four times the third.

25. One integer is 5 more than another. The sum of the two integers is 171. Find the integers.

26. One integer is three times another. If their difference is 72, find the two integers.

27. One integer is 3 more than twice a second integer. Their sum is 114. Find the two integers.

28. The sum of two integers is 57. One of the integers is 7 less than three times the other integer. Find the two integers.

29. The sum of three integers is 39. The second integer is 5 less than the first, and the third integer is twice the first. What are the three integers?

30. The sum of three integers is 49. The first integer is 7 more than twice the second integer, and the third integer is 6 more than the second integer. Find the three integers.

31. The width of a room is $\frac{2}{3}$ of the length. If the room is 18 ft wide, find the length.

32. The length of a field is 40 yards greater than twice the width. If the field is 400 yards long, find the width.

33. Last week, Ralph earned $37 more than he did this week. If he earned $248 this week, how much did he earn last week?

34. Susan made a score of 86 on her mathematics test. If this is 7 points less than her score on the last test, what was her score on the last test?

35. Ann pays a monthly rent of $260. This represents $\frac{2}{7}$ of her monthly income. What is her monthly income?

36. Bob paid $6770 for a new car. This is $\frac{1}{3}$ of his yearly salary. Find his yearly salary.

37. A pair of shoes is on sale for 15% off the original price. If you pay $51 for the shoes, what was the original price?

38. Two-ninths of the people who came for tryouts for the football team will not make the team. If 42 people make the team, how many came for tryouts?

39. A girl is 22 years younger than her mother. The sum of their ages is 44. Find the age of each.

40. A man is 5 years older than three times his son's age. If the total of their ages is 53 years, find the age of each.

41. An 18-foot board is cut into two parts so that one part is 2 feet longer than the other. How long is each part?

42. A couple recently bought a lot and had a house built. The cost of constructing the house was $25,000 more than the cost of the lot. The total cost for both was $80,000. What was the cost of each?

43. H. V. C. High School played 32 basketball games during this season. They won 5 games more than twice the number of games they lost. What is their won-lost record?

44. Matt bought a fish for his aquarium. The total cost was the marked price plus sales tax of 6% of the marked price. If the total cost was $9.01, find the marked price.

45. A furniture store sells a desk for $350. The price is determined by adding the cost and the markup. If the markup is 25% of the cost, find the cost.

46. A mathematics student bought a calculator and textbook for a total cost of $37.95. If the book cost $6.75 more than the calculator, find the cost of each.

47. A woman bought a skirt and blouse for $67.40. The skirt cost $12.50 more than the blouse. Find the cost of each.

48. Lucy bought two boxes of golf balls. She gave the salesperson $40 and received $10.50 in change. What was the cost of the golf balls per box?

49. Lorraine is charged for long distance calls according to the following rule: 65 cents for the first three minutes plus 25 cents for each additional minute. Find the length of a telephone call if the charges were $2.15.

50. Joyce is a real estate agent. Her weekly salary is $400 plus a commission of 2% based on the total sales exceeding $100,000. What were her total sales last week if she earned $900?

3.6 Solving for Any Term in a Formula

Formulas are general rules or principles stated mathematically. In Section 1.8, we discussed several formulas related to geometric figures. There are many formulas in such fields of study as business, economics, medicine, physics, and chemistry as well as mathematics.

Some of these formulas and their meanings are shown here.

Formula	Meaning
1. $I = Prt$	The simple interest (I) earned by investing money is equal to the product of the principal (P) times the rate of interest (r) times the time (t) in years.
2. $C = \dfrac{5}{9}(F - 32)$	Temperature in degrees Celsius (C) equals $\dfrac{5}{9}$ times the difference between the Fahrenheit temperature (F) and 32.
3. $d = rt$	The distance traveled (d) equals the product of the rate of speed (r) and the time (t).
4. $P = 2\ell + 2w$	The perimeter (P) of a rectangle is equal to twice the length (ℓ) plus twice the width (w).
5. $L = 2\pi rh$	The lateral surface area (top and bottom not included) of a cylinder (L) is equal to 2π times the radius (r) of the base and the height (h).

6. $IQ = \dfrac{100M}{C}$ Intelligence quotient (IQ) is calculated by multiplying 100 times mental age (M) as measured by some test and dividing by chronological age (C).

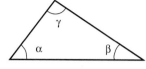

7. $\alpha + \beta + \gamma = 180$ The sum of the angles (α, β, and γ) of a triangle is 180°. **Note:** α, β, and γ are the Greek lowercase letters alpha, beta, and gamma, respectively.

If you know values for all but one variable in a formula, you can substitute those values and find the value of the unknown variable by using the techniques for solving equations discussed in this chapter.

EXAMPLE

1. Given the formula $C = \dfrac{5}{9}(F - 32)$, (a) find C if $F = 212°$ and (b) find F if $C = 20°$.

Solutions

a. $F = 212°$, so substitute 212 for F in the formula.

$$C = \frac{5}{9}(212 - 32) = \frac{5}{9}(180) = 100$$

That is, 212° F is the same as 100° C. Water will boil at 212° F at sea level. This means that, if the temperature is measured in degrees Celsius instead of degrees Fahrenheit, water will boil at 100° C at sea level.

b. $C = 20°$, so substitute 20 for C in the formula.

$$20 = \frac{5}{9}(F - 32) \qquad \text{Now solve for } F.$$

$$\frac{9}{5} \cdot 20 = \frac{9}{5} \cdot \frac{5}{9}(F - 32) \qquad \text{Multiply both sides by } \frac{9}{5}.$$

$$36 = F - 32 \qquad \text{Simplify.}$$

$$68 = F \qquad \text{Add 32 to both sides.}$$

That is, a temperature of 20° C is the same as a comfortable spring day temperature of 68° F.   ■

We say that the formula $d = rt$ is "solved for" d in terms of r and t. Similarly, the formula $A = \dfrac{1}{2}bh$ is solved for A in terms of b and h, and the formula $P = S - C$ (profit is equal to selling price minus cost) is solved for P in terms of S and C. Many times we want to use a certain formula in another form. We want the formula "solved for" some variable other than the one given in terms of the remaining variables. Study the following examples carefully.

EXAMPLES

2. Given $d = rt$, solve for t in terms of d and r. We want to represent the time in terms of distance and rate. We will use this concept later in word problems.

   Solution $d = rt$ Treat r and d as if they were constants.

   $$\frac{d}{r} = \frac{rt}{r}$$ Divide both sides by r.

   $$\frac{d}{r} = t$$ Simplify.

3. Given $P = a + b + c$, solve for a in terms of P, b, and c. This would be a convenient form for the case in which we know the perimeter and two sides of a triangle and want to find the third side.

   Solution $P = a + b + c$ Treat P, b, and c as if they were constants.

   $P - b - c = a + b + c - b - c$ Add $-b - c$ to both sides.
   $P - b - c = a$ Simplify.

4. Given $C = \dfrac{5}{9}(F - 32)$ as in example 1(b), solve for F in terms of C.

   This would give us a formula for finding Fahrenheit temperature given a Celsius temperature value.

   Solution $C = \dfrac{5}{9}(F - 32)$ Treat C as a constant.

   $$\frac{9}{5} \cdot C = \frac{9}{5} \cdot \frac{5}{9}(F - 32)$$ Multiply both sides by $\dfrac{9}{5}$, as in Example 1(b).

   $$\frac{9}{5}C = F - 32$$ Simplify.

   $$\frac{9}{5}C + 32 = F$$ Add 32 to both sides.

   Thus,

   $$F = \frac{9}{5}C + 32 \text{ is solved for } F$$

   $$C = \frac{5}{9}(F - 32) \text{ is solved for } C$$

   These are two forms of the same formula.

5. Suppose you are given $2x + 4y = 10$. (a) Solve first for x in terms of y. (b) Then solve for y in terms of x. This equation is typical of the algebraic equations that we will discuss in Chapter 8.
Solutions

a. Solving for x yields

$$2x + 4y = 10$$ Treat y as a constant.
$$2x + 4y - 4y = 10 - 4y$$ Subtract $4y$ from both sides. (This is the same as adding $-4y$.)
$$2x = 10 - 4y$$ Simplify.
$$\frac{2x}{2} = \frac{10 - 4y}{2}$$ Divide both sides by 2.
$$x = \frac{10}{2} - \frac{4y}{2}$$ Simplify.
$$x = 5 - 2y$$

b. Solving for y yields

$$2x + 4y = 10$$ Treat x as a constant.
$$2x + 4y - 2x = 10 - 2x$$ Subtract $2x$ from both sides.
$$4y = 10 - 2x$$ Simplify.
$$\frac{4y}{4} = \frac{10 - 2x}{4}$$ Divide both sides by 4.
$$y = \frac{10}{4} - \frac{2x}{4}$$ Simplify.
$$y = \frac{5}{2} - \frac{1}{2}x$$

or we can write

$$y = \frac{5 - x}{2}$$ Both forms are correct.

6. Given $3x - y = 15$, solve (a) for x in terms of y and (b) for y in terms of x.
Solutions

a. Solving for x, we have

$$3x - y = 15$$
$$3x = 15 + y$$ Add y to both sides.
$$\frac{1}{3}(3x) = \frac{1}{3}(15 + y)$$ Multiply both sides by $\frac{1}{3}$ (or divide both sides by 3).
$$x = \frac{1}{3} \cdot 15 + \frac{1}{3} \cdot y$$ Simplify using the distributive property.
$$x = 5 + \frac{1}{3}y$$

b. Solving for y, we have

$$3x - y = 15$$

$$-y = 15 - 3x \qquad \text{Subtract } 3x \text{ from both sides.}$$

$$-1(-y) = -1(15 - 3x) \qquad \text{Multiply both sides by } -1 \text{ (or divide both sides by } -1\text{).}$$

$$y = -15 + 3x \qquad \text{Simplify using the distributive property.}$$

7. Given $V = \dfrac{k}{P}$, solve for P in terms of V and k.

Solution $V = \dfrac{k}{P}$

$$V \cdot P = k \qquad \text{Multiply both sides by } P.$$

$$\dfrac{V \cdot P}{V} = \dfrac{k}{V} \qquad \text{Divide both sides by } V.$$

$$P = \dfrac{k}{V}$$
■

Practice Problems   Solve for the indicated variable.

1. $2x - y = 5$; solve for y.   **2.** $2x - y = 5$; solve for x.

3. $A = \dfrac{1}{2}bh$; solve for h.   **4.** $L = 2\pi rh$; solve for r.

5. $P = 2\ell + 2w$; solve for w.

◢ EXERCISES 3.6

Solve for the indicated variable in Exercises 1–56.

1. $P = a + b + c$; solve for b.   **2.** $P = 3s$; solve for s.   **3.** $f = ma$; solve for m.

4. $C = \pi d$; solve for d.   **5.** $A = \ell w$; solve for w.   **6.** $P = R - C$; solve for C.

7. $R = n \cdot p$; solve for n.   **8.** $v = k + gt$; solve for k.   **9.** $I = A - P$; solve for P.

10. $L = 2\pi rh$; solve for h.   **11.** $A = \dfrac{m + n}{2}$; solve for m.   **12.** $W = RI^2t$; solve for R.

13. $P = 4s$; solve for s.   **14.** $C = 2\pi r$; solve for r.   **15.** $d = rt$; solve for t.

16. $P = a + 2b$; solve for a.   **17.** $I = Prt$; solve for t.   **18.** $R = \dfrac{E}{I}$; solve for E.

19. $P = a + 2b$; solve for b.   **20.** $c^2 = a^2 + b^2$; solve for b^2.   **21.** $\alpha + \beta + \gamma = 180$; solve for β.

Answers to Practice Problems   **1.** $y = 2x - 5$ **2.** $x = \dfrac{y + 5}{2}$ **3.** $h = \dfrac{2A}{b}$

4. $r = \dfrac{L}{2\pi h}$ **5.** $w = \dfrac{P - 2\ell}{2}$

22. $A = \dfrac{h}{2}(a + b)$; solve for h.   **23.** $y = mx + b$; solve for x.   **24.** $V = \ell wh$; solve for h.

25. $A = 4\pi r^2$; solve for r^2.   **26.** $V = \pi r^2 h$; solve for h.   **27.** $IQ = \dfrac{100M}{C}$; solve for M.

28. $A = \dfrac{R}{2L}$; solve for R.   **29.** $V = \dfrac{1}{3}\pi r^2 h$; solve for h.   **30.** $A = \dfrac{1}{2}bh$; solve for b.

31. $R = \dfrac{E}{I}$; solve for I.   **32.** $IQ = \dfrac{100M}{C}$; solve for C.   **33.** $A = \dfrac{R}{2L}$; solve for L.

34. $K = \dfrac{mv^2}{2g}$; solve for g.   **35.** $v = k + gt$; solve for t.   **36.** $L = 2\pi rh$; solve for h.

37. $S = 2\pi rh + 2\pi r^2$; solve for h.   **38.** $A = \dfrac{h}{2}(a + b)$; solve for a.   **39.** $S = \dfrac{a}{1 - r}$; solve for r.

40. $K = \dfrac{mv^2}{2g}$; solve for v^2.   **41.** $F = \dfrac{9}{5}C + 32$; solve for C.   **42.** $A = P + Prt$; solve for t.

43. $k = \dfrac{mv^2}{2g}$; solve for m.   **44.** $3x + y = 7$; solve for y.   **45.** $x - y = 3$; solve for y.

46. $x + y = 4$; solve for y.   **47.** $2x + y = 8$; solve for y.   **48.** $3x - y = 14$; solve for y.

49. $x + 2y = 5$; solve for x.   **50.** $-x + 5y = 6$; solve for x.   **51.** $4x - 3y = 9$; solve for y.

52. $2x - 5y + 8 = 0$; solve for x.   **53.** $3x + 8y - 4 = 0$; solve for x.   **54.** $6x - y = -3$; solve for y.

55. $1.2x + 1.5y = 3$; solve for y.   **56.** $2x + 1.7y = 5.1$; solve for y.

3.7 Additional Applications

OBJECTIVE

In this section, you will be learning to solve applied problems by substituting into given formulas and solving the resulting equations.

We have discussed solving equations and working with formulas throughout Chapter 3. In this section, we will present more formulas from a variety of real-life situations, with complete descriptions of the meanings of these formulas. The objectives are to give you more practice with formulas and to illustrate the wide use of formulas. The given information is to be substituted into the appropriate formula and the resulting equation solved for the unknown variable.

EXAMPLE   The lifting force, F, exerted on an airplane wing is found by multiplying some constant, k, times the area, A, of the wing's surface and times the square of the plane's velocity, v. The formula is $F = kAv^2$. Find the force on a plane's wing of area 120 sq ft if k is $\dfrac{4}{3}$ and the plane is traveling 80 miles per hour.

Solution We know that $k = \dfrac{4}{3}$, $A = 120$, and $v = 80$. Substitution gives

$$F = \frac{4}{3} \cdot 120 \cdot 80^2$$

$$= \frac{4}{\cancel{3}} \cdot \overset{40}{\cancel{120}} \cdot 6400$$

$$= 160 \cdot 6400$$

$$F = 1,024,000 \text{ lb}$$

(The force is measured in pounds.) ■

EXERCISES 3.7

In the following problems, read the descriptive information carefully and then substitute the values given in the problem for the corresponding variables in the formulas. Evaluate the resulting expression for the unknown variable.

Velocity

If an object is shot upward with an initial velocity v_0 feet per second, the velocity, v, in feet per second is given by the formula $v = v_0 - 32t$, where t is time in seconds.

1. Find the velocity at the end of 3 seconds if the initial velocity is 144 ft per second.

2. Find the initial velocity of an object if the velocity after 4 seconds is 48 ft per second.

3. An object projected upward with an initial velocity of 106 ft per second has a velocity of 42 ft per second. How many seconds have passed?

Medicine

In nursing, one procedure for determining the dosage for a child is

$$\text{child's dosage} = \frac{\text{age of child in years}}{\text{age of child} + 12} \cdot \text{adult dosage}$$

4. If the adult dosage of a drug is 20 milliliters, how much should a 3-year-old child receive?

5. If the adult dosage of a drug is 340 milligrams, how much should a 5-year-old child receive?

Investment

The amount of money due from investing P dollars is given by the formula $A = P + Prt$, where r is the rate expressed as a decimal and t is the time in years.

6. Find the amount due if $1000 is invested at 6% for 2 years.

7. How long will it take an investment of $600, at a rate of 5%, to be worth $750?

Carpentry

The number, N, of rafters in a roof or studs in a wall can be found by the formula $N = \dfrac{\ell}{d} + 1$, where ℓ is the length of the roof or wall and d is the center-to-center distance from one rafter or stud to the next. ℓ and d must be in the same units.

8. How many rafters will be needed to build a roof 26 ft long if they are placed 2 ft on center?

9. A wall has studs placed 16 in. on center. If the wall is 20 ft long, how many studs are in the wall?

10. How long is a wall if it requires 22 studs placed 16 in. on center?

Cost

The total cost, C, of producing x items can be found by the formula $C = ax + k$, where a is the cost per item and k is the fixed costs (rents, utilities, and so on).

11. Find the cost of producing 30 items if each costs $15 and the fixed costs are $580.

12. It costs $1097.50 to produce 80 dolls per week. If each doll costs $9.50 to produce, find the fixed costs.

13. It costs a company $3.60 to produce a calculator. Last week the total costs were $1308. If the fixed costs are $480 weekly, how many calculators were produced last week?

Depreciation

Many items decrease in value as time passes. This decrease in value is called **depreciation.** One type of depreciation is called **linear depreciation.** The value, V, of an item after t years is given by $V = C - Crt$, where C is the original cost and r is the rate of depreciation expressed as a decimal.

14. If you buy a car for $6000 and depreciate it linearly at a rate of 10% per year, what will be its value after 6 years?

15. A contractor buys a piece of heavy equipment for $20,000. If it cost $25,000 new and is 4 years old, find the rate of depreciation.

In Exercises 16–25, (a) create a formula suggested by the stated relationships and (b) solve as indicated with the given information.

16. The perimeter of a square is the product of 4 with the length of a side. Find the length of a side if the perimeter is 64 meters.

17. The circumference of a circle is the product of 2π and the radius. Find the radius if the circumference is 34π centimeters.

18. The interest on a loan is the product of the amount borrowed (principal) times the rate times the time of the loan. Find the interest on $1100 at 9% borrowed for 3 years.

19. The perimeter of an isosceles triangle (one with 2 sides the same length) is the sum of the base and twice the length of one of the two equal sides. If the perimeter is 72 and the base is 28, what is the length of each of the equal sides?

20. The Celsius temperature can be found by the product of $\dfrac{5}{9}$ with the difference of the Fahrenheit temperature and 32. Find the Celsius temperature, C, if the Fahrenheit temperature is 77 degrees.

21. The area of a circle is the product of π and the square of the radius. If a rotating sprinkler sprays water a distance of 7 meters, how many square meters will it cover in one revolution?

22. The IQ of a person is 100 times the quotient of the mental age and the chronological age. Find the IQ of someone $10\frac{1}{2}$ years old if the mental age is 12 years.

23. In electricity, the resistance of an electrical circuit measured in ohms is the quotient of the volts and the intensity measured in amperes. What is the resistance of a 120-volt circuit with an intensity of 20 amperes?

24. The volume of a cylinder is the product of π, radius squared, and height. Find the height of a cylinder whose volume is 252π cubic centimeters if the radius is 6 centimeters.

25. The area of a trapezoid is $\frac{1}{2}$ the height times the sum of the two bases. If the area of a trapezoid is 108 square centimeters, the height is 12 centimeters, and the length of one base is 8 centimeters, find the length of the other base.

CHAPTER 3 SUMMARY

Key Terms and Formulas

An expression that involves only multiplication and/or division with constants and/or variables is called a **term. A single constant or variable is also a term.** [3.1]

Like terms (or **similar terms**) are terms that are constants or terms that contain the same variables that are of the same power. [3.1]

A value that gives a true statement when substituted for the variable in an equation is called a **solution** to the equation. [3.2]

List of Key Words for determining the appropriate operation:

If a, b, and c are constants and $a \neq 0$, then a **first-degree equation** is an equation that can be written in the form

$$ax + b = c \quad [3.2]$$

Consecutive integers are two integers that differ by 1. [3.5]

Consecutive even integers are even integers that differ by 2. [3.5]

Consecutive odd integers are odd integers that differ by 2. [3.5]

Addition	**Subtraction**	**Multiplication**	**Division**
add	subtract	multiply	divide
sum	difference	product	quotient
plus	minus	times	
more than	less than	twice	
increased by	decreased by	of (with fractions and percent) [3.4]	

Properties and Rules

Addition Property of Equality [3.2]
If the same algebraic expression is added to both sides of an equation, the new equation has the same solutions as the original equation. Symbolically, if A, B, and C are algebraic expressions, then the equations

$$A = B$$

and $\qquad A + C = B + C$

have the same solutions.

Multiplication Property of Equality [3.2]
If both sides of an equation are multiplied by the same nonzero algebraic expression, the new equation has the same solutions as the original solution. Symbolically, if A, B, and C are algebraic expressions, then the equations

$$A = B$$

and $\qquad AC = BC \qquad$ where $C \neq 0$

have the same solutions.

Procedures

To Solve a First-Degree Equation [3.3]

1. Simplify each side of the equation by removing any grouping symbols and combining like terms.
2. Use the addition property of equality to add the opposites of constants and/or variables so that variables are on one side and constants on the other.
3. Use the multiplication property of equality to multiply both sides by the reciprocal of the coefficient of the variable (or divide both sides by the coefficient).
4. Check your answer by substituting it into the original equation.

CHAPTER 3 REVIEW

Combine like terms in Exercises 1–5. [3.1]

1. $3x + 2 + 4x - 1$

2. $-4(x + 3) + 2x$

3. $x + \dfrac{x - 5x}{4}$

4. $\dfrac{2(x + 3x)}{4} + 2x$

5. $2(x^2 - 3x) + 5(2x^2 + 7x)$

Evaluate Exercises 6–10 if $a = 2$, $b = -1$, and $c = 5$. [3.1]

6. $4a^2 - 3a + 2$

7. $3a - 2ab + c$

8. $-b^2 - 7b + 5$

9. $a(4b^2 + 3bc)$

10. $\dfrac{3a^3 - b}{c} + ab^2$

Solve the equations in Exercises 11–25. [3.2–3.3]

11. $x + 3.2 = 1.7$

12. $x - \dfrac{1}{2} = \dfrac{3}{4}$

13. $-8x = 12$

14. $-5x = 20.5$

15. $7x + 4 = -17$

16. $9x - 11 = x + 5$

17. $4x + 3 = 39 - 2x$

18. $5(1 - 2x) = 3x + 57$

19. $2(x + 6) = 6(x + 4)$

20. $5(2x + 3) = 3(x - 4) - 1$

21. $\dfrac{2}{3}x + 5 = 7$

22. $\dfrac{4}{5}x - 3 = 2$

23. $-\dfrac{3}{4}x + 2 = -6$

24. $0.5x + 3 = 0.3x - 5$

25. $4.4 + 0.6x = 1.2 - 0.2x$

Write out in words each expression in Exercises 26–30. [3.4]

26. $3x - 1$

27. $4 - 7y$

28. $5(y + 1)$

29. $2(4n - 1)$

30. $\dfrac{4}{x + 7}$

Write an algebraic expression described by each of the phrases in Exercises 31–35. [3.4]

31. 4 added to three times a number

32. The difference between 9 and twice a number

33. 11 times the difference between a number and 4

34. The number of hours in x days and 5 hours

35. The number of points scored by a basketball team on x field goals (2 points each) and 17 free throws (1 point each)

Solve for each indicated variable in Exercises 36–40. [3.6]

36. $E = mc^2$; solve for m.

37. $F = k \cdot \dfrac{ab}{d^2}$; solve for b.

38. $V = \dfrac{1}{3}\pi r^2 h$; solve for h.

39. $2x + 7y = 8$; solve for y.

40. $5x + 2y = -3$; solve for x.

In Exercises 41–50, set up an equation for each word problem and solve.

41. The perimeter of a rectangle is 240 in. If the length is 85 in., find the width. [3.5]

42. A car will cost \$6400 if purchased on an installment plan. If the monthly payments are \$150 and the down payment is \$1000, how long will it take to pay for the car? [3.5]

43. If twice a certain number is increased by 3, the result is 8 less than three times the number. Find the number. [3.5]

44. Three times a certain number is 10 more than twice the number. Find the number. [3.5]

45. The sum of two consecutive odd integers is 84. Find the integers. [3.5]

46. Find two consecutive integers such that twice the first subtracted from 5 is equal to the second plus 19. [3.5]

47. Sally bought a softball and bat. The bat cost \$9.35 more than the ball. If the total cost was \$16.25, find the cost of each. [3.5]

48. Bill collects limited-edition art prints. He recently ordered a print at a discount of 10% off the original price. However, he had to pay a shipping charge of \$6.50. If the total cost was \$227.00, what was the original price of the print? [3.5]

49. If you buy a car for \$7200 and depreciate it linearly at a rate of 8% per year, what will be its value after 7 years? [3.7]

50. The volume of a right circular cone is one-third of the product of π, radius squared, and height. Find the height of a cone whose volume is 147π cubic inches and whose radius is 7 in. [3.7]

CHAPTER 3 TEST

Combine like terms in Exercises 1 and 2.

1. $6(x^2 - 2x) - (-2x^2 + 5x)$

2. $\dfrac{3(3x - x)}{6} + 2(2x - 5)$

Evaluate Exercises 3 and 4 if $x = -3$, $y = 2$, and $z = -1$.

3. $2x + 3z - xy$

4. $3x^2 - 2xz - y^2$

Solve the equations in Exercises 5–11.

5. $\dfrac{5}{3}x + 1 = -4$

6. $4x - 5 - x = 2x + 5 - x$

7. $8(3 + x) = -4(2x - 6)$

8. $\dfrac{3}{4}x + \dfrac{1}{2} = x - 3$

9. $\dfrac{1}{6}x + \dfrac{2}{3} = x - 1$

10. $4(2x - 1) + 3 = 2(x - 4) - 5$

11. $0.7x + 2 = 0.4x + 8$

12. Write out in words: $4(x - 2)$

13. Write out in words: $3(x + 4) - 7$

14. Write an algebraic expression described by the following phrase: 3 times the sum of twice a number and 5.

15. Write an algebraic expression described by the following phrase: the number of quarts in x gallons and 3 quarts (1 gallon = 4 quarts).

16. Write an algebraic expression described by the following phrase: the annual interest on x dollars if the rate is 9% per year.

Solve for each indicated variable in Exercises 17–19.

17. $S = 2\pi rh$; solve for h.

18. $N = mrt + p$; solve for m.

19. $5x - 3y + 7 = 0$; solve for y.

In Exercises 20–25, set up an equation for each word problem and solve.

20. One number is 5 more than twice another. Their sum is -22. Find the numbers.

21. Find two consecutive integers such that twice the first added to three times the second is equal to 83.

22. A rectangle is 37 in. long and has a perimeter of 122 in. Find the width of the rectangle.

23. The total cost for a shirt was $25.97. The total cost was the marked price plus sales tax of 6% of the marked price. Find the marked price.

24. Quang had a hamburger and chocolate shake for lunch. The hamburger cost 90¢ more than the chocolate shake. If the total cost was $2.80, how much did each cost?

25. The area of a triangle is one-half of the product of the base and the height. (a) Write a formula suggested by this relationship and (b) find the area if the base is 16 meters and the height is 7 meters.

CUMULATIVE REVIEW (3)

Complete the expressions in Exercises 1–3 using the distributive property.

1. $6(4 + 7) = $ _____

2. $2(3x + 1) = $ _____

3. $-3(x + 5) = $ _____

Evaluate the expressions in Exercises 4–6 using the rules for order of operations.

4. $(3^2 - 1) \div 2 \cdot 4 + 2^3$

5. $\dfrac{3}{8} + \dfrac{5}{6} \div \left(\dfrac{1}{2} - \dfrac{1}{7} \right)$

6. $-4^2 + 15 \div 5 \cdot [2 - (-1)]$

Evaluate the expressions in Exercises 7 and 8 if $x = -2$, $y = 3$, and $z = 1$.

7. $3x - xy + 5yz$

8. $2x^2 + xy - y^2$

9. Is $x = -2$ a solution to the equation $3x + 4 = x$?

10. From the product of -6 and -3, subtract the difference between -6 and -3.

11. List the numbers that satisfy the condition $|x| = \dfrac{5}{3}$.

Combine like terms in Exercises 12–14.

12. $3(x + 4) - 5 + x$

13. $\dfrac{3(5x - x)}{4} - 2x - 7$

14. $2(x^2 + 4x) - (x^2 - 5x)$

Solve the equations in Exercises 15–17.

15. $9x - 8 = 4x - 13$

16. $6 - (x - 2) = 5$

17. $10 + 4x = 5x - 3(x + 4)$

Solve for the indicated variable in Exercises 18 and 19.

18. $h = vt - 16t^2$; solve for v.

19. $A = 2\pi r^2 + 2\pi rh$; solve for h.

20. The length of a rectangle is three times the width. If the perimeter is 128 inches, find the width and length.

4 EXPONENTS AND POLYNOMIALS

DID YOU KNOW ?

One of the most difficult problems for students in beginning algebra is to become comfortable with the idea that letters or symbols can be manipulated just like numbers in arithmetic. These symbols may be the cause of "math anxiety."

A great deal of publicity has recently been given to the concept that a large number of people suffer from math anxiety, a painful uneasiness caused by mathematical symbols or a problem-solving situation. Persons affected by math anxiety find it almost impossible to learn mathematics, or they may be able to learn but be unable to apply their knowledge or do well on tests. Persons suffering from math anxiety often develop math avoidance, and they avoid careers, majors, or classes that will require mathematics courses or skills. The sociologist Lucy Sells has determined that mathematics is a critical filter in the job market. Persons who lack quantitative skills are channeled into high-unemployment, low-paying, nontechnical areas.

What causes math anxiety? Researchers are investigating the following hypotheses:

1. a lack of skills that leads to lack of confidence and, therefore, to anxiety;
2. an attitude that mathematics is not useful or significant to society;
3. career goals that seem to preclude mathematics;
4. a self-concept that differs radically from the stereotype of a mathematician;
5. perceptions that parents, peers, or teachers have low expectations for the person in mathematics;
6. social conditioning to avoid mathematics (a particular problem for women).

We hope that you are finding your present experience with algebra successful and that the skills you are acquiring now will enable you to approach mathematical problems with confidence.

Mathematics is the queen of the sciences and arithmetic the queen of mathematics.

Karl F. Gauss (1777–1855)

 **CHAPTER OUTLINE**

*I*n Chapter 4, you will learn the rules of exponents and how exponents can be used to simplify very large and very small numbers. Astronomers and chemists are very familiar with these ideas and the notation used is appropriately called scientific notation. This notation is also used in your hand-held calculator. (Multiply 5,000,000 by 5,000,000 on your calculator and read the results.)

Although not all algebraic expressions are polynomials, the first two courses in algebra may leave this impression. Polynomials and operations with polynomials appear at almost every level of mathematics, which explains their importance in beginning courses. The skills you learn in Chapters 4 and 5 will be needed again and again in any mathematics courses you take in the future.

 **4.1 Exponents**

OBJECTIVE

In this section, you will be learning to simplify expressions by using the properties of integer exponents.

In Section 1.2, an **exponent** was defined as a number that tells how many times a factor occurs in a product. This definition is valid only if the exponents are natural (counting) numbers. In this section, we will develop four properties of exponents that will help in simplifying algebraic expressions and expand your understanding of exponents to include the exponent 0 and negative exponents.

We know that

$$6^2 = 6 \cdot 6 = 36$$

and

$$6^3 = 6 \cdot 6 \cdot 6 = 216$$

Also, we know that

$$x^3 = x \cdot x \cdot x$$

and

$$x^5 = x \cdot x \cdot x \cdot x \cdot x$$

If we want to multiply $6^2 \cdot 6^3$ or $x^3 \cdot x^5$, we could write down all the factors as follows:

$$6^2 \cdot 6^3 = (6 \cdot 6) \cdot (6 \cdot 6 \cdot 6) = 6^5$$

and

$$x^3 \cdot x^5 = (x \cdot x \cdot x) \cdot (x \cdot x \cdot x \cdot x \cdot x) = x^8$$

What do you think would be a simplified form for the product $3^4 \cdot 3^3$? You were right if you thought 3^7. That is, $3^4 \cdot 3^3 = 3^7$.

The preceding discussion, along with the basic concept of whole-number exponents, leads to the following property of exponents.

Property 1 of Exponents   If a is a nonzero number and m and n are integers, then

$$a^m \cdot a^n = a^{m+n}$$

If a variable or constant has no exponent written, it is understood to be 1. For example,

$$y = y^1, \qquad 7 = 7^1, \qquad \text{and} \qquad a = a^1$$

EXAMPLES   Use property 1 of exponents to simplify the following expressions.

1. $x^2 \cdot x^4$

   Solution $x^2 \cdot x^4 = x^{2+4} = x^6$

2. $y \cdot y^6$

   Solution $y \cdot y^6 = y^1 \cdot y^6 = y^{1+6} = y^7$

3. $4^2 \cdot 4$

   Solution $4^2 \cdot 4 = 4^{2+1} = 4^3 = 64$

4. $2^3 \cdot 2^2$

   Solution $2^3 \cdot 2^2 = 2^{3+2} = 2^5 = 32$

5. $(-2)^4(-2)^3$

   Solution $(-2)^4(-2)^3 = (-2)^{4+3} = (-2)^7 = -128$

6. $2y^2 \cdot 3y^9$

   Solution $2y^2 \cdot 3y^9 = 2 \cdot 3 \cdot y^2 \cdot y^9 = 6y^{2+9} = 6y^{11}$

7. $3 \cdot 2^4 \cdot 2$

   Solution $3 \cdot 2^4 \cdot 2 = 3 \cdot 2^{4+1} = 3 \cdot 2^5 = 3 \cdot 32 = 96$

8. $5 \cdot 3 \cdot 3^2$

   Solution $5 \cdot 3 \cdot 3^2 = 5 \cdot 3^{1+2} = 5 \cdot 3^3 = 5 \cdot 27 = 135$ ∎

Property 1 is true if m and n are **integers.** So we need to discuss the meaning of 0 as an exponent and negative exponents.

Study the following patterns of numbers. What do you think are the missing values for 2^0, 3^0, and 5^0?

$2^5 = 32$	$3^5 = 243$	$5^5 = 3125$
$2^4 = 16$	$3^4 = 81$	$5^4 = 625$
$2^3 = 8$	$3^3 = 27$	$5^3 = 125$
$2^2 = 4$	$3^2 = 9$	$5^2 = 25$
$2^1 = 2$	$3^1 = 3$	$5^1 = 5$
$2^0 = \,?$	$3^0 = \,?$	$5^0 = \,?$

Notice that in the column of powers of 2, each number is $\dfrac{1}{2}$ of the preceding number. Since $\dfrac{1}{2} \cdot 2 = 1$, a reasonable guess is that $2^0 = 1$. Similarly, in the column of powers of 3, $\dfrac{1}{3} \cdot 3 = 1$, so $3^0 = 1$ seems reasonable. Also,

$\dfrac{1}{5} \cdot 5 = 1$, so $5^0 = 1$ fits the pattern. In fact, the missing values are $2^0 = 1$, $3^0 = 1$, and $5^0 = 1$.

Another approach to understanding 0 as an exponent involves Property 1. Consider

$$2^0 \cdot 2^3 = 2^{0 + 3} = 2^3 \qquad \text{Using Property 1}$$

and $\qquad\qquad\qquad 1 \cdot 2^3 = 2^3$

So, $2^0 \cdot 2^3 = 1 \cdot 2^3$ and $2^0 = 1$. Similarly,

$$7^0 \cdot 7^2 = 7^{0 + 2} = 7^2$$

and $\qquad\qquad\qquad 1 \cdot 7^2 = 7^2$

So, $7^0 \cdot 7^2 = 1 \cdot 7^2$ and $7^0 = 1$.

This discussion leads directly to Property 2.

Property 2 of Exponents   If *a* is a nonzero number, then

$$a^0 = 1$$

The expression 0^0 is undefined.

EXAMPLES   Simplify the following expressions using Property 2.

9. 10^0

Solution $10^0 = 1$

10. $x^0 \cdot x^3$

Solution $x^0 \cdot x^3 = x^{0 + 3} = x^3$ or $x^0 \cdot x^3 = 1 \cdot x^3 = x^3$

11. 49^0

Solution $49^0 = 1$

12. $(-6)^0$

Solution $(-6)^0 = 1$ ∎

Again, looking at a pattern of powers, we can understand the meaning of negative exponents, as in 2^{-1}, 2^{-2}, 3^{-1}, and 3^{-2}. What do you think the missing values are in the following patterns? (**Hint:** Notice that in the column of powers of 2, each number is $\dfrac{1}{2}$ of the preceding number. In the column of powers of 3, each number is $\dfrac{1}{3}$ of the preceding number.)

$$2^3 = 8 \qquad\qquad 3^3 = 27$$
$$2^2 = 4 \qquad\qquad 3^2 = 9$$
$$2^1 = 2 \qquad\qquad 3^1 = 3$$
$$2^0 = 1 \qquad\qquad 3^0 = 1$$
$$2^{-1} = ? \qquad\qquad 3^{-1} = ?$$
$$2^{-2} = ? \qquad\qquad 3^{-2} = ?$$

The values are $2^{-1} = \dfrac{1}{2}$ and $2^{-2} = \dfrac{1}{2^2} = \dfrac{1}{4}$. Also, $3^{-1} = \dfrac{1}{3}$ and

$3^{-2} = \dfrac{1}{3^2} = \dfrac{1}{9}$. These lead to Property 3.

Property 3 of Exponents   If a is a nonzero number and n is an integer, then

$$a^{-n} = \frac{1}{a^n}$$

EXAMPLES   Simplify each expression so that it contains only positive exponents.

13. 5^{-1}

  Solution $5^{-1} = \dfrac{1}{5^1} = \dfrac{1}{5}$ Using Property 3

14. x^{-3}

  Solution $x^{-3} = \dfrac{1}{x^3}$ Using Property 3

15. $2^{-5} \cdot 2^8$

  Solution $2^{-5} \cdot 2^8 = 2^{-5+8} = 2^3 = 8$ Using Property 1

16. $x^{-9} \cdot x^7$

  Solution $x^{-9} \cdot x^7 = x^{-9+7} = x^{-2} = \dfrac{1}{x^2}$

(Here, Property 1 is used first, then Property 3.)   ∎

Special note: A simplified expression will not contain negative exponents. (In later courses in algebra, this rule will be adjusted to allow for negative exponents in the numerator of a fraction.)

Now we are ready to consider an expression that involves division as well as exponents. For example,

$$\frac{3^3}{3^5} = \frac{\cancel{3} \cdot \cancel{3} \cdot \cancel{3} \cdot 1}{\cancel{3} \cdot \cancel{3} \cdot \cancel{3} \cdot 3 \cdot 3} = \frac{1}{3^2} = \frac{1}{9} \qquad \text{or} \qquad \frac{3^3}{3^5} = 3^{3-5} = 3^{-2} = \frac{1}{3^2} = \frac{1}{9}$$

$$\frac{5^4}{5^2} = \frac{\cancel{5} \cdot \cancel{5} \cdot 5 \cdot 5}{\cancel{5} \cdot \cancel{5} \cdot 1} = \frac{5^2}{1} = 25 \qquad \text{or} \qquad \frac{5^4}{5^2} = 5^{4-2} = 5^2 = 25$$

$$\frac{2^2}{2^7} = \frac{\not{2} \cdot \not{2} \cdot 1}{\not{2} \cdot \not{2} \cdot 2 \cdot 2 \cdot 2 \cdot 2 \cdot 2} = \frac{1}{2^5} = \frac{1}{32} \quad \text{or} \quad \frac{2^2}{2^7} = 2^{2-7} = 2^{-5} = \frac{1}{2^5} = \frac{1}{32}$$

Thus, we are led to Property 4.

Property 4 of Exponents   If a is a nonzero number and m and n are integers, then

$$\frac{a^m}{a^n} = a^{m-n}$$

EXAMPLES   Simplify each expression using the appropriate property of exponents. Remember, a simplified expression does not contain negative exponents.

17. $\dfrac{x^6}{x}$

   Solution $\dfrac{x^6}{x} = x^{6-1} = x^5$

18. $\dfrac{x^6}{x^{-1}}$

   Solution $\dfrac{x^6}{x^{-1}} = x^{6-(-1)} = x^{6+1} = x^7$

19. $\dfrac{y^{-2}}{y^{-7}}$

   Solution $\dfrac{y^{-2}}{y^{-7}} = y^{-2-(-7)} = y^{-2+7} = y^5$

20. $\dfrac{10^{-8}}{10^{-2}}$

   Solution $\dfrac{10^{-8}}{10^{-2}} = 10^{-8-(-2)} = 10^{-8+2} = 10^{-6}$

   $$= \frac{1}{10^6} \quad \text{or} \quad \frac{1}{1,000,000}$$

21. $\dfrac{x^{10} \cdot x^2}{x^{15}}$

   Solution $\dfrac{x^{10} \cdot x^2}{x^{15}} = \dfrac{x^{10+2}}{x^{15}} = \dfrac{x^{12}}{x^{15}} = x^{12-15} = x^{-3} = \dfrac{1}{x^3}$

22. $\dfrac{2^{-5} \cdot 2^8}{2^3}$

   Solution $\dfrac{2^{-5} \cdot 2^8}{2^3} = \dfrac{2^{-5+8}}{2^3} = \dfrac{2^3}{2^3} = 2^{3-3} = 2^0 = 1$

Of special interest are expressions with negative exponents in the denominator, such as $\dfrac{1}{a^{-n}}$. We can write

$$\frac{1}{a^{-n}} = \frac{a^0}{a^{-n}} = a^{0-(-n)} = a^{0+n} = a^n$$

or

$$\frac{1}{a^{-n}} = \frac{1}{\dfrac{1}{a^n}} = 1 \cdot \frac{a^n}{1} = a^n$$

Thus,

$$a^{-n} = \frac{1}{a^n} \qquad \text{and} \qquad \frac{1}{a^{-n}} = a^n$$

In effect, the sign of an exponent is changed whenever a term is moved from numerator to denominator, or vice versa. For example,

$$\frac{4^{-2}}{4^{-5}} = \frac{4^5}{4^2} = 4^{5-2} = 4^3 = 64$$

and

$$\frac{x^6}{x^{-5}} = x^6 \cdot x^5 = x^{6+5} = x^{11}$$

Practice Problems   Simplify each expression.

1. $2^3 \cdot 2^4$ $= 2^7 = 128$

2. $\dfrac{2^3}{2^4}$ $\dfrac{2 \cdot 2 \cdot 2 \cdot 1}{2 \cdot 2 \cdot 2 \cdot 2}$

3. $\dfrac{x^0 \cdot x^7 \cdot x^{-3}}{x^{-2}}$ x^6

4. $\dfrac{10^{-8} \cdot 10^2}{10^{-7}}$ $= 10$

◢ EXERCISES 4.1

Simplify each expression. The final form of expressions with variables should contain only positive exponents. Assume that all variables represent nonzero numbers.

1. $3^2 \cdot 3$ $= 3^3$ 2. $7^2 \cdot 7^3$ $= 7^5$ 3. $8^3 \cdot 8^0$ $= 8^3$ 4. 3^{-1} $\frac{1}{9}$ 5. 4^{-2}

6. $(-5)^{-2}$ 7. 6^{-3} $\frac{1}{216}$ 8. $(-2)^4 \cdot (-2)^0$ 9. $(-4)^3 \cdot (-4)^0$ -64 10. $3 \cdot 2^3$

11. $6 \cdot 3^2$ $= 54$ 12. $-4 \cdot 5^3$ 13. $-2 \cdot 3^3$ 14. $3 \cdot 2^{-3}$ 15. $4 \cdot 3^{-2}$

16. $-3 \cdot 5^{-2}$ 17. $-5 \cdot 2^{-2}$ 18. $x^2 \cdot x^3$ 19. $x^3 \cdot x$ 20. $y^2 \cdot y^0$

21. $y^3 \cdot y^8$ 22. x^{-3} 23. y^{-2} 24. $2x^{-1}$ 25. $5y^{-4}$

26. $x \cdot x^{-1}$ 27. $x \cdot x^{-3}$ 28. $x^0 \cdot x^{-2}$ 29. $y^5 \cdot y^{-2}$ 30. $y^4 \cdot y^{-6}$

$-5 \cdot \frac{1}{4} = \frac{5}{4}$ -54 $\frac{4}{1} \cdot \frac{1}{9} = \frac{4}{9}$

Answers to Practice Problems 1. $2^7 = 128$ 2. $\dfrac{1}{2}$ 3. x^6 4. 10

31. $\dfrac{3^4}{3^2}$ **32.** $\dfrac{7^3}{7}$ **33.** $\dfrac{9^5}{9^2}$ **34.** $\dfrac{10^3}{10^4}$ **35.** $\dfrac{10^2}{10^5}$

36. $\dfrac{2^3}{2^6}$ **37.** $\dfrac{x^4}{x^2}$ **38.** $\dfrac{x^5}{x^3}$ **39.** $\dfrac{x^3}{x}$ **40.** $\dfrac{y^6}{y^4}$

41. $\dfrac{x^7}{x^3}$ **42.** $\dfrac{x^8}{x^3}$ **43.** $\dfrac{x^{-2}}{x^2}$ **44.** $\dfrac{x^{-3}}{x}$ **45.** $\dfrac{x^4}{x^{-2}}$

46. $\dfrac{x^5}{x^{-1}}$ **47.** $\dfrac{x^{-3}}{x^{-5}}$ **48.** $\dfrac{x^{-4}}{x^{-1}}$ **49.** $\dfrac{y^{-2}}{y^{-4}}$ **50.** $\dfrac{y^3}{y^{-3}}$

51. $3x^3 \cdot x^0$ **52.** $3y \cdot y^4$ **53.** $(5x^3)(2x^2)$ **54.** $(3x^2)(3x)$ **55.** $(4x^3)(9x^0)$

56. $(5x^2)(3x^4)$ **57.** $(-2x^2)(7x^3)$ **58.** $(3y^3)(-6y^2)$ **59.** $(-4x^5)(3x)$ **60.** $(6y^4)(5y^5)$

61. $\dfrac{8y^3}{2y^2}$ **62.** $\dfrac{12x^4}{3x}$ **63.** $\dfrac{9y^5}{3y^3}$ **64.** $\dfrac{-10x^5}{2x}$ **65.** $\dfrac{-8y^4}{4y^2}$

66. $\dfrac{12x^6}{-3x^3}$ **67.** $\dfrac{21x^4}{-3x^2}$ **68.** $\dfrac{10 \cdot 10^3}{10^{-3}}$ **69.** $\dfrac{10^4 \cdot 10^{-3}}{10^{-2}}$ **70.** $\dfrac{10 \cdot 10^{-1}}{10^2}$

71. $\dfrac{10^0 \cdot 10^3}{10^5}$ **72.** $\dfrac{x^2 \cdot x^{-3}}{x^4}$ **73.** $\dfrac{x^4 \cdot x^{-2}}{x^{-3}}$ **74.** $\dfrac{y^5 \cdot y^0}{y^{-2}}$ **75.** $\dfrac{y^6 \cdot y^{-2}}{y^4}$

Calculator Problems

Use a calculator to evaluate each expression.

76. $(2.16)^3$ **77.** $(-5.06)^2$ **78.** $(1.6)^{-2}$ **79.** $(6.4)^{-3}$ **80.** $(-14.8)^2(21.3)^2$

◢ 4.2 *More on Exponents and Scientific Notation*

OBJECTIVES

In this section, you will be learning to:

1. Simplify powers of expressions by using the properties of integer exponents.
2. Write decimal numbers in scientific notation.
3. Operate with decimal numbers using scientific notation.

In Section 4.1, we discussed the following four properties of exponents where a is a nonzero real number and m and n are integers.

Property 1: $a^m \cdot a^n = a^{m+n}$

Property 2: $a^0 = 1$

Property 3: $a^{-n} = \dfrac{1}{a^n}$

Property 4: $\dfrac{a^m}{a^n} = a^{m-n}$

In each of these properties, the base of each exponent was one number or variable. Now we will consider expressions such as $(5x)^3$, $(3xy)^4$, $\left(\dfrac{-2}{x}\right)^3$, and $(x^2)^4$, in which the base is a product or a quotient. For example,

$$
\begin{aligned}
(5x)^3 &= (5x) \cdot (5x) \cdot (5x) \\
&= 5 \cdot 5 \cdot 5 \cdot x \cdot x \cdot x \\
&= 5^3 \cdot x^3 \\
&= 125x^3
\end{aligned}
$$

Using the associative and commutative properties of multiplication

$$
\begin{aligned}
(-2y)^5 &= (-2y) \cdot (-2y) \cdot (-2y) \cdot (-2y) \cdot (-2y) \\
&= (-2)(-2)(-2)(-2)(-2) \cdot y \cdot y \cdot y \cdot y \cdot y \\
&= (-2)^5 \cdot y^5 \\
&= -32y^5
\end{aligned}
$$

[Note carefully that in the expression $(-2y)^5$, the negative sign goes with 2, and -2 is treated as a factor of $-2y$.]

In an expression such as $-x^2$, you know that -1 is the understood coefficient of x^2 and the exponent 2 refers only to x as its base. Thus,

$$-x^2 = -1 \cdot x^2.$$

The same is true for an expression such as -7^2 or -2^0. We have

$$-7^2 = -1 \cdot 7^2 = -1 \cdot 49 = -49$$

and

$$-2^0 = -1 \cdot 2^0 = -1 \cdot 1 = -1$$

But if parentheses are used, as in $(-7)^2$ and $(-2)^0$, then the base is the negative number:

$$(-7)^2 = (-7)(-7) = +49$$

and

$$(-2)^0 = 1$$

These examples lead to Property 5.

Property 5 of Exponents   If a and b are nonzero numbers and n is an integer, then

$$(ab)^n = a^n b^n$$

EXAMPLES   Simplify each expression using Property 5.

1. $(5x)^2$

    Solution $(5x)^2 = 5^2 \cdot x^2 = 25x^2$

2. $(xy)^3$

    Solution $(xy)^3 = x^3 \cdot y^3 = x^3 y^3$

3. $(3xy)^4$

    Solution $(3xy)^4 = 3^4 \cdot x^4 \cdot y^4 = 81x^4 y^4$

4. $(-7ab)^2$

    Solution $(-7ab)^2 = (-7)^2 a^2 b^2 = 49a^2 b^2$

5. $(-x)^3$

Solution $(-x)^3 = (-1 \cdot x)^3 = (-1)^3 \cdot x^3 = -1x^3 = -x^3$

6. $(ab)^{-5}$

Solution $(ab)^{-5} = a^{-5} \cdot b^{-5} = \dfrac{1}{a^5} \cdot \dfrac{1}{b^5} = \dfrac{1}{a^5 b^5}$

or, using Property 3 first and then Property 5,

$$(ab)^{-5} = \dfrac{1}{(ab)^5} = \dfrac{1}{a^5 b^5}$$ ∎

Fractions raised to a power can be handled in a similar manner. For example,

$$\left(\dfrac{2}{x}\right)^3 = \dfrac{2}{x} \cdot \dfrac{2}{x} \cdot \dfrac{2}{x} = \dfrac{2 \cdot 2 \cdot 2}{x \cdot x \cdot x} = \dfrac{2^3}{x^3} = \dfrac{8}{x^3}$$

and

$$\left(\dfrac{3x}{y}\right)^4 = \dfrac{3x}{y} \cdot \dfrac{3x}{y} \cdot \dfrac{3x}{y} \cdot \dfrac{3x}{y} = \dfrac{3x \cdot 3x \cdot 3x \cdot 3x}{y \cdot y \cdot y \cdot y}$$

$$= \dfrac{(3x)^4}{y^4} = \dfrac{3^4 x^4}{y^4} = \dfrac{81 x^4}{y^4}$$

Thus, we now have Property 6.

Property 6 of Exponents   If a and b are nonzero numbers and n is an integer, then

$$\left(\dfrac{a}{b}\right)^n = \dfrac{a^n}{b^n}$$

EXAMPLES   Simplify each expression using Property 6.

7. $\left(\dfrac{y}{x}\right)^5$

Solution $\left(\dfrac{y}{x}\right)^5 = \dfrac{y^5}{x^5}$

8. $\left(\dfrac{2a}{b}\right)^4$

Solution $\left(\dfrac{2a}{b}\right)^4 = \dfrac{(2a)^4}{b^4} = \dfrac{2^4 a^4}{b^4} = \dfrac{16 a^4}{b^4}$

9. $\left(\dfrac{2x}{-3y}\right)^2$

Solution $\left(\dfrac{2x}{-3y}\right)^2 = \dfrac{(2x)^2}{(-3y)^2} = \dfrac{2^2 x^2}{(-3)^2 y^2} = \dfrac{4x^2}{9y^2}$ ∎

What happens when a power is raised to a power? For example, if you want to simplify the expressions $(x^2)^3$ and $(2^5)^2$, you might try writing

$$(x^2)^3 = x^2 \cdot x^2 \cdot x^2 = x^{2+2+2} = x^6$$

and $$(2^5)^2 = 2^5 \cdot 2^5 = 2^{5+5} = 2^{10} = 1024$$

However, this technique can be quite time-consuming when the exponent is large, such as in $(3x^2y^3)^{17}$. Property 7 gives a convenient way to handle powers raised to powers.

Property 7 of Exponents	If a is a nonzero number and m and n are integers, then $$(a^m)^n = a^{mn}$$

EXAMPLES   Simplify each expression using Property 7.

10. $(x^2)^4$

　　　Solution　$(x^2)^4 = x^{2 \cdot 4} = x^8$

11. $(3^2)^3$

　　　Solution　$(3^2)^3 = 3^{2 \cdot 3} = 3^6 = 729$

12. $(x^5)^{-2}$

　　　Solution　$(x^5)^{-2} = x^{5(-2)} = x^{-10} = \dfrac{1}{x^{10}}$

　　　or

　　　$$(x^5)^{-2} = \frac{1}{(x^5)^2} = \frac{1}{x^{5 \cdot 2}} = \frac{1}{x^{10}}$$

13. $(y^{-7})^2$

　　　Solution　$(y^{-7})^2 = y^{-7 \cdot 2} = y^{-14} = \dfrac{1}{y^{14}}$

　　　or

　　　$$(y^{-7})^2 = \left(\frac{1}{y^7}\right)^2 = \frac{1^2}{(y^7)^2} = \frac{1}{y^{14}}$$ ∎

　　　Various combinations of all seven properties of exponents are needed to simplify the following examples. There may be more than one correct sequence of steps to follow. **You should apply whichever property you "see" first. The simplified form will be the same in any case.**

EXAMPLES   Simplify each expression using properties of exponents.

14. $\left(\dfrac{-2x}{y^2}\right)^3$

　　　Solution　$\left(\dfrac{-2x}{y^2}\right)^3 = \dfrac{(-2x)^3}{(y^2)^3} = \dfrac{(-2)^3 x^3}{y^6} = \dfrac{-8x^3}{y^6}$

15. $\left(\dfrac{3a^5b}{a^0b^2}\right)^2$

Solution $\left(\dfrac{3a^5b}{a^0b^2}\right)^2 = \left(\dfrac{3a^5}{b^{2-1}}\right)^2 = \dfrac{3^2(a^5)^2}{(b^1)^2} = \dfrac{9a^{10}}{b^2}$ ∎

Another general approach with fractions involving negative exponents is to note that

$$\left(\dfrac{a}{b}\right)^{-n} = \dfrac{a^{-n}}{b^{-n}} = \dfrac{b^n}{a^n} = \left(\dfrac{b}{a}\right)^n$$

EXAMPLE

16. Either of the following approaches can be used to simplify $\left(\dfrac{x^3}{y}\right)^{-4}$.

Solutions

a. $\left(\dfrac{x^3}{y}\right)^{-4} = \left(\dfrac{y}{x^3}\right)^4 = \dfrac{y^4}{(x^3)^4} = \dfrac{y^4}{x^{12}}$

b. $\left(\dfrac{x^3}{y}\right)^{-4} = \dfrac{(x^3)^{-4}}{y^{-4}} = \dfrac{x^{-12}}{y^{-4}} = \dfrac{\dfrac{1}{x^{12}}}{\dfrac{1}{y^4}} = \dfrac{1}{x^{12}} \cdot \dfrac{y^4}{1} = \dfrac{y^4}{x^{12}}$

As was mentioned in Section 4.1, we could have written
$\dfrac{x^{-12}}{y^{-4}} = \dfrac{x^{-12}}{y^{-4}} = \dfrac{y^4}{x^{12}}$ directly since switching between numerator
and denominator changes the sign of the exponent. ∎

Scientific Notation

Problems in physics, chemistry, astronomy, and other scientific and techno-
logical fields sometimes involve very large or very small numbers. For convenience, these numbers can be written in a form called **scientific notation** that
involves exponents. Your calculator also uses this notation.

In scientific notation, all decimal numbers are written as the product
(indicated with a cross, ×) of a number between 1 and 10 and some power
of 10. For example,

$$56{,}000 = 5.6 \times 10^4 \qquad \text{(Note that 5.6 is between 1 and 10.)}$$
$$872{,}000 = 8.72 \times 10^5$$
$$0.0049 = 4.9 \times 10^{-3}$$

To understand this notation and how to find the correct power of 10,
consider the place value system we use for decimal numbers (see Figure 4.1).

$\overline{10^7}\ \overline{10^6}\ \overline{10^5}\ \overline{10^4}\ \overline{10^3}\ \overline{10^2}\ \overline{10^1}\ \overline{10^0}\ \cdot\ \overline{10^{-1}}\ \overline{10^{-2}}\ \overline{10^{-3}}\ \overline{10^{-4}}\ \overline{10^{-5}}\ \overline{10^{-6}}\ \overline{10^{-7}}\ \overline{10^{-8}}$

Values of each position in the decimal system using the powers of 10

Figure 4.1

Write 56,000 with the value of each position indicated:

$$56,000 = \underset{10^4\ 10^3\ 10^2\ 10^1\ 10^0}{5\quad 6\quad 0\quad 0\quad 0}\ .$$

Now, we see that the digit 5 is in the position corresponding to 10^4. That is, if we multiply 5.6×10^4, that will give the value 56,000.

Similarly,

$$872,000 = \underset{10^5\ 10^4\ 10^3\ 10^2\ 10^1\ 10^0}{8\quad 7\quad 2\quad 0\quad 0\quad 0}\ . = 8.72 \times 10^5$$

since the digit 8 is in the 10^5 place value position.

Also,

$$0.0049 = 0\ .\ \underset{10^{-1}\ 10^{-2}\ 10^{-3}\ 10^{-4}}{0\quad 0\quad 4\quad 9} = 4.9 \times 10^{-3}$$

since the digit 4 is in the 10^{-3} place value position.

Your calculator will also give large answers in scientific notation. For example, use your calculator to multiply

	93,000,000 miles	(The distance to the sun from the earth)
times	250,000 miles	(The distance to the moon from the earth)

Your calculator probably shows an expression like

$$2.325\quad 13\qquad \text{or}\qquad 2.325\quad \text{E } 13$$

which is the calculator's version of the scientific notation

$$2.325 \times 10^{13}$$

To see how to arrive at this value using scientific notation and the properties of exponents, we can calculate in the following way:

$$
\begin{aligned}
93,000,000 \times 250,000 &= 9.3 \times 10^7 \times 2.5 \times 10^5 \\
&= 9.3 \times 2.5 \times 10^7 \times 10^5 \\
&= 23.25 \times 10^{7+5} \\
&= 2.325 \times 10^1 \times 10^{12} \\
&= 2.325 \times 10^{13}
\end{aligned}
$$

EXAMPLES

17. Write 8,720,000 in scientific notation.

Solution The digit 8 is in the seventh position to the left of the understood decimal point. In powers of 10, this is the 10^6 position. So,

$$8,720,000 = 8.72 \times 10^6 \qquad \text{(8.72 is between 1 and 10.)}$$

To check, move the decimal point 6 places to the right and get the original number:

$$8.72 \times 10^6 = 8 . \underset{1\ 2\ 3\ 4\ 5\ 6}{7\,2\,0\,0\,0\,0}\ . = 8,720,000.$$

18. Write 0.000000376 in scientific notation.

    Solution The digit 3 is in the seventh position to the right of the decimal point. In powers of 10, this is the 10^{-7} position. So,

    $$0.000000376 = 3.76 \times 10^{-7} \qquad \text{(3.76 is between 1 and 10.)}$$

    To check, move the decimal point 7 places to the left and get the original number:

    $$3.76 \times 10^{-7} = 0 \underbrace{.0 0 0 0 0 0 3}. 7 6$$
    $$7\ 6\ 5\ 4\ 3\ 2\ 1$$
    $$= 0.000000376$$

19. Multiply $300,000 \times 700$ using scientific notation and leave your answer in scientific notation. Verify your answer using your calculator.

    Solution $300,000 \times 700 = 3.0 \times 10^5 \times 7.0 \times 10^2$
    $$= 3 \times 7 \times 10^5 \times 10^2$$
    $$= 21 \times 10^7$$
    $$= 2.1 \times 10^1 \times 10^7$$
    $$= 2.1 \times 10^8$$

20. Simplify the expression $\dfrac{0.0042 \times 0.003}{0.21}$ using scientific notation and leave your answer in scientific notation. Verify your answer using your calculator.

    Solution $\dfrac{0.0042 \times 0.003}{0.21} = \dfrac{4.2 \times 10^{-3} \times 3 \times 10^{-3}}{2.1 \times 10^{-1}}$

    $$= \frac{\overset{2}{\cancel{4.2}} \times 3}{\cancel{2.1}} \times \frac{10^{-6}}{10^{-1}}$$
    $$= 6 \times 10^{-6 + 1}$$
    $$= 6 \times 10^{-5}$$

21. A light-year is the distance traveled by a particle of light in one year. Find the length of a light-year in scientific notation if light travels 186,000 miles per second.

    Solution
    $$60 \text{ sec} = 1 \text{ min}$$
    $$60 \text{ min} = 1 \text{ hr}$$
    $$24 \text{ hr} = 1 \text{ day}$$
    $$365 \text{ days} = 1 \text{ year}$$
    $$1 \text{ light-year} = 186,000 \times 60 \times 60 \times 24 \times 365$$
    $$= 5,865,696,000,000$$
    $$= 5.865696 \times 10^{12} \text{ mi}$$

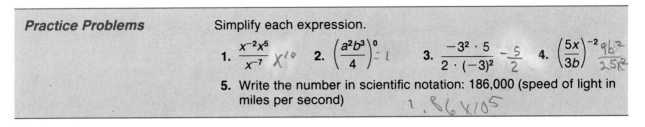

Practice Problems Simplify each expression.

1. $\dfrac{x^{-2}x^5}{x^{-7}}$ x^{10} 2. $\left(\dfrac{a^2b^3}{4}\right)^0$ $=1$ 3. $\dfrac{-3^2 \cdot 5}{2 \cdot (-3)^2}$ $-\dfrac{5}{2}$ 4. $\left(\dfrac{5x}{3b}\right)^{-2}$ $\dfrac{9b^2}{25x^2}$

5. Write the number in scientific notation: 186,000 (speed of light in miles per second) 1.86×10^5

EXERCISES 4.2

Simplify each expression in Exercises 1–40. Assume that all variables represent nonzero numbers.

1. $(6x^3)^2$ $36x^6$ 2. $(-3x^4)^2$ $-9x^8$ 3. $(-3x^2)^3$ $-27x^6$ 4. $(x^2y)^5$ 5. $(4^0xy^2)^3$

6. $-(2x^3y^0)^2$ 7. $(8a^4b)^2$ 8. $(-2^2x^5)^3$ 9. $-(7xy^2)^0$ 10. $(6a^2b^{-3})^{-1}$

11. $(4m^2n^{-3})^{-2}$ 12. $5(x^2y^{-1})^{-2}$ 13. $-2(3x^5y^{-2})^{-3}$ 14. $\left(\dfrac{3x}{y}\right)^3$ 15. $\left(\dfrac{-4x}{y^2}\right)^2$

16. $\left(\dfrac{6m^3}{n^5}\right)^0$ 17. $\left(\dfrac{3x^2}{y^3}\right)^2$ 18. $\left(\dfrac{-2x^2}{y^{-2}}\right)^2$ 19. $\left(\dfrac{x}{y}\right)^{-2}$ 20. $\left(\dfrac{2a}{b}\right)^{-1}$

21. $\left(\dfrac{2x}{y^5}\right)^{-2}$ 22. $\left(\dfrac{3x}{y^{-2}}\right)^{-1}$ 23. $\left(\dfrac{4a^2}{b^{-3}}\right)^{-3}$ 24. $\left(\dfrac{-3}{xy^2}\right)^{-3}$ 25. $\left(\dfrac{5xy^3}{y}\right)^2$

26. $\left(\dfrac{m^2n^3}{mn}\right)^2$ 27. $\left(\dfrac{2ab^3}{b^2}\right)^4$ 28. $\left(\dfrac{-7^2x^2y}{y^3}\right)^{-1}$ 29. $\left(\dfrac{2ab^4}{b^2}\right)^{-3}$ 30. $\left(\dfrac{5x^3y}{y^2}\right)^{-2}$

31. $\left(\dfrac{2x^2y}{y^3}\right)^{-4}$ 32. $\left(\dfrac{x^3y^{-1}}{y^2}\right)^2$ 33. $\left(\dfrac{2a^2b^{-1}}{b^2}\right)^3$ 34. $\left(\dfrac{6y^5}{x^2y^{-2}}\right)^2$ 35. $\left(\dfrac{7x^{-2}y}{xy^{-1}}\right)^2$

36. $\left(\dfrac{3xy^{-1}}{7x^2y}\right)^{-2}$ 37. $\left(\dfrac{2x^2y^{-3}}{3x^{-1}y}\right)^{-1}$ 38. $\left(\dfrac{4x^3y^{-1}}{xy^{-2}}\right)^{-1}$ 39. $\left(\dfrac{3x^3y}{2x^{-2}y}\right)^{-2}$ 40. $\left(\dfrac{2x^{-3}y}{x^2y^{-2}}\right)^{-3}$

Write the numbers in Exercises 41–64 in scientific notation and perform the indicated operations. Leave your answers in scientific notation. Use a calculator to verify your answers.

41. 86,000   42. 927,000   43. 0.0362

44. 0.0061   45. 18,300,000   46. 376,000,000

47. 0.000217   48. 0.00000143   49. 500×9000

50. $35,000 \times 2000$ 51. $410,000 \times 30,000$ 52. $14,000 \times 200,000$

53. 300×0.00015 54. $0.000024 \times 40,000$ 55. 0.0003×0.000025

56. 0.00005×0.00013 57. $\dfrac{3900}{0.003}$ 58. $\dfrac{4800}{12,000}$

59. $\dfrac{125}{50,000}$ 60. $\dfrac{0.0046}{230}$ 61. $\dfrac{0.02 \times 3900}{0.013}$

Answers to Practice Problems 1. x^{10} 2. 1 3. $\dfrac{-5}{2}$ or $-\dfrac{5}{2}$ 4. $\dfrac{9b^2}{25x^2}$

5. 1.86×10^5

62. $\dfrac{0.0084 \times 0.003}{0.21 \times 60}$

63. $\dfrac{0.005 \times 650 \times 3.3}{0.0011 \times 2500}$

64. $\dfrac{5.4 \times 0.003 \times 50}{15 \times 0.0027 \times 200}$

65. The mass of a hydrogen atom is approximately 0.00000000000000000000000167 grams. Write this number in scientific notation.

66. The weight of an atom is expressed in atomic weight units (amu), where 1 amu = 0.0000000000000000000000000016605 kilograms. Express the weight in scientific notation.

67. The distance light travels in one year is called a light-year. A light-year is approximately 5,866,000,000,000 miles. Write this number in scientific notation.

68. The mass of the earth is about 5,980,000,000,000,000,000,000,000,000 grams. Write this number in scientific notation.

69. Light travels approximately 3×10^{10} centimeters per second. How many centimeters would this be per minute? per hour? Express your answers in scientific notation.

70. An atom of gold weighs approximately 3.25×10^{-22} grams. What would be the weight of 2000 atoms of gold? Express your answer in scientific notation.

4.3 Adding and Subtracting Polynomials

OBJECTIVES

In this section, you will be learning to:

1. Name the degree of polynomials.
2. Add polynomials.
3. Subtract polynomials.
4. Simplify expressions by removing grouping symbols and combining like terms.

In Section 3.1, we defined a **term** as an expression that involves only multiplication and/or division with constants and/or variables. Examples of terms are $3x$, $-5y^2$, 17, and $\dfrac{x}{y}$. **Like terms** (or **similar terms**) are terms that contain variables that are of the same power or are constants. We combined like terms using the distributive property. For example,

$$3x + 5x = (3 + 5)x = 8x$$

and

$$14x^3 - 17x^3 = (14 - 17)x^3 = -3x^3$$

A **monomial** is a single term with only whole-number exponents for its variables and no variable in a denominator. The general form of a **monomial in x** is

kx^n **where n is a whole number and k is any number**

n is called the **degree** of the monomial, and k is the **coefficient**.

A monomial may have more than one variable, but only monomials in one variable will be discussed in the remainder of this chapter. More variables will be used in Chapter 5.

In the case of a constant monomial, such as 6, we can write $6 = 6 \cdot 1 = 6 \cdot x^0$. So, we say a nonzero constant is a **monomial of 0 degree**. In the case of 0, we can write $0 = 0x = 0x^2 = 0x^{17}$. Because of all the possible ways of writing 0, we say that 0 is a **monomial of no degree**.

Terms That Are Monomials

$13x^7$	Seventh degree
$\dfrac{1}{2}x^2$	Second degree
$-5y^6$	Sixth degree

Monomials do not have fractional exponents or negative exponents and do not have any variables with positive exponents in a denominator.

Terms That Are Not Monomials

$2x^{-1}$ Negative exponent

$x^{1/2}$ Fractional exponent

$\dfrac{3}{x^2}$ Variable in the denominator

A monomial or algebraic sum of monomials is a **polynomial.** For example, $3x$, $x + 5$, and $a^3 + 5a - 7$ are all polynomials. (**Note:** Since we are dealing only with monomials in one variable, we will also consider only polynomials in one variable. Polynomials such as $5xy^2 + z$ will be discussed later.

The **degree of a polynomial** is the largest of the degrees of its terms after like terms have been combined. Generally, for convenience, a polynomial will be written so that the degrees of its terms decrease from left to right. We say the powers are **descending.** For example,

$$7x^3 - 8 + x^2 - x^4 \qquad \text{Fourth degree}$$

is the same as

$$-x^4 + 7x^3 + x^2 - 8 \qquad \text{Written in descending order, considered the standard form}$$

Some polynomial forms are used so frequently in algebra that they have been given special names, as follows:

Classification of Polynomials		**Examples**
Monomial:	polynomial with one term	$3x$
Binomial:	polynomial with two terms	$x + 5$
Trinomial:	polynomial with three terms	$a^3 + 5a - 7$

EXAMPLES Simplify the following polynomials.

1. $5x^3 + 7x^3$

   Solution $5x^3 + 7x^3 = (5 + 7)x^3 = 12x^3$ Third-degree monomial

2. $5x^3 + 7x^3 - 2x$

   Solution $5x^3 + 7x^3 - 2x = 12x^3 - 2x$ Third-degree binomial

3. $\dfrac{1}{2}y + 3y - \dfrac{2}{3}y^2 - 7$

   Solution

   $$\dfrac{1}{2}y + 3y - \dfrac{2}{3}y^2 - 7 = -\dfrac{2}{3}y^2 + \dfrac{7}{2}y - 7 \qquad \begin{array}{l}\text{Second-degree} \\ \text{trinomial}\end{array}$$

4. $x^2 + 8x - 15 - x^2$

   Solution $x^2 + 8x - 15 - x^2 = 8x - 15$ First-degree binomial ■

Addition with Polynomials

The **sum** of two or more polynomials is found by combining like terms. For example,

$$(x^2 - 5x + 3) + (2x^2 - 8x - 4) + (3x^3 + x^2 - 5)$$
$$= 3x^3 + (x^2 + 2x^2 + x^2) + (-5x - 8x) + (3 - 4 - 5)$$
$$= 3x^3 + 4x^2 - 13x - 6$$

We can also write like terms one beneath the other and add the like terms in each column.

$$
\begin{array}{r}
x^2 - 5x + 3 \\
2x^2 - 8x - 4 \\
3x^3 + x^2 - 5 \\
\hline
3x^3 + 4x^2 - 13x - 6
\end{array}
$$

EXAMPLES

5. Add as indicated.
$$(5x^3 - 8x^2 + 12x + 13) + (-2x^2 - 8) + (4x^3 - 5x + 14)$$
$$= (5x^3 + 4x^3) + (-8x^2 - 2x^2) + (12x - 5x) + (13 - 8 + 14)$$
$$= 9x^3 - 10x^2 + 7x + 19$$

6. Find the sum.
$$
\begin{array}{r}
x^3 - x^2 + 5x \\
4x^3 + 5x^2 - 8x + 9 \\
\hline
5x^3 + 4x^2 - 3x + 9
\end{array}
$$

       ■

Subtraction with Polynomials

If a negative sign is written in front of a polynomial in parentheses, the meaning is the opposite of the entire polynomial. The opposite can be found by changing the sign of every term in the polynomial.

$$-(2x^2 + 3x - 7) = -2x^2 - 3x + 7$$

We can also think of the opposite of a polynomial as -1 times the polynomial.

$$-(2x^2 + 3x - 7) = -1(2x^2 + 3x - 7)$$
$$= -1(2x^2) - 1(3x) - 1(-7)$$
$$= -2x^2 - 3x + 7$$

The result is the same with either approach. So the **difference** between two polynomials can be found by changing the sign of each term of the second polynomial and then combining like terms.

$$(5x^2 - 3x - 7) - (2x^2 + 5x - 8) = 5x^2 - 3x - 7 - 2x^2 - 5x + 8$$
$$= 3x^2 - 8x + 1$$

If the polynomials are written one beneath the other, we change the signs of the terms of the polynomial being subtracted and then combine like terms. Subtract:

$$5x^2 - 3x - 7 \qquad\qquad 5x^2 - 3x - 7$$
$$\underline{-(2x^2 + 5x - 8)} \qquad\qquad \underline{-2x^2 - 5x + 8}$$
$$3x^2 - 8x + 1$$

EXAMPLES

7. Subtract as indicated.

$(9x^4 - 22x^3 + 3x^2 + 10) - (5x^4 + 2x^3 + 5x^2 - x)$
$= 9x^4 - 22x^3 + 3x^2 + 10 - 5x^4 - 2x^3 - 5x^2 + x$
$= 4x^4 - 24x^3 - 2x^2 + x + 10$

8. Find the difference.

$$8x^3 + 5x^2 - 14 \qquad\qquad 8x^3 + 5x^2 + 0x - 14$$
$$\underline{-(-2x^3 + x^2 + 6x)} \qquad\qquad \underline{2x^3 - x^2 - 6x + 0}$$
$$10x^3 + 4x^2 - 6x - 14$$

Write in 0s for missing powers to help with alignment of like terms. ∎

If an expression contains more than one pair of grouping (or inclusion) symbols, such as parentheses ( ), brackets [ ], or braces { }, simplify by removing the innermost pair of symbols first. As examples,

a. $5x - [2x + 3(4 - x) + 1] - 9$
$= 5x - [2x + 12 - 3x + 1] - 9$
$= 5x - [-x + 13] - 9$
$= 5x + x - 13 - 9$
$= 6x - 22$

Work with the parentheses first since they are included inside the brackets.

Brackets second

b. $10 - x + 2[x + 3(x - 5) + 7]$
$= 10 - x + 2[x + 3x - 15 + 7]$
$= 10 - x + 2[4x - 8]$
$= 10 - x + 8x - 16$
$= 7x - 6$

Work with the parentheses first since they are included inside the brackets.

Brackets second

Practice Problems

-3x³ - x² + 4x - 7
- 5x³   - 3x + 4
─────────────────────
-8x³ + x² - 7x + 11

1. Combine like terms and state the degree of the polynomial:
$8x^3 - 3x^2 - x^3 + 5 + 3x^2$

2. Add:
$(15x + 4) + (3x^2 - 9x - 5)$

3. Subtract:
$(-5x^3 - 3x + 4) - (3x^3 - x^2 + 4x - 7)$

4. Simplify:
$2 - [3a - (4 - 7a) + 2a]$

Answers to Practice Problems **1.** $7x^3 + 5$; third degree **2.** $3x^2 + 6x - 1$
3. $-8x^3 + x^2 - 7x + 11$ **4.** $-12a + 6$

EXERCISES 4.3

Simplify the polynomials in Exercises 1–15 and state the degree of the result.

1. $x + 3x$

2. $4x^2 - x + x^2$

3. $x^3 + 3x^2 - 2x$

4. $3x^2 - 8x + 8x$

5. $x^4 - 4x^2 + 2x^2 - x^4$

6. $2 - 6x + 5x - 2$

7. $-x^3 + 6x + x^3 - 6x$

8. $11x^2 - 3x + 2 - 7x^2$

9. $6x^5 + 2x^2 - 7x^5 - 3x^2$

10. $2x^2 - 3x^2 + 2 - 4x^2 - 2 + 5x^2$

11. $4x - 8x^2 + 2x^3 + 8x^2$

12. $2x + 9 - x + 1 - 2x$

13. $5x^2 + 3 - 2x^2 + 1 - 3x^2$

14. $13x^2 - 6x - 9x^2 - 4x$

15. $7x^3 + 3x^2 - 2x^3 + x - 5x^3 + 1$

Add in Exercises 16–35.

16. $(2x^2 + 5x - 1) + (x^2 + 2x + 3)$

17. $(x^2 + 2x - 3) + (x^2 + 5)$

18. $(x^2 + 7x - 7) + (x^2 + 4x)$

19. $(x^2 + 3x - 8) + (3x^2 - 2x + 4)$

20. $(2x^2 - x - 1) + (x^2 + x + 1)$

21. $(3x^2 + 5x - 4) + (2x^2 + x - 6)$

22. $(-2x^2 - 3x + 9) + (3x^2 - 2x + 8)$

23. $(x^2 + 6x - 7) + (3x^2 + x - 1)$

24. $(-4x^2 + 2x - 1) + (3x^2 - x + 2) + (x - 8)$

25. $(8x^2 + 5x + 2) + (-3x^2 + 9x - 4)$

26. $(x^2 + 2x - 1) + (3x^2 - x + 2) + (x - 8)$

27. $(x^3 + 2x - 9) + (x^2 - 5x + 2) + (x^3 - 4x^2 + 1)$

28. $\begin{array}{r} x^2 + 4x - 6 \\ -2x^2 + 3x + 1 \\ \hline \end{array}$

29. $\begin{array}{r} 5x^2 - 3x + 11 \\ -2x^2 + x - 6 \\ \hline \end{array}$

30. $\begin{array}{r} 2x^2 + 4x - 3 \\ 3x^2 - 9x + 2 \\ \hline \end{array}$

31. $\begin{array}{r} x^3 + 3x^2 + x \\ 2x^3 - x^2 + 2x - 4 \\ \hline \end{array}$

? $\begin{array}{r} x^3 + 3x^2 - 4 \\ 7x^2 + 2x + 1 \\ x^3 + x^2 - 6x \\ \hline \end{array}$

33. $\begin{array}{r} x^3 + 5x^2 + 7x - 3 \\ 4x^2 + 3x - 9 \\ 4x^3 + 2x^2 - 2 \\ \hline \end{array}$

34. $\begin{array}{r} 7x^3 + 5x^2 + x - 6 \\ -3x^2 + 4x + 11 \\ -3x^3 - 2x^2 - 5x + 2 \\ \hline \end{array}$

35. $\begin{array}{r} x^3 + 2x^2 - 5 \\ -2x^3 + x - 9 \\ x^3 - 2x^2 + 14 \\ \hline \end{array}$

Subtract in Exercises 36–55.

36. $(2x^2 + 4x + 8) - (x^2 + 3x + 2)$

37. $(3x^2 + 7x - 6) - (x^2 + 2x + 5)$

38. $(x^2 + 8x - 3) - (x^2 + 5x - 7)$

39. $(3x^2 - 5x - 11) - (2x^2 - 4x + 1)$

40. $(x^2 - 9x + 2) - (4x^2 - 3x + 4)$

41. $(2x^2 - x - 10) - (-x^2 + 3x - 2)$

42. $(7x^2 + 4x - 9) - (-2x^2 + x - 9)$

43. $(6x^2 + 11x + 2) - (4x^2 - 2x - 7)$

44. $(10x^2 - 2x + 9) - (-3x^2 + 5x - 2)$

45. $(x^2 - 12x + 3) - (8x^2 - 11x - 6)$

46. $(9x^2 + 6x - 5) - (13x^2 + 6)$

47. $(8x^2 + 9) - (4x^2 - 3x - 2)$

48. $(x^3 + 4x^2 - 7) - (3x^3 + x^2 + 2x + 1)$

49. $\begin{array}{r} 14x^2 - 6x + 9 \\ 8x^2 + x - 9 \\ \hline \end{array}$

50. $\begin{array}{r} 9x^2 - 3x + 2 \\ 4x^2 - 5x - 1 \\ \hline \end{array}$

51. $\begin{array}{r} 11x^2 + 5x - 13 \\ -3x^2 + 5x + 2 \\ \hline \end{array}$

52. $\begin{array}{r} -3x^2 + 7x - 6 \\ 2x^2 - x + 4 \\ \hline \end{array}$

53. $\begin{array}{r} x^3 + 6x^2 - 3 \\ -x^3 + 2x^2 - 3x + 7 \\ \hline \end{array}$

54. $\begin{array}{r} 5x^2 + 8x + 11 \\ -3x^2 + 2x - 4 \\ \hline \end{array}$

55. $\begin{array}{r} 3x^3 + 9x - 17 \\ x^3 + 5x^2 - 2x - 6 \\ \hline \end{array}$

Remove the symbols of inclusion and combine like terms in Exercises 56–70.

56. $5x + 2(x - 3) - (3x + 7)$

57. $-4(x - 6) - (8x + 2)$

58. $11 + [3x - 2(1 + 5x)]$

59. $2x + [9x - 4(3x + 2) - 7]$

60. $8x - [2x + 4(x - 3) - 5]$

61. $17 - [-3x + 6(2x - 3) + 9]$

62. $3x^3 - [5 - 7(x^2 + 2) - 6x^2]$

63. $10x^3 - [8 - 5(3 - 2x^2) - 7x^2]$

64. $(2x^2 + 4) - [-8 + 2(7 - 3x^2) + x]$

65. $-[6x^2 - 3(4 + 2x) + 9] - (x^2 + 5)$

66. $2[3x + (x - 8) - (2x + 5)] - (x - 7)$

67. $-3[-x + (10 - 3x) - (8 - 3x)] + (2x - 1)$

68. $(x^2 - 1) + x[4 + (3 - x)]$

69. $x(x - 5) + [6x - x(4 - x)]$

70. $x(2x + 1) - [5x - x(2x + 3)]$

4.4 Multiplying Polynomials

OBJECTIVE

In this section, you will be learning to multiply polynomials.

We have multiplied terms such as $5x^2 \cdot 3x^4 = 15x^6$. Also, we have applied the distributive property to expressions such as $5(2x + 3) = 10x + 15$. Now we will use both of these procedures to multiply polynomials. We will discuss first the product of a monomial with a polynomial of two or more terms; second, the product of two binomials; and third, the product of a binomial with a polynomial of more than two terms.

Using the distributive property $a(b + c) = ab + ac$ with multiplication indicated on the left, we can find the product of a monomial with a polynomial of two or more terms as follows:

$$5x(2x + 3) = 5x \cdot 2x + 5x \cdot 3 = 10x^2 + 15x$$
$$3x^2(4x - 1) = 3x^2 \cdot 4x + 3x^2(-1) = 12x^3 - 3x^2$$
$$-4a^5(a^2 - 8a + 5) = -4a^5 \cdot a^2 - 4a^5(-8a) - 4a^5(5)$$
$$= -4a^7 + 32a^6 - 20a^5$$

Now suppose that we want to multiply two binomials, say, $(x + 3)(x + 7)$. We will apply the distributive property in the following way with multiplication indicated on the right of the parentheses.

Compare $\qquad (x + 3)(x + 7)$

to $\qquad (a + b)c$

Think of $(x + 7)$ as taking the place of c. Thus,

$$(a + b)c = ac + bc$$

takes the form $\quad (x + 3)(x + 7) = x(x + 7) + 3(x + 7)$

Completing the products on the right, using the distributive property twice again, gives

$$(x + 3)(x + 7) = x(x + 7) + 3(x + 7)$$
$$= x^2 + x \cdot 7 + 3x + 3 \cdot 7$$
$$= x^2 + 7x + 3x + 21$$
$$= x^2 + 10x + 21$$

In the same manner,

$$(x + 2)(3x + 4) = x(3x + 4) + 2(3x + 4)$$
$$= x \cdot 3x + x \cdot 4 + 2 \cdot 3x + 2 \cdot 4$$
$$= 3x^2 + 4x + 6x + 8$$
$$= 3x^2 + 10x + 8$$

Similarly,

$$(2x - 1)(x^2 + x - 5)$$
$$= 2x(x^2 + x - 5) - 1(x^2 + x - 5)$$
$$= 2x \cdot x^2 + 2x \cdot x + 2x(-5) - 1 \cdot x^2 - 1 \cdot x - 1(-5)$$
$$= 2x^3 + 2x^2 - 10x - x^2 - x + 5$$
$$= 2x^3 + x^2 - 11x + 5$$

One quick way to check if your products are correct is to substitute some convenient number for x into the original two factors and into the product. Choose any nonzero number that you like. The values of both expressions should be the same. For example, let $x = 1$. Then

$$(x + 2)(3x + 4) = (1 + 2)(3 \cdot 1 + 4) = (3)(7) = 21$$
$$3x^2 + 10x + 8 = 3 \cdot 1^2 + 10 \cdot 1 + 8 = 3 + 10 + 8 = 21$$

The product $(x + 2)(3x + 4) = 3x^2 + 10x + 8$ seems to be correct. We could double check by letting $x = 5$.

$$(x + 2)(3x + 4) = (5 + 2)(3 \cdot 5 + 4) = (7)(19) = 133$$
$$3x^2 + 10x + 8 = 3 \cdot 5^2 + 10 \cdot 5 + 8 = 75 + 50 + 8 = 133$$

Convinced? This is just a quick check, however, and is not foolproof unless you try more numbers than the degree of the product.

The product of two polynomials can also be found by writing one polynomial under the other. **The distributive property is applied by multiplying each term of one polynomial by each term of the other.** Writing in the usual manner and multiplying from the left, we have

$$(2x^2 + 3x - 4)(3x + 7)$$
$$= (2x^2 + 3x - 4)3x + (2x^2 + 3x - 4)7$$
$$= 2x^2 \cdot 3x + 3x \cdot 3x - 4 \cdot 3x + 2x^2 \cdot 7 + 3x \cdot 7 - 4 \cdot 7$$
$$= 6x^3 + 9x^2 - 12x + 14x^2 + 21x - 28$$
$$= 6x^3 + 23x^2 + 9x - 28$$

Now, writing one under the other and applying the distributive property, we obtain

$$
\begin{array}{r}
2x^2 + 3x - 4 \\
3x + 7 \\
\hline
14x^2 + 21x - 28
\end{array}
$$

Multiply by $+7$.

$$
\begin{array}{r}
2x^2 + 3x - 4 \\
3x + 7 \\
\hline
14x^2 + 21x - 28 \\
6x^3 + 9x^2 - 12x \phantom{{}- 28} \\
\hline
\end{array}
$$

Multiply by $3x$.
Align the like terms so that they can be easily combined.

$(3a+4)(2a^2+a+5)$

$6a^3+3a^2+15a+8a^2+4a+20$

$a^3+11a^2+19a+20$

$$2x^2 + 3x - 4$$
$$\underline{ 3x + 7}$$
$$14x^2 + 21x - 28$$
$$\underline{6x^3 + 9x^2 - 12x}$$
$$6x^3 + 23x^2 + 9x - 28 \qquad \text{Combine like terms.}$$

EXAMPLES Find each product.

1. $-4x(x^2 - 3x + 12)$

   Solution $-4x(x^2 - 3x + 12) = -4x \cdot x^2 - 4x(-3x) - 4x \cdot 12$
   $$= -4x^3 + 12x^2 - 48x$$

$3a+4$

$2a^2+a+5$

3

$2a^2+a+5$

$\underline{ 3a+4}$

$8a^2+4a+20$

$a^3 \quad 3a+15a$

$\underline{a^3+11a^2+19a+20}$

$(x-5)(x+2)(x-1)$

$x^2+2x-5x-10(x-1)$

$x^2-3x-10(x-1)$

$x^3-x^2-3x^2+3x-10x+10$

$x^3-4x^2-7x+10$

2. $(3a + 4)(2a^2 + a + 5)$

   Solution

   $(3a + 4)(2a^2 + a + 5) = (3a + 4)2a^2 + (3a + 4)a + (3a + 4)5$
   $$= 6a^3 + 8a^2 + 3a^2 + 4a + 15a + 20$$
   $$= 6a^3 + 11a^2 + 19a + 20$$

3. $7y^2 - 3y + 2$
   $ 2y + 3$

   Solution
   $$7y^2 - 3y + 2$$
   $$\underline{ 2y + 3}$$
   $$+ 21y^2 - 9y + 6 \qquad \text{Multiply by 3.}$$
   $$\underline{14y^3 - 6y^2 + 4y} \qquad \text{Multiply by } 2y.$$
   $$14y^3 + 15y^2 - 5y + 6 \qquad \text{Combine like terms.}$$

4. $(x - 5)(x + 2)(x - 1)$

   Solution First multiply $(x - 5)(x + 2)$, then multiply this result by $(x - 1)$.

   $$(x - 5)(x + 2) = x(x + 2) - 5(x + 2)$$
   $$= x^2 + 2x - 5x - 10$$
   $$= x^2 - 3x - 10$$

   $$(x^2 - 3x - 10)(x - 1) = (x^2 - 3x - 10)x + (x^2 - 3x - 10)(-1)$$
   $$= x^3 - 3x^2 - 10x - x^2 + 3x + 10$$
   $$= x^3 - 4x^2 - 7x + 10 \qquad \blacksquare$$

Practice Problems	Find each product.

1. $2x(3x^2 + x - 1)$ 　　 2. $(x + 3)(x - 7)$

3. $(x - 1)(x^2 + x - 4)$ 　　 4. $(x + 1)(x - 2)(x + 2)$

$6x^3+2x^2-2x$

$x^3+x^2-4x-x^2-x+4$

x^3-5x+4

$x^2-7x+3x-21$

$x^2-4x-21$

$x^2-2x+x-2$

Answers to Practice Problems 1. $6x^3 + 2x^2 - 2x$ 2. $x^2 - 4x - 21$
3. $x^3 - 5x + 4$ 4. $x^3 + x^2 - 4x - 4$

$x^3-4x-4+x^2$

$x^2-2x+x-2$

$x^2-x-2(x+2)$

$x^3+2x^2-x^2-2x-2x-4$

EXERCISES 4.4

Multiply as indicated in Exercises 1–45 and simplify if possible.

1. $-3x^2(-2x^3)$ **2.** $4x^5(x^2)$ **3.** $5x(-4x^2)$

4. $9x^3(2x)$ **5.** $-1(3x^2 + 2x)$ **6.** $-7(2x^2 + 3x)$

7. $-4x(x^2 + 1)$ **8.** $x^2(x^2 + 4x)$ **9.** $7x^2(x^2 + 2x - 1)$

10. $5x^3(5x^2 - x + 2)$ **11.** $-x(x^3 + 5x - 4)$ **12.** $-2x^4(x^3 - x^2 + 2x)$

13. $3x(2x + 1) - 2(2x + 1)$ **14.** $x(3x + 4) + 7(3x + 4)$ **15.** $3x(3x - 5) + 5(3x - 5)$

16. $6x(x - 1) + 5(x - 1)$ **17.** $5x(-2x + 7) - 2(-2x + 7)$

18. $x(x^2 + 1) - 1(x^2 + 1)$ **19.** $x(x^2 + 3x + 2) + 2(x^2 + 3x + 2)$

20. $4x(x^2 - x + 1) + 3(x^2 - x + 1)$ **21.** $(x + 4)(x - 3)$

22. $(x + 7)(x - 5)$ **23.** $(x + 6)(x - 8)$ **24.** $(x + 2)(x - 4)$

25. $(x - 2)(x - 1)$ **26.** $(x - 7)(x - 8)$ **27.** $3(x + 4)(x - 5)$

28. $-4(x + 6)(x - 7)$ **29.** $x(x + 3)(x + 8)$ **30.** $x(x - 4)(x - 7)$

31. $(2x + 1)(x - 4)$ **32.** $(3x - 1)(x + 4)$ **33.** $(6x - 1)(x + 3)$

34. $(3x + 5)(3x - 5)$ **35.** $(2x + 3)(2x - 3)$ **36.** $(8x + 15)(x + 1)$

37. $(4x + 1)(4x + 1)$ **38.** $(5x - 2)(5x - 2)$ **39.** $(x + 3)(x^2 - x + 4)$

40. $(2x + 1)(x^2 - 7x + 2)$ **41.** $3x + 7$ **42.** $x^2 + 3x + 1$

                                           $\underline{x - 5}$ $\underline{5x - 9}$

43. $8x^2 + 3x - 2$ **44.** $2x^2 + 3x + 5$ **45.** $6x^2 - x + 8$

        $\underline{\quad - 2x + 7}$ $\underline{x^2 + 2x - 3}$ $\underline{2x^2 + 5x + 6}$

Find the product in Exercises 46–70 and simplify if possible. Check by letting $x = 2$.

46. $(3x - 4)(x + 2)$ **47.** $(x + 6)(4x - 7)$ **48.** $(2x + 5)(x - 1)$ **49.** $(5x - 3)(x + 4)$

50. $(2x + 1)(3x - 8)$ **51.** $(x - 2)(3x + 8)$ **52.** $(7x + 1)(x - 2)$ **53.** $(3x + 7)(2x - 5)$

54. $(2x + 3)(2x + 3)$ **55.** $(5x + 2)(5x + 2)$ **56.** $(x + 3)(x^2 - 4)$ **57.** $(x^2 + 2)(x - 4)$

58. $(2x + 7)(2x - 7)$ **59.** $(3x - 4)(3x + 4)$ **60.** $(x + 1)(x^2 - x + 1)$

61. $(x - 2)(x^2 + 2x + 4)$ **62.** $(7x - 2)(7x - 2)$ **63.** $(5x - 6)(5x - 6)$

64. $(2x + 3)(x^2 - x - 1)$ **65.** $(3x + 1)(x^2 - x + 9)$ **66.** $(x + 1)(x + 2)(x + 3)$

67. $(x - 1)(x - 2)(x - 3)$ **68.** $(x^2 + x - 1)(x^2 - x + 1)$ **69.** $(x^2 + x + 2)(x^2 + x - 2)$

70. $(x^2 + 3x + 2)^2$

4.5 Special Products of Binomials

In Section 4.4, we emphasized the use of the distributive property in finding products of polynomials. This approach is correct and serves to provide a solid understanding of multiplication. However, there is a more efficient way to multiply binomials used by most students. After some practice, you can probably find these products mentally.

The FOIL Method

Consider finding the product $(2x + 3)(5x + 2)$ using the distributive property, then investigating the positions of the terms in the product.

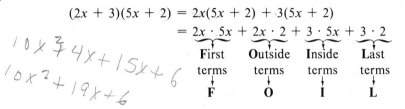

$$(2x + 3)(5x + 2) = 2x(5x + 2) + 3(5x + 2)$$

$$= \underbrace{2x \cdot 5x}_{\substack{\text{First} \\ \text{terms} \\ \downarrow \\ \textbf{F}}} + \underbrace{2x \cdot 2}_{\substack{\text{Outside} \\ \text{terms} \\ \downarrow \\ \textbf{O}}} + \underbrace{3 \cdot 5x}_{\substack{\text{Inside} \\ \text{terms} \\ \downarrow \\ \textbf{I}}} + \underbrace{3 \cdot 2}_{\substack{\text{Last} \\ \text{terms} \\ \downarrow \\ \textbf{L}}}$$

This technique, illustrated by the following diagramed equations, is called the **FOIL method** of mentally multiplying two binomials.

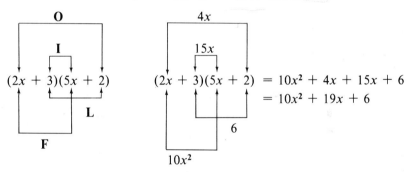

The same layout is shown again to help you get a mental image of the process.

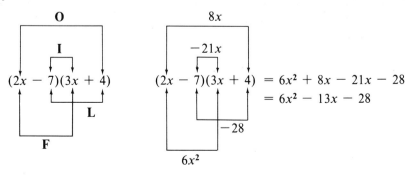

EXAMPLES   Use the FOIL method to find each product of binomials.

1. $(2x + 1)(4x + 3)$

Solution

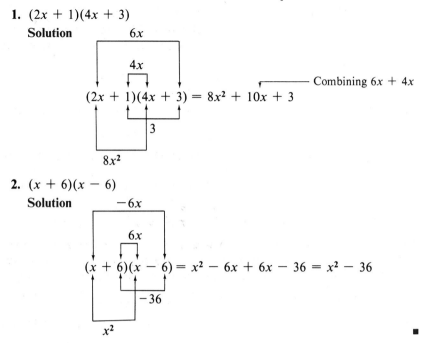

2. $(x + 6)(x - 6)$

Solution

$$(x + 6)(x - 6) = x^2 - 6x + 6x - 36 = x^2 - 36$$

∎

The Difference of Two Squares:
$(x + a)(x - a) = x^2 - a^2$

In Example 2, the middle terms, $-6x$ and $+6x$, are opposites of each other and their sum is 0. Therefore, the resulting product has only two terms.

$$(x + 6)(x - 6) = x^2 - 36$$

The simplified product is in the form of a difference and both terms are squares. Thus, we have the following special case called the **difference of two squares.**

$$(x + a)(x - a) = x^2 - a^2$$

When the two binomials are in the form of the sum and difference of the same two terms, the product will always be the difference of the squares of the terms. In such a case, we can write the answer directly with no calculations.

The numbers

$$0, 1, 4, 9, 16, 25, 36, 49, 64, \text{ and so on}$$

are called **perfect squares** since they are the squares of integers.

EXAMPLES   Find each product.

3. $(x + 5)(x - 5)$

Solution   The two binomials represent the sum and difference of x and 5. So the product is the difference of their squares.

$$(x + 5)(x - 5) = x^2 - 5^2 = x^2 - 25$$

4. $(8x + 3)(8x - 3)$

Solution $(8x + 3)(8x - 3) = (8x)^2 - (3)^2 = 64x^2 - 9$

5. $(x^3 + 2)(x^3 - 2)$

Solution $(x^3 + 2)(x^3 - 2) = (x^3)^2 - 2^2 = x^6 - 4$ ■

Perfect Square Trinomials:
$(x + a)^2 = x^2 + 2ax + a^2$
$(x - a)^2 = x^2 - 2ax + a^2$

We now want to consider the case where the two binomials being multiplied are the same. That is, we want to consider the **square of a binomial.**

As the following discussion concludes, there is a pattern that, after some practice, allows us to go directly to the product.

$$(x + 5)^2 = (x + 5)(x + 5) = x^2 + 5x + 5x + 25$$
$$= x^2 + 2 \cdot 5x + 25$$
$$= x^2 + 10x + 25$$
$$(x + 7)^2 = (x + 7)(x + 7) = x^2 + 2 \cdot 7x + 49$$
$$= x^2 + 14x + 49$$
$$(x + 10)^2 = (x + 10)(x + 10) = x^2 + 20x + 100$$

and the basic pattern is

$$(x + a)^2 = x^2 + 2ax + a^2$$

The expression $x^2 + 2ax + a^2$ is called a **perfect square trinomial.**

One interesting device for remembering the result of squaring a binomial is the square shown in Figure 4.2.

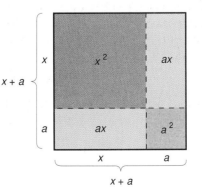

For the area of the square,

$$(x + a)^2 = x^2 + ax + ax + a^2$$
$$= x^2 + 2ax + a^2$$

Figure 4.2

Another **perfect square trinomial** results if an expression of the form $(x - a)$ is squared. The sign between x and a is $-$ instead of $+$.

$$(x - 5)(x - 5) = x^2 - 5x - 5x + 25$$
$$= x^2 - 2 \cdot 5x + 25$$
$$= x^2 - 10x + 25$$
$$(x - 7)(x - 7) = x^2 - 2 \cdot 7x + 49$$
$$= x^2 - 14x + 49$$
$$(x - 10(x - 10) = x^2 - 20x + 100$$

In general,

$$(x - a)^2 = x^2 - 2ax + a^2$$

EXAMPLES Find the following products.

6. $(2x + 3)^2$

Solution The pattern for squaring a binomial gives

$$(2x + 3)^2 = (2x)^2 + 2 \cdot 3 \cdot 2x + (3)^2$$
$$= 4x^2 + 12x + 9$$

7. $(5x - 1)^2$

Solution $(5x - 1)^2 = (5x)^2 - 2(1)(5x) + (1)^2$
$$= 25x^2 - 10x + 1$$

8. $(9 - x)^2$

Solution $(9 - x)^2 = (9)^2 - 2(9)(x) + x^2$
$$= 81 - 18x + x^2$$

9. $(y^3 + 1)^2$

Solution $(y^3 + 1)^2 = (y^3)^2 + 2(1)(y^3) + 1^2$
$$= y^6 + 2y^3 + 1$$ ■

Common Error

Many beginning algebra students make the following error:

$(x + a)^2 = x^2 + a^2$
$(x + 6)^2 = x^2 + 36$ ⟶ **WRONG**

Avoid this error by remembering that the square of a binomial is a trinomial.

$(x + 6)^2 = x^2 + 2 \cdot 6x + 36$
$= x^2 + 12x + 36$ ⟵——— **RIGHT**

Practice Problems   Find the indicated products.

1. $(x + 10)(x - 10)$ **2.** $(x + 3)^2$

3. $(2x - 1)(x + 3)$ **4.** $(2x - 5)^2$

5. $(x^2 + 4)(x^2 - 3)$

◢ EXERCISES 4.5

Write the products of Exercises 1–30 and identify those that are the difference of two squares or perfect square trinomials.

1. $(x + 3)(x - 3)$ **2.** $(x - 7)^2$ **3.** $(x - 5)^2$ **4.** $(x + 4)(x + 4)$

5. $(x - 6)(x + 6)$ **6.** $(x + 9)(x - 9)$ **7.** $(x + 8)(x + 8)$ **8.** $(x + 12)(x - 12)$

9. $(2x + 3)(x - 1)$ **10.** $(3x + 1)(2x + 5)$ **11.** $(3x - 4)^2$ **12.** $(5x + 2)(5x - 2)$

13. $(2x + 1)(2x - 1)$ **14.** $(3x + 1)^2$ **15.** $(3x - 2)(3x - 2)$ **16.** $(4x + 5)(4x - 5)$

17. $(3 + x)^2$ **18.** $(8 - x)(8 - x)$ **19.** $(5 - x)(5 - x)$ **20.** $(11 - x)(11 + x)$

21. $(5x - 9)(5x + 9)$ **22.** $(4 - x)^2$ **23.** $(2x + 7)(2x + 7)$ **24.** $(3x + 2)^2$

25. $(9x + 2)(9x - 2)$ **26.** $(6x + 5)(6x - 5)$ **27.** $(5x^2 + 2)(2x^2 - 3)$ **28.** $(4x^2 + 7)(2x^2 + 1)$

29. $(1 + 7x)^2$ **30.** $(2 - 5x)^2$

Write the indicated products in Exercises 31–58.

31. $(x + 2)(5x + 1)$ **32.** $(7x - 2)(x - 3)$ **33.** $(4x - 3)(x + 4)$ **34.** $(x + 11)(x - 8)$

35. $(3x - 7)(x - 6)$ **36.** $(x + 7)(2x + 9)$ **37.** $(5 + x)(5 + x)$ **38.** $(3 - x)(6 - x)$

39. $(x^2 + 1)(x^2 - 1)$ **40.** $(x^2 + 5)(x^2 - 5)$ **41.** $(x^2 + 3)(x^2 + 3)$ **42.** $(x^2 - 4)^2$

43. $(x^3 - 2)^2$ **44.** $(x^3 + 8)(x^3 - 8)$ **45.** $(x^2 - 6)(x^2 + 9)$ **46.** $(x^2 + 3)(x^2 - 5)$

47. $\left(x + \dfrac{2}{3}\right)\left(x - \dfrac{2}{3}\right)$ **48.** $\left(x - \dfrac{1}{2}\right)\left(x + \dfrac{1}{2}\right)$ **49.** $\left(x + \dfrac{3}{4}\right)\left(x - \dfrac{3}{4}\right)$ **50.** $\left(x + \dfrac{3}{8}\right)\left(x - \dfrac{3}{8}\right)$

51. $\left(x + \dfrac{3}{5}\right)\left(x + \dfrac{3}{5}\right)$ **52.** $\left(x + \dfrac{4}{3}\right)\left(x + \dfrac{4}{3}\right)$ **53.** $\left(x - \dfrac{5}{6}\right)^2$ **54.** $\left(x - \dfrac{2}{7}\right)^2$

55. $\left(x + \dfrac{1}{4}\right)\left(x - \dfrac{1}{2}\right)$ **56.** $\left(x - \dfrac{1}{5}\right)\left(x + \dfrac{2}{3}\right)$ **57.** $\left(x + \dfrac{1}{3}\right)\left(x + \dfrac{1}{2}\right)$ **58.** $\left(x - \dfrac{4}{5}\right)\left(x - \dfrac{3}{10}\right)$

Calculator Problems

Write the indicated products in Exercises 59–71.

59. $(x + 1.4)(x - 1.4)$ **60.** $(x - 2.1)(x + 2.1)$ **61.** $(x - 2.5)^2$

62. $(x + 1.7)^2$ **63.** $(x + 2.15)(x - 2.15)$ **64.** $(x + 1.36)(x - 1.36)$

65. $(x + 1.24)^2$ **66.** $(x - 2.45)^2$ **67.** $(1.42x + 9.6)^2$

68. $(0.46x - 0.71)^2$ **69.** $(11.4x + 3.5)(11.4x - 3.5)$ **70.** $(2.5x + 11.4)(1.3x - 16.9)$

71. $(12.6x - 6.8)(7.4x + 15.3)$

Answers to Practice Problems **1.** $x^2 - 100$ **2.** $x^2 + 6x + 9$ **3.** $2x^2 + 5x - 3$
4. $4x^2 - 20x + 25$ **5.** $x^4 + x^2 - 12$

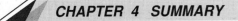 **CHAPTER 4 SUMMARY**

Key Terms and Formulas

In **scientific notation,** decimal numbers are written as the product of a number between 1 and 10 and some power of 10. [4.2]

The general form of a **monomial in x** is

kx^n where n is a whole number and k is any number [4.3]

n is called the **degree** of the monomial and k is the **coefficient.** [4.3]

A nonzero constant is a **monomial of 0 degree.** [4.3]

Zero is a monomial of **no degree.** [4.3]

A monomial or algebraic sum of monomials is a **polynomial.** [4.3]

The **degree of a polynomial** is the largest of the degrees of its terms after like terms have been combined. [4.3]

Classification of Polynomials [4.3]

Monomial:	polynomial with one term
Binomial:	polynomial with two terms
Trinomial:	polynomial with three terms

Difference of Two Squares [4.5]
$$(x + a)(x - a) = x^2 - a^2$$

Perfect Square Trinomials [4.5]
$$(x + a)^2 = x^2 + 2ax + a^2$$
$$(x - a)^2 = x^2 - 2ax + a^2$$

Properties and Rules

Properties of Exponents

For nonzero real numbers a and b and integers m and n,

1. $a^m \cdot a^n = a^{m+n}$ [4.1]

2. $a^0 = 1$ [4.1]

3. $a^{-n} = \dfrac{1}{a^n}$ [4.1]

4. $\dfrac{a^m}{a^n} = a^{m-n}$ [4.1]

5. $(ab)^n = a^n b^n$ [4.2]

6. $\left(\dfrac{a}{b}\right)^n = \dfrac{a^n}{b^n}$ [4.2]

7. $(a^m)^n = a^{mn}$ [4.2]

The expression 0^0 is undefined. [4.1]

Procedures

The **sum** of two or more polynomials is found by combining like terms. [4.3]

The **difference** of two polynomials is found by changing the sign of each term of the second polynomial and then combining like terms. [4.3]

The **product** of two polynomials can be found by applying the distributive property in the form
$(a + b)c = ac + bc$. [4.4]

The **FOIL method** can be used to find the product of two binomials. [4.5]

CHAPTER 4 REVIEW

Simplify each expression in Exercises 1–20. Expressions with variables should contain positive exponents in the answers. [4.1–4.2]

1. $6^4 \cdot 6^3$

2. $(-3)^3 \cdot (-3)^2$

3. $\dfrac{7^4}{7^5}$

4. $\dfrac{5^3 \cdot 5^0}{5^4}$

5. $y^3 \cdot y^2$

6. $y^{-3} \cdot y^4$

7. $\dfrac{2x^3}{x^{-1}}$

8. $\dfrac{2y^{-5}}{3y^{-7}}$

9. $\dfrac{x^0 \cdot x^{-4}}{x^3}$

10. $\dfrac{4x^3}{2x^{-2}x^4}$

11. $(4x^2y)^3$

12. $(7x^5y^{-2})^2$

13. $\left(\dfrac{6x^2}{y^5}\right)^2$

14. $\dfrac{x^{-3}}{x^3y^2}$

15. $(a^{-3}b^2)^{-2}$

16. $\left(\dfrac{3xy^4}{x^2}\right)^{-1}$

17. $\left(\dfrac{2x^2y^0}{x^{-1}y}\right)^{-3}$

18. $\left(\dfrac{8x^{-3}y^2}{xy^{-1}}\right)^0$

19. $\left(\dfrac{x^3y^{-1}}{xy^{-2}}\right)^2$

20. $\left(\dfrac{3^{-1}x^3y^{-1}}{x^{-1}y^2}\right)^2$

Write each of the expressions in Exercises 21–25 in scientific notation. [4.2]

21. 4,270,000

22. 0.00023

23. 0.0015×4200

24. $\dfrac{840}{0.00021}$

25. $\dfrac{0.005 \times 66}{0.011 \times 250}$

Simplify the polynomials in Exercises 26–31 and state the degree of the result. [4.3]

26. $7x^2 - 2x^2 + 1$

27. $3x + 4x^2 + 6x$

28. $-x^2 + 5x + 2x^2 - x$

29. $9x - x^3 + 3x^2 - x + x^3$

30. $-x^3 + 4x^2 - 3 + x - x^2 + 1$

31. $8x - 7x^2 + x^2 - x^3 - 6x - 4$

Add or subtract as indicated in Exercises 32–39. [4.3]

32. $(7x^2 + 2x - 1) + (8x - 4)$

33. $(6x^2 - 5x + 3) + (-4x^2 - 3x + 9)$

34. $(-2x^2 - 11x + 1) + (x^3 + 3x - 7) + (2x - 1)$

35. $(4x^2 + 2x - 7) - (5x^2 + x - 2)$

36. $(x^3 + 4x^2 - x) - (-2x^3 + 6x + 3)$

37. $(6x^2 + x - 10) - (x^3 - x^2 + x - 4)$

38. $(x^2 + 2x + 6) + (5x^2 - x - 2) - (8x + 3)$

39. $(2x^2 - 5x - 7) - (3x^2 - 4x + 1) + (x^2 - 9)$

Remove the symbols of inclusion in Exercises 40–45 and combine like terms. [4.3]

40. $3x(x + 4) - 5(x^2 + 3x)$

41. $-2[7x - (2x + 5) + 3]$

42. $6x - [9 - 2(3x - 1) + 7x]$

43. $2x^2 + x[4x - (8x - 3) + x(2x + 7)]$

44. $5 - 3[4x - (7 - 2x)]$

45. $3x + 2[x - 3(x + 5) + 4(x - 1)]$

Find the products in Exercises 46–55 and identify those products that are the difference of squares or perfect square trinomials. [4.4–4.5]

46. $-3x(x^2 - 4)$

47. $x^2(x - x^3)$

48. $(x + 6)(x - 6)$

49. $(x + 4)(x - 3)$

50. $(3x + 7)(3x + 7)$

51. $(2x - 1)(2x + 1)$

52. $(x^2 + 5)(x^2 - 5)$

53. $(x^2 - 2)(x^2 - 2)$

54. $\left(x + \dfrac{2}{5}\right)\left(x - \dfrac{2}{5}\right)$

55. $\left(x + \dfrac{5}{8}\right)\left(x + \dfrac{5}{8}\right)$

Find the products in Exercises 56–65 and check by letting $x = 3$. [4.4]

56. $-1(x^2 - 5x + 2)$

57. $3x(x^2 + 2x - 1)$

58. $(3x + 2)(4x - 7)$

59. $(2x - 9)(x + 4)$

60. $(x - 6)(5x + 3)$

61. $(3x - 4)(2x + 3)$

62. $(x^2 + 4)(x^2 - 4)$

63. $(x - 6)(x^2 + 8)$

64. $(4x + 1)(x^2 - x)$

65. $(5x - 2)(x^2 + x - 2)$

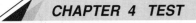

CHAPTER 4 TEST

Simplify each expression in Exercises 1–6 so that it has only positive exponents.

1. $(4x^4)(-3x^2)$

2. $\dfrac{8x^5 \cdot x^2}{4x^{-2}}$

3. $(5x^{-2}y^3)^2$

4. $\dfrac{3x^{-2}}{9x^{-3}y^2}$

5. $\left(\dfrac{4xy^2}{x^3}\right)^{-1}$

6. $\left(\dfrac{2x^0y^3}{x^{-1}y}\right)^2$

Write the expressions in Exercises 7 and 8 in scientific notation.

7. 230×0.005

8. $\dfrac{65 \times 0.012}{150}$

Simplify the polynomials in Exercises 9 and 10 and state the degree of the result.

9. $3x + 4x^2 - x^3 + 4x^2 + x^3$

10. $2x^2 + 3x - x^3 + x^2 - 1$

Perform the indicated operations and simplify the results in Exercises 11–16.

11. $(x^2 + 3x - 8) + (2x^2 - x + 1)$

12. $(-x^2 + 9x - 6) + (3x^2 - 2)$

13. $(5x^3 - 2x + 7) + (-x^2 + 8x - 2)$

14. $(4x^2 + 3x - 1) - (6x^2 + 2x + 5)$

15. $(x^2 + 3x + 9) - (-6x^2 - 11x + 5)$

16. $(3x^3 - 2x) - (4x^2 + 3x - 8)$

Remove the symbols of inclusion in Exercises 17–19 and combine like terms.

17. $7x + [2x - 3(4x + 1) + 5]$

18. $12x - 2[5 - (7x + 1) + 3x]$

19. $4x - [(6x + 7) - (3x + 4)] - 6$

Find the product in Exercises 20–25.

20. $5x^2(-3x^2 + 9x)$

21. $(5x + 4)(5x - 4)$

22. $(4 - 3x)^2$

23. $3x(x - 7)(2x + 5)$

24. $(x^2 + 9)(x^2 - 9)$

25. $\left(x - \dfrac{4}{5}\right)^2$

CUMULATIVE REVIEW (4)

Use the distributive property to complete the expression in Exercises 1 and 2.

1. $3x + 45 = 3(\quad)$

2. $6x + 16 = 2(\quad)$

List the whole numbers that are factors of the number in Exercises 3 and 4.

3. 54

4. 24

For each pair of numbers in Exercises 5–8, find two factors of the first number whose sum is the second number.

5. 32, 18

6. 36, −13

7. −56, 10

8. −48, −8

Combine like terms in Exercises 9 and 10.

9. $2(3x^2 + 5x) - (x^2 - 6x)$

10. $4(2x - 3) - 3(x + 2)$

Solve the equations in Exercises 11 and 12.

11. $7(4 - x) = 3(x + 4)$

12. $-2(5x + 1) + 2x = 4(x + 1)$

Simplify each expression in Exercises 13 and 14 so that it has only positive exponents.

13. $\dfrac{12x^4 \cdot x^3}{3x^2}$

14. $\dfrac{9x^5y^4}{27xy^2}$

Find the product in Exercises 15–18.

15. $2x^2(7x + 9)$

16. $(3x + 8)(3x - 8)$

17. $(2x - 9)^2$

18. $(4x + 3)(2x - 5)$

19. Find two consecutive even integers whose sum is 134.

20. The length of a rectangle is 9 centimeters more than its width. The perimeter is 82 centimeters. Find the dimensions of the rectangle.

5 FACTORING POLYNOMIALS AND SOLVING EQUATIONS

◢ DID YOU KNOW ?

You have noticed by now that almost every algebraic skill somehow relates to equation solving and applied problems. The emphasis on equation solving has always been a part of classical algebra.

In Italy, during the Renaissance, it was the custom for one mathematician to challenge another mathematician to an equation-solving contest. A large amount of money, often in gold, was supplied by patrons or sponsoring cities as the prize. At that time, it was important not to publish equation-solving methods, since mathematicians could earn large amounts of money if they could solve problems that their competitors could not. Equation-solving techniques were passed down from a mathematician to an apprentice, but they were never shared.

A Venetian mathematician, Niccolò Fontana (1500?–57), known as Tartaglia, "the stammerer," discovered how to solve third-degree or cubic equations. At that time, everyone could solve first- and second-degree equations and special kinds of equations of higher degree. Tartaglia easily won equation-solving contests simply by giving his opponents third-degree equations to solve.

Tartaglia planned to keep his method secret, but, after receiving a pledge of secrecy, he gave his method to Girolamo Cardano (1501–76). Cardano broke his promise by publishing one of the first successful Latin algebra texts, *Ars Magna,* "The Great Art." In it, he included not only Tartaglia's solution to the third-degree equations but also a pupil's (Ferrari) discovery of the general solution to fourth-degree equations. Until recently, Cardano received credit for discovering both methods.

It was not until 300 years later that it was shown that there are no general algebraic methods for solving fifth- or higher-degree equations. Thus, a great deal of time and energy has gone into developing the methods of equation solving that you are learning.

Algebra is the intellectual instrument which has been created for rendering clear the quantitative aspect of the world.

Alfred North Whitehead (1861–1947)

CHAPTER OUTLINE

*F*actoring is the reverse of multiplication. That is, to factor polynomials, you need to remember how you multiplied them. In this way, the concept of factoring is built on your previous knowledge and skills with multiplication. For example, if you are given a product, such as $x^2 - a^2$ (the difference of two squares), you must recall the factors of $(x + a)$ and $(x - a)$ from your work in multiplying polynomials in Chapter 4.

Studying mathematics is a building process with each topic dependent on previous topics with a few new ideas added each time. The equations and applications in Chapter 5 involve many of the concepts studied earlier, yet you will find them a step higher and more interesting and more challenging.

5.1 Greatest Common Factor and Solving Equations

OBJECTIVES

In this section, you will be learning to:

1. Find the quotients of two monomials.

2. Factor polynomials by finding the greatest common monomial factor.

3. Factor polynomials by grouping.

4. Solve equations by factoring common monomials.

In Chapter 4, we developed techniques using the distributive property and the FOIL method for multiplying polynomials. The result of multiplication is called the **product.** The expressions, or numbers, being multiplied are called **factors** of the product. For example,

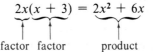

$$2x(x + 3) = 2x^2 + 6x$$

factor factor   product

Now we want to reverse this process. That is, given the product, we want to find the factors. This is called **factoring** and is very useful in solving equations and simplifying algebraic fractions. In this chapter, we will discuss several techniques for factoring based on the skills developed for multiplying polynomials in Chapter 4.

Common Monomial Factors

From our work with exponents, we know the property that $\dfrac{a^m}{a^n} = a^{m-n}$. This property, along with reducing fractions and/or dividing numbers, can be used when dividing two monomials. For example,

$$\frac{35x^8}{5x^2} = 7x^6 \quad \text{and} \quad \frac{16a^5}{-8a} = -2a^4$$

To divide a polynomial by a monomial, a procedure that will be discussed in more detail in Chapter 6, we divide each term in the polynomial by the monomial. For example,

$$\frac{8x^3 - 14x^2 + 10x}{2x} = \frac{8x^3}{2x} - \frac{14x^2}{2x} + \frac{10x}{2x}$$
$$= 4x^2 - 7x + 5$$

This concept of dividing each term by a monomial is part of finding a monomial factor of a polynomial. Finding the greatest common monomial factor in a polynomial means to **choose the monomial with the highest degree and the largest integer coefficient that will divide into each term of the polynomial.**

This monomial will be one factor, and the sum of the various quotients will be the other factor. For example, factor

$$8x^3 - 14x^2 + 10x$$

We have already seen that $2x$ will divide into each term. Note carefully that this division yields **integer** coefficients. The factors of $8x^3 - 14x^2 + 10x$ are $2x$ and $4x^2 - 7x + 5$. We say that $2x$ is **factored out** and we write

$$8x^3 - 14x^2 + 10x = 2x(4x^2 - 7x + 5)$$

Now factor $24x^6 - 12x^4 - 18x^3$. Look closely at each term. Each term is divisible by several different monomials, such as x, x^2, $3x^2$, $2x^3$, and $6x^3$. We want the one with the largest coefficient and the largest exponent, namely, $6x^3$. Thus, we factor out $6x^3$ as follows:

$$24x^6 - 12x^4 - 18x^3 = 6x^3 \cdot 4x^3 + 6x^3(-2x) + 6x^3(-3)$$
$$= 6x^3(4x^3 - 2x - 3)$$

$$\left(\text{Note: } \frac{24x^6}{6x^3} = 4x^3, \quad \frac{-12x^4}{6x^3} = -2x, \quad \frac{-18x^3}{6x^3} = -3 \right)$$

With practice, all this division can be done mentally.

If all the terms are negative or if the leading term (the term of highest degree) is negative, we will choose the greatest common factor to be negative. This will leave a positive coefficient for the first term in parentheses. We can also think of factoring out a monomial as using the distributive property in a sort of reverse sense.

$$ab + ac = a(b + c) \qquad \text{where } a \text{ is the greatest common factor of the two terms } ab \text{ and } ac$$

EXAMPLES   Factor each polynomial by finding the greatest common monomial factor.

1. $x^3 - 7x$
   Solution $x^3 - 7x = x \cdot x^2 + x(-7) = x(x^2 - 7)$

2. $5x^3 - 15x^2$
   Solution $5x^3 - 15x^2 = 5x^2 \cdot x + 5x^2(-3) = 5x^2(x - 3)$

3. $2x^4 - 3x^2 + 1$
   Solution $2x^4 - 3x^2 + 1$ No common monomial factor

4. $-4a^5 + 2a^3 - 6a^2$

    Solution $-4a^5 + 2a^3 - 6a^2 = 2a^2(-2a^3 + a - 3)$

        or, factoring $-2a^2$,

            $-4a^5 + 2a^3 - 6a^2 = -2a^2(2a^3 - a + 3)$

(Here a negative term was factored out because the leading term was negative. Either answer is correct, but the second is preferred.)   ■

    A polynomial may be **in more than one variable.** For example, $5x^2y + 10xy^2$ is in the two variables x and y. Thus, a common monomial factor may have more than one variable.

$$5x^2y + 10xy^2 = 5xy \cdot x + 5xy \cdot 2y$$
$$= 5xy(x + 2y)$$

Similarly,

$$4xy^3 - 2x^2y^2 + 8xy^2 = 2xy^2 \cdot 2y + 2xy^2(-x) + 2xy^2 \cdot 4$$
$$= 2xy^2(2y - x + 4)$$

$$\left(\textbf{Note: } \frac{4xy^3}{2xy^2} = 2y, \quad \frac{-2x^2y^2}{2xy^2} = -x, \quad \frac{8xy^2}{2xy^2} = 4 \right)$$

EXAMPLES   Factor each polynomial by finding the greatest common monomial factor.

5. $4ax^3 + 4ax$

    Solution $4ax^3 + 4ax = 4ax(x^2 + 1)$ Note that $4ax = 1 \cdot 4ax$.

6. $3x^2y^2 - 6xy^2$

    Solution $3x^2y^2 - 6xy^2 = 3xy^2(x - 2)$

7. $14by^3 + 7b^2y - 21by^2$

    Solution $14by^3 + 7b^2y - 21by^2 = 7by(2y^2 + b - 3y)$

8. $-2x^2y^3 + 4xy^4$

    Solution $-2x^2y^3 + 4xy^4 = -2xy^3(x - 2y)$ ■

Factoring by Grouping

Consider the expression

$$y(x + 4) + 2(x + 4)$$

as the sum of two "terms," $y(x + 4)$ and $2(x + 4)$. Each of these "terms" has the common factor $(x + 4)$. Factoring out this common factor using the distributive property gives

$$y(x + 4) + 2(x + 4) = (x + 4)(y + 2)$$

Similarly,

$$3(x - 2) - a(x - 2) = (x - 2)(3 - a)$$

Now, we see that the product

$$(x + 3)(y + 5) = (x + 3)y + (x + 3)5$$
$$= xy + 3y + 5x + 15$$

has four terms and none of the terms are like terms; therefore, none can be combined. Factoring polynomials with four or more terms can **sometimes** be accomplished by grouping the terms and using the distributive property, as in the following examples.

EXAMPLES   Factor each polynomial by grouping.

9. $xy + 5x + 3y + 15$

Solution

$xy + 5x + 3y + 15 = (xy + 5x) + (3y + 15)$ Group terms that have a common monomial factor.

$= x(y + 5) + 3(y + 5)$ Use the distributive property.

$= (y + 5)(x + 3)$ Note that $y + 5$ is a common factor.

10. $ax + ay + bx + by$

Solution

$ax + ay + bx + by = (ax + ay) + (bx + by)$ Group.

$= a(x + y) + b(x + y)$ Use the distributive property.

$= (x + y)(a + b)$ Use the distributive property again.

Some other grouping may yield the same results. For example,

$$ax + ay + bx + by = (ax + bx) + (ay + by)$$
$$= x(a + b) + y(a + b)$$
$$= (a + b)(x + y)$$

11. $x^2 - xy - 5x + 5y$

Solution $x^2 - xy - 5x + 5y = (x^2 - xy) + (-5x + 5y)$
$$= x(x - y) + 5(-x + y)$$

This does not work because $x - y \neq -x + y$. Try factoring -5 instead of $+5$ from the last two terms.

$$x^2 - xy - 5x + 5y = (x^2 - xy) + (-5x + 5y)$$
$$= x(x - y) - 5(x - y)$$
$$= (x - y)(x - 5) \qquad \text{Success!}$$

12. $x^2 + ax + 3x + 3y$

Solution $x^2 + ax + 3x + 3y = x(x + a) + 3(a + y)$
But $x + a \neq x + y$ and there is no common factor.
So, $x^2 + ax + 3x + 3y$ is **not factorable**.   ■

Practice Problems

Factor each polynomial by finding the greatest common monomial factor.

1. $2x^2 - 4$ **2.** $-5x^2 - 5x$

3. $7ax^2 - 7ax$ **4.** $9x^2y^2 + 12x^2y - 6x^2$

5. Factor by grouping: $5x + 35 - xy - 7y$

Solving Equations by Factoring

To solve an equation in which a product of polynomials is equal to 0, we can let each factor in turn equal 0 to find all possible solutions. The reason is that a product is 0 only if at least one of the factors is 0.

$$\text{If } a \cdot b = 0, \text{ then } a = 0 \text{ or } b = 0$$

EXAMPLES Solve the following equations by getting 0 on one side and factoring the other side. Then set each factor equal to 0 and solve.

13. $3x^2 = 6x$

    Solution

$3x^2 - 6x = 0$	Get 0 on one side.
$3x(x - 2) = 0$	Factor.
$3x = 0 \quad \text{or} \quad x - 2 = 0$	Set each factor equal to 0.
$x = 0 \qquad\qquad x = 2$	Solve each equation.

    The solutions are 0 and 2.

14. $5x^2 - 45x = 0$

    Solution

$5x(x - 9) = 0$	Factor. (One side is already 0.)
$5x = 0 \quad \text{or} \quad x - 9 = 0$	Set each factor equal to 0.
$x = 0 \qquad\qquad x = 9$	Solve each equation.

    The solutions are 0 and 9.

15. If the side of a square is x, its area is x^2. Suppose that the area of a particular square is numerically the same as 25 times its perimeter. What is the length of one side of the square?

    Solution Let $x =$ length of one side

$$x^2 = \text{area of square}$$
$$4x = \text{perimeter of square}$$

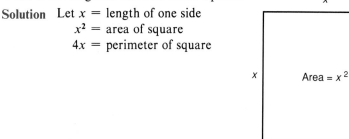

Answers to Practice Problems **1.** $2(x^2 - 2)$ **2.** $-5x(x + 1)$ **3.** $7ax(x - 1)$
4. $3x^2(3y^2 + 4y - 2)$ **5.** $(x + 7)(5 - y)$

Thus, the equation is

$$x^2 = 25(4x)$$
$$x^2 = 100x \qquad \text{Simplify.}$$
$$x^2 - 100x = 0 \qquad \text{One side must be 0.}$$
$$x(x - 100) = 0 \qquad \text{Factor.}$$
$$x = 0 \quad \text{or} \quad x = 100$$

The solution is 100. The value of 0 is a solution to the equation but it must be discarded because it does not make sense in the problem. The length of a side of a square cannot be 0. ■

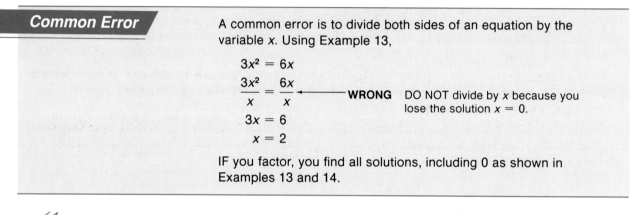

Common Error

A common error is to divide both sides of an equation by the variable x. Using Example 13,

$$3x^2 = 6x$$
$$\frac{3x^2}{x} = \frac{6x}{x} \longleftarrow \textbf{WRONG} \quad \text{DO NOT divide by } x \text{ because you lose the solution } x = 0.$$
$$3x = 6$$
$$x = 2$$

IF you factor, you find all solutions, including 0 as shown in Examples 13 and 14.

EXERCISES 5.1

Simplify the expressions in Exercises 1–10. Assume that no denominator is equal to 0.

1. $\dfrac{x^3}{x}$ **2.** $\dfrac{y^6}{y^4}$ **3.** $\dfrac{x^7}{x^3}$ **4.** $\dfrac{x^8}{x^3}$ **5.** $\dfrac{-8y^3}{2y^2}$

6. $\dfrac{12x^2}{2x}$ **7.** $\dfrac{9x^5}{3x^2}$ **8.** $\dfrac{-10x^5}{2x}$ **9.** $\dfrac{4x^3y^2}{2xy}$ **10.** $\dfrac{21x^4y^3}{-3xy^2}$

Factor each of the polynomials in Exercises 11–38 by finding the greatest common monomial factor.

11. $11x - 121$

12. $14x + 21$

13. $16y^3 + 12y$

14. $4ax - 8ay$

15. $-8a - 16b$

16. $-3x^2 + 6x$

17. $-6ax + 9ay$

18. $10x^2y - 25xy$

19. $16x^4y - 14x^2y$

20. $18y^2z^2 - 2yz$

21. $-14x^2y^3 + 14x^2y$

22. $8y^2 - 32y + 8$

23. $5x^2 - 15x - 5$

24. $ad^2 + 10ad + 25a$

25. $8mx - 12my + 4mz$

26. $36tx - 45ty + 24tz$

27. $34x^5 - 51x^4 + 17x^3$

28. $-56x^4 - 98x^3 - 35x^2$

29. $15x^7 + 24x^6 - 32x^4$

30. $50axb^2 - 2axc^2$

31. $a^2x^2z - y^2z$

32. $9axy^3 - 9axy$

33. $7x^4y^6 + 28x^2y^4$

34. $36x^2y^2 + 12x^2y - 15x^2$

35. $15xy^2 - 20x^2y^3 - 25x^5y^7$

36. $-3x^2y^2 - 6x^3y^3 - 9xy$

37. $-16x^5y - 15x^4y + 3x^2y$

38. $22x^3z + 11x^2z - 33xz$

Factor each of the polynomials in Exercises 39–52 by grouping.

39. $bx + b + cx + c$

40. $3x + 3y + ax + ay$

41. $x^3 + 3x^2 + 6x + 18$

42. $2z^3 - 14z^2 + 3z - 21$

43. $x^2 - 4x + 6xy - 24y$

44. $3x + 3y - bx - by$

45. $5xy + yz - 20x - 4z$

46. $x - 3xy + 2z - 6zy$

47. $24y + 2xz - 3yz - 16x$

48. $10xy - 2y^2 + 7yz - 35xz$

49. $x^4 - 3x^3 + 7x - 21$

50. $16y^3 + 56y^2 + 10y + 35$

51. $4x^4 - 6x^3 - 14x^2 + 21x$

52. $10x^4 - 25x^3 - 18x^2 + 45x$

Solve the equations in Exercises 53–64 by factoring.

53. $x^2 - 7x = 0$

54. $x^2 + 8x = 0$

55. $2x^2 + 12x = 0$

56. $3x^2 - 15x = 0$

57. $4x^2 = 16x$

58. $5x^2 = -25x$

59. $-3x^2 + 6x = 0$

60. $18x - 4x^2 = 0$

61. $3x^2 + 2x = 0$

62. $2x^2 - 7x = 0$

63. $4x^2 = 3x$

64. $5x^2 = -2x$

65. The area of a square is numerically the same as six times its perimeter. What is the length of the side of the square?

66. The area of a square is numerically the same as three times its perimeter. What is the length of the sides of the square?

67. The square of a number is equal to 8 times the number. Find the number.

68. Four times the square of a number is equal to 12 times the number. Find the number.

69. The product of two consecutive even integers is equal to 8 times the smaller integer. Find the integers.

70. The product of two consecutive integers is equal to 11 times the smaller integer. Find the integers.

5.2 *Factoring Special Products and Solving Equations*

OBJECTIVES

In this section, you will be learning to:

1. Factor the differences of two squares.
2. Factor perfect square trinomials.
3. Complete the square of trinomials by determining the missing terms that make incomplete trinomials perfect squares.
4. Solve equations by factoring.

In Chapter 4, we discussed the following three special products.

I. $(x + a)(x - a) = x^2 - a^2$ Difference of two squares

II. $(x + a)^2 = x^2 + 2ax + a^2$ Perfect square trinomial

III. $(x - a)^2 = x^2 - 2ax + a^2$ Perfect square trinomial

Factoring Special Products

Now we want to factor products of these types so that we can solve equations and the related word problems.

Suppose that the product polynomial $x^2 - 9$ is given. If we recognize this expression as the difference of two squares, then the factors are known.

$$x^2 - 9 = (x + 3)(x - 3)$$

Similarly,

$$x^2 - y^2 = (x + y)(x - y)$$
$$x^2 - 25 = (x + 5)(x - 5)$$
$$36x^2 - 49 = (6x + 7)(6x - 7)$$

If the polynomial is a perfect square trinomial with 1 as the coefficient of x^2, then the last term must be a perfect square and the middle coefficient must be 2 times the term that was squared (or -2 times the term that was squared). For example,

$$x^2 + 6x + 9 = (x + 3)^2; \quad \text{here } 9 = 3^2 \text{ and } 6 = 2 \cdot 3.$$
$$x^2 - 14x + 49 = (x - 7)^2; \quad \text{here } 49 = 7^2 \text{ and } -14 = -2(7).$$

Recognizing the form of the polynomial is the key to factoring the three special products discussed here. Sometimes the form may be disguised by a common monomial factor or by a rearrangement of the terms. **Always look for a common monomial factor first.** For example,

$$5x^2y - 20y = 5y(x^2 - 4) \qquad \text{Factoring out } 5y$$
$$= 5y(x + 2)(x - 2) \qquad \text{Difference of two squares}$$

EXAMPLES   Factor completely each of the following polynomials.

1. $y^2 - 10y + 25$

   Solution

   $$y^2 - 10y + 25 = (y - 5)^2 \qquad \text{Perfect square trinomial}$$

2. $6a^2b - 6b$

   Solution

   $$6a^2b - 6b = 6b(a^2 - 1) \qquad \text{Common monomial factor}$$
   $$= 6b(a + 1)(a - 1) \qquad \text{Difference of two squares}$$

3. $3x^2 + 12 + 12x$

   Solution

   $$3x^2 + 12 + 12x = 3(x^2 + 4 + 4x) \qquad \text{Common monomial factor}$$
   $$= 3(x^2 + 4x + 4) \qquad \text{Rearranging terms}$$
   $$= 3(x + 2)^2 \qquad \text{Perfect square trinomial}$$

4. $a^6 - 100$ Even powers, such as a^6, can always be treated as squares. $a^6 = (a^3)^2$

   Solution

   $$a^6 - 100 = (a^3 + 10)(a^3 - 10) \qquad \text{Difference of two squares} \quad ∎$$

Procedures to Factor
Special Products

1. Look for a common monomial factor.
2. Check the number of terms:
   a. Two terms—Is it the difference of two squares?
   b. Three terms—Is it a perfect square trinomial?
3. Check the possibility of factoring any of the factors.

Completing the Square

Closely related to factoring special products is the procedure of **completing the square.** This procedure helps in understanding the nature of perfect square trinomials and we will use it again in Chapter 10 in solving second-degree equations. The procedure involves adding a square term to a binomial so that the resulting trinomial is a perfect square trinomial, thus "completing the square." For example,

$$x^2 + 10x + \underline{} = ()^2$$

The middle coefficient, 10, is twice the number that is to be squared. So, by taking half this coefficient and squaring the result, we will have the missing constant.

$$x^2 + 10x + \underline{} = ()^2$$

$$x^2 + 10x + \underline{\ 25\ } = (x + 5)^2; \quad \frac{1}{2}(10) = 5 \quad \text{and} \quad 5^2 = 25$$

For $x^2 + 18x$, we get

$$x^2 + 18x + \underline{} = ()^2$$

$$x^2 + 18x + \underline{\ 81\ } = (x + 9)^2; \quad \frac{1}{2}(18) = 9 \quad \text{and} \quad 9^2 = 81$$

EXAMPLES Complete the square as indicated.

5. $y^2 + 20y + \underline{} = ()^2$

Solution Since $\frac{1}{2}(20) = 10$ and $10^2 = 100$, then

$$y^2 + 20y + \underline{\ 100\ } = (y + 10)^2$$

6. $x^2 + 3x + \underline{} = ()^2$

Solution Since $\frac{1}{2}(3) = \frac{3}{2}$ and $\left(\frac{3}{2}\right)^2 = \frac{9}{4}$, then

$$x^2 + 3x + \underline{\ \frac{9}{4}\ } = \left(x + \frac{3}{2}\right)^2$$

(**Note:** We introduce fractions in this situation because they seem to cause considerable difficulty later on for many students.)

7. $a^2 - 5a + \underline{} = ()^2$

Solution Since $\frac{1}{2}(-5) = -\frac{5}{2}$ and $\left(\frac{-5}{2}\right)^2 = \frac{25}{4}$, then

$$a^2 - 5a + \underline{\ \frac{25}{4}\ } = \left(a - \frac{5}{2}\right)^2.$$

■

To help in understanding the relationships between the terms in a perfect square trinomial, we look at completing the square from a slightly different perspective. Suppose that the middle x-term is missing and we want to find this term so that the resulting trinomial is the square of a binomial. For example,

$$x^2 + \underline{\hspace{1cm}} + 36 = (\quad)^2$$

From the formulas,

$$(x + a)^2 = x^2 + \underline{2ax} + a^2$$
$$(x - a)^2 = x^2 - \underline{2ax} + a^2$$

we know that the middle (missing) term is to have coefficient $2a$ (or $-2a$) and, in this case, $a^2 = 36 = 6^2$ and $2a = 2 \cdot 6 = 12$. So,

$$x^2 + \underline{12x} + 36 = (x + 6)^2$$

Similarly, for

$$x^2 - \underline{\hspace{1cm}} + 16 = (\quad)^2$$

we get

$$x^2 - \underline{8x} + 16 = (x - 4)^2 \qquad 16 = 4^2 \text{ and } -2 \cdot 4 = -8$$

EXAMPLE   Complete the square as indicated.

8. $x^2 - \underline{\hspace{1cm}} + \dfrac{9}{16} = (\quad)^2$

   Solution   Since $\dfrac{9}{16} = \left(\dfrac{3}{4}\right)^2$ and $-2 \cdot \dfrac{3}{4} = -\dfrac{3}{2}$, then

   $$x^2 - \underline{\dfrac{3}{2}x} + \dfrac{9}{16} = \left(x - \dfrac{3}{4}\right)^2 \qquad \blacksquare$$

Practice Problems

Factor completely.

1. $x^2 - 16$ 2. $25 - x^2$
3. $3y^2 + 6y + 3$ 4. $y^{10} - 81$

Complete the square.

5. $x^2 + 6x + \underline{\hspace{1cm}} = (\quad)^2$ 6. $x^2 + \underline{\hspace{1cm}} + 49 = (\quad)^2$

Answers to Practice Problems   **1.** $(x + 4)(x - 4)$ **2.** $(5 + x)(5 - x)$ **3.** $3(y + 1)^2$
4. $(y^5 + 9)(y^5 - 9)$ **5.** $x^2 + 6x + \underline{9} = (x + 3)^2$
6. $x^2 + \underline{14x} + 49 = (x + 7)^2$

Solving Equations by Factoring

The technique for solving equations by factoring special products is the same as that discussed in Section 5.1. First, get 0 on one side of the equation. Then factor and set each nonconstant factor equal to 0. Remember,

$$\text{If } a \cdot b = 0, \text{ then } a = 0 \text{ or } b = 0.$$

EXAMPLES Solve the following equations by getting 0 on one side and factoring the other side. Then set each nonconstant factor equal to 0 and solve.

9. $4x^2 = 36$

Solution

$4x^2 - 36 = 0$	Get 0 on one side.
$4(x^2 - 9) = 0$	Factor out a common monomial.
$4(x + 3)(x - 3) = 0$	Factor the difference of two squares.
$x + 3 = 0 \quad \text{or} \quad x - 3 = 0$	Set each nonconstant factor equal to 0.
$x = -3 \qquad\qquad x = 3$	(Note that the constant factor 4 has no effect on the solutions.)

The solutions are -3 and 3.

10. $x^2 + 81 = 18x$

Solution

$x^2 + 81 - 18x = 0$	Get 0 on one side.
$x^2 - 18x + 81 = 0$	Rearrange the terms.
$(x - 9)^2 = 0$	Factor the perfect square trinomial.
$(x - 9)(x - 9) = 0$	Here, the same factor appears twice.
$x - 9 = 0$	
$x = 9$	We call the solution a **double solution** or a **double root** because the two factors are the same.

The solution is 9. ∎

◢ EXERCISES 5.2

Factor completely each of the polynomials in Exercises 1–34.

1. $x^2 - 1$ **2.** $x^2 - 4$ **3.** $x^2 - 49$ **4.** $x^2 - 144$

5. $x^2 + 4x + 4$ **6.** $x^2 + 10x + 25$ **7.** $x^2 - 12x + 36$ **8.** $x^2 - 20x + 100$

9. $16x^2 - 9$ **10.** $4x^2 - 49$ **11.** $9x^2 - 1$ **12.** $36x^2 - 1$

13. $25 - 4x^2$ **14.** $16 - 9x^2$ **15.** $x^2 - 14x + 49$ **16.** $x^2 - 2x + 1$

17. $x^2 + 6x + 9$ **18.** $x^2 + 8x + 16$ **19.** $3x^2 - 27y^2$ **20.** $27x^2 - 48y^2$

21. $x^3 - xy^2$ **22.** $12ax^2 - 75ay^2$ **23.** $x^2 - \dfrac{1}{4}$ **24.** $x^2 - \dfrac{4}{9}$

25. $x^2 - \dfrac{9}{16}$ **26.** $x^2 - \dfrac{25}{49}$ **27.** $x^4 - 1$ **28.** $x^4 - 16y^4$

29. $2x^2 - 32x + 128$ **30.** $3x^2 - 30x + 75$ **31.** $ay^2 + 2ay + a$ **32.** $ax^2 + 18ax + 81a$

33. $4x^2y^2 - 24xy^2 + 36y^2$ **34.** $2ax^2 - 44ax + 242a$

In Exercises 35–50, complete the square by adding the correct term, then factor as indicated.

35. $x^2 - 6x + \underline{\quad} = (\quad)^2$ **36.** $x^2 - 12x + \underline{\quad} = (\quad)^2$

37. $x^2 + 4x + \underline{\quad} = (\quad)^2$ **38.** $x^2 + 20x + \underline{\quad} = (\quad)^2$

39. $x^2 + \underline{\quad} + 16 = (\quad)^2$ **40.** $x^2 - \underline{\quad} + 49 = (\quad)^2$

41. $x^2 - \underline{\quad} + 81 = (\quad)^2$ **42.** $x^2 + \underline{\quad} + 121 = (\quad)^2$

43. $x^2 + x + \underline{\quad} = (\quad)^2$ **44.** $x^2 - 7x + \underline{\quad} = (\quad)^2$

45. $x^2 - 9x + \underline{\quad} = (\quad)^2$ **46.** $x^2 + 3x + \underline{\quad} = (\quad)^2$

47. $x^2 + \underline{\quad} + \dfrac{25}{4} = (\quad)^2$ **48.** $x^2 - \underline{\quad} + \dfrac{121}{4} = (\quad)^2$

49. $x^2 - \underline{\quad} + \dfrac{9}{4} = (\quad)^2$ **50.** $x^2 + \underline{\quad} + \dfrac{81}{4} = (\quad)^2$

Solve the equations in Exercises 51–64 by factoring.

51. $x^2 - 100 = 0$ **52.** $x^2 - 121 = 0$ **53.** $3x^2 - 75 = 0$

54. $5x^2 - 45 = 0$ **55.** $x^2 + 8x + 16 = 0$ **56.** $x^2 + 14x + 49 = 0$

57. $3x^2 = 18x - 27$ **58.** $5x^2 = 10x - 5$ **59.** $4x^2 - 49 = 0$

60. $16x^2 - 25 = 0$ **61.** $2x^2 = 8x - 8$ **62.** $6x^2 = 60x - 150$

63. $x^2 + 3x + \dfrac{9}{4} = 0$ **64.** $x^2 + \dfrac{2}{3}x + \dfrac{1}{9} = 0$

65. The area of a square is 81 square centimeters. How long is each side of the square?

66. The length of a rectangle is three times the width. If the area is 48 square feet, find the length and width of the rectangle.

67. The length of a rectangle is five times the width. If the area is 180 square inches, find the length and width of the rectangle.

68. One number is eight more than another. Their product is -16. What are the numbers?

69. One number is 10 more than another. If their product is -25, find the numbers.

70. Find two numbers whose product is -49 if one number is fourteen less than the other.

◢ 5.3 Factoring Trinomials

In this section, we will investigate two methods for factoring trinomials in the general form

$$ax^2 + bx + c$$

where the trinomial may or may not be a perfect square trinomial. The two methods are the **trial-and-error method** and the **ac-method.**

Trial-and-Error Method

The key to this discussion is the FOIL method of multiplication that we studied in Section 4.5. We will use this method in a reverse sense since the product will be given and the object is to find factors. Reviewing the FOIL method of multiplication, we can find the product

$$(2x + 5)(3x + 1) = 6x^2 + 17x + 5$$

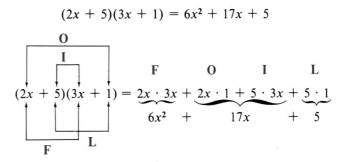

$$(2x + 5)(3x + 1) = \underbrace{2x \cdot 3x}_{6x^2} + \underbrace{2x \cdot 1 + 5 \cdot 3x}_{17x} + \underbrace{5 \cdot 1}_{5}$$

F   The product of the **first** two terms is $6x^2$.

$\left.\begin{array}{l}\mathbf{O}\\\mathbf{I}\end{array}\right\}$ The sum of the **inner** and **outer** products is $17x$.

L   The product of the **last** two terms is 5.

First we consider the problem of factoring a trinomial such as

$$x^2 + 7x + 12$$

where the coefficient a is 1. We know here that $F = x^2 = x \cdot x$ and that $L = 12$.

$$x^2 + 7x + 12 = (x \qquad)(x \qquad)$$
$$L = 12$$
$$F = x^2$$

So, the only problem is to find the factors of 12 whose sum is 7. The factors are 4 and 3 since $4 \cdot 3 = 12$ and $4 + 3 = 7$.

$$O = 4x$$
$$I = 3x$$
$$x^2 + 7x + 12 = (x + 3)(x + 4); \qquad 4x + 3x = 7x$$
$$L = 12$$
$$F = x^2$$

Similarly, to factor

$$x^2 - 5x - 24$$

we need only find factors of -24 whose sum is -5. The factors are -8 and $+3$ since $(-8)(+3) = -24$ and $-8 + 3 = -5$.

$$x^2 - 5x - 24 = (x + 3)(x - 8); \qquad -8x + 3x = -5x$$

Now consider the problem of factoring

$$6x^2 + 31x + 5$$

as a product of two binomials.

$$6x^2 + 31x + 5 = (\qquad)(\qquad)$$

$$L = +5$$
$$F = 6x^2$$

We know that $F = 6x^2$

and $6x^2 = 3x \cdot 2x$

and $6x^2 = 6x \cdot x$

We also know that $L = +5$

and $+5 = (+1)(+5)$

and $+5 = (-1)(-5)$

So, by the **trial-and-error method:**

1. List all the possible combinations of factors of $6x^2$ and $+5$ in their respective **F** and **L** positions.
2. Check the sum of the products in the **O** and **I** positions until you find the sum to be $31x$.
3. If none of these sums is $31x$, the trinomial is not factorable.

Note: After some practice, much of the trial time can be shortened and the method will become reasonably efficient.

$$\overset{\text{F}}{\underset{\text{L}}{\big|}}$$

a. $(3x + 1)(2x + 5)$

b. $(3x + 5)(2x + 1)$

c. $(6x + 1)(x + 5)$

d. $(6x + 5)(x + 1)$

e. $(3x - 1)(2x - 5)$

f. $(3x - 5)(2x - 1)$

g. $(6x - 1)(x - 5)$

h. $(6x - 5)(x - 1)$

We really don't need to check these four because the **O** and **I** sum would have to be negative, and we are looking for $+31x$.

Now, for the possibilities with positive constants, we need to check the sums of the outer (**O**) and inner (**I**) products to find $+31x$.

a. $(3x + 1)(2x + 5);\qquad 15x + 2x = 17x$

b. $(3x + 5)(2x + 1);\qquad 3x + 10x = 13x$

c. $(6x + 1)(x + 5);\qquad 30x + x = 31x$

We have found the correct combination of factors so we need not try $(6x + 5)(x + 1)$. So,

$$6x^2 + 31x + 5 = (6x + 1)(x + 5)$$

To help limit the trial-and-error search, **look at the constant term.**

1. If the sign of the constant term is positive $(+)$, the signs in both factors will be the same, either both positive $(+)$ or both negative $(-)$.

2. If the sign of the constant term is negative $(-)$, the signs in the factors will differ, one positive $(+)$ and one negative $(-)$.

EXAMPLES   Factor using the trial-and-error method.

1. $x^2 - 3x - 40$

   Solution $x^2 - 3x - 40 = (x - 8)(x + 5)$ The trial-and-error method is particularly easy to use when 1 is the coefficient of x^2. Here we need to search only for factors of -40 whose sum is -3.

2. $x^2 + 9x + 20$

   Solution $x^2 + 9x + 20 = (x + 5)(x + 4)$ Again 1 is the coefficient of x^2 and we have $5 \cdot 4 = 20$ and $5 + 4 = 9$.

3. Factor $6x^2 - 31x + 5$.

   Solution   Since the middle term is $-31x$ and the constant is $+5$, we know that the two factors of 5 must both be negative, -5 and -1.

   $$6x^2 - 31x + 5 = (6x - 1)(x - 5); \qquad -30x - x = -31x$$

4. Factor $2x^2 + 12x + 10$ completely.

   Solution $2x^2 + 12x + 10 = 2(x^2 + 6x + 5)$ First find any common monomial factor.

   $$= 2(x + 5)(x + 1); \qquad x + 5x = 6x$$

   Special Note: To factor completely means to find factors of the polynomial none of which are themselves factorable. Thus, $2x^2 + 12x + 10 = (2x + 10)(x + 1)$ is **not** factored completely since $2x + 10 = 2(x + 5)$. We could write

   $$2x^2 + 12x + 10 = (2x + 10)(x + 1)$$
   $$= 2(x + 5)(x + 1) \qquad \blacksquare$$

   Finding the greatest common monomial factor first generally makes the problem easier. The trial-and-error method may seem difficult at first, but with practice you will learn to "guess" better and to eliminate certain combinations quickly. For example, to factor $10x^2 + x - 2$, do we use $10x$ and x or

$5x$ and $2x$; and for -2, do we use -2 and $+1$ or $+2$ and -1? The terms $5x$ and $2x$ are more likely candidates since they are closer together than $10x$ and x and the middle term is small, $1x$. So,

$(5x + 1)(2x - 2)$	$-10x + 2x = -8x$	Reject.
$(5x - 1)(2x + 2)$	$+10x - 2x = 8x$	Reject.
$(5x + 2)(2x - 1)$	$-5x + 4x = -x$	Reject.
$(5x - 2)(2x + 1)$	$5x - 4x = x$	Accept!

$$10x^2 + x - 2 = (5x - 2)(2x + 1)$$

Not all polynomials are factorable. For example, no matter what combinations we try, $3x^2 - 3x + 4$ will not have two binomial factors with integer coefficients. This polynomial is **irreducible (or not factorable); it cannot be factored as a product of polynomials with integer coefficients.**

   The sum of two squares $a^2 + b^2$ is an important polynomial that is not factorable. For example, $x^2 + 4$ is not factorable. There are no factors with integer coefficients whose product is $x^2 + 4$.

EXAMPLES   Factor completely. Look first for the greatest common monomial factor.

   5. $4x^2 + 100$

   Solution $4x^2 + 100 = 4(x^2 + 25)$ $x^2 + 25$ is not factorable.

   6. $6x^3 - 8x^2 + 2x$

   Solution

   $$6x^3 - 8x^2 + 2x = 2x(3x^2 - 4x + 1) = 2x(3x - 1)(x - 1)$$

   7. $2x^2 + x - 6$

   Solution $2x^2 + x - 6 = (2x - 3)(x + 2)$

   8. $x^2 + x + 1$

   Solution $x^2 + x + 1 = x^2 + x + 1$ This trinomial is not factorable.

■

ac-*Method (Grouping)*

The *ac*-method is very systematic. It also involves the technique of grouping that we discussed in Section 5.1. We refer to trinomials in the form

$$ax^2 + bx + c \qquad \text{where } a, b, \text{ and } c \text{ are integers.}$$

   Consider the problem of factoring $3x^2 + 17x + 10$ where $a = 3$, $b = 17$, and $c = 10$.

Analysis of Factoring by the ac-Method	$3x^2 + 17x + 10$	$ax^2 + bx + c$
	Step 1 Multiply $3 \cdot 10 = 30$.	Multiply $a \cdot c$.
	Step 2 Find two integers whose product is 30 and whose sum is 17. (In this case, $2 \cdot 15 = 30$ and $2 + 15 = 17$.)	Find two integers whose product is ac and whose sum is b. If this is not possible, then the trinomial is **not factorable.**
	Step 3 Rewrite the middle term ($17x$) using 2 and 15 as coefficients. $3x^2 + 17x + 10$ $= 3x^2 + 2x + 15x + 10$	Rewrite the middle term (bx) using the two numbers found in Step 2 as coefficients.
	Step 4 Factor by grouping the first two terms and the last two terms. $3x^2 + 2x + 15x + 10$ $= x(3x + 2) + 5(3x + 2)$	Factor by grouping the first two terms and the last two terms.
	Step 5 Factor out the common binomial factor $(3x + 2)$. Thus, $3x^2 + 17x + 10$ $= 3x^2 + 2x + 15x + 10$ $= x(3x + 2) + 5(3x + 2)$ $= (3x + 2)(x + 5)$	Factor out the common binomial factor to find two binomial factors of the trinomial $ax^2 + bx + c$.

EXAMPLES

9. Factor $x^2 - 3x - 28$ using the *ac*-method.

Solution $a = 1$, $b = -3$, and $c = -28$

Step 1   Find the product $a \cdot c$: $1(-28) = -28$.

Step 2   Find the two integers whose product is -28 and whose sum is -3:

$$(-7)(+4) = -28 \quad \text{and} \quad -7 + 4 = -3$$

Step 3   Rewrite $-3x$ as $-7x + 4x$, giving

$$x^2 - 3x - 28 = x^2 - 7x + 4x - 28$$

Step 4   Factor by grouping:

$$x^2 - 3x - 28 = x^2 - 7x + 4x - 28$$
$$= x(x - 7) + 4(x - 7)$$

Step 5   Factor out the common binomial factor $(x - 7)$:

$$x^2 - 3x - 28 = x(x - 7) + 4(x - 7) = (x - 7)(x + 4)$$

10. Factor $12x^3 - 26x^2 + 12x$ using the *ac*-method.

Solution   First factor out the greatest common factor $2x$:

$$12x^3 - 26x^2 + 12x = 2x(6x^2 - 13x + 6)$$

Now factor the trinomial $6x^2 - 13x + 6$ with $a = 6$, $b = -13$, and $c = 6$.

Step 1   Find the product $a \cdot c$: $6(6) = 36$.

Step 2   Find two integers whose product is 36 and whose sum is -13. (**Note:** This may take some time and experimentation. We do know that both numbers must be negative since the product is positive and the sum is negative.)

$$(-9)(-4) = +36 \qquad \text{and} \qquad -9 + (-4) = -13$$

Steps 3 and 4   Factor by grouping:

$$6x^2 - 13x + 6 = 6x^2 - 9x - 4x + 6$$
$$= 3x(x - 3) - 2(x - 3)$$

[**Note:** -2 is factored from the last two terms so that there will be a common binomial factor $(x - 3)$.]

Step 5   Factor out the common binomial factor $(x - 3)$:

$$6x^2 - 13x + 6 = 6x^2 - 9x - 4x + 6$$
$$= 3x(x - 3) - 2(x - 3)$$
$$= (x - 3)(3x - 2)$$

Thus, for the original expression,

$$12x^3 - 26x^2 + 12x = 2x(6x^2 - 13x + 6)$$
$$= 2x(x - 3)(3x - 2)$$

A Summary of Procedures to Follow in Factoring

1. Look for a common factor.
2. Check the number of terms:
   a. Two terms
      (1) difference of two squares?—factorable
      (2) sum of two squares?—not factorable
   b. Three terms
      (1) perfect square trinomial?
      (2) use trial-and-error method?
      (3) use the *ac*-method?
   c. Four terms
      (1) group terms with a common factor
3. Check the possibility of factoring any of the factors.

Remember that the product of the factors must always equal the original polynomial. This is a good way to check your work.

Practice Problems   Factor completely.

1. $3x^2 + 7x - 6$ **2.** $x^2 + 25$

3. $3x^2 + 15x + 18$ **4.** $10x^2 - 41x - 18$

5. $x^2 + 11x + 28$

◢ EXERCISES 5.3

For Exercises 1–6, first list **all** possible combinations of the form $(ax + b)(cx + d)$ suggested by the FOIL method, then choose the one correct pair of factors.

1. $x^2 + 5x + 6$ **2.** $x^2 - 6x + 8$ **3.** $2x^2 - 3x - 5$ **4.** $3x^2 - 4x - 7$

5. $6x^2 + 11x + 5$ **6.** $4x^2 - 11x + 6$

Factor as completely as possible Exercises 7–70.

7. $x^2 - 3x + 2$ **8.** $x^2 + 5x + 6$ **9.** $x^2 - 3x - 10$ **10.** $x^2 - 11x + 10$

11. $x^2 - 13x - 14$ **12.** $x^2 - 12x + 36$ **13.** $x^2 + 8x + 64$ **14.** $x^2 + 2x + 3$

15. $2x^2 - x - 1$ **16.** $2y^2 + 3y + 1$ **17.** $4t^2 - 3t - 1$ **18.** $2x^2 - 3x - 2$

19. $5a^2 - a - 6$ **20.** $3a^2 + 4a + 1$ **21.** $7x^2 + 5x - 2$ **22.** $8x^2 - 10x - 3$

23. $4x^2 + 23x + 15$ **24.** $6x^2 + 23x + 21$ **25.** $x^2 + 6x - 16$ **26.** $9x^2 - 3x - 20$

27. $6x^2 - 19x + 10$ **28.** $4b^2 - 4b + 1$ **29.** $3x^2 - 7x + 2$ **30.** $7x^2 - 11x - 6$

31. $9x^2 - 6x + 1$ **32.** $4x^2 - 25$ **33.** $4x^2 + 25$ **34.** $x^2 + 81$

35. $12y^2 - 7y - 12$ **36.** $x^2 - 46x + 45$ **37.** $5x^2 + 45$ **38.** $4x^2 + 64$

39. $4a^2 - 11a + 6$ **40.** $6m^2 - 73m + 12$ **41.** $x^2 + x + 1$ **42.** $x^2 + 2x + 2$

43. $16x^2 - y^2$ **44.** $3x^2 - 11x - 4$ **45.** $64x^2 - 48x + 9$ **46.** $9x^2 - 12x + 4$

47. $6x^2 + 2x - 20$ **48.** $12y^2 - 15y + 3$ **49.** $10x^2 + 35x + 30$ **50.** $24y^2 + 4y - 4$

51. $18x^2 - 8y^2$ **52.** $15y^2 - 10y - 40$ **53.** $7x^4 - 5x^3 + 3x^2$ **54.** $12x^2 - 60x + 75$

55. $-12m^2 + 22m + 4$ **56.** $32y^2 + 50z^2$ **57.** $6x^3 + 9x^2 - 6x$ **58.** $-5y^2 + 40y - 60$

59. $9x^3y^3 + 9xy$ **60.** $30a^3 + 51a^2 + 9a$ **61.** $12x^3 - 108x^2 + 243x$

62. $48x^2y - 354xy + 126y$ **63.** $48xy^3 - 100xy^2 + 48xy$ **64.** $24a^2x^2 + 72a^2x + 48a^2$

65. $21y^4 - 98y^3 + 56y^2$ **66.** $72a^3 - 306a^2 + 189a$ **67.** $ax + ay + 3x + 3y$

68. $2x^2 + 2y^2 + ax^2 + ay^2$ **69.** $5x^2 - 5y^2 + bx^2 - by^2$ **70.** $ax + bx - 4a - 4b$

◢ 5.4 Solving Quadratic Equations by Factoring

OBJECTIVE

In this section, you will be learning to solve quadratic equations by factoring.

In this section, we will summarize the work we have done in Sections 5.1, 5.2, and 5.3, introduce some new terminology, and solve more equations.

The emphasis has been on second-degree polynomials. Generally, the difference of two squares, perfect square trinomials, the trial-and-error method, and the *ac*-method have been related to expressions in which the term of highest

Answers to Practice Problems   1. $(3x - 2)(x + 3)$ **2.** Not factorable
3. $3(x + 2)(x + 3)$ **4.** $(5x + 2)(2x - 9)$ **5.** $(x + 7)(x + 4)$

degree was an x^2 term. Such second-degree polynomials are called quadratics and they play a major role in algebra. We will continue to emphasize them throughout the first two courses in algebra.

Equations of the form

$$ax^2 + bx + c = 0 \qquad (a \neq 0)$$

are called **quadratic equations.**

The following list summarizes what we know about solving quadratic equations.

To Solve a Quadratic Equation by Factoring	1. Add or subtract terms as necessary so that 0 is on one side of the equation and the equation is in the form $$ax^2 + bx + c = 0 \qquad (a \neq 0)$$ 2. If there are any fractional coefficients, multiply each term by the least common denominator so that all coefficients will be integers. 3. Factor the quadratic, if possible. 4. Set each nonconstant factor equal to 0 and solve for the unknown. 5. Remember, there will never be more than two solutions.

EXAMPLES Solve the following quadratic equations by factoring.

1. $x^2 - 7x + 12 = 0$

　　Solution 　　$x^2 - 7x + 12 = 0$

　　　　　　　$(x - 4)(x - 3) = 0$ 　　　　　$(-4)(-3) = +12$ and
　　　　　　　　　　　　　　　　　　　　　　$(-4) + (-3) = -7$

　　　　$x - 4 = 0$ 　　or 　　$x - 3 = 0$

　　　　　　$x = 4$ 　　　　　　　$x = 3$

2. $4x^2 - 4x = 24$

　　Solution 　　　　　$4x^2 - 4x = 24$

　　　　　　　　$4x^2 - 4x - 24 = 0$ 　　　One side must be 0.

　　　　　　　$4(x^2 - x - 6) = 0$ 　　　4 is a common monomial
　　　　　　　　　　　　　　　　　　　　factor.

　　　　　　$4(x - 3)(x + 2) = 0$ 　　　The constant factor 4 can
　　　　　　　　　　　　　　　　　　　　never be 0 and does not
　　　　　　　　　　　　　　　　　　　　affect the solution.

　　　　$x - 3 = 0$ 　　or 　　$x + 2 = 0$

　　　　　　$x = 3$ 　　　　　　　$x = -2$

3. $\dfrac{2x^2}{15} - \dfrac{x}{3} = -\dfrac{1}{5}$

Solution $\dfrac{2x^2}{15} - \dfrac{x}{3} = \dfrac{-1}{5}$

$15 \cdot \dfrac{2x^2}{15} - \dfrac{x}{3} \cdot 15 = -\dfrac{1}{5} \cdot 15$ Multiply each term by 15, the LCM of the denominators, to get integer coefficients.

$2x^2 - 5x = -3$ Simplify.

$2x^2 - 5x + 3 = 0$ One side must be 0.

$(2x - 3)(x - 1) = 0$ Factor by the trial-and-error method or the *ac*-method.

$2x - 3 = 0 \qquad \text{or} \qquad x - 1 = 0$

$2x = 3 \qquad\qquad\qquad\quad x = 1$

$x = \dfrac{3}{2}$

The following two examples illustrate situations in which the first step is to multiply expressions so that like terms can be combined. Then the quadratic equations can be solved by factoring.

EXAMPLES

4. $(x + 5)^2 = 36$

Solution $(x + 5)^2 = 36$

$x^2 + 10x + 25 = 36$ $(x + 5)^2$ gives a perfect square trinomial.

$x^2 + 10x - 11 = 0$ One side must be 0.

$(x + 11)(x - 1) = 0$

$x + 11 = 0 \qquad \text{or} \qquad x - 1 = 0$

$x = -11 \qquad\qquad\qquad x = 1$

5. $3x(x - 1) = 2(5 - x)$

Solution $3x(x - 1) = 2(5 - x)$

$3x^2 - 3x = 10 - 2x$ Multiply using the distributive property.

$3x^2 - 3x + 2x - 10 = 0$ One side must be 0.

$3x^2 - x - 10 = 0$ Simplify.

$(3x + 5)(x - 2) = 0$ Factor using the trial-and-error method or the *ac*-method.

$3x + 5 = 0 \qquad \text{or} \qquad x - 2 = 0$

$x = -\dfrac{5}{3} \qquad\qquad\qquad x = 2$

In the next example, we show that factoring can be used to solve equations of degree higher than second degree. There will be more than two factors. But, just as with quadratics, if the product is 0, then the solutions are found by setting each factor equal to 0.

EXAMPLE

6. $2x^3 - 4x^2 - 6x = 0$

Solution $2x^3 - 4x^2 - 6x = 0$

$2x(x^2 - 2x - 3) = 0$ Factor out the common monomial $2x$.

$2x(x - 3)(x + 1) = 0$

$2x = 0$ or $x - 3 = 0$ or $x + 1 = 0$ Set each factor

$x = 0$ $x = 3$ $x = -1$ equal to 0.   ∎

Practice Problems

Solve each equation by factoring.

1. $x^2 - 6x = 0$ **2.** $6x^2 - x - 1 = 0$
3. $(x - 2)^2 - 25 = 0$ **4.** $x^3 - 8x^2 + 16x = 0$

◿ EXERCISES 5.4

Solve the equations in Exercises 1–64.

1. $x^2 - 5x + 6 = 0$ **2.** $x^2 + 3x - 10 = 0$ **3.** $x^2 - 5x - 14 = 0$
4. $x^2 + 3x - 28 = 0$ **5.** $x^2 - 6x + 9 = 0$ **6.** $x^2 + 8x + 12 = 0$
7. $x^2 - 25 = 0$ **8.** $x^2 + 10x + 25 = 0$ **9.** $2x^2 - 4x = 0$
10. $3x^2 + 9x = 0$ **11.** $x^3 - 3x^2 = 4x$ **12.** $x^3 + 7x^2 = -12x$
13. $5x^2 + 15x = 0$ **14.** $x^2 - 11x = -18$ **15.** $2x^2 - 24 = 2x$
16. $4x^2 - 12x = 0$ **17.** $x^2 + 8 = 6x$ **18.** $x^2 = x + 30$
19. $2x^2 + 2x - 24 = 0$ **20.** $9x^2 + 63x + 90 = 0$ **21.** $2x^2 - 5x - 3 = 0$
22. $2x^2 - x - 6 = 0$ **23.** $3x^2 - 4x - 4 = 0$ **24.** $3x^2 - 8x + 5 = 0$
25. $2x^2 - 7x - 4 = 0$ **26.** $4x^2 + 8x + 3 = 0$ **27.** $3x^2 + 2x - 8 = 0$
28. $6x^2 + 7x + 2 = 0$ **29.** $4x^2 - 12x + 9 = 0$ **30.** $25x^2 - 60x + 36 = 0$
31. $(x - 1)^2 = 4$ **32.** $(x - 3)^2 = 1$ **33.** $(x + 5)^2 = 9$
34. $(x + 4)^2 = 16$ **35.** $(x + 4)(x - 1) = 6$ **36.** $(x - 5)(x + 3) = 9$
37. $(x + 2)(x - 4) = 27$ **38.** $(x + 2)(x + 4) = -1$ **39.** $x(x + 7) = 3(x + 4)$
40. $x(x + 9) = 6(x + 3)$ **41.** $3x(x + 1) = 2(x + 1)$ **42.** $2x(x - 1) = 3(x - 1)$
43. $x(2x + 1) = 6(x + 2)$ **44.** $3x(x + 3) = 2(2x - 1)$ **45.** $5x^2 = 8x$
46. $3x^2 = 15x$ **47.** $9x^2 = 36$ **48.** $4x^2 - 16 = 0$
49. $5x^2 + 10x + 5 = 0$ **50.** $2x^2 = 4x + 6$ **51.** $8x^2 + 32 = 32x$

Answers to Practice Problems   **1.** $x = 0, x = 6$ **2.** $x = \dfrac{1}{2}, x = -\dfrac{1}{3}$
3. $x = 7, x = -3$ **4.** $x = 0, x = 4$

52. $-6x^2 + 18x + 24 = 0$

53. $\dfrac{x^2}{3} - 2x + 3 = 0$

54. $\dfrac{x^2}{9} = 1$

55. $\dfrac{x^2}{5} - x - 10 = 0$

56. $\dfrac{2}{3}x^2 + 2x - \dfrac{20}{3} = 0$

57. $\dfrac{x^2}{8} + x + \dfrac{3}{2} = 0$

58. $\dfrac{x^2}{6} - \dfrac{1}{2}x - 3 = 0$

59. $x^2 - x + \dfrac{1}{4} = 0$

60. $x^2 + \dfrac{7}{6}x + \dfrac{1}{3} = 0$

61. $x^3 + 8x = 6x^2$

62. $x^3 = x^2 + 30x$

63. $6x^3 + 7x^2 = -2x$

64. $3x^3 = 8x - 2x^2$

◢ 5.5 Applications

OBJECTIVES

In this section, you will be learning to solve word problems by writing quadratic equations that can be factored and solved.

Whether or not word problems cause you difficulty depends a great deal on your personal experiences and general reasoning abilities. These abilities are developed over a long period of time. A problem that is easy for you, possibly because you have had experience in a particular situation, might be quite difficult for a friend, and vice versa.

Most problems do not say specifically to add, subtract, multiply, or divide. You are to know from the nature of the problem what to do. You are to ask yourself, "What information is given? What am I trying to find? What tools, skills, and abilities do I need to use?"

Word problems should be approached in an orderly manner. Have an "attack plan."

Attack Plan for Word Problems

1. Read the problem carefully at least twice.
2. Decide what is asked for and assign a variable or variable expression to the unknown quantities.
3. Organize a chart, or a table, or a diagram relating all the information provided.
4. Form an equation. (Possibly a formula of some type is necessary.)
5. Solve the equation.
6. Check your solution with the wording of the problem to be sure it makes sense.

Several types of problems lead to quadratic equations. The problems in this section are set up so that the equations can be solved by factoring. More general problems and approaches to solving quadratic equations are discussed in Chapter 10.

EXAMPLES

1. One number is four more than another, and the sum of their squares is 296. What are the numbers?

   Solution   Let x = smaller number
   $x + 4$ = larger number

   $$x^2 + (x + 4)^2 = 296 \qquad \text{Add the squares.}$$
   $$x^2 + x^2 + 8x + 16 = 296$$
   $$2x^2 + 8x - 280 = 0$$
   $$2(x^2 + 4x - 140) = 0$$
   $$2(x + 14)(x - 10) = 0 \qquad \text{The constant factor 2 does not affect the solution.}$$

   $$x + 14 = 0 \qquad \text{or} \qquad x - 10 = 0$$
   $$x = -14 \qquad\qquad\qquad x = 10$$
   $$x + 4 = -10 \qquad\qquad\quad x + 4 = 14$$

   There are two sets of answers to the problem, 10 and 14 or -14 and -10.

   Check $10^2 + 14^2 = 100 + 196 = 296$

   and $(-14)^2 + (-10)^2 = 196 + 100 = 296$

2. In an orange grove, there are 10 more trees in each row than there are rows. How many rows are there if there are 96 trees in the grove?

   Solution   Let r = number of rows
   $r + 10$ = number of trees per row

   $$r(r + 10) = 96$$
   $$r^2 + 10r = 96$$
   $$r^2 + 10r - 96 = 0$$
   $$(r - 6)(r + 16) = 0$$

   $$r - 6 = 0 \qquad \text{or} \qquad r + 16 = 0$$
   $$r = 6 \qquad\qquad\qquad \cancel{r = -16}$$

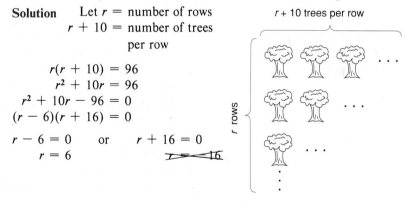

   ($r = -16$ is called an extraneous solution because -16 does not fit the conditions of the problem even though -16 is a solution to the equation.)

   There are 6 rows in the grove ($6 \cdot 16 = 96$ trees).

3. A rectangle has an area of 135 square meters and a perimeter of 48 meters. What are the dimensions of the rectangle?

Solution   The area of a rectangle is the product of its length and width ($A = \ell w$). The perimeter of a rectangle is given by $P = 2\ell + 2w$. Since the perimeter is 48 meters, then the length plus the width must be 24 meters.

Let w = width
$24 - w$ = length

$$w(24 - w) = 135 \qquad \text{Area} = \text{width times length.}$$
$$24w - w^2 = 135$$
$$0 = w^2 - 24w + 135$$
$$0 = (w - 9)(w - 15)$$

$w - 9 = 0$	or	$w - 15 = 0$
$w = 9$		$w = 15$
$24 - w = 15$		$24 - w = 9$

The width is 9 meters and the length is 15 meters ($9 \cdot 15 = 135$).

4. A man wants to build a block wall shaped like a rectangle along three sides of his property. If 180 feet of fencing are needed and the area of the lot is 4000 square feet, what are the dimensions of the lot?

Solution   Let x = one of two equal sides
$180 - 2x$ = third side

The product of width times length gives the area.

$$x(180 - 2x) = 4000$$
$$180x - 2x^2 = 4000$$

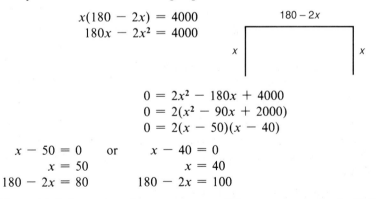

$$0 = 2x^2 - 180x + 4000$$
$$0 = 2(x^2 - 90x + 2000)$$
$$0 = 2(x - 50)(x - 40)$$

$x - 50 = 0$	or	$x - 40 = 0$
$x = 50$		$x = 40$
$180 - 2x = 80$		$180 - 2x = 100$

From this information, there are two possible answers: the lot is 50 feet by 80 feet, or the lot is 40 feet by 100 feet.

5. The sum of the squares of two positive consecutive odd integers is 202. Find the integers.

Solution   Let n = first odd integer
$n + 2$ = next consecutive odd integer

$$n^2 + (n + 2)^2 = 202$$
$$n^2 + n^2 + 4n + 4 = 202$$
$$2n^2 + 4n - 198 = 0$$
$$2(n^2 + 2n - 99) = 0$$
$$2(n - 9)(n + 11) = 0$$

$n = 9$ or   ~~$n = -11$~~   The problem asked for positive
$n + 2 = 11$ integers.

The first integer is 9 and the next consecutive odd integer is 11
$(9^2 + 11^2 = 81 + 121 = 202)$.   ∎

Special Note: Many of the problems in Exercises 5.5 relate to geometric figures. These figures and related formulas are discussed in detail in Section 1.5.

EXERCISES 5.5

Determine a quadratic equation for each of the following problems in Exercises 1–40. Then solve the equation.

1. The square of an integer is equal to seven times the integer. Find the integer.

2. The square of an integer is equal to twice the integer. Find the integer.

3. The square of a positive integer is equal to the sum of the integer and twelve. Find the integer.

4. If the square of a positive integer is added to three times the integer, the result is 28. Find the integer.

5. One number is seven more than another. Their product is 78. Find the numbers.

6. One positive number is three more than twice another. If the product is 27, find the numbers.

7. If the square of a positive integer is added to three times the number, the result is 54. Find the number.

8. One number is six more than another. The difference between their squares is 132. What are the numbers?

9. The difference between two positive integers is 8. If the smaller is added to the square of the larger, the sum is 124. Find the numbers.

10. One number is three less than twice another. The sum of their squares is 74. Find the numbers.

11. One number is five less than another. The sum of their squares is 97. Find the numbers.

12. Find a positive integer such that the product of the integer with a number three less than the integer is equal to the integer increased by thirty-two.

13. The product of two consecutive positive integers is 72. Find the integers.

14. Find two consecutive integers whose product is 110.

15. Find two consecutive positive integers such that the sum of their squares is 85.

16. Find two consecutive positive integers such that the square of the second integer added to four times the first is equal to 41.

17. Find two consecutive positive integers such that the difference between their squares is 17.

18. The product of two consecutive odd integers is 63. Find the integers.

19. The product of two consecutive even integers is 120. Find the integers.

20. The product of two consecutive even integers is 168. Find the integers.

21. The length of a rectangle is twice the width. The area is 72 square inches. Find the length and width of the rectangle.

22. The length of a rectangle is three times the width. If the area is 147 square centimeters, find the length and width.

23. The length of a rectangular yard is 12 meters greater than the width. If the area of the yard is 85 square meters, find the length and width.

24. The length of a rectangle is three centimeters greater than the width. The area is 108 square centimeters. Find the length and width.

25. The width of a rectangle is 4 feet less than the length. The area is 117 square feet. Find the length and width.

26. The altitude of a triangle is 4 feet less than the base. The area of the triangle is 16 square feet. Find the length of the base and altitude.

27. The base of a triangle exceeds the altitude by 5 meters. If the area is 42 square meters, find the length of the base and altitude.

28. The base of a triangle is 15 inches greater than the altitude. If the area is 63 square inches, find the length of the base.

29. The base of a triangle is 6 feet less than the altitude. The area is 56 square feet. Find the length of the altitude.

30. The perimeter of a rectangle is 32 inches. The area of the rectangle is 48 square inches. Find the dimensions.

31. The area of a rectangle is 104 square centimeters. If the perimeter is 42 centimeters, find the length and width.

32. The area of a rectangle is 24 square centimeters. If the perimeter is 20 centimeters, find the length and width.

33. The perimeter of a rectangle is 40 meters and the area is 96 square meters. Find the dimensions.

34. An orchard has 140 orange trees. The number of rows exceeds the number of trees per row by 13. How many trees are there in each row?

35. One formation for a drill team is rectangular. The number of members in each row exceeds the number of rows by 3. If there is a total of 108 members in the formation, how many rows are there?

36. A theater can seat 144 people. The number of rows is 7 less than the number of seats in each row. How many rows of seats are there?

37. The length of a rectangle is 7 centimeters greater than the width. If 4 centimeters are added to both the length and width, the new area would be 98 square centimeters. Find the dimensions of the original rectangle.

38. The width of a rectangle is 5 meters less than the length. If 6 meters are added to both the length and width, the new area will be 300 square meters. Find the dimensions of the original rectangle.

39. Suzie is going to fence a rectangular flower garden in her back yard. She has 50 feet of fencing, and she plans to use the house as one side of the garden. If the area is 300 square feet, what are the dimensions of the flower garden?

40. A rancher is going to build a corral with 52 yards of fencing. He is planning to use the barn as one side of the corral. If the area is 320 square yards, what are the dimensions?

◢ 5.6 *Additional Applications*

OBJECTIVES

In this section, you will be learning to:

1. Solve applied formulas.
2. Solve the resulting quadratic equations.

The following applications provide practice with equations found in real-life settings. The formulas are provided without any discussion of how or why they fit the described situation. Further studies in other fields will show how some of these formulas are generated.

Volume of a Cylinder

The volume of a cylinder is given by the formula

$$V = \pi r^2 h$$

where V is the volume

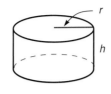

$\pi = 3.14$ (3.14 is an approximation for π)

r is the radius

h is the height

1. Find the volume of a cylinder with a radius of 6 in. and a height of 20 in.

2. Find the height of a cylinder if the volume is 282.6 in. and the radius is 3 in.

3. A cylinder has a height of 14 in. and a volume of 1099 cu in. Find the radius.

4. Find the radius of a cylinder whose volume is 2512 cu cm and whose height is 8 cm.

Load on a Wooden Beam

The safe load L of a horizontal wooden beam supported at both ends is expressed by the formula

$$L = \frac{kbd^2}{\ell}$$

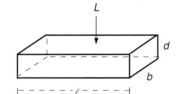

where L is expressed in pounds

b is the breadth or width in inches

d is the depth in inches

ℓ is the length in inches

k, a constant, depends on the grade of the beam and is expressed in pounds per square inch

5. What is the safe maximum load of a white pine beam 180 in. long, 3 in. wide, and 4 in. deep if $k = 3000$ lb per sq in.?

6. Find the constant k if a beam 144 in. long, 2 in. wide, and 6 in. deep supports a maximum load of 1100 lb.

7. A solid oak beam is required to support a load of 12,000 lb. It can be no more than 8 in. deep and is 192 in. long. For this grade of oak, $k = 6000$ lb per sq in. How wide should the beam be?

8. The safe maximum load of a white pine beam 4 in. wide and 150 in. long is 2880 lb. For white pine, $k = 3000$ lb per sq in. Find the depth of the beam.

9. A Douglas fir beam is required to support a load of 20,000 lb. For Douglas fir, $k = 4800$ lb per sq in. If the beam is 6 in. wide and 144 in. long, what is the minimum depth for the beam to support the required load?

Height of a Projectile

The equation

$$h = -16t^2 + v_0 t$$

gives the height h, in feet, that a body will be above the earth at time t, in seconds, if it is projected upward with an initial velocity v_0, in feet per second. (The initial velocity, v_0 is the velocity when $t = 0$.)

10. Find the height of an object 3 seconds after it has been projected upward at a rate of 56 feet per second.

11. Find the height of an object 5 seconds after it has been projected upward at a rate of 120 feet per second.

12. A ball is thrown upward with a velocity of 144 feet per second. When will it strike the ground?

13. An object is projected upward at a rate of 160 feet per second. Find the time when it is 384 feet above the ground.

14. An object is projected upward at a rate of 96 feet per second. Find the time when it is 144 feet above the ground.

Electrical Power

The power output of a generator armature is given by the equation

$$P_o = E_g I - r_g I^2$$

where P_o is measured in kilowatts

       E_g is measured in volts

       r_g is measured in ohms

       I is measured in amperes

15. Find I if $P_o = 120$ kilowatts, $E_g = 16$ volts, and $r_g = \dfrac{1}{2}$ ohm.

16. Find I if $P_o = 180$ kilowatts, $E_g = 22$ volts, and $r_g = \dfrac{2}{3}$ ohm.

Consumer Demand

The demand for a product is the number of units of the product that consumers are willing to buy when the market price is p dollars. The consumers' total expenditure for the product is found by multiplying the price times the demand.

17. When fishing reels are priced at p dollars, local consumers will buy $36 - p$ fishing reels. What is the price if total sales were $320?

18. A manufacturer can sell $100 - 2p$ lamps at p dollars each. If the receipts from the lamps total $1200, what is the price of the lamps?

Consumer Demand

The demand for a certain commodity is given by

$$D = -20p^2 + ap + 1200 \text{ units per month}$$

where p is the selling price

       a is a constant

19. Find the selling price if 1120 units are sold and $a = 60$.

20. Find the selling price if 1860 units are sold and $a = 232$.

◢ 5.7 Additional Factoring Review

Factoring is a basic skill necessary for success in algebra. Exercises 1–66 are provided as extra practice in this most important skill.

Factor completely.

1. $m^2 + 7m + 6$

2. $a^2 - 4a + 3$

3. $x^2 + 11x + 18$

4. $y^2 + 8y + 15$

5. $x^2 - 100$

6. $n^2 - 8n + 12$

7. $m^2 - m - 6$

8. $y^2 - 49$

9. $a^2 + 2a + 24$ 10. $x^2 + 12x + 35$ 11. $64x^2 - 1$ 12. $x^2 - 4x - 21$

13. $x^2 + 10x + 25$ 14. $x^2 + 3x - 10$ 15. $x^2 + 9x - 36$ 16. $x^2 + 17x + 72$

17. $x^2 + 13x + 36$ 18. $2y^2 - 24y + 70$ 19. $5x^2 - 70x + 240$ 20. $7x^2 + 14x - 168$

21. $200 + 20x - 4x^2$ 22. $64 + 49x^2$ 23. $3x^2 - 147$ 24. $x^3 - 4x^2 - 12x$

25. $3x^3 + 15x^2 + 18x$ 26. $112x - 2x^2 - 2x^3$ 27. $16x^3 - 100x$ 28. $2x^2 - 3x + 1$

29. $3x^2 - 17x + 10$ 30. $2x^2 + 7x + 3$ 31. $4x^2 - 14x + 6$ 32. $6x^2 - 11x + 4$

33. $12x^2 - 32x + 5$ 34. $12x^2 + x - 6$ 35. $6x^2 + x - 35$ 36. $4x^2 - 18x + 20$

37. $8x^2 + 6x - 35$ 38. $12x^2 + 5x - 3$ 39. $20x^2 - 21x - 54$ 40. $150x^2 - 96$

41. $12x^2 - 60x - 75$ 42. $14 + 11x - 15x^2$ 43. $24 + x - 3x^2$ 44. $21x^2 - x - 10$

45. $8x^2 - 22x + 15$ 46. $63x^2 - 40x - 12$ 47. $20x^2 + 9x - 20$ 48. $35x^2 - x - 6$

49. $18x^2 - 15x + 2$ 50. $12x^2 - 47x + 11$ 51. $252x - 175x^3$ 52. $12x^3 + 2x^2 + 70x$

53. $21x^3 - 13x^2 + 2x$ 54. $36x^3 + 120x^2 + 100x$ 55. $36x^3 + 21x^2 - 30x$

56. $63x - 3x^2 - 30x^3$ 57. $16x^3 - 52x^2 + 22x$ 58. $24x^3 - 4x^2 - 160x$

59. $75 + 10x - 120x^2$ 60. $144x^3 - 10x^2 - 50x$ 61. $xy + 3y - 4x - 12$

62. $2xz + 10x + z + 5$ 63. $x^2 + 2xy - 6x - 12y$ 64. $2y^2 + 6yz + 5y + 15z$

65. $x^3 - 8x^2 - 5x + 40$ 66. $2x^3 - 14x^2 - 3x + 21$

CHAPTER 5 SUMMARY

Key Terms and Formulas

Finding the greatest common monomial factor in a polynomial means to choose the monomial with the highest degree and the largest integer coefficient that will divide into each term of the polynomial. [5.1]

Special forms:

I. $(x^2 - a^2) = (x + a)(x - a)$
   Difference of two squares

II. $x^2 + 2ax + a^2 = (x + a)^2$
   Perfect square trinomial

III. $x^2 - 2ax + a^2 = (x - a)^2$
   Perfect square trinomial [5.2]

Adding a square term to a binomial so that the resulting trinomial is a perfect square trinomial is called **completing the square.** [5.2]

A polynomial is **irreducible** (or **not factorable**) if it cannot be factored as a product of polynomials with integer coefficients. A polynomial that is the **sum of two squares** $(x^2 + a^2)$ is not factorable. [5.3]

Equations of the form

$$ax^2 + bx + c = 0 \qquad (a \neq 0)$$

are called **quadratic equations.** [5.4]

Properties and Rules

If $a \cdot b = 0$, then $a = 0$ or $b = 0$. [5.2]

Procedures

A Summary of Procedures to Follow in Factoring   [5.3]
1. Look for a common factor.
2. Check the number of terms:
   a. Two terms
      (1) difference of two squares?—factorable
      (2) sum of two squares?—not factorable
   b. Three terms
      (1) perfect square trinomial?
      (2) use trial-and-error method?
      (3) use the *ac*-method?
   c. Four terms
      (1) group terms with a common factor
3. Check the possibility of factoring any of the factors.

To Solve a Quadratic Equation by Factoring   [5.4]
1. Add or subtract terms as necessary so that 0 is on one side of the equation and the equation is in the form

$$ax^2 + bx + c = 0 \quad (a \neq 0)$$

2. If there are any fractional coefficients, multiply each term by the least common denominator so that all coefficients will be integers.
3. Factor the quadratic, if possible.
4. Set each nonconstant factor equal to 0 and solve for the unknown.
5. Remember, there will never be more than two solutions.

Attack Plan for Word Problems   [5.5]
1. Read the problem carefully at least twice.
2. Decide what is asked for and assign a variable or variable expression to the unknown quantities.
3. Organize a chart, or a table, or a diagram relating all the information provided.
4. Form an equation. (Possibly a formula of some type is necessary.)
5. Solve the equation.
6. Check your solution with the wording of the problem to be sure it makes sense.

CHAPTER 5 REVIEW

Simplify the expressions in Exercises 1–5. Assume that no denominator is equal to 0.   [5.1]

1. $\dfrac{x^5}{x^2}$ **2.** $\dfrac{x^4}{x}$ **3.** $\dfrac{4x^3}{2x^2}$ **4.** $\dfrac{-24x^4y^2}{3x^2y}$ **5.** $\dfrac{36x^3y^5}{9xy^4}$

Factor completely Exercises 6–30.   [5.1–5.3]

6. $5x - 10$ **7.** $-12x^2 - 16x$ **8.** $16x^2y - 24xy$
9. $10x^4 - 25x^3 + 5x^2$ **10.** $4x^2 - 1$ **11.** $y^2 - 20y + 100$
12. $81x^2 - 4y^2$ **13.** $4x^2 - 12x + 9$ **14.** $ac^2 + a^2b^2 + ad$
15. $3x^2 - 48y^2$ **16.** $x^2 - 7x - 18$ **17.** $5x^2 + 40x + 80$
18. $5x^2 + 17x + 6$ **19.** $x^2 + x - 30$ **20.** $x^2 + x + 3$
21. $25x^2 + 20x + 4$ **22.** $3x^2 + 5x + 2$ **23.** $2x^3 - 20x^2 + 50x$
24. $4x^3 + 4xz^2$ **25.** $6x^2 - x - 2$ **26.** $8x^3 - 10x^2 - 12x$
27. $x^4 + 3x^2 - 28$ **28.** $x^4 - 2x^2 - 8$ **29.** $xy + 3x + 2y + 6$
30. $ax - 2a - 2b + bx$

Complete the square by adding the correct constant in Exercises 31–35 so that the trinomials in each will factor as indicated.   [5.2]

31. $x^2 - 4x + \underline{\hspace{1.5em}} = ($ $)^2$ **32.** $x^2 + 18x + \underline{\hspace{1.5em}} = ($ $)^2$
33. $x^2 - 8x + \underline{\hspace{1.5em}} = ($ $)^2$ **34.** $x^2 - \underline{\hspace{1.5em}} + 25 = ($ $)^2$
35. $x^2 + \underline{\hspace{1.5em}} + \dfrac{25}{4} = ($ $)^2$

Solve the equations in Exercises 36–54. [5.4]

36. $(x - 7)(x + 1) = 0$

37. $x(3x + 5) = 0$

38. $2x(x + 5)(x - 2) = 0$

39. $21x - 3x^2 = 0$

40. $x^2 + 8x + 12 = 0$

41. $x^2 = 3x + 28$

42. $\dfrac{x^2}{3} - 3 = 0$

43. $x^3 + 5x^2 - 6x = 0$

44. $x^2 - 55 = 6x$

45. $\dfrac{1}{4}x^2 + x - 15 = 0$

46. $15 - 12x - 3x^2 = 0$

47. $x^3 + 14x^2 + 49x = 0$

48. $8x = 12x + 2x^2$

49. $6x^2 = 24x$

50. $5x^2 - 7x - 6 = 0$

51. $4x^2 + 9x - 9 = 0$

52. $(4x - 3)(x + 1) = 2$

53. $2x(x - 2) = x + 25$

54. $\dfrac{3}{2}x^2 - \dfrac{7}{4}x - 5 = 0$

55. The difference between two positive numbers is 9. If the smaller is added to the square of the larger, the result is 147. Find the numbers. [5.5]

56. The length of a rectangle exceeds the width by 5 cm. If both dimensions were increased by 3 cm, the area would be increased by 96 square centimeters. Find the dimensions of the original rectangle. [5.5]

57. Find two consecutive whole numbers such that the sum of their squares is equal to 145. [5.5]

58. The length of a rectangle is 1 foot more than twice the width. If the area is 78 square feet, find the length and width. [5.5]

59. The base of a triangle is 5 inches longer than the altitude. If the area is 187 square inches, find the length of the base and altitude. [5.5]

60. A ball is thrown upward with an initial velocity of 88 feet per second. When will it be 120 feet above the ground? ($h = -16t^2 + v_0 t$.) [5.6]

CHAPTER 5 TEST

Simplify the expressions in Exercises 1 and 2. Assume that no denominator is 0.

1. $\dfrac{16x^4}{8x}$

2. $\dfrac{42x^3y^3}{-6x^3y}$

Factor completely each of the polynomials in Exercises 3–12.

3. $20x^3y^2 + 18x^3y - 15x^4y$

4. $x^2 - 9x + 20$

5. $x^2 + 14x + 49$

6. $36x^2 - 1$

7. $6x^2 + x - 5$

8. $3x^2 + x - 24$

9. $16x^2 - 25y^2$

10. $2x^3 - x^2 - 3x$

11. $6x^2 - 13x + 6$

12. $2xy - 3y + 14x - 21$

Add the correct term in Exercises 13–15 so that the trinomials factor as indicated.

13. $x^2 - 10x + \underline{\hspace{1cm}} = (\quad)^2$

14. $x^2 + 16x + \underline{\hspace{1cm}} = (\quad)^2$

15. $x^2 + \underline{\hspace{1cm}} + 144 = (\quad)^2$

Solve the equations in Exercises 16–22.

16. $(x + 2)(3x - 5) = 0$

17. $x^2 - 7x - 8 = 0$

18. $-3x^2 = 18x$

19. $\dfrac{2x^2}{5} - 6 = \dfrac{4x}{5}$

20. $4x^2 + 17x - 15 = 0$

21. $8x^2 - 2x - 15 = 0$

22. $(2x - 7)(x + 1) = 6x - 19$

Solve the equations in Exercises 23–25.

23. One number is 10 less than five times another number. Their product is 120. Find the numbers.

24. The product of two consecutive positive integers is 342. Find the two integers.

25. The length of a rectangle is 7 centimeters less than twice the width. If the area of the rectangle is 165 square centimeters, find the length and width.

CUMULATIVE REVIEW (5)

Find the LCM for each set of numbers in Exercises 1 and 2.

1. 20, 12, 24

2. $8x^2$, $14x^2y$, $21xy$

Perform the indicated operation in Exercises 3–6. Reduce all answers to lowest terms.

3. $\dfrac{7}{12} + \dfrac{9}{16}$

4. $\dfrac{11}{15a} - \dfrac{5}{12a}$

5. $\dfrac{6x}{25} \cdot \dfrac{5}{4x}$

6. $\dfrac{40}{92} \div \dfrac{2}{15x}$

Simplify each of the expressions in Exercises 7 and 8 so that it has only positive exponents.

7. $\dfrac{-4x^2y}{12xy^{-3}}$

8. $\dfrac{21xy^2}{3x^{-1}y}$

Simplify each of the expressions in Exercises 9 and 10.

9. $2(4x + 3) + 5(x - 1)$

10. $2x(x + 5) - (x + 1)(x - 3)$

Perform the indicated operations in Exercises 11–16.

11. $(x^2 + 3x - 1) + (2x^2 - 8x + 4)$

12. $(-2x^2 + 6x - 7) + (2x^2 - x - 1)$

13. $(2x^2 - 5x + 3) - (x^2 - 2x + 8)$

14. $(x + 1) - (4x^2 + 3x - 2)$

15. $(2x - 7)(x + 4)$

16. $-(x + 6)(3x - 1)$

Factor each expression as completely as possible in Exercises 17–20.

17. $8x^2 - 20x$

18. $6x^2 - 96$

19. $2x^2 - 15x + 18$

20. $6x^2 - x - 12$

DID YOU KNOW ?

Pythagoras (c. 550 B.C.) was a Greek mathematician who founded a secret brotherhood whose objective was to investigate music, astronomy, and mathematics, especially geometry. The Pythagoreans believed that all physical phenomena were explainable in terms of "arithmos," the basic properties of whole numbers and their ratios.

In Chapter 6, you will be studying rational numbers (fractions), numbers written as the ratio of two integers. The Pythagorean Society attached mystical significance to these rational numbers. Their investigations of musical harmony revealed that a vibrating string must be divided exactly into halves, thirds, fourths, fifths, and so on, to produce tones in harmony with the string vibrating as a whole.

The recognition that the chords which are pleasing to the ear correspond to exact divisions of the vibrating string by whole numbers stimulated Pythagoras to propose that all physical properties could be described using rational numbers. The Society attempted to compute the orbits of the planets by relating them to the musical intervals. The Pythagoreans thought that as the planets moved through space they produced music, the music of the spheres.

Unfortunately, a scandalous idea soon developed within the Pythagorean Society. It became apparent that there were some numbers that could

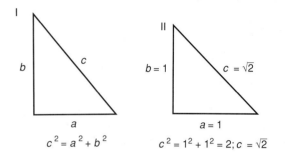

$$c^2 = a^2 + b^2 \qquad c^2 = 1^2 + 1^2 = 2; c = \sqrt{2}$$

not be represented as the ratio of two whole numbers. This shocking idea came about when the Pythagoreans investigated their master's favorite theorem, the Pythagorean Theorem:

Given a right triangle, the length of the hypotenuse squared is equal to the sum of the lengths squared of the other two sides (as illustrated in the diagrams). The hypotenuse of the second triangle is some number that, when squared, gives two.

It can be proven (and it was by the Pythagoreans) that no such rational number exists. Imagine the amazement within the Brotherhood. Legend has it that the Pythagoreans attempted to keep the "irrational" numbers a secret. Hippaes, the brother who first told an outsider about the new numbers, was supposedly expelled from the Society and punished by death for his unfaithfulness.

Number rules the Universe.
The Pythagoreans

◢ CHAPTER OUTLINE

*R*ational expressions are fractions. So, if you know how to work with fractions in arithmetic, you will find the same techniques used in Chapter 6. You will need to find a common denominator to add (or subtract) rational expressions, and division is accomplished by multiplying by the reciprocal of the divisor. The basic difference is that the numerators and denominators are polynomials and the skills of multiplying and factoring that you learned in Chapters 4 and 5 will be needed. Keep in mind, throughout Chapter 6, that you are following the same rules you learned for fractions in arithmetic and some of the difficult-looking expressions may seem considerably easier.

◢ 6.1 Division with Polynomials

OBJECTIVES

In this section, you will be learning to:

1. Divide polynomials by monomials.

2. Divide polynomials by binomials by using long division.

Fractions, such as $\dfrac{135}{8}$ and $\dfrac{7}{8}$, in which the numerator and denominator are integers are called **rational numbers.** Fractions in which the numerator and denominator are polynomials are called **rational expressions.** (No denominator can be 0.)

Rational numbers: $\dfrac{2}{3}, \dfrac{-5}{16}, \dfrac{22}{7}, \dfrac{0}{6}, \dfrac{3}{4}$, and $-\dfrac{3}{4}$

Rational expressions: $\dfrac{x}{x^2 + 1}, \dfrac{x^2 + 5x + 6}{x^2 + 7x + 12}, \dfrac{x^2 + 2x + 1}{x}$, and $\dfrac{1}{5x}$

In this section, we want to treat a rational expression as an indicated division problem. From this basis, there are two situations to consider:

1. the denominator (divisor) is a monomial, or
2. the denominator (divisor) is not a monomial.

Monomial Denominators

We know from arithmetic that the sum of fractions with the same denominator can be written as a single fraction by adding the numerators and using the common denominator. For example,

$$\frac{3}{a} + \frac{2b}{a} + \frac{c}{a} = \frac{3 + 2b + c}{a}$$

Reversing this process (with a monomial in the denominator), we can write

$$\frac{3x^3 + 6x^2 + 9x}{3x} = \frac{3x^3}{3x} + \frac{6x^2}{3x} + \frac{9x}{3x} = x^2 + 2x + 3$$

Each term in the numerator is divided by the denominator. Similarly,

$$\frac{x^2 + 2x + 1}{x} = \frac{x^2}{x} + \frac{2x}{x} + \frac{1}{x} = x + 2 + \frac{1}{x}$$

and $$\frac{3xy + 6xy^2 - 18}{2y} = \frac{3xy}{2y} + \frac{6xy^2}{2y} - \frac{18}{2y} = \frac{3x}{2} + 3xy - \frac{9}{y}$$

EXAMPLES   Divide each polynomial by the monomial denominator by writing a sum of fractions. Reduce each fraction, if possible.

1. $$\frac{8x^2 - 14x + 1}{2}$$

    Solution $$\frac{8x^2 - 14x + 1}{2} = \frac{8x^2}{2} - \frac{14x}{2} + \frac{1}{2} = 4x^2 - 7x + \frac{1}{2}$$

2. $$\frac{12x^3 + 3x^2 - 9x}{3x^2}$$

    Solution $$\frac{12x^3 + 3x^2 - 9x}{3x^2} = \frac{12x^3}{3x^2} + \frac{3x^2}{3x^2} - \frac{9x}{3x^2}$$

    $$= 4x + 1 - \frac{3}{x}$$

3. $$\frac{10x^2y + 25xy + 3y^2}{5xy^2}$$

    Solution $$\frac{10x^2y + 25xy + 3y^2}{5xy^2} = \frac{10x^2y}{5xy^2} + \frac{25xy}{5xy^2} + \frac{3y^2}{5xy^2}$$

    $$= \frac{2x}{y} + \frac{5}{y} + \frac{3}{5x}$$ ∎

The Division Algorithm

In arithmetic, the process (or series of steps) that we follow in dividing two numbers is called the **division algorithm** (or **long division**). By this division algorithm, we can find $135 \div 8$ as follows:

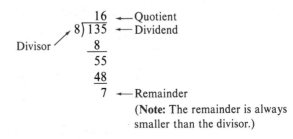

Check: $8 \cdot 16 + 7 = 128 + 7 = 135$

We can also write the division in fraction form and the remainder over the divisor, giving a mixed number.

$$135 \div 8 = \frac{135}{8} = 16\frac{7}{8}$$

In algebra, the division algorithm with polynomials is quite similar. If we divide one polynomial by another, the quotient will be another polynomial with a remainder. Symbolically, if we have $P \div D$, then

$$P = Q \cdot D + R$$

or

$$\frac{P}{D} = Q + \frac{R}{D}$$

where P is the dividend, D is the divisor, Q is the quotient, and R is the remainder. **The remainder must be of smaller degree than the divisor. If the remainder is 0, then the divisor and quotient are factors of the dividend.**

In the following two examples, the **division algorithm** is illustrated in a step-by-step form with first-degree divisors. Each step is explained in detail and should be studied carefully.

EXAMPLES

4. $\dfrac{3x^2 - 5x + 4}{x + 2}$ or $(3x^2 - 5x + 4) \div (x + 2)$

Solution

	Steps	Explanation
Step 1	$x + 2 \overline{)3x^2 - 5x + 4}$	Write both polynomials in order of descending powers. If any powers are missing, fill in with 0s.
Step 2	$\begin{array}{r} 3x \\ x + 2 \overline{)3x^2 - 5x + 4} \end{array}$	Mentally divide $3x^2$ by x. $\dfrac{3x^2}{x} = 3x$. Write $3x$ above $3x^2$.
Step 3	$\begin{array}{r} 3x \\ x + 2 \overline{)3x^2 - 5x + 4} \\ 3x^2 + 6x \end{array}$	Multiply $3x$ times $(x + 2)$, and write the terms under the like terms in the dividend.
Step 4	$\begin{array}{r} 3x \\ x + 2 \overline{)3x^2 - 5x + 4} \\ \mp 3x^2 \mp 6x \\ -11x \end{array}$	Subtract $3x^2 + 6x$ by changing signs and adding.
Step 5	$\begin{array}{r} 3x \\ x + 2 \overline{)3x^2 - 5x + 4} \\ \mp 3x^2 \mp 6x \\ -11x + 4 \end{array}$	Bring down the 4.

	Steps	**Explanation**

Step 6

$$\begin{array}{r} 3x - 11 \\ x + 2 \overline{)\, 3x^2 - 5x + 4\,} \\ \underline{\mp 3x^2 \mp 6x} \\ -11x + 4 \end{array}$$

Mentally divide $-11x$ by x.

$\dfrac{-11x}{x} = -11$. Write -11 in the quotient directly above $-11x$.

Step 7

$$\begin{array}{r} 3x - 11 \\ x + 2 \overline{)\, 3x^2 - 5x + 4\,} \\ \underline{\mp 3x^2 \mp 6x} \\ -11x + 4 \\ -11x - 22 \end{array}$$

Multiply -11 times $(x + 2)$, and write the terms under like terms in the expression $-11x + 4$.

Step 8

$$\begin{array}{r} 3x - 11 \\ x + 2 \overline{)\, 3x^2 - 5x + 4\,} \\ \underline{\mp 3x^2 \mp 6x} \\ -11x + 4 \\ \underline{\pm 11x \pm 22} \\ 26 \end{array}$$

Subtract $-11x - 22$ by changing signs and adding. The remainder is 26.

Step 9   **Check:** Multiply the divisor and quotient, and add the remainder; the result should be the dividend.

$$(3x - 11)(x + 2) + 26 = 3x^2 + 6x - 11x - 22 + 26$$
$$= 3x^2 - 5x + 4$$

The answer can also be written in the form $Q + \dfrac{R}{D}$ as

$$3x - 11 + \frac{26}{x + 2}$$

5. $\dfrac{4x^3 - 7x - 5}{2x + 1}$ or $(4x^3 - 7x - 5) \div (2x + 1)$

This example will be done in fewer steps. Note that $0x^2$ is inserted so that like terms will be aligned.

Solution

	Steps	**Explanation**

Step 1 $2x + 1 \overline{)\, 4x^3 + 0x^2 - 7x - 5\,}$

x^2 is a missing power, so $0x^2$ is supplied.

Step 2

$$\begin{array}{r} 2x^2 \\ 2x + 1 \overline{)\, 4x^3 + 0x^2 - 7x - 5\,} \\ \underline{4x^3 + 2x^2} \end{array}$$

$\dfrac{4x^3}{2x} = 2x^2$

$2x^2(2x + 1) = 4x^3 + 2x^2$

Step 3

$$\begin{array}{r} 2x^2 \\ 2x + 1 \overline{)\, 4x^3 + 0x^2 - 7x - 5\,} \\ \underline{\mp 4x^3 \mp 2x^2} \\ -2x^2 - 7x \end{array}$$

Subtract $4x^3 + 2x^2$ and bring down $-7x$.

Steps	Explanation

Step 4
$$2x + 1 \overline{)\begin{array}{l} 2x^2 - x \\ 4x^3 + 0x^2 - 7x - 5 \end{array}}$$
$$\mp 4x^3 \mp 2x^2$$
$$ -2x^2 - 7x$$
$$ -2x^2 - x$$

$\dfrac{-2x^2}{2x} = -x$

Multiply $-x$ times $(2x + 1)$.

Step 5
$$2x + 1 \overline{)\begin{array}{l} 2x^2 - x - 3 \\ 4x^3 + 0x^2 - 7x - 5 \end{array}}$$
$$\mp 4x^3 \mp 2x^2$$
$$ 2x^2 - 7x$$
$$ \pm 2x^2 \pm x$$
$$ -6x - 5$$
$$ \pm 6x \pm 3$$
$$ -2$$

Continue dividing, using the same procedure. The remainder is -2.

Step 6   **Check:** $(2x + 1)(2x^2 - x - 3) - 2$
$$= 4x^3 - 2x^2 - 6x + 2x^2 - x - 3 - 2$$
$$= 4x^3 - 7x - 5$$

The answer can also be written in the form

$$2x^2 - x - 3 + \frac{-2}{2x + 1} \quad \text{or} \quad 2x^2 - x + 3 - \frac{2}{2x + 1} \ \blacksquare$$

The division algorithm can be applied whenever the degree of the dividend, P, is greater than or equal to the degree of the divisor, D. As was stated earlier, the remainder, R, must be of smaller degree than the divisor, D.

EXAMPLE

6. Use the division algorithm (or long division) to divide:
    $(x^3 - 3x^2 - 5x - 8) \div (x^2 + 2x + 3)$
    Solution

$$x^2 + 2x + 3 \overline{)\begin{array}{l} x - 5 \\ x^3 - 3x^2 - 5x - 8 \end{array}}$$
$$\mp x^3 \mp 2x^2 \mp 3x$$
$$ -5x^2 - 8x - 8$$
$$ \pm 5x^2 \pm 10x \pm 15$$
$$ 2x + 7$$

Quotient

Remainder is first-degree, smaller degree than the divisor.

$$Q + \frac{R}{D} = x - 5 + \frac{2x + 7}{x^2 + 2x + 3}$$

Check $(x^2 + 2x + 3)(x - 5) + (2x + 7)$
$$= x^3 + 2x^2 + 3x - 5x^2 - 10x - 15 + 2x + 7$$
$$= x^3 - 3x^2 - 5x - 8 \qquad\qquad\qquad \blacksquare$$

Practice Problems

1. Express the quotient as a sum of fractions:

$$\frac{4x^2 + 6x + 1}{2x}$$

Use the division algorithm to divide.

2. $(3x^2 + 8x - 4) \div (x + 2)$

3. $(x^3 + 4x^2 - 5) \div (x^2 + x - 1)$

EXERCISES 6.1

Express each quotient in Exercises 1–19 as a sum of fractions and simplify if possible.

1. $\dfrac{4x^2 + 8x + 3}{4}$

2. $\dfrac{6x^2 - 10x + 1}{2}$

3. $\dfrac{10x^2 - 15x - 3}{5}$

4. $\dfrac{9x^2 - 12x + 5}{3}$

5. $\dfrac{2x^2 + 5x}{x}$

6. $\dfrac{8x^2 - 7x}{x}$

7. $\dfrac{x^2 + 6x - 3}{x}$

8. $\dfrac{-2x^2 - 3x + 8}{x}$

9. $\dfrac{4x^2 + 6x - 3}{2x}$

10. $\dfrac{3x^3 - 2x^2 + x}{x^2}$

11. $\dfrac{6x^3 - 9x^2 - 3x}{3x^2}$

12. $\dfrac{5x^2y - 10xy^2 - 3y}{5xy}$

13. $\dfrac{7x^2y^2 + 21xy^3 - 11y^3}{7xy^2}$

14. $\dfrac{12x^3y + 6x^2y^2 - 3xy}{6x^2y}$

15. $\dfrac{3x^2y - 8xy^2 - 4y^3}{4xy}$

16. $\dfrac{2x^3y^2 - 6x^2y^3 + 15xy}{3xy^2}$

17. $\dfrac{5x^2y^2 - 8xy^2 + 16xy}{8xy^2}$

18. $\dfrac{3x^3y - 14x^2y - 7xy}{7x^2y}$

19. $\dfrac{8x^3y^2 - 9x^2y^3 + 5xy}{9xy^2}$

Divide in Exercises 20–48 using the long division procedure, and check your answer.

20. $119 \div 17$

21. $278 \div 23$

22. $326 \div 64$

23. $437 \div 59$

24. $(x^2 + 3x + 2) \div (x + 2)$

25. $(x^2 - 5x - 6) \div (x + 3)$

26. $(y^2 + 8y + 15) \div (y + 4)$

27. $(a^2 - 2a - 15) \div (a - 2)$

28. $(x^2 - 7x - 18) \div (x + 5)$

29. $(y^2 - y - 42) \div (y - 4)$

30. $(4a^2 - 21a + 2) \div (a - 6)$

31. $(5y^2 + 14y - 7) \div (y + 5)$

32. $(8x^2 + 10x - 4) \div (2x + 3)$

33. $(8c^2 + 2c - 14) \div (2c + 3)$

34. $(6x^2 + x - 4) \div (2x - 1)$

35. $(10m^2 - m - 6) \div (5m - 3)$

36. $(x^2 - 6) \div (x + 2)$

37. $(x^2 + 3x) \div (x + 5)$

38. $(2x^3 + 4x^2 - x + 1) \div (x - 3)$

39. $(y^3 - 9y^2 + 26y - 24) \div (y - 2)$

40. $(3t^3 + 10t^2 + 6t + 3) \div (3t + 1)$

41. $(12a^3 - 3a^2 + 4a + 1) \div (4a - 1)$

Answers to Practice Problems   1. $2x + 3 + \dfrac{1}{2x}$ 2. $3x + 2 - \dfrac{8}{x + 2}$

3. $x + 3 + \dfrac{-2x - 2}{x^2 + x - 1}$

42. $(2x^3 + x^2 - 6) \div (x + 4)$
44. $(x^3 - 8) \div (x - 2)$
46. $(x^3 + 7x^2 + x - 2) \div (x^2 - x + 1)$
48. $(x^3 + 3x^2 + 1) \div (x^2 + 2x + 3)$

43. $(x^3 + 2x^2 - 5) \div (x - 5)$
45. $(x^3 + 27) \div (x + 3)$
47. $(2x^3 - x + 3) \div (x^2 - 2)$

6.2 *Multiplication and Division*

OBJECTIVES

In this section, you will be learning to:

1. Reduce rational expressions to lowest terms.
2. Multiply rational expressions.
3. Divide rational expressions.

Before we proceed further with operations with rational expressions, we want to emphasize that the rules are essentially the same as those in arithmetic with fractions. That is, adding, subtracting, multiplying and dividing with rational expressions are operations involving factoring and common denominators just as with fractions in arithmetic. The following rules for arithmetic with fractions were reviewed in Chapter 1.

Summary of Arithmetic Rules for Fractions

A **fraction** is a number that can be written in the form $\frac{a}{b}$ and means $a \div b$ where $b \neq 0$. No denominator can be 0.

The Fundamental Principle: $\frac{a}{b} = \frac{a \cdot k}{b \cdot k}$ where $k \neq 0$

The **reciprocal** of $\frac{a}{b}$ is $\frac{b}{a}$ and $\frac{a}{b} \cdot \frac{b}{a} = 1$.

Multiplication: $\frac{a}{b} \cdot \frac{c}{d} = \frac{a \cdot c}{b \cdot d}$

Division: $\frac{a}{b} \div \frac{c}{d} = \frac{a}{b} \cdot \frac{d}{c}$

Addition: $\frac{a}{b} + \frac{c}{b} = \frac{a + c}{b}$

Subtraction: $\frac{a}{b} - \frac{c}{b} = \frac{a - c}{b}$

Reducing Rational Expressions

To reduce a rational expression, factor both the numerator and denominator and use the Fundamental Principle to "divide out" any common **factors.** For example,

fraction: $\frac{15}{35} = \frac{\cancel{5} \cdot 3}{\cancel{5} \cdot 7} = \frac{3}{7}$

rational expression: $\frac{x^2 + 5x + 6}{x^2 + 7x + 12} = \frac{\cancel{(x + 3)}(x + 2)}{\cancel{(x + 3)}(x + 4)} = \frac{x + 2}{x + 4}$

(**Note:** Since no denominator can be 0, $x + 3 \neq 0$ and $x + 4 \neq 0$ or $x \neq -3$ and $x \neq -4$. This includes all fractions in the original expression.)

Common Error

The key word when reducing is **factor.** Many students incorrectly "divide out" terms that are not factors.

"Divide out" only common factors.

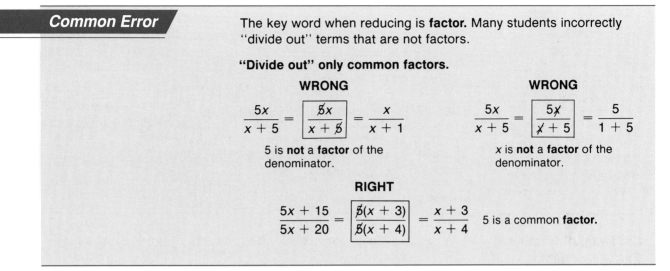

	WRONG			**WRONG**	

$$\frac{5x}{x + 5} = \boxed{\frac{\cancel{5}x}{x + \cancel{5}}} = \frac{x}{x + 1} \qquad\qquad \frac{5x}{x + 5} = \boxed{\frac{5\cancel{x}}{\cancel{x} + 5}} = \frac{5}{1 + 5}$$

5 is **not** a **factor** of the denominator. x is **not** a **factor** of the denominator.

RIGHT

$$\frac{5x + 15}{5x + 20} = \boxed{\frac{\cancel{5}(x + 3)}{\cancel{5}(x + 4)}} = \frac{x + 3}{x + 4} \qquad \text{5 is a common } \textbf{factor.}$$

EXAMPLES Reduce each rational expression to lowest terms. State any restrictions on the variable because **no denominator can be 0.**

1. $\dfrac{2x + 4}{3x + 6}$

Solution $\dfrac{2x + 4}{3x + 6} = \dfrac{2\cancel{(x + 2)}}{3\cancel{(x + 2)}} = \dfrac{2}{3}$ Since $x + 2 \neq 0$, $x \neq -2$.

2. $\dfrac{x^2 - 16}{x - 4}$

Solution $\dfrac{x^2 - 16}{x - 4} = \dfrac{(x + 4)\cancel{(x - 4)}}{\cancel{x - 4}} = x + 4$ Since $x - 4 \neq 0$, $x \neq 4$.

3. $\dfrac{a}{a^2 - 5a}$

Solution

$$\frac{a}{a^2 - 5a} = \frac{\overset{1}{\cancel{a}}}{\cancel{a}(a - 5)} = \frac{1}{a - 5} \quad (a \neq 0, 5)$$ Note that these restrictions are determined before reducing.

This example illustrates the importance of writing 1 in the numerator if all the factors divide out. 1 is an understood factor.

4. $\dfrac{3 - y}{y - 3}$

Solution $\dfrac{3 - y}{y - 3} = \dfrac{-1(-3 + y)}{y - 3} = \dfrac{-1\cancel{(y - 3)}}{\cancel{y - 3}} = -1 \quad (y \neq 3)$ ∎

In Example 4, the result was -1. The two expressions $3 - y$ and $y - 3$ are opposites of each other. That is,

$$-1(3 - y) = -3 + y = y - 3$$

When opposites are divided, the quotient is -1.

$$\dfrac{-8}{+8} = -1, \qquad \dfrac{14}{-14} = -1, \qquad \dfrac{x - 5}{5 - x} = \dfrac{\overset{1}{\cancel{(x - 5)}}}{-1\cancel{(x - 5)}} = \dfrac{1}{-1} = -1$$

In general, for $b \neq a$, $(a - b)$ and $(b - a)$ are opposites and

$$\dfrac{a - b}{b - a} = \dfrac{a - b}{-1(a - b)} = -1$$

Now we want to clarify the following fact about negative fractions. Without changing the value of the fraction, the negative sign can be placed

1. in front of the fraction, or
2. with the numerator, or
3. with the denominator.

For example,

$$-\dfrac{6}{2} = -3, \qquad \dfrac{-6}{2} = -3, \qquad \text{and} \qquad \dfrac{6}{-2} = -3$$

Thus,

$$-\dfrac{6}{2} = \dfrac{-6}{2} = \dfrac{6}{-2}$$

We can make the following general statements about the placement of negative signs.

For integers a and b, $b \neq 0$,

$$-\dfrac{a}{b} = \dfrac{-a}{b} = \dfrac{a}{-b}$$

For polynomials P and Q, $Q \neq 0$,

$$-\dfrac{P}{Q} = \dfrac{-P}{Q} = \dfrac{P}{-Q}$$

At this stage of the discussion, we will assume that no denominator is 0. You should keep in mind that there are certain restrictions on the variable, but these restrictions will not be stated.

EXAMPLES   Simplify the following expressions.

5. $-\dfrac{-2x + 6}{x^2 - 3x}$

Solution $-\dfrac{-2x + 6}{x^2 - 3x} = \dfrac{-(-2x + 6)}{x^2 - 3x} = \dfrac{2x - 6}{x^2 - 3x}$

$$= \dfrac{2(x - 3)}{x(x - 3)} = \dfrac{2}{x}$$

6. $-\dfrac{5 - 10x}{6x - 3}$

Solution $-\dfrac{5 - 10x}{6x - 3} = -\dfrac{5(1 - 2x)}{3(2x - 1)} = -\dfrac{-5}{3} = \dfrac{5}{3}$

or

$$-\dfrac{5 - 10x}{6x - 3} = -\dfrac{-(5 - 10x)}{6x - 3} = \dfrac{-1(5)(1 - 2x)}{3(2x - 1)} = \dfrac{5}{3}$$

or

$$-\dfrac{5 - 10x}{6x - 3} = \dfrac{5 - 10x}{-(6x - 3)} = \dfrac{5(1 - 2x)}{-1(3)(2x - 1)} = \dfrac{-5}{-3} = \dfrac{5}{3}$$ ∎

Multiplying Rational Expressions

To **multiply** any two rational expressions, multiply the numerators and multiply the denominators, keeping the expressions in factored form. Then "divide out" any common factors.

If *P, Q, R,* and *S* are polynomials with *Q, S* ≠ 0, then

$$\frac{P}{Q} \cdot \frac{R}{S} = \frac{P \cdot R}{Q \cdot S}$$

EXAMPLES   Multiply and reduce if possible.

7. $\dfrac{x + 3}{x} \cdot \dfrac{x - 3}{x + 5}$

Solution $\dfrac{x + 3}{x} \cdot \dfrac{x - 3}{x + 5} = \dfrac{(x + 3)(x - 3)}{x(x + 5)} = \dfrac{x^2 - 9}{x(x + 5)}$

(There are no common factors in the numerator and denominator.)

8. $\dfrac{x + 5}{7x} \cdot \dfrac{49x^2}{x^2 - 25}$

     Solution $\dfrac{x + 5}{7x} \cdot \dfrac{49x^2}{x^2 - 25} = \dfrac{\overset{7x}{\cancel{49x^2}(\cancel{x + 5})}}{\cancel{7x}(\cancel{x + 5})(x - 5)} = \dfrac{7x}{x - 5}$

9. $\dfrac{x^2 + 5x + 6}{3x + 6} \cdot \dfrac{x^2 - 4}{x^2 - 2x - 8}$

     Solution

$$\dfrac{x^2 + 5x + 6}{3x + 6} \cdot \dfrac{x^2 - 4}{x^2 - 2x - 8} = \dfrac{(\cancel{x + 2})(x + 3)(\cancel{x + 2})(x - 2)}{3(\cancel{x + 2})(x - 4)(\cancel{x + 2})}$$

$$= \dfrac{(x + 3)(x - 2)}{3(x - 4)}$$

$$= \dfrac{x^2 + x - 6}{3(x - 4)} \qquad \blacksquare$$

As shown in Examples 7 and 9, we will multiply the factors in the numerator yet leave the denominator in factored form. This form is not necessary, but it is useful for adding and subtracting, as we will see in Section 6.3.

Dividing Rational Expressions

To **divide** any two rational expressions, multiply by the **reciprocal** of the divisor.

If P, Q, R, and S are polynomials with Q, R, $S \neq 0$, then

$$\dfrac{P}{Q} \div \dfrac{R}{S} = \dfrac{P}{Q} \cdot \dfrac{S}{R}$$

EXAMPLES   Divide and reduce if possible.

10. $\dfrac{a^2 - 49}{12a^2} \div \dfrac{a^2 + 8a + 7}{18a}$

     Solution $\dfrac{a^2 - 49}{12a^2} \div \dfrac{a^2 + 8a + 7}{18a} = \dfrac{a^2 - 49}{12a^2} \cdot \dfrac{18a}{a^2 + 8a + 7}$

$$= \dfrac{(\cancel{a + 7})(a - 7) \cdot \cancel{6} \cdot 3 \cdot \cancel{a}}{\cancel{6} \cdot 2 \cdot \underset{a}{\cancel{a^2}}(\cancel{a + 7})(a + 1)}$$

$$= \dfrac{3(a - 7)}{2a(a + 1)}$$

$$= \dfrac{3a - 21}{2a(a + 1)}$$

11. $\dfrac{2x^2 + 3x - 2}{x^2 + 3x + 2} \div \dfrac{1 - 2x}{x - 2}$

Solution

$$\dfrac{2x^2 + 3x - 2}{x^2 + 3x + 2} \div \dfrac{1 - 2x}{x - 2} = \dfrac{2x^2 + 3x - 2}{x^2 + 3x + 2} \cdot \dfrac{x - 2}{1 - 2x}$$

$$= \dfrac{\overset{-1}{(2x - 1)}(x + 2)(x - 2)}{(x + 1)(x + 2)(1 - 2x)}$$

$$= \dfrac{-1(x - 2)}{x + 1}$$

$$= \dfrac{-x + 2}{x + 1}$$ ∎

Practice Problems

Perform the indicated operations and simplify. Assume that no denominator is 0.

1. $\dfrac{2y^2 - 16y}{6y^2 + 7y - 3} \cdot \dfrac{2y^2 + 11y + 12}{y^2 - 9y + 8}$

2. $\dfrac{a - b}{b - a} \div \dfrac{a^2 + 2ab + b^2}{a^2 + ab}$

3. $\dfrac{x^2 + 3x - 4}{x^2 - 1} \div \dfrac{x^2 + 6x + 8}{x + 1}$

EXERCISES 6.2

Reduce Exercises 1–20. State any restrictions on the variable.

1. $\dfrac{3x - 6}{6x + 3}$

2. $\dfrac{5x + 20}{6x + 24}$

3. $\dfrac{x}{x^2 - 4x}$

4. $\dfrac{4 - 2x}{2x - 4}$

5. $\dfrac{7x + 14}{x + 2}$

6. $\dfrac{3x^2 - 3x}{2x - 2}$

7. $\dfrac{5 + 3x}{3x + 5}$

8. $\dfrac{4xy + y^2}{3y^2 + 2y}$

9. $\dfrac{4 - 4x^2}{4x^2 - 4}$

10. $\dfrac{4x - 8}{(x - 2)^2}$

11. $\dfrac{2x + 6}{(x + 3)^2}$

12. $\dfrac{x^2 - 4}{2x + 4}$

13. $\dfrac{x^2 + 7x + 10}{x^2 - 25}$

14. $\dfrac{x^2 - 3x - 10}{x^2 - 7x + 10}$

15. $\dfrac{x^2 - 3x - 18}{x^2 + 6x + 9}$

16. $\dfrac{8x^2 + 6x - 9}{16x^2 - 9}$

17. $\dfrac{x^2 - 5x + 6}{8x - 2x^3}$

18. $\dfrac{6x^2 - 11x + 3}{4x^2 - 12x + 9}$

19. $\dfrac{16x^2 + 40x + 25}{4x^2 + 13x + 10}$

20. $\dfrac{6x^2 - 11x + 4}{3x^2 - 7x + 4}$

Answers to Practice Problems   1. $\dfrac{2y^2 + 8y}{(3y - 1)(y - 1)}$ **2.** $\dfrac{-a}{a + b}$ **3.** $\dfrac{1}{x + 2}$

Perform the indicated operations in Exercises 21–60. Assume that no denominator is zero.

21. $\dfrac{2x - 4}{3x + 6} \cdot \dfrac{3x}{x - 2}$

22. $\dfrac{5x + 20}{2x} \cdot \dfrac{4x}{2x + 4}$

23. $\dfrac{4x^2}{x^2 + 3x} \cdot \dfrac{x^2 - 9}{2x - 2}$

24. $\dfrac{x^2 - x}{x - 1} \cdot \dfrac{x + 1}{x}$

25. $\dfrac{x^2 - 4}{x} \cdot \dfrac{3x}{x + 2}$

26. $\dfrac{x - 3}{15x} \div \dfrac{3x - 9}{30x^2}$

27. $\dfrac{x^2 - 1}{5} \div \dfrac{x^2 + 2x + 1}{10}$

28. $\dfrac{x - 5}{3} \div \dfrac{x^2 - 25}{6x + 30}$

29. $\dfrac{10x^2 - 5x}{6x^2 + 12x} \div \dfrac{2x - 1}{x^2 + 2x}$

30. $\dfrac{x^2 + 2x - 8}{4x} \div \dfrac{2x^2 + 5x + 2}{3x^2}$

31. $\dfrac{x^2 + x}{x^2 + 2x + 1} \cdot \dfrac{x^2 - x - 2}{x^2 - 1}$

32. $\dfrac{x^2 - x - 6}{x^2 - 4} \cdot \dfrac{x^2 - 25}{x^2 + 2x - 15}$

33. $\dfrac{x + 3}{x^2 + 3x - 4} \cdot \dfrac{x^2 + x - 2}{x + 2}$

34. $\dfrac{x^2 + 5x + 6}{2x + 4} \cdot \dfrac{5x}{x + 3}$

35. $\dfrac{6x^2 - 7x - 3}{x^2 - 1} \cdot \dfrac{x - 1}{2x - 3}$

36. $\dfrac{x^2 - 9}{2x^2 + 7x + 3} \cdot \dfrac{2x^2 + 11x + 5}{x^2 - 3x}$

37. $\dfrac{x^2 - 8x + 15}{x^2 - 9x + 14} \cdot \dfrac{7 - x}{x^2 + 4x - 21}$

38. $\dfrac{x^2 - 16x + 39}{6 + x - x^2} \cdot \dfrac{4x + 8}{x + 1}$

39. $\dfrac{3x^2 - 7x + 2}{1 - 9x^2} \cdot \dfrac{3x + 1}{x - 2}$

40. $\dfrac{16 - x^2}{x^2 + 2x - 8} \cdot \dfrac{4 - x^2}{x^2 - 2x - 8}$

41. $\dfrac{4x^2 - 1}{x^2 - 16} \div \dfrac{2x + 1}{x^2 - 4x}$

42. $\dfrac{4x^2 - 13x + 3}{16x^2 - 4x} \div \dfrac{x^2 - 6x + 9}{8x^2}$

43. $\dfrac{x^2 + x - 6}{2x^2 + 6x} \div \dfrac{x^2 - 5x + 6}{8x^2}$

44. $\dfrac{x^2 - 4}{x^2 - 5x + 6} \div \dfrac{x^2 + 3x + 2}{x^2 - 2x - 3}$

45. $\dfrac{2x^2 - 5x - 12}{x^2 - 10x + 24} \div \dfrac{4x^2 - 9}{x^2 - 9x + 18}$

46. $\dfrac{2x^2 - 7x + 3}{x^2 - 3x + 2} \div \dfrac{x^2 - 6x + 9}{x^2 - 4x + 3}$

47. $\dfrac{6x^2 - x - 2}{12x^2 + 5x - 2} \div \dfrac{4x^2 - 1}{8x^2 - 6x + 1}$

48. $\dfrac{8x^2 + 6x - 9}{8x^2 - 26x + 15} \div \dfrac{4x^2 + 12x + 9}{16x^2 + 18x - 9}$

49. $\dfrac{3x^2 + 11x + 6}{4x^2 + 16x + 7} \div \dfrac{3x^2 - x - 2}{2x^2 - x - 28}$

50. $\dfrac{2x^2 + 5x - 3}{5x^2 + 17x - 12} \div \dfrac{2x^2 + 3x - 2}{5x^2 + 7x - 6}$

51. $\dfrac{x^2 - 16}{x^2 - 4x} \cdot \dfrac{x^2}{x + 4} \cdot \dfrac{2x^2 - 2x}{x - 1}$

52. $\dfrac{10x^2 + 3x - 1}{6x^2 + x - 2} \cdot \dfrac{2x^2 - x}{2x^2 - x - 1} \cdot \dfrac{3x + 2}{5x^2 - x}$

53. $\dfrac{x^2 - 3x}{x^2 + 2x + 1} \cdot \dfrac{x - 3}{x^2 - 9} \cdot \dfrac{x^2 - 3x - 18}{x - 6}$

54. $\dfrac{x^2 + 2x - 3}{2x^2 + 3x - 2} \cdot \dfrac{3x + 5}{2x - 3} \div \dfrac{3x^2 + 2x - 5}{2x^2 + x - 6}$

55. $\dfrac{2x^2 + 5x - 3}{x^2 + 2x - 3} \cdot \dfrac{x^2 + 3x - 4}{4x^2 - 1} \div \dfrac{x^2 + 4x}{x^3 + 5x^2}$

56. $\dfrac{6x^2 + 7x - 3}{2x^2 + 5x + 3} \div \dfrac{x^2 + 5x + 6}{4x^2 + 3x - 1} \cdot \dfrac{x^2 + 2x - 3}{4x^2 + 7x - 2}$

57. $\dfrac{x^2 + 4x + 3}{x^2 + 8x + 7} \div \dfrac{x^2 - 7x - 8}{35 + 12x + x^2} \div \dfrac{x^2 + 8x + 15}{x^2 - 9x + 8}$

58. $\dfrac{12x^2}{x^2 - 1} \div \dfrac{4x^2 + 4x - 3}{2x^2 - 5x + 3} \div \dfrac{6x^2 - 9x}{1 - 4x^2}$

59. $\dfrac{2x^2 - 7x + 3}{36x^2 - 1} \div \dfrac{6x^2 + 5x + 1}{2x + 1} \div \dfrac{2x^2 - 5x - 3}{18x^2 + 3x - 1}$

60. $\dfrac{12x^2 - 8x - 15}{4x^2 - 8x + 3} \cdot \dfrac{6x^2 - 13x + 6}{4x^2 + 5x + 1} \div \dfrac{18x^2 + 3x - 10}{8x^2 - 2x - 1}$

6.3 Addition and Subtraction

OBJECTIVES

In this section, you will be learning to:

1. Add rational expressions.
2. Subtract rational expressions.

Addition

To add rational expressions with a common denominator, proceed just as with fractions: add the numerator and use the common denominator. For example,

$$\frac{5}{x + 1} + \frac{6}{x + 1} = \frac{5 + 6}{x + 1} = \frac{11}{x + 1}$$

Sometimes the sum can be reduced.

$$\frac{x^2}{x + 1} + \frac{2x + 1}{x + 1} = \frac{x^2 + 2x + 1}{x + 1} = \frac{(x + 1)^2}{x + 1} = x + 1$$

For polynomials P, Q, and R, with $Q \neq 0$,

$$\frac{P}{Q} + \frac{R}{Q} = \frac{P + R}{Q}$$

EXAMPLES   Find the indicated sums and reduce if possible.

1. $\dfrac{x^2 + 5}{x + 5} + \dfrac{6x}{x + 5}$

Solution $\dfrac{x^2 + 5}{x + 5} + \dfrac{6x}{x + 5} = \dfrac{x^2 + 5 + 6x}{x + 5} = \dfrac{x^2 + 6x + 5}{x + 5}$

$$= \frac{(x + 5)(x + 1)}{x + 5} = x + 1$$

2. $\dfrac{3}{x^2 + 3x + 2} + \dfrac{2x + 1}{x^2 + 3x + 2}$

Solution $\dfrac{3}{x^2 + 3x + 2} + \dfrac{2x + 1}{x^2 + 3x + 2} = \dfrac{3 + 2x + 1}{x^2 + 3x + 2}$

$$= \frac{2x + 4}{(x + 2)(x + 1)} = \frac{2(x + 2)}{(x + 2)(x + 1)} = \frac{2}{x + 1}$$ ∎

The rational expressions added in Examples 1 and 2 had common denominators. To add expressions with different denominators, we need to find the least common multiple (LCM) of the denominators. The LCM was discussed in Section 1.2. We list the procedure here for polynomials and use this in finding the common denominator for rational expressions.

To Find the LCM for a Set of Polynomials	**1.** Completely factor each polynomial (including prime factors for numerical factors). **2.** Form the product of all factors that appear, using each factor the most number of times it appears in any one polynomial.

EXAMPLES   Find the indicated sums by finding the LCM of the denominators and changing each expression to an equivalent expression with that LCM as denominator.

3. $\dfrac{5}{x^2} + \dfrac{3}{x^2 - 4x}$

Solution $\left. \begin{array}{l} x^2 = x^2 \\ x^2 - 4x = x(x - 4) \end{array} \right\}$ $\text{LCM} = x^2(x - 4)$

So, $x^2(x - 4)$ is the common denominator. Multiply the numerator and denominator of each fraction so that it has denominator $x^2(x - 4)$.

$$\frac{5}{x^2} + \frac{3}{x^2 - 4x} = \frac{5(x - 4)}{x^2(x - 4)} + \frac{3 \cdot x}{x(x - 4) \cdot x}$$

$$= \frac{5x - 20}{x^2(x - 4)} + \frac{3x}{x^2(x - 4)}$$

$$= \frac{5x - 20 + 3x}{x^2(x - 4)} = \frac{8x - 20}{x^2(x - 4)}$$

4. $\dfrac{5}{x^2 - 1} + \dfrac{4}{(x - 1)^2}$

Solution $\left. \begin{array}{l} x^2 - 1 = (x + 1)(x - 1) \\ (x - 1)^2 = (x - 1)^2 \end{array} \right\}$ $\text{LCM} = (x + 1)(x - 1)^2$

$$\frac{5}{x^2 - 1} + \frac{4}{(x - 1)^2} = \frac{5(x - 1)}{(x + 1)(x - 1)(x - 1)} + \frac{4(x + 1)}{(x - 1)^2(x + 1)}$$

$$= \frac{5x - 5 + 4x + 4}{(x - 1)^2(x + 1)} = \frac{9x - 1}{(x - 1)^2(x + 1)}$$ ■

Subtraction

To subtract one rational expression from another with the same denominator, simply subtract the numerators and use the common denominator. Since the numerators will be polynomials, **a good idea is to put both numerators in parentheses so that all changes in signs will be done correctly.**

For polynomials P, Q, and R with $Q \neq 0$,

$$\frac{P}{Q} - \frac{R}{Q} = \frac{P - R}{Q}$$

EXAMPLES Find the indicated differences and reduce if possible.

5. $\dfrac{3}{2x-4} - \dfrac{5}{2x-4}$

Solution $\dfrac{3}{2x-4} - \dfrac{5}{2x-4} = \dfrac{3-5}{2x-4} = \dfrac{-2}{2x-4}$

$$= \dfrac{\overset{-1}{\cancel{-2}}}{\cancel{2}(x-2)} = \dfrac{-1}{x-2}$$

6. $\dfrac{x+4}{x+3} - \dfrac{5x+1}{x+3}$

Solution $\dfrac{x+4}{x+3} - \dfrac{5x+1}{x+3} = \dfrac{(x+4)-(5x+1)}{x+3}$

$$= \dfrac{x+4-5x-1}{x+3} = \dfrac{-4x+3}{x+3} \quad ■$$

Common Error

Many beginning students make a **mistake** in subtracting fractions by not subtracting the entire numerator. They make a **mistake** similar to the following mistake:

WRONG

$$\dfrac{7}{x+5} - \dfrac{2-x}{x+5} = \boxed{\dfrac{7-2-x}{x+5}}$$

By using parentheses, you can avoid such mistakes.

RIGHT

$$\dfrac{7}{x+5} - \dfrac{2-x}{x+5} = \boxed{\dfrac{7-(2-x)}{x+5}} = \dfrac{7-2+x}{x+5} = \dfrac{5+x}{x+5}$$

$$= \dfrac{x+5}{x+5} = 1$$

As with addition, if the rational expressions do not have the same denominator, find the LCM of the denominators and change each fraction to an equivalent fraction with the LCM as denominator.

EXAMPLE

7. $\dfrac{2x}{x^2 - 9} - \dfrac{1}{x^2 + 7x + 12}$

Solution

$$\left.\begin{array}{l} x^2 - 9 = (x + 3)(x - 3) \\ x^2 + 7x + 12 = (x + 4)(x + 3) \end{array}\right\} \quad \text{LCM} = (x + 3)(x - 3)(x + 4)$$

$$\dfrac{2x}{x^2 - 9} - \dfrac{1}{x^2 + 7x + 12}$$

$$= \dfrac{2x(x + 4)}{(x + 3)(x - 3)(x + 4)} + \dfrac{-1(x - 3)}{(x + 4)(x + 3)(x - 3)} \qquad \begin{array}{l}\text{Place the} \\ \text{negative sign in} \\ \text{the numerator} \\ \text{and add.}\end{array}$$

$$= \dfrac{2x^2 + 8x - x + 3}{(x + 3)(x - 3)(x + 4)} = \dfrac{2x^2 + 7x + 3}{(x + 3)(x - 3)(x + 4)}$$

$$= \dfrac{(2x + 1)(x + 3)}{(x + 3)(x - 3)(x + 4)} = \dfrac{2x + 1}{(x - 3)(x + 4)} \qquad \blacksquare$$

Whenever the denominators are opposites of each other or contain factors that are opposites of each other, it is important to use -1 as a factor or to multiply both the numerator and denominator by -1. This will simplify the work considerably.

EXAMPLE

8. $\dfrac{10}{x - 2} + \dfrac{8}{2 - x}$

Solution   Note that $x - 2$ and $2 - x$ are opposites and

$$x - 2 = -1(2 - x)$$

So, if we multiply

$$\dfrac{8}{(2 - x)} \cdot \dfrac{(-1)}{(-1)} = \dfrac{-8}{x - 2}$$

the fractions will have the same denominator and we do not have to find an LCM.

$$\dfrac{10}{x - 2} + \dfrac{8}{2 - x} = \dfrac{10}{x - 2} + \dfrac{8}{(2 - x)} \cdot \dfrac{(-1)}{(-1)}$$

$$= \dfrac{10}{x - 2} + \dfrac{-8}{x - 2}$$

$$= \dfrac{10 - 8}{x - 2} = \dfrac{2}{x - 2}$$

Do not try to reduce this last fraction, since 2 is **not a factor** of the denominator. $\blacksquare$

Practice Problems   Perform the indicated operations and reduce if possible.

1. $\dfrac{5}{x-1} - \dfrac{4+x}{x-1}$ 2. $\dfrac{5}{x+1} + \dfrac{10x}{x^2+4x+3}$

3. $\dfrac{1}{x^2+x} + \dfrac{4}{x^2} - \dfrac{2}{x^3-x}$ 4. $\dfrac{x}{2x-1} - \dfrac{2}{1-2x}$

EXERCISES 6.3

Perform the indicated operations and reduce if possible in Exercises 1–60.

1. $\dfrac{4x}{x+2} + \dfrac{8}{x+2}$

2. $\dfrac{x-1}{x+4} + \dfrac{x+9}{x+4}$

3. $\dfrac{3x-1}{2x-6} + \dfrac{x-11}{2x-6}$

4. $\dfrac{2x+5}{4(x+1)} + \dfrac{-x+3}{4x+4}$

5. $\dfrac{x^2}{x^2+2x+1} + \dfrac{-2x-3}{x^2+2x+1}$

6. $\dfrac{2x-1}{x^2-x-6} + \dfrac{1-x}{x^2-x-6}$

7. $\dfrac{-(3x+2)}{x^2-7x+6} + \dfrac{4x-4}{x^2-7x+6}$

8. $\dfrac{x+1}{x^2-2x-3} + \dfrac{-(3x-5)}{x^2-2x-3}$

9. $\dfrac{x^2+2}{x^2-4} + \dfrac{-(4x-2)}{x^2-4}$

10. $\dfrac{2x+5}{2x^2-x-1} + \dfrac{-(4x+2)}{2x^2-x-1}$

11. $\dfrac{3x+2}{4x-2} - \dfrac{x+2}{4x-2}$

12. $\dfrac{2x+1}{x^2-x-6} - \dfrac{x-1}{x^2-x-6}$

13. $\dfrac{2x^2-3x+1}{x^2-3x-4} - \dfrac{x^2+x+6}{x^2-3x-4}$

14. $\dfrac{x^2-x-2}{x^2-4} - \dfrac{x^2+x-2}{x^2-4}$

15. $\dfrac{3}{x-3} - \dfrac{2}{2x-6}$

16. $\dfrac{2x}{3x+6} + \dfrac{5}{2x+4}$

17. $\dfrac{8}{5x-10} - \dfrac{6}{3x-6}$

18. $\dfrac{7}{4x-20} - \dfrac{1}{3x-15}$

19. $\dfrac{2}{x} + \dfrac{1}{x+4}$

20. $\dfrac{8x+3}{x+1} + \dfrac{x-1}{x}$

21. $\dfrac{x}{x+4} + \dfrac{x}{x-4}$

22. $\dfrac{5}{x-2} + \dfrac{x}{x+3}$

23. $\dfrac{3}{2-x} + \dfrac{6}{x-2}$

24. $\dfrac{5}{2x-3} + \dfrac{2}{3-2x}$

25. $\dfrac{x}{5-x} + \dfrac{2x+3}{x-5}$

26. $\dfrac{x-1}{3x-1} + \dfrac{4}{x+2}$

27. $\dfrac{x}{x-1} - \dfrac{1}{x+2}$

28. $\dfrac{x+2}{x+3} + \dfrac{4}{3-x}$

29. $\dfrac{x+1}{x+4} + \dfrac{2x}{4-x}$

30. $\dfrac{8}{x} - \dfrac{x+1}{x-6}$

31. $\dfrac{8}{x-2} - \dfrac{4}{x+2}$

32. $\dfrac{8}{2x+2} + \dfrac{3}{3x+3}$

33. $\dfrac{x}{4x-8} - \dfrac{3x+1}{3x-6}$

34. $\dfrac{7x-1}{x^2-25} + \dfrac{4}{x+5}$

35. $\dfrac{2x+3}{x^2+4x-5} - \dfrac{4}{x-1}$

36. $\dfrac{2x-3}{x+1} - \dfrac{x+1}{2x-2}$

Answers to Practice Problems   1. -1 2. $\dfrac{15}{x+3}$ 3. $\dfrac{5x^2-3x-4}{x^2(x+1)(x-1)}$ 4. $\dfrac{x+2}{2x-1}$

37. $\dfrac{3x}{6 + x} - \dfrac{2x}{x^2 - 36}$

38. $\dfrac{x}{7 + x} - \dfrac{7 - 13x}{x^2 - 49}$

39. $\dfrac{2x + 1}{x - 7} + \dfrac{3x}{x^2 - 8x + 7}$

40. $\dfrac{x - 4}{x - 2} - \dfrac{x - 7}{2x - 10}$

41. $\dfrac{x + 1}{x + 2} - \dfrac{x + 2}{2x + 6}$

42. $\dfrac{3x - 4}{x^2 - x - 20} - \dfrac{2}{x - 5}$

43. $\dfrac{x + 2}{x^2 - 16} - \dfrac{x + 1}{2x - 8}$

44. $\dfrac{7}{x - 9} - \dfrac{x - 1}{3x + 6}$

45. $\dfrac{1}{x^2 + x - 2} - \dfrac{1}{x^2 - 1}$

46. $\dfrac{4}{x^2 + x - 6} + \dfrac{4}{x^2 + 5x + 6}$

47. $\dfrac{2x}{x^2 + x - 12} + \dfrac{3x}{x^2 - 9}$

48. $\dfrac{x}{x^2 + 4x - 21} + \dfrac{1 - x}{x^2 + 8x + 7}$

49. $\dfrac{x - 3}{x^2 + 4x + 4} - \dfrac{3x}{x^2 + 3x + 2}$

50. $\dfrac{x - 1}{2x^2 + 3x - 2} + \dfrac{x}{2x^2 - 3x + 1}$

51. $\dfrac{x}{x^2 - 16} - \dfrac{3x}{x^2 + 5x + 4}$

52. $\dfrac{2x}{x^2 - 2x - 15} - \dfrac{5}{x^2 - 6x + 5}$

53. $\dfrac{x - 2}{2x^2 + 5x - 3} + \dfrac{2x - 5}{2x^2 - 9x + 4}$

54. $\dfrac{6}{x^2 - 4x - 12} + \dfrac{x + 1}{x^2 - 3x - 18}$

55. $\dfrac{x}{x^2 + 3x - 10} + \dfrac{3x}{x^2 - 4}$

56. $\dfrac{2x + 3}{x^2 - 6x - 7} + \dfrac{x - 1}{x^2 - 5x - 14}$

57. $\dfrac{3x}{x - 4} + \dfrac{7x}{x + 4} - \dfrac{x + 3}{x^2 - 16}$

58. $\dfrac{x}{x + 3} - \dfrac{x + 1}{x - 3} + \dfrac{x^2 + 4}{x^2 - 9}$

59. $\dfrac{1}{x - 2} - \dfrac{1}{x - 1} + \dfrac{x}{x^2 - 3x + 2}$

60. $\dfrac{2}{x^2 - 9} - \dfrac{3}{x^2 - 4x + 3} + \dfrac{x - 1}{x^2 + 2x - 3}$

◢ 6.4 Complex Algebraic Fractions

A fraction that contains various combinations of addition, subtraction, multiplication, and division with rational expressions is called a **complex algebraic fraction.** Examples of complex algebraic fractions are

$$\dfrac{\dfrac{1}{x} + \dfrac{1}{y}}{x + y}, \qquad \dfrac{\dfrac{1}{x + 2} - \dfrac{1}{x}}{1 + \dfrac{2}{x}}, \qquad \text{and} \qquad \dfrac{4 - x}{x + 3} + \dfrac{x}{x + 3} \div \dfrac{x}{x - 3}$$

In an expression such as $\dfrac{\dfrac{1}{x} + \dfrac{1}{y}}{x + y}$, the large fraction bar is a symbol of inclusion. The expression could be written as follows:

$$\dfrac{\dfrac{1}{x} + \dfrac{1}{y}}{x + y} = \left(\dfrac{1}{x} + \dfrac{1}{y}\right) \div (x + y)$$

Similarly, $$\dfrac{\dfrac{1}{x + 2} - \dfrac{1}{x}}{1 + \dfrac{2}{x}} = \left(\dfrac{1}{x + 2} - \dfrac{1}{x}\right) \div \left(1 + \dfrac{2}{x}\right)$$

Simplify by working with the numerator and denominator separately as if they were in parentheses. Make the numerator a single fraction and the denominator a single fraction. Then divide.

a. $$\dfrac{\dfrac{1}{x} + \dfrac{1}{y}}{x + y} = \dfrac{\dfrac{1 \cdot y}{x \cdot y} + \dfrac{1 \cdot x}{y \cdot x}}{\dfrac{x + y}{1}}$$

$$= \dfrac{\dfrac{y + x}{xy}}{\dfrac{x + y}{1}} = \dfrac{\cancel{y + x}}{xy} \cdot \dfrac{1}{\cancel{x + y}} = \dfrac{1}{xy}$$

b. $$\dfrac{\dfrac{1}{x + 2} - \dfrac{1}{x}}{1 + \dfrac{2}{x}} = \dfrac{\dfrac{1 \cdot x}{(x + 2)x} - \dfrac{1(x + 2)}{x(x + 2)}}{\dfrac{x}{x} + \dfrac{2}{x}}$$

$$= \dfrac{\dfrac{x - (x + 2)}{x(x + 2)}}{\dfrac{x + 2}{x}} = \dfrac{\dfrac{x - x - 2}{x(x + 2)}}{\dfrac{x + 2}{x}}$$

$$= \dfrac{-2}{\cancel{x}(x + 2)} \cdot \dfrac{\cancel{x}}{x + 2} = \dfrac{-2}{(x + 2)^2}$$

A second method is to find the LCM of the denominators in the fractions in both the original numerator and the original denominator, and then multiply both the numerator and the denominator by this LCM.

a. $$\dfrac{\dfrac{1}{x} + \dfrac{1}{y}}{x + y} = \dfrac{\dfrac{1}{x} + \dfrac{1}{y}}{\dfrac{x + y}{1}} \qquad \left.\begin{array}{c} x \\ y \\ 1 \end{array}\right\} \; \text{LCM} = xy$$

$$= \dfrac{\left(\dfrac{1}{x} + \dfrac{1}{y}\right)xy}{\left(\dfrac{x + y}{1}\right)xy} = \dfrac{\dfrac{1}{x} \cdot xy + \dfrac{1}{y} \cdot xy}{(x + y)xy}$$

$$= \dfrac{\cancel{y + x}}{(\cancel{x + y})xy} = \dfrac{1}{xy}$$

b. $\dfrac{\dfrac{1}{x+2} - \dfrac{1}{x}}{1 + \dfrac{2}{x}}$

$$= \dfrac{\left(\dfrac{1}{x+2} - \dfrac{1}{x}\right) \cdot x(x+2)}{\left(1 + \dfrac{2}{x}\right) \cdot x(x+2)} \qquad \left.\begin{array}{c} x \\ x+2 \end{array}\right\} \text{ LCM} = x(x+2)$$

$$= \dfrac{\dfrac{1}{\cancel{x+2}} \cdot x(\cancel{x+2}) - \dfrac{1}{\cancel{x}} \cdot \cancel{x}(x+2)}{1 \cdot x(x+2) + \dfrac{2}{\cancel{x}} \cdot \cancel{x}(x+2)}$$

$$= \dfrac{x - (x+2)}{x(x+2) + 2(x+2)} = \dfrac{x - x - 2}{(x+2)(x+2)} = \dfrac{-2}{(x+2)^2}$$

Each of the techniques just described is valid. Sometimes one is easier to use than the other, but the choice is up to you.

EXAMPLES

1. Simplify the following complex algebraic fraction: $\dfrac{\dfrac{1}{x+y} - \dfrac{1}{x-y}}{\dfrac{2y}{x^2 - y^2}}$

Solution

$$\dfrac{\dfrac{1}{x+y} - \dfrac{1}{x-y}}{\dfrac{2y}{x^2 - y^2}} = \dfrac{\dfrac{1(x-y)}{(x+y)(x-y)} - \dfrac{1(x+y)}{(x-y)(x+y)}}{\dfrac{2y}{x^2 - y^2}}$$

$$= \dfrac{\dfrac{(x-y) - (x+y)}{(x-y)(x+y)}}{\dfrac{2y}{x^2 - y^2}}$$

$$= \dfrac{x - y - x - y}{(x-y)(x+y)} \cdot \dfrac{x^2 - y^2}{2y}$$

$$= \dfrac{\overset{-1}{\cancel{-2y}}}{\cancel{(x-y)(x+y)}} \cdot \dfrac{\cancel{(x-y)(x+y)}}{\cancel{2y}} = -1$$

Or use the technique of multiplying the numerator and the denominator by the LCM of the denominators of the various fractions.

$$\frac{\dfrac{1}{x+y} - \dfrac{1}{x-y}}{\dfrac{2y}{x^2-y^2}} = \frac{\left(\dfrac{1}{x+y} - \dfrac{1}{x-y}\right)(x+y)(x-y)}{\left(\dfrac{2y}{x^2-y^2}\right)(x+y)(x-y)}$$

$$= \frac{\dfrac{1}{x+y}(x+y)(x-y) - \dfrac{1}{x-y}(x+y)(x-y)}{\dfrac{2y(x+y)(x-y)}{(x+y)(x-y)}}$$

$$= \frac{(x-y)-(x+y)}{2y}$$

$$= \frac{x-y-x-y}{2y} = \frac{-2y}{2y} = -1$$

2. In an expression such as

$$\frac{4-x}{x+3} + \frac{x}{x+3} \div \frac{x}{x-3}$$

the rules for order of operations indicate that the division is to be done first.

Solution $\dfrac{4-x}{x+3} + \dfrac{x}{x+3} \div \dfrac{x}{x-3} = \dfrac{4-x}{x+3} + \dfrac{x}{x+3} \cdot \dfrac{x-3}{x}$

$$= \frac{4-x}{x+3} + \frac{x-3}{x+3}$$

$$= \frac{4-x+x-3}{x+3}$$

$$= \frac{1}{x+3}$$

■

Practice Problems   Simplify the following expressions.

1. $\dfrac{\dfrac{1}{x}}{1 + \dfrac{1}{x}}$

2. $\dfrac{1 + \dfrac{3}{x-3}}{x - \dfrac{x^2}{x-3}}$

Answers to Practice Problems   1. $\dfrac{1}{x+1}$ **2.** $-\dfrac{1}{3}$

EXERCISES 6.4

Simplify the complex algebraic fractions in Exercises 1–32.

1. $\dfrac{\dfrac{4}{5}}{\dfrac{7}{10}}$
2. $\dfrac{\dfrac{5}{6}}{\dfrac{2}{3}}$
3. $\dfrac{\dfrac{1}{2}+\dfrac{1}{3}}{\dfrac{5}{6}-\dfrac{1}{4}}$
4. $\dfrac{\dfrac{3}{4}-\dfrac{7}{8}}{\dfrac{1}{3}-\dfrac{1}{6}}$
5. $\dfrac{1+\dfrac{1}{3}}{2+\dfrac{2}{3}}$
6. $\dfrac{2+\dfrac{5}{7}}{3-\dfrac{2}{7}}$

7. $\dfrac{\dfrac{3x}{5}}{\dfrac{x}{5}}$
8. $\dfrac{\dfrac{x}{2y}}{\dfrac{3x}{y}}$
9. $\dfrac{\dfrac{4}{xy^2}}{\dfrac{6}{x^2y}}$
10. $\dfrac{\dfrac{x}{2y^2}}{\dfrac{5x^2}{6y}}$
11. $\dfrac{\dfrac{8x^2}{5y}}{\dfrac{x}{10y^2}}$
12. $\dfrac{\dfrac{15y^2}{2x^3}}{\dfrac{8y}{}}$

13. $\dfrac{\dfrac{12x^3}{7y^4}}{\dfrac{3x^5}{2y}}$
14. $\dfrac{\dfrac{9x^2}{5y^3}}{\dfrac{3xy}{10}}$
15. $\dfrac{\dfrac{3x}{x+1}}{\dfrac{x-2}{x+1}}$
16. $\dfrac{\dfrac{x-3}{x+4}}{\dfrac{x-2}{x+4}}$
17. $\dfrac{\dfrac{x+3}{2x}}{\dfrac{2x-1}{4x^2}}$
18. $\dfrac{\dfrac{x-2}{6x}}{\dfrac{x+3}{3x^2}}$

19. $\dfrac{\dfrac{x^2-9}{x}}{x-3}$
20. $\dfrac{\dfrac{x^2-3x+2}{x-4}}{x-2}$
21. $\dfrac{\dfrac{x+2}{x-2}}{\dfrac{4x^2+8x}{x^2-4}}$
22. $\dfrac{\dfrac{x-1}{x+3}}{\dfrac{x+2}{x^2+2x-3}}$
23. $\dfrac{\dfrac{1}{x}-\dfrac{1}{3x}}{\dfrac{x+6}{x^2}}$

24. $\dfrac{\dfrac{1}{3}+\dfrac{1}{x}}{\dfrac{1}{2}-\dfrac{1}{x}}$
25. $\dfrac{1+\dfrac{1}{x}}{1-\dfrac{1}{x^2}}$
26. $\dfrac{\dfrac{3}{x}-\dfrac{6}{x^2}}{\dfrac{1}{x}-\dfrac{2}{x^2}}$
27. $\dfrac{\dfrac{1}{x}-\dfrac{1}{y}}{\dfrac{y}{x^2}-\dfrac{1}{y}}$
28. $\dfrac{\dfrac{x}{y}-\dfrac{1}{3}}{\dfrac{6}{y}-\dfrac{2}{x}}$

29. $\dfrac{2-\dfrac{4}{x}}{\dfrac{x^2-4}{x^2+x}}$
30. $\dfrac{\dfrac{1}{x}}{1-\dfrac{1}{x-2}}$
31. $\dfrac{x-\dfrac{2}{x+1}}{x+\dfrac{x-3}{x+1}}$
32. $\dfrac{x-\dfrac{2x-3}{x-2}}{2x-\dfrac{x+3}{x-2}}$

Write each of the expressions as a single fraction reduced to lowest terms in Exercises 33–40.

33. $\dfrac{1}{x+1}-\dfrac{3}{2x}\cdot\dfrac{4x}{x+1}$

34. $\dfrac{4}{x}-\dfrac{2}{x^2-2x}\cdot\dfrac{x-2}{5}$

35. $\left(\dfrac{8}{x}-\dfrac{3}{4x}\right)\div\dfrac{4x+5}{x}$

36. $\left(\dfrac{2}{x}+\dfrac{5}{x-3}\right)\cdot\dfrac{2x-6}{x}$

37. $\dfrac{x}{x-1}-\dfrac{3}{x-1}\cdot\dfrac{x+2}{x}$

38. $\dfrac{x+3}{x+2}+\dfrac{x}{x+2}\div\dfrac{x^2}{x-3}$

39. $\dfrac{x-1}{x+4}+\dfrac{x-6}{x^2+3x-4}\div\dfrac{x-4}{x-1}$

40. $\dfrac{x}{x+3}-\dfrac{3}{x-5}\cdot\dfrac{x^2-3x-10}{x-2}$

6.5 Solving Equations Involving Rational Expressions

In this section, you will be learning to:

1. Solve equations involving rational expressions.
2. Solve proportions.
3. Solve word problems using proportions.

We have solved equations involving fractions with constants in the denominators (Section 3.3). For example, to solve

$$\frac{x+1}{3} + \frac{x}{2} = 1$$

we would multiply both sides of the equation by 6, the LCM of the denominators. The procedure looks like this:

$$\frac{x+1}{3} + \frac{x}{2} = 1$$

$$6\left(\frac{x+1}{3} + \frac{x}{2}\right) = 1 \cdot 6 \qquad \text{Multiply both sides by } \mathbf{6.}$$

$$\overset{2}{6}\left(\frac{x+1}{3}\right) + \overset{3}{6}\left(\frac{x}{2}\right) \text{p}= 1 \cdot 6 \qquad \text{Multiply each fraction by } \mathbf{6.}$$

$$2x + 2 + 3x = 6$$

$$5x = 4$$

$$x = \frac{4}{5}$$

If the denominators have variables, we follow the same procedure:

1. Find the LCM of the denominators.
2. Multiply both sides of the equation by this LCM and simplify by using the distributive property.
3. Solve the resulting equation.
4. Check to see that no solution makes a denominator 0.

EXAMPLES First state any restrictions on the variable, then solve each of the following equations.

1. $\dfrac{3}{x} + \dfrac{1}{2} = \dfrac{7}{x}$

Solution

$$\frac{3}{x} + \frac{1}{2} = \frac{7}{x} \qquad\qquad x \neq 0 \text{ and LCM} = 2x.$$

$$2x\left(\frac{3}{x} + \frac{1}{2}\right) = 2x\left(\frac{7}{x}\right) \qquad \text{Multiply both sides by } 2x.$$

$$2x\left(\frac{3}{x}\right) + 2x\left(\frac{1}{2}\right) = 2x\left(\frac{7}{x}\right) \qquad \text{Multiply each fraction by } 2x.$$

$$6 + x = 14 \qquad\qquad \text{Simplify.}$$

$$x = 8$$

Now, $8 \neq 0$ and checking, we have

$$\frac{3}{8} + \frac{1}{2} \overset{?}{=} \frac{7}{8}$$

$$\frac{3}{8} + \frac{4}{8} \overset{?}{=} \frac{7}{8}$$

$$\frac{7}{8} = \frac{7}{8}$$

So, $x = 8$ is the solution.

2. $\dfrac{3}{x - 2} + 2 = \dfrac{18}{x + 1}$

Solution

$$\frac{3}{x - 2} + 2 = \frac{18}{x + 1} \qquad\qquad \begin{array}{l} x \neq 2,\, -1 \text{ and} \\ \text{LCM} = (x - 2)(x + 1). \end{array}$$

$$(x - 2)(x + 1)\left[\frac{3}{x - 2} + 2\right] = \frac{18}{x + 1}(x - 2)(x + 1) \qquad \begin{array}{l}\text{Multiply both sides by} \\ (x - 2)(x - 1).\end{array}$$

$$(x - 2)(x + 1)\left(\frac{3}{x - 2}\right) + 2(x - 2)(x + 1) = \left(\frac{18}{x + 1}\right)(x - 2)(x + 1) \qquad \begin{array}{l}\text{Multiply each fraction,} \\ \text{including 2, by} \\ (x - 2)(x + 1).\end{array}$$

$$3(x + 1) + 2(x^2 - x - 2) = 18(x - 2)$$
$$3x + 3 + 2x^2 - 2x - 4 = 18x - 36$$
$$2x^2 + x - 1 = 18x - 36$$
$$2x^2 - 17x + 35 = 0$$
$$(2x - 7)(x - 5) = 0$$
$$2x - 7 = 0 \qquad \text{or} \qquad x - 5 = 0$$
$$x = \frac{7}{2} \qquad\qquad\qquad x = 5$$

There are two solutions: $\dfrac{7}{2}$ and 5. (Both will check.) (Note that neither of these is 2 or -1, the restricted values.)

3. $\dfrac{4}{x-4} - \dfrac{x}{x-4} = 1$

Solution $\dfrac{4}{x-4} - \dfrac{x}{x-4} = 1$ $x \neq 4$ and
LCM $= x - 4$.

$$(x-4)\left[\dfrac{4}{x-4} - \dfrac{x}{x-4}\right] = 1(x-4)$$

Multiply both
sides by $x - 4$.

$$(x-4)\left(\dfrac{4}{x-4}\right) - (x-4)\left(\dfrac{x}{x-4}\right) = 1(x-4)$$

$$4 - x = x - 4$$
$$8 = 2x$$
$$4 = x$$

But $x \neq 4$ was the restriction because this value will give 0 denominators. Therefore, **this equation has no solution.**   ■

Proportions

A **ratio** is a comparison by division of two numbers. We write ratios in the form *a:b* or $\dfrac{a}{b}$. For example, suppose the ratio of women to men faculty in a certain department at a university is 3 to 2. This ratio can be written 3:2 or $\dfrac{3}{2}$ and does not necessarily mean that there are only 5 people in the department. There are many ratios (fractions) that reduce to $\dfrac{3}{2}$. If there are 15 people in the department, then there are 9 women and 6 men because the ratios $\dfrac{9}{6}$ and $\dfrac{3}{2}$ are equal: $\dfrac{9}{6} = \dfrac{3}{2}$.

A **proportion** is an equation that states two ratios are equal. Such equations can involve rational expressions. Our purpose here is to show how such proportions can evolve from word problems and to solve these proportions.

The following format can be used in identifying the terms of a proportion.

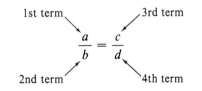

a and *d* (the first and fourth terms) are called the **extremes.**
b and *c* (the second and third terms) are called the **means.**

We use the following fact to solve proportions.

In a true proportion, the product of the extremes is equal to the product of the means.

$$\text{If } \frac{a}{b} = \frac{c}{d} \text{ , then } a \cdot d = b \cdot c.$$

For example,

$$\frac{6}{10} = \frac{9}{15} \text{ because } 6 \cdot 15 = 10 \cdot 9.$$

EXAMPLE   Solve the following proportion.

4. $\dfrac{4}{x - 5} = \dfrac{2}{x + 3}$ **Note:** $x \neq 5, -3.$

Solution $4(x + 3) = 2(x - 5)$ Product of extremes equals
$4x + 12 = 2x - 10$ product of means.
$4x + 12 - 2x - 12 = 2x - 10 - 2x - 12$
$2x = -22$
$x = -11$ ■

Proportions arise naturally in everyday problems. For example, if tires are on sale at 2 for \$75 but you need 5 new tires, what would you pay for the 5 tires? Setting up a proportion gives:

$$\frac{2 \ tires}{75 \ dollars} = \frac{5 \ tires}{x \ dollars} \qquad \textbf{or} \qquad \frac{75 \ dollars}{x \ dollars} = \frac{2 \ tires}{5 \ tires}$$

Numerators agree in type   Numerators correspond
and   and
denominators agree in type.   denominators correspond.

Solving, we have

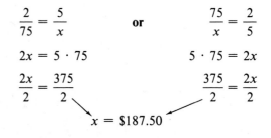

$$\frac{2}{75} = \frac{5}{x} \qquad \textbf{or} \qquad \frac{75}{x} = \frac{2}{5}$$

$$2x = 5 \cdot 75 \qquad\qquad 5 \cdot 75 = 2x$$

$$\frac{2x}{2} = \frac{375}{2} \qquad\qquad \frac{375}{2} = \frac{2x}{2}$$

$$x = \$187.50$$

The cost of 5 tires would be \$187.50.

Be very careful with the units in the setup of a proportion. One of the following conditions must be true:

1. the numerators agree in type and the denominators agree in type, or

2. the numerators correspond and the denominators correspond.

Any other arrangement will give a wrong answer. For example,

$$\frac{75 \text{ dollars}}{x \text{ dollars}} = \frac{5 \text{ tires}}{2 \text{ tires}} \quad \text{is} \quad \textbf{WRONG}$$

because 75 dollars does not correspond to 5 tires.

EXAMPLES

5. On an architect's scale drawing of a building, $\frac{1}{2}$ inch represents 12 feet. What does 3 inches represent?

Solution $$\frac{\dfrac{1}{2} \text{ inch}}{12 \text{ feet}} = \frac{3 \text{ inches}}{x \text{ feet}}$$

$$\frac{1}{2}x = 3 \cdot 12$$

$$2 \cdot \frac{1}{2}x = 36 \cdot 2$$

$$x = 72$$

Three inches represents 72 feet.

6. Making a statistical analysis, Mike finds 3 defective computer disks in a sample of 20 disks. If this ratio is consistent, how many bad disks does he expect to find in an order of 2400?

Solution $$\frac{3 \text{ defective disks}}{20 \text{ disks}} = \frac{x \text{ defective disks}}{2400 \text{ disks}}$$

$$3 \cdot 2400 = 20x$$

$$\frac{7200}{20} = \frac{20x}{20}$$

$$360 = x$$

He expects to find 360 defective disks. (**Note:** He probably will return the order to the manufacturer.)   ■

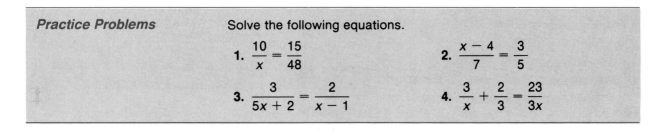

Practice Problems

Solve the following equations.

1. $\dfrac{10}{x} = \dfrac{15}{48}$

2. $\dfrac{x-4}{7} = \dfrac{3}{5}$

3. $\dfrac{3}{5x+2} = \dfrac{2}{x-1}$

4. $\dfrac{3}{x} + \dfrac{2}{3} = \dfrac{23}{3x}$

Answers to Practice Problems **1.** $x = 32$ **2.** $x = \dfrac{41}{5}$ **3.** $x = -1$ **4.** $x = 7$

EXERCISES 6.5

First state any restriction on the variable, then solve each equation or proportion in Exercises 1–38.

1. $\dfrac{4x}{3} - \dfrac{3}{4} = \dfrac{5x}{6}$

2. $\dfrac{5x}{8} - \dfrac{3}{4} = \dfrac{5}{16}$

3. $\dfrac{x-2}{3} - \dfrac{x-3}{5} = \dfrac{13}{15}$

4. $\dfrac{2+x}{4} - \dfrac{5x-2}{12} = \dfrac{8-2x}{5}$

5. $\dfrac{4}{5x} = \dfrac{2}{25}$

6. $\dfrac{2}{15} = \dfrac{8}{3x}$

7. $\dfrac{21}{x+4} = \dfrac{7}{8}$

8. $\dfrac{5}{x+3} = \dfrac{6}{11}$

9. $\dfrac{2}{3x} = \dfrac{1}{4} - \dfrac{1}{6x}$

10. $\dfrac{3}{8x} - \dfrac{7}{10} = \dfrac{1}{5x}$

11. $\dfrac{1}{x} - \dfrac{8}{21} = \dfrac{3}{7x}$

12. $\dfrac{3}{4x} = \dfrac{7}{8x} + \dfrac{2}{3}$

13. $\dfrac{2x}{x-1} = \dfrac{3}{2}$

14. $\dfrac{x+7}{x-4} = \dfrac{5}{6}$

15. $\dfrac{10}{x} = \dfrac{5}{x-2}$

16. $\dfrac{8}{x-3} = \dfrac{12}{2x-3}$

17. $\dfrac{6}{x-4} = \dfrac{5}{x+7}$

18. $\dfrac{4}{x-6} = \dfrac{11}{5x+2}$

19. $\dfrac{5}{x-6} - 5 = \dfrac{2x}{x-6}$

20. $\dfrac{2x}{x-2} + 3 = \dfrac{9}{x-2}$

21. $3 + \dfrac{2x}{x-3} = \dfrac{4}{x-3}$

22. $\dfrac{12}{x-4} = 6 + \dfrac{3x}{x-4}$

23. $\dfrac{x}{x+3} + \dfrac{2}{x} = 1$

24. $\dfrac{15}{x} + \dfrac{3x-7}{x+2} = 3$

25. $\dfrac{x}{x+3} + \dfrac{1}{x+2} = 1$

26. $\dfrac{2x}{x-4} + \dfrac{2}{x+1} = 2$

27. $\dfrac{x}{x-4} - \dfrac{4}{2x-1} = 1$

28. $\dfrac{x}{x-2} - \dfrac{x+1}{x+4} = 1$

29. $\dfrac{2}{4x-1} + \dfrac{1}{x+1} = \dfrac{3}{x+1}$

30. $\dfrac{x-2}{x+4} - \dfrac{3}{2x+1} = \dfrac{x-7}{x+4}$

31. $\dfrac{x}{x-4} - \dfrac{12x}{x^2+x-20} = \dfrac{x-1}{x+5}$

32. $\dfrac{x-2}{x-3} + \dfrac{x-3}{x-2} = \dfrac{2x^2}{x^2-5x+6}$

33. $\dfrac{12}{x+1} + 1 = \dfrac{5}{x-1}$

34. $\dfrac{8}{x+3} - 3 = \dfrac{1}{x-2}$

35. $\dfrac{x}{x-3} + \dfrac{8}{x^2-2x-3} = \dfrac{-2}{x+1}$

36. $\dfrac{x}{x+4} + \dfrac{8x}{x^2-16} = \dfrac{1}{x-4}$

37. $\dfrac{1}{x+2} + \dfrac{1}{x+5} = \dfrac{5}{4}$

38. $\dfrac{3}{x-1} - \dfrac{5}{x+4} = \dfrac{11}{14}$

Set up proportions and solve the word problems in Exercises 39–50.

39. The local sales tax is 6 cents for each dollar. What is the price of an item if the sales tax is 51 cents?

40. The property tax rate is $10.87 on every $100 of assessed valuation. Find the property tax on a house assessed at $38,000.

41. On a map, each inch represents $7\frac{1}{2}$ miles. What is the distance represented by 6 inches?

42. An elementary school has a ratio of 1 teacher for each 24 children. If the school presently has 21 teachers, how many students are enrolled?

43. At the Bright-As-Day light-bulb plant, 3 out of each 100 bulbs produced are defective. If the daily production is 4800 bulbs, how many are defective?

44. A floor plan is drawn to scale where 1 inch represents 4 feet. What size will the drawing be for a room that is 26 feet by 32 feet?

45. Pete is averaging 17 hits for each 50 times at bat. If he maintains this average, how many "at bats" will he need in order to get 204 hits?

46. Every 50 pounds of a certain alloy contains 6 pounds of copper. To have 27 pounds of copper, how many pounds of alloy will you need?

47. Instructions for Never-Ice Antifreeze state that 4 quarts of antifreeze are needed for each 10 quarts of radiator capacity. If Sal's car has a 22-quart radiator, how many quarts of antifreeze will it need?

48. The negative of a rectangular-shaped picture is $1\frac{1}{2}$ in. by 2 in. If the largest paper available is 11 in. long, what would be the maximum width of the enlargement?

49. A flagpole casts a 30-foot shadow at the same time as a 6-foot man casts a $4\frac{1}{2}$-foot shadow. How tall is the flagpole?

50. In a group of 30 people, there are 3 left-handed people. How many left-handed people would you expect to find in a group of 180 people?

6.6 Applications

In Section 6.5, we solved equations involving rational expressions (fractions) and some word problems using proportions. In this section, we continue to deal with rational expressions in solving a variety of applications involving numbers and proportions, time $\left(\text{using the relationship } d = rt \text{ in the form } t = \dfrac{d}{r} \right)$, and work.

More on Proportions

Continuing our discussion of proportions from Section 6.5, we are interested in situations where a total amount is known and this total is to be separated into two parts according to some ratio. The key idea here is to recognize the relationship between the total and the two parts. In general,

if $\qquad\qquad\qquad T =$ total amount

and $\qquad\qquad\qquad x =$ one part

then $\qquad\qquad T - x =$ second part

EXAMPLES

1. The sum of two integers is 104 and they are in the ratio of 8 to 5. Find the two numbers.

Solution   Here, we know the total sum of the two numbers is 104. So,

let $\qquad\qquad n =$ one number

and $\qquad 104 - n =$ second number

Then set up a proportion using the ratio 8 to 5.

$$\frac{n}{104 - n} = \frac{8}{5}$$

$$5n = 8(104 - n)$$

$$5n = 832 - 8n$$

$$\frac{13n}{13} = \frac{832}{13}$$

$$n = 64$$

$$104 - n = 40$$

The two numbers are 64 and 40.

$$\left(\text{\textbf{Checking,} we get } 64 + 40 = 104 \text{ and } \frac{64}{40} = \frac{8 \cdot 8}{8 \cdot 5} = \frac{8}{5}.\right)$$

2. Suppose that an artist expects that for every 9 special brushes he orders, 7 will be good and 2 will be defective. If he orders 45 brushes, how many will he expect to be defective?

Solution Let n = number of defective brushes
  $45 - n$ = number of good brushes (Total $- n$)

Then, using the ratio of good to defective,

$$\frac{7}{2} = \frac{45 - n}{n}$$

$$7n = 2(45 - n)$$

$$7n = 90 - 2n$$

$$9n = 90$$

$$n = 10$$

He expects 10 defective brushes.   ∎

Time $\left(\textbf{d = rt } or \textbf{ t} = \dfrac{\textbf{d}}{\textbf{r}}\right)$

Another application of rational expressions and proportions involves the formula $d = rt$ (distance equals rate times time). The same formula can be written in three forms:

$$d = rt; \qquad r = \frac{d}{t}; \qquad t = \frac{d}{r}.$$

The last formula on the right indicates that time can be represented as a ratio of distance divided by rate.

EXAMPLES

3. Janice can run 32 kilometers (about 19 miles) in the same amount of time that her twin sister, Jill, can run 24 kilometers. If Janice runs 2 kilometers per hour faster than Jill, how fast does each girl run?

   Solution   Let r = Jill's rate

   and $r + 2$ = Janice's rate (She runs faster than Jill.)

	Rate	Time = $\dfrac{\text{distance}}{\text{rate}}$	Distance
Jill	r	$\dfrac{24}{r}$	24
Janice	$r + 2$	$\dfrac{32}{r + 2}$	32

   Using the formula $t = \dfrac{d}{r}$ and the fact that their times are equal, we can write the equation:

   $$\left.\begin{array}{c}\text{Jill's}\\\text{time}\end{array}\right\} \longrightarrow \frac{24}{r} = \frac{32}{r + 2} \longleftarrow \left\{\begin{array}{c}\text{Janice's}\\\text{time}\end{array}\right.$$

   $$24(r + 2) = 32r$$
   $$24r + 48 = 32r$$
   $$48 = 8r$$
   $$6 = r$$

   So,   Jill's rate = r = 6 km/hr
   and   Janice's rate = $r + 2$ = 8 km/hr.

4. A small plane can fly at 120 mph in still air. If it can fly 490 miles with a tailwind in the same time that it can fly 350 miles against a headwind, what is the speed of the wind?

   Solution   Let w = speed of the wind

	Rate	Time = $\dfrac{\text{distance}}{\text{rate}}$	Distance
Tailwind	$120 + w$	$\dfrac{490}{120 + w}$	490
Headwind	$120 - w$	$\dfrac{350}{120 - w}$	350

Using the formula $t = \dfrac{d}{r}$ and the fact that the times are equal, the equation is

$$\frac{490}{120 + w} = \frac{350}{120 - w}$$

$$490(120 - w) = 350(120 + w)$$

$$58{,}800 - 490w = 42{,}000 + 350w$$

$$16{,}800 = 840w$$

$$20 = w$$

The speed of the wind is 20 mph.   ∎

Work

Problems involving "work" can be very sophisticated and require calculus and physics. The problems we will be concerned with relate to the time involved to complete the work on a particular job. These problems involve only the idea of what fraction of the job is done in one unit of time (hours, minutes, days, weeks, and so on). For example, if a man can dig a ditch in 4 hours, what part (of the ditch-digging job) did he do in one hour? The answer is $\dfrac{1}{4}$. If the job took 5 hours, he would do $\dfrac{1}{5}$ in one hour. If the job took x hours, he would do $\dfrac{1}{x}$ in one hour.

EXAMPLES

5. Mike can clean his family's pool in 2 hours. His younger sister, Stacey, can do it in 3 hours. If they work together, how long will it take them to clean the pool?

   Solution   Let x = number of hours working together. Then

	Hours	Part in 1 hour
Mike	2	$\dfrac{1}{2}$
Stacey	3	$\dfrac{1}{3}$
Together	x	$\dfrac{1}{x}$

   $$\underbrace{\text{part done in}}_{\frac{1}{2}} \; + \; \underbrace{\text{part done in}}_{\frac{1}{3}} \; = \; \underbrace{\text{part done in}}_{\frac{1}{x}}$$
   $$\text{1 hr by Mike} \qquad \text{1 hr by Stacey} \qquad \text{1 hr together}$$

   $$\frac{1}{2} \; + \; \frac{1}{3} \; = \; \frac{1}{x}$$

$$\frac{1}{2}(6x) + \frac{1}{3}(6x) = \frac{1}{x}(6x)$$

Multiply each term on both sides of the equation by **6x,** the LCM of the denominators.

$$3x + 2x = 6$$

$$5x = 6$$

$$x = \frac{6}{5} \text{ hr}$$

Together they can clean the pool in $\frac{6}{5}$ hours, or 1 hour, 12 minutes.

(Note that this answer seems reasonable since it is less time than either person would take working alone.)

6. A man was told that his new Jacuzzi® pool would fill through an inlet valve in 3 hours. He knew something was wrong when the pool took 8 hours to fill. He found he had left the drain valve open. How long will it take to drain the pool?

Solution   Let t = time to drain pool. (**Note:** In this case, the inlet and outlet valves work against each other.)

	Hours	Part in 1 hour
Inlet	3	$\frac{1}{3}$
Outlet	t	$\frac{1}{t}$
Together	8	$\frac{1}{8}$

$$\underbrace{\text{part filled}}_{\text{by inlet}} - \underbrace{\text{part emptied}}_{\text{by outlet}} = \underbrace{\text{part filled}}_{\text{together}}$$

$$\frac{1}{3} - \frac{1}{t} = \frac{1}{8}$$

$$\frac{1}{3}(24t) - \frac{1}{t}(24t) = \frac{1}{8}(24t)$$

$$8t - 24 = 3t$$

$$5t = 24$$

$$t = \frac{24}{5}$$

The pool will drain in $\frac{24}{5}$ hours, or 4 hours, 48 minutes.   ∎

EXERCISES 6.6

1. The ratio of two numbers is 7 to 9. If the sum of the numbers is 48, find the numbers.

2. The ratio of two numbers is 3 to 7. If the sum of the numbers is 60, find the numbers.

3. In a mathematics class, the ratio of women to men is 6 to 5. If there are 44 students in the class, how many men are there?

4. In a certain class, the ratio of success to failure is 7 to 2. How many can be expected to successfully complete the class if 36 people are enrolled?

5. An office has two copiers. One can make 8 copies while the other makes 5. If both are used to make 650 copies, how many are made by each machine?

6. In an iodine-alcohol mixture, there are 3 parts of iodine for each 8 parts of alcohol. If you want a total of 220 milliliters of mixture, how much iodine and alcohol are needed?

7. The denominator of a fraction is two more than the numerator. If both numerator and denominator are increased by three, the result is $\frac{5}{6}$. Find the original fraction.

8. The denominator of a fraction is one more than twice the numerator. If both numerator and denominator are increased by seven, the result is $\frac{3}{4}$. Find the original fraction.

9. Julie travels 15 miles per hour faster than Derek. She can travel 180 miles in the same amount of time that he can travel 135 miles. Find the two rates.

	Rate	Time	Distance
Derek	x		135
Julie	$x + 15$		180

10. An airliner travels 1820 miles in the same amount of time it takes a private plane to travel 520 miles. If the airliner's average speed is 52 mph more than three times the average speed of the private plane, find both rates.

	Rate	Time	Distance
Private plane	x		520
Airliner	$x + 52$		1820

11. An airplane can travel 320 mph in still air. If it travels 690 miles with the wind in the same length of time it travels 510 miles against the wind, what is the speed of the wind?

12. A boat can travel 5 miles downstream in the same length of time it travels 3 miles upstream. If the rate of the current is 3 mph, find the speed of the boat in still water.

13. After traveling 750 miles, the pilot of an airliner increased the speed by 20 mph and continued on for another 780 miles. If the same amount of time was spent at each rate, find the rates.

14. Mr. Snyder had a sales meeting scheduled 80 miles from his home. After traveling the first 30 miles at a rather slow rate due to traffic, he found that, in order to make the meeting on time, he had to increase his speed by 25 mph. If he traveled the same length of time at each rate, find the rates.

15. An airliner leaves Phoenix traveling at an average rate of 420 mph. Thirty minutes later, a second airliner leaves Phoenix traveling along a parallel route. If the second plane overtakes the first at a point 1680 miles from Phoenix, find its rate. (**Hint:** The difference in the times is $\frac{1}{2}$ hour.)
[See table on page 217.]

	Rate	Time	Distance
First airliner	420		1680
Second airliner	x		1680

16. Mrs. Green leaves El Paso on a business trip to Los Angeles, traveling at 56 mph. Fifteen minutes later, Mr. Green discovers that she had forgotten her briefcase and, jumping into their second car, he begins to pursue her. What was his average speed if he caught her 84 miles outside of El Paso? (**Hint:** The difference in times is $\frac{1}{4}$ hour.)

	Rate	Time	Distane
Mrs. Green	56		84
Mr. Green	x		84

17. A boat traveled 9 miles downstream and returned. The total trip took $2\frac{1}{4}$ hours. If the speed of the current was 3 mph, find the speed of the boat in still water.

18. A boat travels 15 miles upstream and returns. If the boat travels 14 mph in still water, find the speed of the current if the total trip takes 2 hours, 20 minutes.

19. Miles rides the ski lift to the top of Snowy Peak, a distance of $1\frac{3}{4}$ miles. He then skies directly down the slope. If he skis five times as fast as the lift travels and the total trip takes 45 minutes, find the rate at which he skis.

20. Bonnie lives 45 miles from town. Recently she went to town and returned home. She traveled 6 mph faster returning home than she did going to town. If her total travel time was 1 hour, 35 minutes, find the two rates.

21. Tom can fix a car in 6 hours. Ellen can do it in 4 hours. How long would it take them working together?

22. One company can clear a parcel of land in 25 hours. Another company requires 30 hours to do the job. If both are hired, how long will it take them to clear the parcel working together?

23. One pipe can fill a tank in 15 minutes. The tank can be drained by another pipe in 45 minutes. If both pipes are opened, how long will it take to fill the tank?

24. One pipe can fill a tank three times as fast as another. When both pipes are used, it takes $1\frac{1}{2}$ hours to fill the tank. How long does it take each pipe alone?

25. Working together, Rick and Rod can clean the snow from the driveway in 20 minutes. It would have taken Rick, working alone, 36 minutes. How long would it have taken Rod alone?

26. A carpenter and his partner can put up a patio cover in $3\frac{3}{7}$ hours. If the partner needs 8 hours to complete the patio alone, how long will it take the carpenter working alone?

27. It takes Maria twice as long to complete a research project as it takes Judy to do the same project. If they both work on it, it takes them 1 hour. How long would it take each of them working alone?

28. It takes Bob four hours longer to repair a car than it takes Ken. Working together, they can complete the job in $1\frac{1}{2}$ hours. How long would it take each of them working alone?

29. A power plant has two coal-burning boilers. If both boilers are in operation, a load of coal is burned in 6 days. If only one boiler is used, a load of coal will last the new "fuel-efficient" boiler 5 days longer than the old boiler. How long will a load of coal last each boiler?

30. Working together, Alice and Sharon can prepare the company's sales report in $2\frac{2}{5}$ hours. Working alone, Sharon would need two hours longer to prepare the report than Alice would. How long would it take each of them working alone?

6.7 Additional Applications

In this section, you will be solving applied problems.

Gears

If a gear with T_1 teeth is meshed with a gear having T_2 teeth, the numbers of revolutions of the two gears are related by the proportion

$$\frac{T_1}{T_2} = \frac{R_2}{R_1} \qquad \text{or} \qquad T_1 R_1 = T_2 R_2$$

where R_1 and R_2 are the numbers of revolutions.

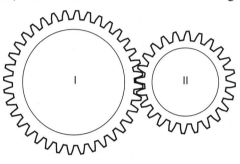

1. A gear having 96 teeth is meshed with a gear having 27 teeth. Find the speed (in revolutions per minute) of the small gear if the large one is traveling at 108 rpm.

2. What size gear (how many teeth) is needed to turn a pulley on a feed grinder at 24 revolutions per minute if it is driven by a 15-tooth gear turning at 56 revolutions per minute?

3. What size gear (how many teeth) is needed to turn the reel on a harvester at 20 revolutions per minute if it is driven by a 12-tooth gear turning at 60 revolutions per minute?

Lifting Force

The lifting force P in pounds exerted by the atmosphere on the wings of an airplane is related to the area A of the wings in square feet and the speed of the plane V in miles per hour by the formula

$$P = kAV^2 \qquad \text{where } k \text{ is the constant of proportionality.}$$

(**Hint for Problems 4–9:** First substitute the given information to find the value for k, the constant of proportionality. Then answer the question using the given change in conditions.)

4. If the lift is 9600 lb for a wing area of 120 sq ft and a speed of 80 mph, find the lift of the same airplane at a speed of 100 mph.

5. If a lift is 12,000 lb for a wing area of 110 sq ft and a speed of 90 mph, what would be the necessary wing area to attain the same lift at 80 mph?

6. The lift for a wing of area 280 sq ft is 34,300 lb when the plane is going 210 mph. What is the lift if the speed is decreased to 180 mph?

Pressure

Boyle's Law states that if the temperature of a gas sample remains the same, the pressure of the gas is related to the volume by the formula

$$P = \frac{k}{V} \qquad \text{where } k \text{ is the constant of proportionality.}$$

7. The temperature of a gas remains the same. The pressure of the gas is 16 lb/sq ft when the volume is 300 cu ft. What will be the pressure if the gas is compressed into 25 cu ft?

8. A pressure of 1600 lb/sq ft is exerted by 2 cu ft of air in a cylinder. If a piston is pushed into the cylinder until the pressure is 1800 lb/sq ft, what will be the volume of the air?

9. If 1000 cu ft of gas exerting a pressure of 140 lb/sq ft must be placed in a container which has a capacity of 200 cu ft, what is the new pressure exerted by the gas?

Medicine

In order to administer the medication prescribed by a doctor, a nurse must often figure the dosage in relation to the drugs available. The following formula is sometimes used:

Labeling Information   **Prescribed Dosage**

$$\frac{\text{stated dosage}}{\text{stated amount of solution}} = \frac{\text{desired dosage}}{\text{amount of solution needed}}$$

10. If a doctor orders 1.5 mg of reserpine, how many milliliters of a solution labeled 2.5 mg in 1 ml should be given?

11. The doctor orders 450,000 units of penicillin. The dose on hand is in a solution labeled 3,000,000 units per 10 ml. How many milliliters should be administered?

12. A doctor orders 25 units of insulin. How many milliliters of a solution labeled 40 units per milliliter should be given?

Concrete Mix

The ingredients in concrete are cement, sand, and gravel mixed in water. The proportions of each ingredient, except water, are expressed as

cement:sand:gravel

13. Find the number of sacks of cement (one cubic foot per sack), cubic feet of sand, and cubic feet of gravel needed to make 81 cubic feet of 1:3:5 concrete.

14. At the cost of $1.30 per sack, determine the cost of the cement required to make 60 cubic feet of concrete using a 1:3:4 mix (1 sack = 1 cubic foot).

Lever

If a lever is balanced, with weight on opposite sides of the balance points, then the following proportion exists:

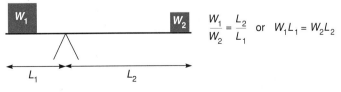

$$\frac{W_1}{W_2} = \frac{L_2}{L_1} \quad \text{or} \quad W_1 L_1 = W_2 L_2$$

(Ignore the weight of the lever.)

15. How much weight can be raised at one end of an 8-ft bar by the downward force of 60 lb when the balance point is $\dfrac{1}{2}$ ft from the unknown weight?

16. A force of 40 lb at one end of a 5-ft bar is to balance 160 lb at the other end of the bar.

Ignoring the weight of the bar, how far from the 40-lb weight should a hole be drilled in the bar for a bolt to serve as a balance point?

17. Where should the balance point of a 12-ft bar be located if a 120-lb force is to raise a load weighing 960 lb?

Electricity

The resistance, R (in ohms), in a wire is given by the formula

$$R = \frac{kL}{d^2} \qquad \text{where } L \text{ is the length of the wire and } d \text{ is the diameter.}$$

18. The resistance of a wire 500 ft long and with a diameter of 0.01 in. is 20 ohms. What is the resistance of a wire 1500 ft long and with a diameter of 0.02 in.?

19. The resistance of a wire 100 ft long and 0.01 in. in diameter is 8 ohms. What is the resistance of a piece of the same type of wire with a length of 150 ft and a diameter of 0.015 in.?

20. The resistance is 2.6 ohms when the diameter of a wire is 0.02 in. and the wire is 10 ft long. Find the resistance of the same type of wire with a diameter of 0.01 in. and a length of 5 ft.

◢ CHAPTER 6 SUMMARY

Key Terms and Formulas

Rational expressions are fractions in which the numerator and denominator are polynomials.   [6.1]

A fraction that contains various combinations of addition, subtraction, multiplication, and division with rational expressions is called a **complex algebraic fraction.**   [6.4]

A **ratio** is a comparison by division of two numbers, usually written as $a{:}b$ or $\dfrac{a}{b}$.   [6.5]

A **proportion** is an equation that states two ratios are equal:

$$\frac{a}{b} = \frac{c}{d} \quad [6.5]$$

Properties and Rules

If P, Q, R, and S are polynomials with Q, $S \neq 0$, then

1. $\dfrac{P}{Q} \cdot \dfrac{R}{S} = \dfrac{P \cdot R}{Q \cdot S}$ [6.2]

2. $\dfrac{P}{Q} \div \dfrac{R}{S} = \dfrac{P}{Q} \cdot \dfrac{S}{R}$ [6.2]

3. $\dfrac{P}{Q} + \dfrac{R}{Q} = \dfrac{P + R}{Q}$ [6.3]

4. $\dfrac{P}{Q} - \dfrac{R}{Q} = \dfrac{P - R}{Q}$ [6.3]

In general, for $b \neq a$, $(a - b)$ and $(b - a)$ are opposites and

$$\frac{a - b}{b - a} = \frac{a - b}{-1(a - b)} = -1 \quad [6.2]$$

For integers a and b, $b \neq 0$,

$$-\frac{a}{b} = \frac{-a}{b} = \frac{a}{-b} \quad [6.2]$$

For polynomials P and Q, $Q \neq 0$,

$$-\frac{P}{Q} = \frac{-P}{Q} = \frac{P}{-Q} \quad [6.2]$$

In a true proportion, the product of the extremes is equal to the product of the means.

$$\text{If } \frac{a}{b} = \frac{c}{d}, \text{ then } a \cdot d = b \cdot c. \quad [6.5]$$

Procedures

To Divide a Polynomial by a Polynomial [6.1]
The **division algorithm** with polynomials is the procedure followed when dividing a polynomial by a polynomial: For $P \div D$,

$$P = Q \cdot D + R \quad \text{or} \quad \frac{P}{D} = Q + \frac{R}{D}$$

where P is the dividend, D is the divisor, Q is the quotient, and R is the remainder. **The remainder must be of smaller degree than the divisor.**

To Find the LCM for a Set of Polynomials [6.3]
1. Completely factor each polynomial (including prime factors for numerical factors).
2. Form the product of all factors that appear, using each factor the most number of times it appears in any one polynomial.

To Solve an Equation Involving Rational Expressions [6.5]
1. Find the LCM of the denominators.
2. Multiply both sides of the equation by this LCM and simplify by using the distributive property.
3. Solve the resulting equation.
4. Check to see that no solution makes a denominator 0.

◢◢ CHAPTER 6 REVIEW

Express each quotient in Exercises 1–5 as a sum of fractions and simplify if possible. [6.1]

1. $\dfrac{6x^2 + 3x - 5}{3}$

2. $\dfrac{8x^2 - 14x + 6}{2x}$

3. $\dfrac{3x^3y + 10x^2y^2 + 5xy^3}{5x^2y}$

4. $\dfrac{7x^2y^2 - 21x^2y^3 + 8x^2y^4}{7x^2y^2}$

5. $\dfrac{20x^3y^2 - 15x^2y^3 + 10xy^4}{10xy^2}$

Divide Exercises 6–15 using the long division procedue and check your answers. [6.1]

6. $378 \div 37$

7. $792 \div 61$

8. $(x^2 + 7x - 18) \div (x + 9)$

9. $(x^2 - 14x - 16) \div (x + 1)$

10. $(4x^2 + 5x - 2) \div (4x - 3)$

11. $(2x^3 + 5x^2 + 7) \div (x + 3)$

12. $\dfrac{4x^2 + 5x + 9}{x - 2}$

13. $\dfrac{x^2 + 9}{x + 3}$

14. $(6x^3 - 5x^2 + 1) \div (3x + 2)$

15. $(x^3 + 64) \div (x + 4)$

Reduce each fraction in Exercises 16–23 to its lowest terms. State any restrictions on the variable. [6.2]

16. $\dfrac{x}{x^2 + x}$

17. $\dfrac{4 - x}{3x - 12}$

18. $\dfrac{2x + 6}{3x + 9}$

19. $\dfrac{x^2 + 3x}{x^2 + 7x + 12}$

20. $\dfrac{x^2 + 2x - 15}{2x^2 - 12x + 18}$

21. $\dfrac{x^2 - 9x + 20}{16 - x^2}$

22. $\dfrac{8x^2 - 2x - 3}{4x^2 + x - 3}$

23. $\dfrac{2x^2 - 5x^3}{10x^2 - 4x}$

Perform the indicated operations in Exercises 24–42. Assume that no denominator is 0.   [6.2–6.3]

24. $\dfrac{x^2}{x + y} - \dfrac{y^2}{x + y}$

25. $\dfrac{x^2 + x - 7}{x^2 + 3x - 10} + \dfrac{x^2 + 5x - 13}{x^2 + 3x - 10}$

26. $\dfrac{3x}{x + 2} \cdot \dfrac{x^2 + 2x}{x^2}$

27. $\dfrac{3x}{x + 2} \div \dfrac{4x}{x + 2}$

28. $\dfrac{2y + 4}{5} \cdot \dfrac{35}{6y + 12}$

29. $\dfrac{7}{x} + \dfrac{9}{x - 4}$

30. $\dfrac{4}{y + 2} - \dfrac{4}{y + 3}$

31. $\dfrac{4x}{3x + 6} - \dfrac{x}{x + 1}$

32. $\dfrac{4x}{x - 4} \div \dfrac{12x^2}{x^2 - 16}$

33. $\dfrac{x^2 + 3x + 2}{x + 3} \div \dfrac{x + 1}{3x^2 + 6x}$

34. $\dfrac{2x + 1}{x^2 + 5x - 6} \cdot \dfrac{x^2 + 6x}{x}$

35. $\dfrac{5}{x - 5} + \dfrac{x}{25 - x^2}$

36. $\dfrac{4}{x + 6} + \dfrac{8x}{x^2 + x - 6}$

37. $\dfrac{8}{x^2 + x - 6} + \dfrac{2x}{x^2 - 3x + 2}$

38. $\dfrac{x}{x^2 + 3x - 4} - \dfrac{x + 1}{x^2 - 1}$

39. $\dfrac{x^2 + 3x - 10}{x^2 - 9} \cdot \dfrac{x^2 - x - 6}{x^2 + 7x + 10}$

40. $\dfrac{x + 1}{x^2 + 4x + 4} \div \dfrac{x^2 - x - 2}{x^2 - 2x - 8}$

41. $\dfrac{7 - x}{7x^3} \div \dfrac{7x^2 + 48x - 7}{x^3 + 7x^2} \cdot \dfrac{7x^2 - 50x + 7}{x^2 - 7x}$

42. $\dfrac{3}{x + 2} + \dfrac{2x}{x^2 - 3x - 10} - \dfrac{1}{x^2 - 2x - 15}$

Simplify the complex algebraic fractions in Exercises 43–47.   [6.4]

43. $\dfrac{1 - \dfrac{2}{3}}{\dfrac{1}{4} + \dfrac{5}{6}}$

44. $\dfrac{\dfrac{3}{x} + \dfrac{1}{6x}}{\dfrac{7}{3x}}$

45. $\dfrac{\dfrac{2x}{x - 4}}{\dfrac{x}{x^2 - 3x - 4}}$

46. $\dfrac{\dfrac{1}{x} - \dfrac{1}{x^2}}{\dfrac{1}{x} + \dfrac{1}{x^2}}$

47. $\dfrac{\dfrac{4}{3x} + \dfrac{1}{6x}}{\dfrac{1}{x^2} - \dfrac{1}{2x}}$

Solve each proportion or equation in Exercises 48–53.   [6.5]

48. $\dfrac{4}{7} = \dfrac{x}{91}$

49. $\dfrac{5}{8} = \dfrac{9}{x}$

50. $\dfrac{3}{x} = \dfrac{5}{x + 3}$

51. $\dfrac{x - 2}{x + 4} = \dfrac{3}{7}$

52. $\dfrac{3x}{x + 2} + \dfrac{2}{x - 4} = 3$

53. $\dfrac{x}{x - 4} + \dfrac{5}{x} = \dfrac{4}{x^2 - 4x}$

Set up a proportion or other equation for each problem in Exercises 54–62 and then solve the equation.

54. A set of house plans is drawn to scale, where $\dfrac{1}{4}$ inch represents 1 foot. What will be the size of the drawing of a room that is 18 feet by 24 feet?   [6.5]

55. During basketball season, Red averaged 5 successful free throws for each 7 attempted. If he made 115 free throws, how many did he attempt?   [6.5]

56. Two numbers are in a ratio of 2 to 5. If their sum is 84, find the numbers.   [6.5]

57. Patti can travel 125 miles on her bike in the same amount of time that Bart can travel 100 miles. Patti's rate is 5 mph faster than Bart's. Find the two rates.   [6.6]

58. A family travels 18 miles downriver and returns. It takes 8 hours to make the round trip. Their rate in still water is twice the rate of the current. Find the rate of the current.   [6.6]

59. Sonny needs 5 hours to complete the yardwork. His wife, Loni, needs 4 hours to do the work. How long will it take if they work together?   [6.6]

60. A contractor hires two bulldozers to clear a tract of land. One works twice as fast as the other. It takes them 3 days to clear the land working together. How long would it have taken each of them working alone?   [6.6]

61. If the lift is 16,000 lb for a wing area of 180 sq ft and a speed of 120 mph, find the lift of the same plane at a speed of 150 mph $(P = kAv^2)$. [6.7]

62. The resistance of a wire 250 ft long and with a diameter of 0.01 in. is 10 ohms. What is the resistance of a piece of the same type of wire with a length of 300 ft and a diameter of 0.02 in. $\left(R = \dfrac{kL}{d^2}\right)$? [6.7]

CHAPTER 6 TEST

Express each quotient in Exercises 1 and 2 as a sum of fractions and simplify if possible.

1. $\dfrac{4x^3 + 3x^2 - 6x}{2x^2}$

2. $\dfrac{5x^2y + 6x^2y^2 + 3xy^3}{3x^2y}$

Divide Exercises 3–5 using long division.

3. $(x^2 + 5x - 36) \div (x - 4)$

4. $(2x^2 - 9x - 20) \div (2x + 3)$

5. $(x^3 - 27) \div (x - 3)$

6. Reduce and state any restrictions on the variable: $\dfrac{2x^2 + 5x - 3}{2x^2 - x}$

Reduce the fractions in Exercises 7 and 8. Assume that no denominator is 0.

7. $\dfrac{8x^2 - 2x - 3}{4x^2 + x - 3}$

8. $\dfrac{15x^4 - 18x^3}{36x - 30x^2}$

Perform the indicated operations in Exercises 9–14 and reduce your answer. Assume that no denominator is 0.

9. $\dfrac{16x^2 + 24x + 9}{3x^2 + 14x + 8} \cdot \dfrac{21x + 14}{4x^2 + 3x}$

10. $\dfrac{3x - 3}{3x^2 - x - 2} \div \dfrac{24x^2 - 6}{6x^2 + x - 2}$

11. $\dfrac{3}{x^2 - 1} + \dfrac{7}{x^2 - x - 2}$

12. $\dfrac{8x}{x^2 - 25} - \dfrac{3}{x^2 + 3x - 10}$

13. $\dfrac{1}{x - 1} + \dfrac{2x}{x^2 - x - 12} - \dfrac{5}{x^2 - 5x + 4}$

14. $\dfrac{x^2 + 8x + 15}{x^2 + 6x + 5} \cdot \dfrac{x - 3}{2x} \div \dfrac{x + 3}{x^2}$

Simplify the complex fractions in Exercises 15 and 16.

15. $\dfrac{\dfrac{1}{x} - \dfrac{1}{x^2}}{1 - \dfrac{1}{x^2}}$

16. $\dfrac{\dfrac{6}{x - 5}}{\dfrac{x + 18}{x^2 - 2x - 15}}$

Solve the equations in Exercises 17–20.

17. $\dfrac{9}{3 - x} = \dfrac{8}{2x + 1}$

18. $\dfrac{3}{4} = \dfrac{2x + 5}{5 - x}$

19. $\dfrac{3}{x} - \dfrac{2}{x + 1} = \dfrac{5}{2x}$

20. $\dfrac{2x}{x - 1} - \dfrac{3}{x^2 - 1} = \dfrac{2x + 5}{x + 1}$

Set up proportions or other equations for Exercises 21–25 and then solve the equations.

21. A 20-foot flagpole casts a shadow 12 feet long. If a man 6 feet tall is standing beside the flagpole, how long is his shadow?

22. Eighty percent of a mathematics class received a passing grade on the last exam. If 32 people passed, how many are in the class?

23. Two numbers are in a ratio of 5 to 7. Their sum is 156. Find the numbers.

24. Lisa can travel 228 miles in the same time that Kim travels 168 miles. If Lisa's speed is 15 mph faster than

25. A plumber can complete a job in 2 hours. If an apprentice helps, it takes only $1\frac{1}{3}$ hours. How long would it take the apprentice working alone?

CUMULATIVE REVIEW (6)

1. Find 4.2% of 85.

2. Find 12.5% of 40.

Find the products in Exercises 3 and 4.

3. $(3x + 7)(x - 4)$

4. $(2x - 5)^2$

5. In the formula $A = p + prt$, solve for t.

Simplify each expression in Exercises 6 and 7.

6. $-4(x + 2) + 5(2x - 3)$

7. $x(x + 7) - (x + 1)(x - 4)$

Factor each expression in Exercises 8 and 9.

8. $4x^2 + 16x + 15$

9. $2x^2 + 6x - 20$

Simplify each expression in Exercises 10 and 11.

10. $\dfrac{32x^3y^2}{4xy^2}$

11. $\dfrac{15xy^{-1}}{3x^{-2}y^{-3}}$

12. Reduce to lowest terms: $\dfrac{8x^3 + 4x^2}{2x^2 - 5x - 3}$

Perform the indicated operation in Exercises 13–15.

13. $\dfrac{2x^2 - 8}{5x + 5} \cdot \dfrac{7x + 7}{x^2 - 4x - 12}$

14. $\dfrac{5}{2x + 3} + \dfrac{3}{x - 1}$

15. $\dfrac{3x}{x - 7} - \dfrac{4}{x - 5}$

Solve each of the equations in Exercises 16–18.

16. $4(x + 3) - 7 = -2(3x + 1) - 3$

17. $(x + 4)(x - 5) = 10$

18. $\dfrac{7}{2x - 1} = \dfrac{3}{x + 6}$

19. An airplane can travel 1035 miles in the same time that a train travels 270 miles. The speed of the plane is 50 mph more than three times the speed of the train. Find the speed of each.

20. A man can wax his car three times as fast as his daughter can. Together they can complete the job in 2 hours. How long does it take each of them working alone?

7 APPLICATIONS, INCLUDING INEQUALITIES

◢ DID YOU KNOW ?

The way in which individuals approach problem-solving situations has itself been a subject of research. In 1944, the French mathematician Jacques Hadamard wrote *An Essay on the Psychology of Invention in the Mathematical Field.* One of Hadamard's more interesting ideas was that the subconscious mind continues to work on a difficult unsolved problem even while the conscious mind is involved in other activities. Hadamard cites instances of scientists experiencing "flashes," in which the complete solution to a problem came to them when they were not involved actively in their research. Some scientists even report instances of thinking carefully about a problem before going to sleep and then waking with knowledge of the solution to the problem.

Hadamard points out, however, that it is impossible for the unconscious mind to assist in problem solving unless the conscious mind has done the necessary work of preparation. When Isaac Newton was asked how he discovered gravity, he is reported to have said, "by constantly thinking it over." Hadamard claims that sudden inspirations never happen except after voluntary efforts at consciously solving the problem. This means that solutions to problems are never chance happenings, but rather are the result of hard work, both conscious and unconscious.

As you use the attack plan for word problems, you will see that there is a definite method for the preparatory work in solving word problems. Your conscious mind can be made to work along certain lines of attack that are most likely to succeed. **The important idea is that the problems must be actively attacked before any inspiration can occur.**

The stages of invention: 1. preparation, 2. incubation, 3. illumination, 4. verification.

Jacques Hadamard (1865–1963)

◢ CHAPTER OUTLINE

*R*ead and reread and reread until you understand the problem. Do not start to work a word problem until you understand what is being asked. Be sure you understand all the words. (Vocabulary is crucial in mathematics and science.) Read each sentence completely. Speed reading is not an advantage in algebra because each word is important. Take your time and be sure you understand what is being asked before you start to write any equation.

◢ 7.1 Review of Solving Equations

OBJECTIVE

In this section, you will be reviewing the methods for solving first-degree equations and quadratic equations.

In Chapter 3, we developed the techniques for solving first-degree equations. First-degree equations can be written in the form $ax + b = c$, where a, b, and c are constants and $a \neq 0$. The basic procedures for solving first-degree equations are listed in Section 3.3 and again here for convenient reference.

To Solve a First-Degree Equation

1. Simplify each side of the equation by removing any grouping symbols and combining like terms. (In some cases, you may want to multiply each term by an expression to clear fractions or decimal coefficients.)
2. Use the addition property of equality to add the opposites of constants and/or variables so that variables are on one side and constants on the other.
3. Use the multiplication property of equality to multiply both sides by the reciprocal of the coefficient of the variable (or divide both sides by the coefficient).
4. Check your answer by substituting it into the original equation.

Remember, the objective is to get the variable by itself on one side of the equation.

EXAMPLES Solve the following equations.

1. $3x + 14 = x - 2(x + 1)$

   Solution

$3x + 14 = x - 2(x + 1)$	Write the equation.
$3x + 14 = x - 2x - 2$	Use the distributive property to remove parentheses.
$3x + 14 = -x - 2$	Simplify.
$4x + 14 = -2$	Add x to both sides.
$4x = -16$	Add -14 to both sides.
$x = -4$	Divide both sides by 4.

2. $1 + 2x + 3 - 3x = 20 - x + 6x$

$\qquad$ **Solution** $\quad 1 + 2x + 3 - 3x = 20 - x + 6x \qquad$ Write the equation.

$$4 - x = 20 + 5x \qquad \text{Simplify.}$$

$$4 = 20 + 6x \qquad \text{Add } x \text{ to both sides.}$$

$$-16 = 6x \qquad \text{Add } -20 \text{ to both sides.}$$

$$-\frac{8}{3} = x \qquad \text{Divide both sides by 6 and reduce.}$$

3. $\dfrac{3x}{4} - 7 = -1$

$\qquad$ **Solution** $\quad \dfrac{3x}{4} - 7 = -1 \qquad$ Write the equation.

$$\frac{3x}{4} = 6 \qquad \text{Add } +7 \text{ to both sides.}$$

$$3x = 24 \qquad \text{Multiply both sides by 4.}$$

$$x = 8 \qquad \text{Divide both sides by 3 and reduce.} \qquad \blacksquare$$

Since $\dfrac{3x}{4} = \dfrac{3}{4} \cdot \dfrac{x}{1} = \dfrac{3}{4}x$, we could solve an equation such as $\dfrac{3x}{4} = 6$ in one step by multiplying both sides by $\dfrac{4}{3}$, the reciprocal of $\dfrac{3}{4}$, as follows:

$$\frac{3x}{4} = 6$$

$$\left(\frac{4}{3} \cdot \frac{3}{4} \right)x = \frac{4}{3} \cdot 6$$

$$x = 8$$

Example 3 can also be solved by first multiplying both sides by 4 instead of adding $+7$ to both sides first. In this procedure, however, we must be sure to **multiply each term by 4 on both sides of the equation.**

$$\frac{3x}{4} - 7 = -1 \qquad \text{Write the equation.}$$

$$4\left(\frac{3x}{4} - 7 \right) = 1 \cdot 4 \qquad \text{Multiply both sides by 4.}$$

$$\frac{3x}{4} \cdot 4 - 7 \cdot 4 = -1 \cdot 4 \qquad \text{Multiply each term by 4.}$$

$$3x - 28 = -4 \qquad \text{Simplify.}$$

$$3x = 24 \qquad \text{Add } +28 \text{ to both sides.}$$

$$x = 8 \qquad \text{Divide both sides by 3 and reduce.}$$

If an equation contains fractions, you should multiply each term by the LCM (least common multiple) of the denominators. Generally, this makes the equation easier to solve. **No denominator can have a value of 0.** So, if you multiply by a variable, be sure to check your answers.

EXAMPLES Solve the following equations.

4. $\dfrac{3}{2x} + \dfrac{1}{x} = 1$

Solution $\dfrac{3}{2x} + \dfrac{1}{x} = 1$ Here the LCM of the denominators is $2x$ ($x \neq 0$).

$2x\left(\dfrac{3}{2x} + \dfrac{1}{x}\right) = 2x \cdot 1$ Multiply both sides by **2x.**

$2x \cdot \dfrac{3}{2x} + 2x \cdot \dfrac{1}{x} = 2x \cdot 1$ Multiply each term by **2x.**

$3 + 2 = 2x$ The result is a first-degree equation.

$5 = 2x$

$\dfrac{5}{2} = \dfrac{2x}{2}$

$\dfrac{5}{2} = x$

5. $0.3x - 0.5x + 0.4 = 0.2(x + 1)$

Solution $0.3x - 0.5x + 0.4 = 0.2(x + 1)$ Here the equation is first-degree with decimal coefficients.

$10(0.3x) - 10(0.5x) + 10(0.4) = 10(0.2)(x + 1)$ Multiply each term on both sides by **10** so that the coefficients will be integers.

$3x - 5x + 4 = 2(x + 1)$

$-2x + 4 = 2x + 2$

$-2x - 2x = 2 - 4$

$-4x = -2$

$\dfrac{-4x}{-4} = \dfrac{-2}{-4}$

$x = \dfrac{1}{2}$

or $x = 0.5$ ∎

In Chapter 5, we solved quadratic equations using factoring techniques and the following fact about products.

If $a \cdot b = 0$, then $a = 0$ or $b = 0$.

EXAMPLE Solve the following equation.

6. $5x^2 + x = 6$

> **Solution**
> $$5x^2 + x = 6$$
> $$5x^2 + x - 6 = 0 \qquad \text{One side must be 0.}$$
> $$(5x + 6)(x - 1) = 0 \qquad \text{Factor.}$$

$$5x + 6 = 0 \qquad \text{or} \qquad x - 1 = 0$$
$$5x = -6 \qquad\qquad\qquad x = 1$$
$$x = -\frac{6}{5}$$

There are two solutions: $-\dfrac{6}{5}$ and 1.   ∎

In Chapter 6, we solved equations involving rational expressions by multiplying each side of the equation by the least common multiple (LCM) of the denominators.

EXAMPLES Solve the following equations.

7. $\dfrac{2}{x} + \dfrac{2}{x + 3} = 1$

> **Solution**
> $$\frac{2}{x} + \frac{2}{x + 3} = 1 \qquad \begin{array}{l}\text{LCM} = x(x + 3). \\ \text{Also, } x \neq 0 \text{ and} \\ x \neq -3.\end{array}$$
>
> $$x(x + 3)\left[\frac{2}{x} + \frac{2}{x + 3}\right] = 1 \cdot x(x + 3) \qquad \begin{array}{l}\text{Multiply} \\ \text{both sides by} \\ x(x + 3).\end{array}$$

$$\frac{2}{x} \cdot x(x + 3) + \frac{2}{\cancel{x + 3}} \cdot x(\cancel{x + 3}) = 1 \cdot x(x + 3) \qquad \begin{array}{l}\text{Multiply} \\ \text{each term by} \\ x(x + 3).\end{array}$$

$$2(x + 3) + 2x = x(x + 3) \qquad \text{Simplify.}$$
$$2x + 6 + 2x = x^2 + 3x$$
$$0 = x^2 - x - 6 \qquad \begin{array}{l}\text{The result is a} \\ \text{quadratic.}\end{array}$$
$$0 = (x + 2)(x - 3) \qquad \text{Factor.}$$

$$x + 2 = 0 \qquad \text{or} \qquad x - 3 = 0$$
$$x = -2 \qquad\qquad\qquad x = 3$$

There are two solutions: -2 and 3. Both are valid solutions since neither gives a denominator value of 0. The two values for x that could not be solutions are 0 and -3.

8. $\dfrac{x}{x-2} + \dfrac{x-6}{x(x-2)} = \dfrac{5x}{x-2} - \dfrac{10}{x-2}$

Solution $\dfrac{x}{x-2} + \dfrac{x-6}{x(x-2)} = \dfrac{5x}{x-2} - \dfrac{10}{x-2}$ LCM $= x(x-2)$.
Also, $x \neq 0$ and
$x \neq 2$.

Now multiply each term by $x(x-2)$.

$$\dfrac{x}{x-2} \cdot x(x-2) + \dfrac{x-6}{x(x-2)} \cdot x(x-2) = \dfrac{5x}{x-2} \cdot x(x-2) - \dfrac{10}{x-2} \cdot x(x-2)$$

$$x^2 + x - 6 = 5x^2 - 10x \qquad \text{Simplify.}$$
$$0 = 4x^2 - 11x + 6$$
$$0 = (4x - 3)(x - 2)$$
$$4x - 3 = 0 \qquad \text{or} \qquad x - 2 = 0$$
$$4x = 3 \qquad\qquad\qquad x = 2$$
$$x = \dfrac{3}{4}$$

The only solution is $x = \dfrac{3}{4}$; 2 is **not** a solution since no denominator can be 0.   ∎

Practice Problems

Solve the following equations.

1. $9x = 11x + 11$

2. $x + 17 = 2x + 7 - 3x$

3. $2x^2 = x + 1$

4. $0.3a + 0.6a = a + 0.007$

5. $\dfrac{5}{8}x + 1 = x - \dfrac{1}{6}$

6. $\dfrac{x-5}{x} + \dfrac{7}{x} = x$

◢ EXERCISES 7.1

Solve the following equations in Exercises 1–80.

1. $3x + 6 = x - 4$ **2.** $x + 9 = 4x - 3$ **3.** $7 - x = 2x - 8$ **4.** $11 - 2x = -3 + 5x$

5. $-3x - 5 = 7 - 5x$ **6.** $7x - 2 = -2x + 7$ **7.** $4x + 3 = 2x + 8$ **8.** $3x + 8 = -3x + 14$

9. $18x - 4 = 3x + 26$ **10.** $7x + 9 = 4x + 36$ **11.** $3x + 14 = 29 - 2x$ **12.** $8x + 5 = 5x - 6$

13. $11x - 3 = 6 + 5x$ **14.** $10x + 13 = 4x - 8$ **15.** $2 - 5x = 7x - 4$ **16.** $4x + 9 = 5 - 2x$

17. $\dfrac{5x}{6} - 1 = \dfrac{3}{2}$ **18.** $\dfrac{3x}{4} + \dfrac{1}{2} = -3$ **19.** $\dfrac{x}{3} - \dfrac{6}{7} = \dfrac{x}{21}$ **20.** $\dfrac{2x}{5} + \dfrac{2}{3} = \dfrac{x}{3}$

21. $\dfrac{3x}{14} - \dfrac{x}{7} = -\dfrac{1}{2}$ **22.** $\dfrac{4x}{3} - \dfrac{3}{4} = \dfrac{x}{6}$ **23.** $\dfrac{x}{4} - \dfrac{x}{3} = \dfrac{1}{12}$ **24.** $\dfrac{3x}{2} + \dfrac{x}{6} = -2$

Answers to Practice Problems   **1.** $x = -\dfrac{11}{2}$ **2.** $x = -5$ **3.** $x = 1$ or $x = -\dfrac{1}{2}$

4. $a = -0.07$ **5.** $x = \dfrac{28}{9}$ **6.** $x = 2$ or $x = -1$

25. $\dfrac{x}{3} - \dfrac{x}{7} = 4$

26. $2x - \dfrac{6x}{5} = \dfrac{9}{10}$

27. $\dfrac{x}{2} - \dfrac{2x}{3} = \dfrac{3}{4} + \dfrac{x}{3}$

28. $\dfrac{x}{4} - \dfrac{1}{2} = \dfrac{x}{6} - \dfrac{x}{3}$

29. $-0.15x = 9 - 0.6x$

30. $7 - 0.23x = 2.6 + 0.32x$

31. $0.4x - 2.2 = 0.48x + 5$

32. $0.2x - 0.04 = 0.15x + 0.02$

33. $3 - 0.2x = 0.05x - 1.25$

34. $15.6 + 0.4x = 1.15x - 8.4$

35. $3(2 - x) - 2x = -7$

36. $2x - 3(x + 2) = 10$

37. $-(x + 5) = 2x + 4$

38. $3(2x - 1) = 5 - x$

39. $3x + 8 = -3(2x - 3)$

40. $6(3x + 1) = 5(1 - 2x)$

41. $4(6 - x) = -2(3x + 1)$

42. $x - (2x + 5) = 7 - (4 - x) + 10$

43. $\dfrac{2}{3x} = \dfrac{1}{4} - \dfrac{1}{6x}$

44. $\dfrac{x - 4}{x} + \dfrac{3}{x} = 0$

45. $\dfrac{3}{8x} - \dfrac{7}{10} = \dfrac{1}{5x}$

46. $\dfrac{1}{x} - \dfrac{8}{21} = \dfrac{3}{7x}$

47. $\dfrac{5}{9x} + \dfrac{3}{2} = \dfrac{1}{2x}$

48. $\dfrac{3}{4x} - \dfrac{1}{2} = \dfrac{7}{8x} + \dfrac{1}{6}$

49. $\dfrac{2}{3x} + \dfrac{3}{4} = \dfrac{1}{6x} + \dfrac{3}{2}$

50. $\dfrac{5}{10x} - \dfrac{2}{5} = \dfrac{4}{5x} + \dfrac{1}{2}$

51. $3x^2 - 15x = 0$

52. $8x^2 + 24x = 0$

53. $5x^2 + 9x = 0$

54. $6x^2 - 14x = 0$

55. $x^2 - 6x - 7 = 0$

56. $x^2 + x - 56 = 0$

57. $x^2 + 3x = 10$

58. $2x^2 - 9x + 9 = 0$

59. $5x^2 - x - 6 = 0$

60. $4x^2 + 23x + 15 = 0$

61. $5x^2 - 7x - 6 = 0$

62. $8x^2 - 10x = 3$

63. $3x^2 = 7x - 2$

64. $3x^2 = 11x + 4$

65. $6x^2 = 20 - 2x$

66. $12x^2 = 10x + 12$

67. $\dfrac{3}{x - 6} = \dfrac{5}{x}$

68. $\dfrac{3}{x - 5} = \dfrac{-2}{x}$

69. $\dfrac{1}{x - 1} = \dfrac{2}{x - 2}$

70. $\dfrac{7}{x - 3} = \dfrac{6}{x - 4}$

71. $\dfrac{x}{x + 3} + \dfrac{1}{x + 2} = 1$

72. $\dfrac{1}{3} + \dfrac{1}{x + 4} = \dfrac{1}{x}$

73. $\dfrac{5}{x + 3} - \dfrac{2}{x + 1} = \dfrac{1}{4}$

74. $\dfrac{4}{x} - \dfrac{2}{x + 1} = \dfrac{4}{3}$

75. $\dfrac{2}{x + 2} + \dfrac{5}{x - 2} = -3$

76. $\dfrac{6}{x + 4} + \dfrac{2}{x - 3} = -1$

Calculator Problems

77. $0.035x - 0.04 = 0.02x + 0.32$

78. $0.0031 - 0.012x = 0.024x - 0.0185$

79. $0.361x - 1.036 = 0.127x + 1.868$

80. $0.147x + 1.651 = 0.099x - 0.269$

7.2 First-Degree Inequalities and Applications

In Section 2.1, we discussed inequalities (using the symbols $<$, $>$, $\leq$, and $\geq$) and number lines. We stated that certain inequalities, such as $-2 < +5$, are true while others, such as $-3.1 \geq 0$, are false. We also graphed integers and rational numbers (fractions) on number lines. The set of numbers $\left\{ -\dfrac{3}{2}, -1, 0, 2 \right\}$ is graphed in Figure 7.1.

Figure 7.1

In this section, we will encounter inequalities with variables and learn to solve these inequalities and graph the solutions. There is only one solution to a first-degree equation such as $x + 2 = 7$. That is, $x = 5$ is the one and only solution. However, an inequality, such as $x + 2 < 7$, may have an infinite number of solutions. In fact, any number less than $(<)5$ is a solution.

Before we discuss the mechanics of solving first-degree inequalities, a comment concerning number lines is appropriate. The numbers that correspond to the points on a line are called **real numbers,** and number lines are called **real number lines.** In Chapter 2, we mentioned **rational numbers** (which include integers and fractions with integers as numerator and denominator) and **irrational numbers.** Together, these two types of numbers make up the **real number system.**

Irrational numbers, such as $\sqrt{2}$, $\sqrt{3}$, $-\sqrt{7}$, and π will be discussed in some detail in Chapter 9. The important idea here is that such numbers do indeed correspond to points on a number line, and when we consider inequalities, such as $x < 8$, all real numbers less than 8 are included (Figure 7.2).

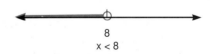

$$8$$
$$x < 8$$

Figure 7.2

Remember, for every real number, there is one corresponding point on a line, and for every point on a line, there is one corresponding real number.

Graphing the solutions to inequalities involves the following steps:

1. Putting a small circle on one point.
   a. An open circle indicates that the point is **not** included.
   b. A shaded circle indicates that the point is included.
2. Shading the portion of the number line corresponding to the solutions.

EXAMPLES   Graph the numbers that satisfy each of the following inequalities.

1. $x < 2$
  Solution

2. $x \leq 2$
  Solution

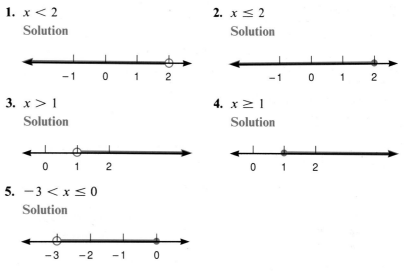

3. $x > 1$
  Solution

4. $x \geq 1$
  Solution

5. $-3 < x \leq 0$
  Solution

This case represents the numbers satisfying two inequalities at the same time:

$$-3 < x \quad \textbf{and} \quad x \leq 0$$

6. $x < -1$ **or** $x > 1$
  Solution

This case represents a combination (or union) of two intervals. The word "or" is necessary, and there is no way to write the two intervals together as there is in Example 5.   ■

The graphs (or sets of numbers) shown in Examples 1–5 are all **intervals** of real numbers. Example 5 represents a combination of two intervals called an **intersection.** Example 6 represents a combination of two intervals called a **union.**

An inequality that can be written in one of the forms

$$ax + b < c \text{ (or } ax + b > c\text{)} \qquad \text{or} \qquad ax + b \leq c \text{ (or } ax + b \geq c\text{)}$$

where *a*, *b*, and *c* are constants and $a \neq 0$ is called a **first-degree inequality.** We are now interested in developing the techniques for solving first-degree inequalities or combinations of first-degree inequalities and graphing their solutions on number lines.

The following basic properties of inequalities are needed. Similar properties for equations were discussed in Chapter 3.

Addition Property of Inequalities	If A, B, and C are algebraic expressions, then the inequalities
	$$A < B \quad \text{and} \quad A + C < B + C$$
	have the same solutions.

Multiplication Property of Inequalities	If A, B, and C are algebraic expressions, then
	1. $A < B$ and $AC < BC$ have the same solutions if $C > 0$.
	2. $A < B$ and $AC > BC$ have the same solutions if $C < 0$.

Solving an inequality, such as $2x + 1 < 7$, is similar to solving a first-degree equation. The object is to get the variable by itself on one side of the inequality. The important difference involves multiplying or dividing by negative numbers. **Multiplying or dividing both sides of an inequality by a negative number reverses the sense of the inequality.** That is, "less than" becomes "greater than," and vice versa. For example,

a.	$5 < 7$	5 is **less than** 7.
	$-2(5) \downarrow -2(7)$	Multiply both numbers by -2.
	$-10 > -14$	Now -10 is **greater than** -14.
b.	$6 \geq -12$	6 is **greater than or equal to** -12.
	$\dfrac{6}{-6} \downarrow \dfrac{-12}{-6}$	Divide both numbers by -6.
	$-1 \leq 2$	-1 is **less than or equal to** 2.
c.	$-2x > 6$	$-2x$ is **greater than** 6.
	$\dfrac{-2x}{-2} < \dfrac{6}{-2}$	Divide both sides by -2 and reverse the inequality.
	$x < -3$	The solution is "x is **less than** -3."

To Solve a First-Degree Inequality	**1.** Simplify each side of the inequality by removing any parentheses and combining like terms.
	2. Add constants and/or variables to or subtract them from both sides of the inequality so that variables are on one side and constants on the other.
	3. Divide both sides by the coefficient (or multiply them by the reciprocal) of the variable, and **reverse the sense of the inequality if this coefficient is negative.**
	4. A quick (and generally satisfactory) check is to select any one number in your solution and substitute it into the original inequality.

EXAMPLES   Solve the following inequalities and graph the solutions.

7. $5x + 4 \le -1$

Solution

$5x + 4 \le -1$	Write the inequality.
$5x + 4 - \mathbf{4} \le -1 - \mathbf{4}$	Add $-\mathbf{4}$ to both sides.
$5x \le -5$	Simplify.
$\dfrac{5x}{\mathbf{5}} \le \dfrac{-5}{\mathbf{5}}$	Divide both sides by $\mathbf{5}$.
$x \le -1$	Simplify.

(**Note:** The closed dot means that -1 is included.)

Check   As a quick check, pick any number less than or equal to -1, say, -4, and substitute it into the original inequality. If a false statement results, then you have made a mistake.

$$5(-4) + 4 \le -1$$
$$-16 \le -1 \qquad \text{True statement}$$

8. $x + 1 > 3 - 2x$

Solution

$x + 1 > 3 - 2x$	Write the inequality.
$x + 1 + \mathbf{2x} > 3 - 2x + \mathbf{2x}$	Add $\mathbf{2x}$ to both sides.
$3x + 1 > 3$	Simplify.
$3x + 1 - \mathbf{1} > 3 - \mathbf{1}$	Add $-\mathbf{1}$ to both sides.
$3x > 2$	Simplify.
$\dfrac{3x}{\mathbf{3}} > \dfrac{2}{\mathbf{3}}$	Divide both sides by $\mathbf{3}$.
$x > \dfrac{2}{3}$	Simplify.

(**Note:** The open circle means that $\dfrac{2}{3}$ is not included in the solution.)

9. $-1 < -2(x + 1) \le 3$

Solution   Note: Here we are actually solving two inequalities at once: $-1 < -2(x + 1)$ **and** $-2(x + 1) \le 3$. We want to find those numbers that satisfy **both** inequalities.

$-1 < -2(x + 1) \le 3$	Write the expression.
$-1 < -2x - 2 \le 3$	Use the distributive property.
$-1 + \mathbf{2} < -2x - 2 + \mathbf{2} \le 3 + \mathbf{2}$	Add $\mathbf{2}$ to all three expressions.
$1 < -2x \le 5$	Simplify.
$\dfrac{1}{-\mathbf{2}} > \dfrac{-2x}{-\mathbf{2}} \ge \dfrac{5}{-\mathbf{2}}$	Divide each expression by $-\mathbf{2}$ and reverse the sense of each inequality.
$-\dfrac{1}{2} > x \ge -\dfrac{5}{2}$	Simplify.

Or, rewriting with the smaller number on the left,

$$-\frac{5}{2} \le x < -\frac{1}{2}$$

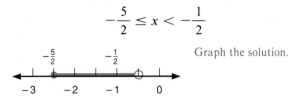

Graph the solution.

Check   For a quick check, substitute $x = -2$.

$$-1 < -2(-2 + 1) \le 3$$
$$-1 < -2(-1) \le 3$$
$$-1 < 2 \le 3 \qquad \text{True statement}$$

While a check does **not** guarantee there are no errors, it is a good indicator.   ■

We are generally familiar with the use of equations to solve word problems. In the following two examples, we show that solving inequalities can be related to real-world problems.

EXAMPLES

10. A physics student has grades of 85, 98, 93, and 90 on four examinations. If he must average 90 or better to receive an A for the course, what score can he receive on the final exam and get an A?

    Solution   Let x = score on final exam. The average is found by adding the scores and dividing by 5.

    $$\frac{85 + 98 + 93 + 90 + x}{5} \ge 90$$

    $$\frac{366 + x}{5} \ge 90$$

    $$\cancel{5}\left(\frac{366 + x}{\cancel{5}}\right) \ge 5 \cdot 90$$

    $$366 + x \ge 450$$

    $$x \ge 450 - 366$$

    $$x \ge 84$$

    If the student scores 84 or more on the final exam, he will average 90 or more and receive an A in physics.

11. Ellen is going to buy 30 stamps, some 15¢ and some 25¢. If she has $5.30, what is the maximum number of 25¢ stamps she can buy?

    Solution   Let $\quad\quad x$ = number of 25¢ stamps

    then $\quad 30 - x$ = number of 15¢ stamps

    Ellen cannot spend more than $5.30.

$$0.25x + 0.15(30 - x) \le 5.30$$
$$25x + 15(30 - x) \le 530$$
$$25x + 450 - 15x \le 530$$
$$10x + 450 \le 530$$
$$10x \le 530 - 450$$
$$10x \le 80$$
$$x \le 8$$

Multiply by 100 to get integer coefficients.

Ellen can buy at most 8 25¢ stamps if she buys 30 stamps.   ■

Practice Problems

1. Graph the numbers that satisfy the inequalities
   $x \le 2$ or $x > 5$.

Solve the inequalities and graph the solutions.

2. $7 + x < 3$

3. $\dfrac{x}{2} + 1 \ge \dfrac{x}{3}$

4. $-5 \le 2x + 1 < 9$

EXERCISES 7.2

Graph the numbers that satisfy the inequalities in Exercises 1–10.

1. $-2 \le x < 3$ **2.** $-4 \le x \le -1$ **3.** $x > 2$ or $x \le -4$ **4.** $x \le 3$ or $x > 6$

5. $x \le 1$ and $x > -2$ **6.** $x > 2$ or $x \le -5$ **7.** $x < \dfrac{10}{3}$ and $x \ge 1$ **8.** $x \ge 6$ or $x < \dfrac{13}{4}$

9. $x < -1$ or $x > \dfrac{1}{3}$ **10.** $x \ge 3$ and $x \le 10$

Solve the inequalities in Exercises 11–50 and graph the solutions.

11. $8 - x < 4$ **12.** $3 - 2a < 21$ **13.** $3x + 14 < 0$ **14.** $3x + 8 < 4$

15. $4x + 5 \ge -6$ **16.** $2x + 3 > -8$ **17.** $3x + 2 > 2x - 1$ **18.** $3y + 2 \le y + 8$

19. $y - 6 \le 4 - y$ **20.** $4x - 2 < 6x + 6$ **21.** $3x - 5 > 3 - x$ **22.** $3y - 1 \ge 11 - 3y$

23. $5y + 6 < 2y - 2$ **24.** $4 - 2x < 5 + x$ **25.** $4 + x > 1 - x$ **26.** $x - 6 > 3x + 5$

27. $\dfrac{x}{4} + 1 \le 5 - \dfrac{x}{4}$ **28.** $\dfrac{x}{2} - 1 \le \dfrac{5x}{2} - 3$ **29.** $\dfrac{x}{3} - 2 > 1 - \dfrac{x}{3}$ **30.** $\dfrac{5x}{3} + 2 > \dfrac{x}{3} - 1$

31. $-(x + 5) \le 2x + 4$ **32.** $-3(2x - 5) \le 3(x - 1)$

33. $x - (2x + 5) \ge 7 - (4 - x) + 10$ **34.** $x - 3(4 - x) + 5 \ge -2(3 - 2x) - x$

Answers to Practice Problems **1.**

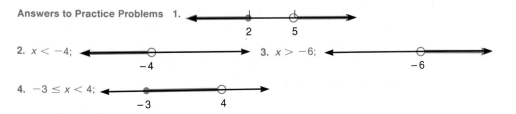

2. $x < -4$;

3. $x > -6$;

4. $-3 \le x < 4$;

35. $-5 < 1 - x < 3$ **36.** $2 < 2x - 4 \le 8$ **37.** $1 < 3x - 2 < 4$ **38.** $-5 < 4x + 1 < 7$
39. $-5 \le 5x + 3 < 9$ **40.** $-1 < 3 - 2x < 6$

41. To receive a B grade, a student must average 80 or more but less than 90. If John received a B in the course and had five grades of 94, 78, 91, 86, and 87 before taking the final exam, what were the possible grades for his final if there were 100 points possible?

42. The range for a C grade is 70 or more but less than 80. Before taking the final exam, Clyde had grades of 59, 68, 76, 84, and 69. If the final exam is counted as two tests, what is the minimum grade he could make on the final to receive a C? If there were 100 points possible, could he receive a B grade if an average of at least 80 points was required?

43. The temperature of a mixture in a chemistry experiment varied from 15° Celsius to 65° Celsius. What is this temperature expressed in degrees Fahrenheit? $\left[C = \dfrac{5}{9}(F - 32). \right]$

44. The temperature at Braver Lake ranged from a low of 23° F to a high of 59° F. What is the equivalent range of temperatures in degrees Celsius? $\left[F = \dfrac{9}{5}C + 32. \right]$

45. The sum of the lengths of any two sides of a triangle must be greater than the third side. If a triangle has one side that is 17 cm and a second side that is 1 cm less than twice the third side, what are the smallest possible lengths for the second and third sides?

46. The sum of four times a number and 21 is greater than 45 and less than 73. What are the possible values for the number?

47. In order for Chuck to receive a B in his mathematics class, he must have a total of at least 400 points. If he has scores of 72, 68, 85, and 89, what scores can he make on the final and receive a B? (The maximum possible score on the final is 100.)

48. In Exercise 47, if the final exam counted twice, could Chuck receive a grade of A in the class if it takes 540 points for an A? Assume that the maximum possible score on the final is 100 points.

49. The Concert Hall has 400 seats. For a concert, the admission will be $6.00 for adults and $3.50 for students. If the expense of producing the concert is $1825, what is the least number of adult tickets that must be sold to realize a profit if all seats are taken?

50. The Pep Club is selling candied apples to raise money. The price per apple is 25¢ until Friday, when they will sell for 20¢. If the Pep Club sells 200 apples, what is the minimum number they must sell at 25¢ each in order to raise at least $47.50?

7.3 Applications: Distance and Geometry

Distance

As was discussed in Section 5.5, the difficulty or ease with which you solve a particular problem depends on many factors, including your personal experiences and general reasoning abilities. For example, suppose you were given the following problem:

"A car travels 170 miles in 3 hours. What was the average speed?"

The problem does not say directly to MULTIPLY, DIVIDE, ADD, or SUB-TRACT. You must know that rate multiplied by time equals distance, or $r \cdot t = d$. You are given the distance (170 miles) and the time (3 hours). You are to find the average speed. The tool you need is the formula $r \cdot t = d$. We used this formula in Section 6.6.

Let r = average speed. Then,

$$3 \cdot r = 170$$
$$r = 56\frac{2}{3} \text{ mph}$$

EXAMPLES

1. A man leaves on a business trip, and at the same time his wife takes their children to visit their grandparents. The cars, traveling in opposite directions, are 360 miles apart at the end of 3 hours. If the man's average speed is 10 mph more than his wife's, what is her average speed?

   Solution Let x = average speed of wife. Then

	rate	· time =	distance
Wife	x	3	$3x$
Man	$(x + 10)$	3	$3(x + 10)$

   $$\underbrace{\text{distance for wife}} + \underbrace{\text{distance for man}} = \underbrace{\text{distance apart}}$$

   $$3x \ + \ 3(x + 10) \ = \ 360$$
   $$3x + 3x + 30 \ = \ 360$$
   $$6x \ = \ 330$$
   $$x \ = \ 55 \text{ mph}$$

   The wife's average speed is 55 mph.

2. Two trains, A and B, are 540 kilometers apart and travel toward each other on parallel tracks. Train A travels at 40 kilometers per hour, and train B travels at 50 kilometers per hour. In how many hours will they meet?

   Solution Let x = time. Then

	rate	· time	= distance
Train A	40	x	$40x$
Train B	50	x	$50x$

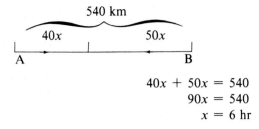

$$40x + 50x = 540$$
$$90x = 540$$
$$x = 6 \text{ hr}$$

The trains will meet in 6 hours.

3. Arno can paddle his canoe 4 mph in still water. After paddling downstream for 3.5 hours, Arno takes 4.5 hours to return to the same place he started by paddling upstream. What is the rate of the current?

Solution Let c = rate of the current. Then

	rate	$\cdot$ time	= distance
Downstream	$4 + c$	3.5	$3.5(4 + c)$
Upstream	$4 - c$	4.5	$4.5(4 - c)$

$$3.5(4 + c) = 4.5(4 - c) \quad \text{The two distances are equal.}$$
$$14 + 3.5c = 18 - 4.5c$$
$$4.5c + 3.5c = 18 - 14$$
$$8c = 4$$
$$c = 0.5$$

The rate of the current is 0.5 mph. ∎

Geometry

In Section 1.5, we discussed several formulas related to geometric figures. When a word problem involves a geometric figure, one of these formulas will be used with the understanding that you are familiar with the formula. You should review these formulas at this time.

EXAMPLE

4. A rectangle with a perimeter of 140 meters has a length that is 20 meters less than twice the width. Find the dimensions of the rectangle.

Solution Draw a diagram and use the formula $P = 2\ell + 2w$.

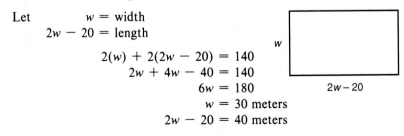

Let $\quad w = \text{width}$
$\quad 2w - 20 = \text{length}$

$$2(w) + 2(2w - 20) = 140$$
$$2w + 4w - 40 = 140$$
$$6w = 180$$
$$w = 30 \text{ meters}$$
$$2w - 20 = 40 \text{ meters}$$

The width is 30 meters and the length is 40 meters. ∎

EXERCISES 7.3

1. The length of a rectangle is 5 times the width. The perimeter of the rectangle is 96 cm. Find the dimensions of the rectangle.

2. The length of a rectangle is 11 ft more than the width. If the perimeter is 90 ft, what are the dimensions?

3. The perimeter of a rectangular parcel of land is 720 ft. The length is 60 ft less than twice the width. What are the dimensions of the parcel?

4. The length of a rectangle is twice the width. If the perimeter is 96 cm, what are the dimensions of the rectangle?

5. The length of a rectangle is 18 meters more than the width. Find the dimensions of the rectangle if the perimeter is 104 meters.

6. The perimeter of a rectangle is 242 meters. The length is 1 meter more than twice the width. Find the length and width of the rectangle.

7. The perimeter of a triangle is 51 meters. If the second side is twice the first side and the third side is 1 meter longer than the second side, how long is each side? (**Hint:** $P = a + b + c$ where a, b, and c are the sides of a triangle.)

8. The length of side b of a triangle is 4 cm more than the length of a side a. The length of side c is 1 cm less than twice the length of side a. If the perimeter is 43 cm, find the length of each side.

9. The second side of a triangle is three meters more than twice the first. The third side is 11 m less than the sum of the lengths of the other two sides. If the perimeter is 49 m, what is the length of each side?

10. An isosceles triangle is a triangle with two sides equal. The perimeter of an isosceles triangle is 51 cm. If the third side is 5 cm less than twice the length of one of the equal sides, how long is each side?

11. Two sides of a triangle are the same length. The third side is 2 centimeters less than the sum of the other two sides. If the perimeter is 22 centimeters, find the length of each side.

12. The ratio of the width of a rectangle to the length of a rectangle is 5 to 9. If the perimeter of the rectangle is 84 cm, find the length and width of the rectangle. (**Hint:** $5a$ and $9a$ are in the ratio of 5 to 9.)

13. Two trains leave Kansas City at the same time. One train travels east and the other travels west. The speed of the west-bound train is 5 mph greater than the speed of the east-bound train. After 6 hours, they are 510 miles apart. Find the rate of each train. Assume the trains travel in a straight line in directly opposite directions. **Hint:** Complete the following chart.

	rate	· time	= distance
West-bound	$x + 5$		
East-bound	x		

14. Steve travels 4 times as fast as Fred. Traveling in opposite directions, they are 105 miles apart after 3 hours. Find their rates of travel.

	rate	· time	= distance
Fred	x	3	
Steve	$4x$	3	

15. Sue travels 5 mph less than twice as fast as June. Starting at the same point and traveling in the same direction, they are 80 miles apart after 4 hours. Find their speeds.

16. Mary and Linda live 324 miles apart. They start at the same time and travel toward each other. Mary's speed is 8 mph greater than Linda's. If they meet in 3 hours, find their speeds.

17. Two planes leave from points 1860 miles apart at the same time and travel toward each other— at slightly different altitudes, of course. If the rates are 220 mph and 400 mph, how soon will they meet? **Hint:** Complete the following chart.

	rate	· time	= distance
Plane 1	220	x	
Plane 2	400	x	

18. Two buses leave Ocarche at the same time traveling in opposite directions. One bus travels at 55 mph, and the other at 59 mph. How soon will they be 285 miles apart?

19. A motor boat crossed a lake traveling 8 mph and returned along the same route at 12 mph. If it took $\frac{3}{4}$ of an hour less for the return trip, how far was the distance across the lake?

	rate	· time	= distance
Across	8		x
Back again	12		x

20. A jogger runs into the countryside at a rate of 10 mph. He returns along the same route at 6 mph. If the total trip took 1 hour, 36 minutes, how far did he jog?

21. A cyclist traveled to her destination at an average rate of 15 mph. By traveling 3 mph faster, she took 30 minutes less to return. What distance did she travel each way?

22. An airliner's average speed is $3\frac{1}{2}$ times the average speed of a private plane. Two hours after they leave the same airport at the same time, traveling in the same direction, they are 580 miles apart. What is the average speed of each plane? (**Hint:** Since they are traveling in the same direction, the distance between them will be the difference of their distances.)

23. A plane left Denver for Hawaii flying at 480 mph. Thirty minutes later, a second plane followed, traveling at 520 mph. How long will it take the second plane to overtake the first?

24. A train left Strong City traveling 48 mph. Forty-five minutes later, a second train left traveling in the opposite direction at 60 mph. How long will it take for the trains to be 279 miles apart?

25. The length of a rectangle is 10 meters more than one-half the width. If the perimeter is 44 meters, what is the length and the width?

26. The length of a rectangle is 1 meter less than twice the width. If each side is increased by 4 meters, the perimeter will be 116 meters. Find the length and the width of the original rectangle.

27. The River Queen tour boat can travel 12 mph in still water. After traveling for 3 hr downstream, it takes 5 hr to return. What is the rate of the current?

28. An airplane can travel 360 mph in still air. Traveling against the wind, it takes 5 hours to reach the destination. The return trip, with the wind, takes 4 hours. Find the speed of the wind.

29. The length of a rectangle is three times the width. If each side is increased by 4 cm, the perimeter will be doubled. Find the length and the width of the original rectangle.

30. A farmer has 160 meters of chain link fencing to build a rectangular corral. If he uses the barn for one of the longer sides, what will be the dimensions of the corral if the length is 5 meters less than three times the width?

31. Jane rides her bike to Blue Lake and returns. Going to the lake, she averages 12 mph. On the return trip, she averages 10 mph. If the total trip takes 5.5 hours, how far is it to Blue Lake?

32. Carol has 7 hours to spend on a hike up a mountain trail and back again. She can hike up the trail at an average of 1.5 mph and can hike down at an average of 2 mph. How far up the trail can she hike before turning back?

 **7.4 Applications: Interest and Mixture**

Interest

People in business know several formulas involving the principal (amount of money invested), rate (percent or rate of interest), and interest (the actual profit or interest earned). These formulas can depend on such related topics as the way the interest is paid on a loan (monthly, daily, yearly, and so on), whether or not there are penalties for early payment of a loan, or escalation clauses if an investment is particularly profitable.

In this section, we will use only the basic formula that calculates simple interest on an annual basis: $P \cdot R = I$, or principal times rate equals interest. This is a special case of the formula $P \cdot R \cdot T = I$ with $T = 1$.

EXAMPLES

1. A man invests in a certain bond yielding 9% interest and then invests $500 in a high-risk stock yielding 12%. After one year, his total interest from the two investments is $240. What amount did he invest in the bond?

   Solution   Let $P =$ principal invested at 9%. Then

	principal	rate	interest
Bond	P	0.09	$0.09P$
High-risk stock	500	0.12	0.12(500)

$$\underbrace{\text{interest on bond}} + \underbrace{\text{interest on stock}} = \underbrace{\text{total income}}$$

$$0.09P + 0.12(500) = 240$$
$$0.09P + 60 = 240$$
$$0.09P = 180$$
$$\frac{0.09P}{0.09} = \frac{180}{0.09}$$
$$P = \$2000$$

   He invested $2000 in the bond yielding 9% interest.

2. A woman has $7000. She decides to separate her funds into two investments. One yields an interest of 6%, and the other, 10%. If she wants an annual income from the investments to be $580, how should she split the money?

   Solution   Since we know that the total to be invested is $7000, if one investment is x, the other must be $7000 − x.

   Let $\quad\quad\quad x =$ amount invested at 10%
   $$7000 - x = \text{amount invested at 6\%}$$

	principal	$\cdot$ rate =	interest
10% investment	x	0.10	$0.10x$
6% investment	$(7000 - x)$	0.06	$0.06(7000 - x)$

$$\underbrace{\text{interest on}}_{0.10x} + \underbrace{\text{interest on}}_{0.06(7000 - x)} = \underbrace{\text{total}}_{580}$$
$$\text{10% investment} \quad \text{6% investment} \quad \text{income}$$

$$0.10x + 0.06(7000 - x) = 580$$
$$10x + 6(7000 - x) = 58{,}000 \quad \text{Multiply each term by 100 to eliminate decimals.}$$

$$10x + 42{,}000 - 6x = 58{,}000$$
$$4x = 16{,}000$$
$$x = \$4000 \ @ \ 10\%$$
$$7000 - x = \$3000 \ @ \ 6\%$$

She should invest $4000 at 10% and $3000 at 6%.   ■

Mixture

Problems involving mixtures occur in physics and chemistry and in such places as a candy store or a tobacco shop. Two or more items of a different percentage of concentration of a chemical such as salt, chlorine, or antifreeze are to be mixed; or two or more types of tobacco are to be mixed to form a final mixture that satisfies certain conditions of percentage of concentration.

The basic plan is to write an equation that deals only with one part of the mixture (such as the salt in the mixture). The following examples explain how this can be accomplished.

EXAMPLES

3. A particular experiment in chemistry demands a 10% solution of acid. If the lab assistant has 9 ounces of a 5% solution on hand, how much acid should be added to get the 10% solution? (**Hint:** Write an equation that deals only with amounts of acid.)

   Solution Let $x =$ amount of acid to be added. Then

	amount of solution	$\cdot$ percent acid	= amount of acid
Original solution	9	0.05	0.05(9)
Added solution	x	1.00	$1.00(x)$
Final solution	$(x + 9)$	0.10	$0.10(x + 9)$

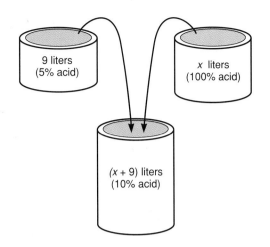

$$\underbrace{\begin{array}{c}\text{acid in}\\9\text{ oz}\end{array}}_{} + \underbrace{\begin{array}{c}\text{acid}\\\text{added}\end{array}}_{} = \underbrace{\begin{array}{c}\text{acid in final}\\\text{solution}\end{array}}_{}$$

$$
\begin{array}{rll}
0.05(9) + 1.00(x) &= 0.10(x + 9) & \\
5(9) + 100(x) &= 10(x + 9) & \text{Multiply each term by 100.}\\
45 + 100x &= 10x + 90 & \\
90x &= 45 & \\
x &= \dfrac{45}{90} & \\
x &= 0.5 \text{ oz of acid} &
\end{array}
$$

Check $\underbrace{\begin{array}{c}\text{acid in}\\9\text{ oz}\end{array}}_{} + \underbrace{\begin{array}{c}\text{acid}\\\text{added}\end{array}}_{} = \underbrace{\begin{array}{c}\text{acid in}\\\text{final mix}\end{array}}_{}$

$$
\begin{array}{rcl}
0.05(9) + 0.5 &\overset{?}{=}& 0.10(0.5 + 9)\\
0.45 + 0.5 &\overset{?}{=}& 0.10(9.5)\\
0.95 &=& 0.95
\end{array}
$$

The 10% solution can be had by adding 0.5 oz of acid.

4. How many gallons of a 20% salt solution should be mixed with a 30% salt solution to produce 50 gallons of 23% solution? (**Hint:** Write an equation that deals only with amounts of salt.)

   Solution Let x = amount of 20% solution
   $50 - x$ = amount of 30% solution

   Note: Since the total number of gallons is known, one amount is found by **subtracting** the other amount from the total.

	amount of solution	. percent salt =	amount of salt
20% solution	x	0.20	$0.20x$
30% solution	$50 - x$	0.30	$0.30(50 - x)$
23% solution	50	0.23	$0.23(50)$

$$\underbrace{\text{salt in 20\% solution}} + \underbrace{\text{salt in 30\% solution}} = \underbrace{\text{salt in 23\% solution}}$$

$$
\begin{aligned}
0.20x \quad + \quad 0.30(50 - x) &= 0.23(50) \\
20x \quad + \quad 30(50 - x) &= 23(50) \\
20x \quad + \quad 1500 - 30x &= 1150 \\
-10x &= 1150 - 1500 \\
-10x &= -350 \\
x &= 35 \text{ gal of 20\% solution}
\end{aligned}
$$

Check $\underbrace{\text{salt in 20\% solution}} + \underbrace{\text{salt in 30\% solution}} = \underbrace{\text{salt in 50 gal}}$

$$
\begin{aligned}
0.20(35) \quad + \quad 0.30(50 - 35) &\overset{?}{=} .23(50) \\
7.0 + 0.30(15) &\overset{?}{=} 11.5 \\
7.0 + 4.5 &\overset{?}{=} 11.5 \\
11.5 &= 11.5
\end{aligned}
$$

Thirty-five gallons of the 20% solution should be added to 15 gallons of the 30% solution. ∎

<hr>

⟁ EXERCISES 7.4

1. Amy receives $273 annually from two investments. If she has $1500 invested at 11%, how much does she have invested at 9%?

principal · rate = interest

11% investment	1500	0.11	
9% investment	x	0.09	

2. A company has $12,000 invested in a project yielding a 10% return. How much should it invest in a project having a 14% return to earn $4000 annually from both investments?

principal · rate = interest

10% investment	12,000	0.10	
14% investment	x	0.14	

3. Mr. Jackson invested $900 at 11% interest. How much did he invest at 9% if his yearly interest from both investments is $202.50?

4. Mildred has money in two savings accounts. One rate is 8% and the other is 10%. If she has $200 more in the 10% account, how much is invested at 8% if the total interest is $101?

5. Money is invested at two rates. One rate is 9% and the other is 13%. If there is $700 more invested at 9%, find the amount invested at each rate if the annual interest is $239.

6. Frank has half of his investments in stock paying an 11% dividend and the other half in a debentured stock paying 13% interest. If his total annual interest is $840, how much does he have invested?

7. Betty invested some of her money at 12% interest. She invested $300 more than twice that amount at 10%. How much is invested at each rate if her income is $318 annually?

8. GFA invested some money in a development yielding 24% and $9000 less in a development yielding 18%. If the first investment produces $2820 more per year than the second, how much is invested in each development?

9. Judy invests a certain amount of money at 7% annual interest and three times that amount at 8%. If her annual income is $232.50, how much does she have invested at each rate?

10. Norman has a certain amount of money invested at 5% annual interest and $500 more than twice that amount invested in bonds yielding 7%. His total income is $187. How much does he have invested at each rate?

11. How many ounces of a 15% hydrochloric acid solution should be mixed with 24 ounces of a 10% solution to make a 12% solution?

	amount of solution	. percent acid	= amount of acid
15% acid	x	0.15	
10% acid	24	0.10	
Final solution	$x + 24$	0.12	

12. Sixty liters of a 30% acid solution is to be reduced to a 20% acid solution by adding water. How much water should be added?

	amount of solution	. percent acid	= amount of acid
30% acid	60	0.30	
Water	x	0.00	
Final solution	$x + 60$	0.20	

13. A pharmacist wishes to reduce 36 ounces of a 10% iodine in alcohol solution to a 3% solution of iodine in alcohol. How many ounces of alcohol will he need? (**Hint:** The amount of iodine is the same in both the original solution and the final solution.)

14. How many gallons of water must be added to 4 gallons of a 30% salt brine to produce a 20% brine?

15. How many quarts of pure antifreeze must be added to 16 quarts of a 30% antifreeze solution to produce a 50% solution? (**Hint:** Pure antifreeze is a 100% solution.)

16. In preparing for an experiment in chemistry, the lab assistant noticed that he needed a 10% acid solution. If he has 20 ounces of a 12% acid solution, how much water must he add to reduce it to a 10% solution? (**Hint:** The amount of acid is the same in both the original solution and the final solution.)

17. A disinfectant contains 15% alcohol. How many liters of pure alcohol must be added to 48 liters of the disinfectant to obtain a solution that is 32% alcohol?

18. A chemist has 25 liters of a 2% acid solution. How many liters of a 5% acid solution should be added to the original solution to obtain a 3% acid solution?

19. How many pounds of milk that is 4% butterfat must be added to 200 pounds of milk that is 3% butterfat to produce milk that is 3.2% butterfat?

20. Determine the number of pounds of a 50% protein supplement that must be added to 2000 pounds of grain that is 3% protein content in order to make a hog ration that is 15% protein.

21. A total of $6000 is invested, part at 8% and the remainder at 12%. How much is invested at each rate if the annual interest is $620?

22. Mrs. Brown has $12,000 invested. Part is invested at 9% and the remainder at 11%. If the interest from the 9% investment exceeds the interest from the 11% investment by $380, how much is invested at each rate?

23. Eight thousand dollars is to be invested, part at 15% and the remainder at 12%. If the annual income from the 15% investment exceeds the income from the 12% investment by $66, how much is invested at each rate?

24. On an investment of $9500, Bill lost 3% on one part and earned 6% on the remainder. If his net annual receipts were $282, how much was each investment?

25. A manufacturer has received an order for 24 tons of a 60% copper alloy. His stock contains only 80% copper alloy and 50% copper alloy. How much of each will be needed to fill the order? (**Hint:** 24 tons is the total amount of alloy.)

26. A tobacco shop wants 50 ounces of tobacco that is 24% of a rare Turkish blend. If a 30% Turkish blend and a 20% Turkish blend are mixed, how much of each blend will be needed?

27. How many pounds each of a 12% zinc alloy and a 30% zinc alloy must be used to produce 90 pounds of a 22% zinc alloy?

28. To meet the government's specifications, an alloy must be 65% aluminum. How many pounds each of a 70% aluminum alloy and a 54% aluminum alloy will be needed to produce 640 pounds of the 65% aluminum alloy?

◢ CHAPTER 7 SUMMARY

Key Terms and Formulas

A first-degree inequality can be written in the form

$$ax + b < c \quad \text{(or } ax + b > c)$$

or $ax + b \leq c$ (or $ax + b \geq c$) where a, b, and c are constants and $a \neq 0$. [7.2]

Properties and Rules

Addition Property of Inequalities [7.2]
If A, B, and C are algebraic expressions, then the inequalities

$$A < B \quad \text{and} \quad A + C < B + C$$

have the same solutions.

Multiplication Property of Inequalities [7.2]
If A, B, and C are algebraic expressions, then

   1. $A < B$ and $AC < BC$ have the same solutions if $C > 0$.

   2. $A < B$ and $AC > BC$ have the same solutions if $C < 0$.

Multiplying or dividing both sides of an inequality by a negative number reverses the sense of the inequality. [7.2]

For every real number, there is one corresponding point on a line, and for every point on a line, there is one corresponding real number. [7.2]

Procedures

To Solve a First-Degree Equation [7.1]
   1. Simplify each side of the equation by removing any grouping symbols and combining like terms. (In some cases, you may want to multiply each term by an expression to clear fractions or decimal coefficients.)

   2. Use the addition property of equality to add the opposites of constants and/or variables so that variables are on one side and constants on the other.

3. Use the multiplication property of equality to multiply both sides by the reciprocal of the coefficient of the variable (or divide both sides by the coefficient).

4. Check your answer by substituting it into the original equation.

To Solve Quadratic Equations   [7.1]

Use factoring techniques and the following fact about products.

$$\text{If } a \cdot b = 0, \text{ then } a = 0 \quad \text{or} \quad b = 0.$$

To Solve a First-Degree Inequality   [7.2]

1. Simplify each side of the inequality by removing any parentheses and combining like terms.

2. Add constants and/or variables to or subtract them from both sides of the inequality so that variables are on one side and constants on the other.

3. Divide both sides by the coefficient (or multiply them by the reciprocal) of the variable, and **reverse the sense of the inequality if this coefficient is negative.**

4. A quick (and generally satisfactory) check is to select any one number in your solution and substitute it into the original inequality.

Some formulas related to word problems are

$r \cdot t = d$	Rate times time equals distance.
$P \cdot R = I$	Principal times rate of interest equals interest.
$P = 2\ell + 2w$	Perimeter of a rectangle equals twice the length plus twice the width.

Other formulas related to geometric figures are listed in Section 1.5   [7.3]

Problems with mixture involve writing an equation that deals with only one quantity in the mixture.   [7.4]

CHAPTER 7 REVIEW

Find the solution for the equations in Exercises 1–20.   [7.1]

1. $4x + 1 = 17$

2. $\dfrac{x}{2} + 9 = 6$

3. $6x + 2 = 3x - 1$

4. $\dfrac{5x}{3} + 2 = 12$

5. $\dfrac{2x}{5} + 3 = -3$

6. $2x + 7 = 5(x - 1)$

7. $37 - 4x = 20 + (x + 12)$

8. $2(x + 6) + 14 = 5(x + 2) - 18$

9. $\dfrac{1}{2}(x - 7) = \dfrac{1}{3}(2x + 5)$

10. $x - \dfrac{x}{5} = 10 - \dfrac{x}{2}$

11. $\dfrac{2}{5}(x - 4) = \dfrac{3}{4}x + 2$

12. $\dfrac{x}{3} + 1 = \dfrac{x + 4}{5}$

13. $\dfrac{3}{2x} = \dfrac{4}{5} + \dfrac{1}{2x}$

14. $\dfrac{2}{3x} = \dfrac{1}{2} + \dfrac{1}{6x}$

15. $1.3x + 11 = 0.5x + 3.8$

16. $2.4x - 4 = 0.8(x + 3)$

17. $3x^2 + 8x = 0$

18. $4x^2 - 5x - 6 = 0$

19. $\dfrac{7}{x + 3} = \dfrac{5}{x - 3}$

20. $\dfrac{4}{x + 1} + \dfrac{2}{x - 3} = \dfrac{-2}{3}$

Graph the numbers that satisfy the inequalities in Exercises 21–27.   [7.2]

21. $-5 < x < 7$

22. $-1 \le x \le 4$

23. $x < -2$ or $x \ge 5$

24. $x < -2$ or $x > \dfrac{4}{5}$

25. $x < \dfrac{7}{8}$ and $x > -3$

26. $2 \le x < 14$

27. $-3 < x \le 2.7$

Solve the inequalities and graph the solutions in Exercises 28–40.   [7.2]

28. $3x + 2 \le 8$

29. $4x - 7 > 9$

30. $\dfrac{x}{3} + 1 > 2$

31. $3x + 5 \le 6$

32. $2x + 3 \geq 4x + 5$ **33.** $5x + 1 \geq 2x + 6$ **34.** $5x + 3 \geq 2x + 15$ **35.** $2x - 5 < 3x + 2$

36. $2(x - 7) < 4(2x + 3)$ **37.** $7x + 4 \leq \frac{1}{2}(5x - 1)$ **38.** $x - (4 - 2x) \leq 2(x + 6) + 1$

39. $-2 \leq 4x + 7 \leq 3$ **40.** $-6 < 4 - 2x \leq 1$

41. A man has a rectangular parcel of land whose length is 40 meters less than 10 times its width. The perimeter is 250 meters. What are the dimensions of the parcel? [7.3]

42. The sum of two numbers is 30. If $\frac{3}{4}$ of the smaller is less than $\frac{1}{2}$ of the larger one, find the smaller number. [7.2]

43. A man has three times as much money invested at 6% as he has invested at 4%. If his yearly income is $1430, how much is invested at each rate? [7.4]

44. Mark paddles his canoe downstream for $1\frac{1}{2}$ hours. The return trip takes 6 hours. If the speed of the current is 3 mph, how fast does Mark row in still water? (**Hint:** If r is his rate in still water, $r + 3$ is his rate downstream, and $r - 3$ is his rate upstream.) [7.3]

45. Minh's father is 11 years more than three times as old as Minh. If the father's age is between 47 and 62 years inclusive, what is the range for Minh's age? [7.2]

46. A meat market has 20 pounds of ground beef that is 40% fat. How many pounds of extra lean ground beef that is only 15% fat would be needed to obtain ground beef that is 25% fat? [7.4]

47. Twice a certain number increased by 26 is greater than 4 times the number increased by 2. Find those numbers that satisfy this condition. [7.2]

48. The length of a rectangle is 7 centimeters more than twice the width. If both are increased by 3 centimeters, the area will be increased by 84 square centimeters. What are the length and width of the rectangle? [7.3]

49. Lisa can travel 228 miles in the same time that Soo travels 168 miles. If Lisa's speed is 15 mph faster than Soo's, find their rates. [7.3]

50. How many liters each of a 40% acid solution and a 55% acid solution must be used to produce 60 liters of a 45% acid solution? [7.4]

CHAPTER 7 TEST

Solve the equations in Exercises 1–8.

1. $5x - 3 = 4 - 2x$ **2.** $\frac{5x}{6} + 4 = 9$ **3.** $-2(5 - 3x) = 6 - (4x - 3)$

4. $0.4x - 6 = 8 - 0.3x$ **5.** $6x^2 - 9x = 0$ **6.** $(2x + 1)(x - 1) = 5$

7. $\frac{5}{2x} - \frac{11}{3x} = \frac{1}{6}$ **8.** $\frac{x}{x - 2} + \frac{3}{x - 1} = 1$

Graph the numbers that satisfy the inequalities in Exercises 9–11.

9. $x < -4$ or $x \geq 1.5$ **10.** $x < \frac{3}{2}$ and $x > -1$ **11.** $-\frac{3}{4} < x \leq 3$

Solve the inequalities in Exercises 12–16.

12. $4x - 5 > 12 - 5x$ **13.** $-2 < 3 - 5x < 6$ **14.** $\frac{2x}{5} - 3 \leq \frac{3}{10} - \frac{x}{3}$

15. $-2(5 - 3x) > 6 - (4x - 3)$ **16.** $2x - (7 - 3x) \geq 2(x + 4) + 6$

Set up equations for Exercises 17–25 and solve them.

17. The length of a rectangle is 9 feet more than twice its width. The perimeter of the rectangle is 66 feet. Find the dimensions.

18. Two automobiles leave Phoenix at the same time and travel in opposite directions. After two hours, they are 234 miles apart. If one automobile travels 13 mph faster than the other, find the rate of each.

19. Lynda invested some money yielding 18% and $1200 more at a yield of 12%. If the 18% investment produces $156 more each year than the second, how much was invested at each rate?

20. The sum of three times a number and 11 is greater than 20 and less than or equal to 47. What are the possible values for the number?

21. How many gallons of gasoline priced at $1.10 a gallon should be mixed with 300 gallons of gasoline priced at $1.00 a gallon to obtain a mixture worth $1.04 a gallon?

22. Alvin has two investments yielding a total annual interest of $185.60. The amount invested at 8% is $320 less than twice the amount invested at 6%. How much is invested at each rate?

23. Denny drove his truck to the garage for service and returned home on his bicycle. He averaged 50 mph in his truck and 12 mph on his bicycle. If the total travel time was 62 minutes, how far is it from his home to the garage?

24. The range for a B grade is 75 or more but less than 85. Eric has grades of 73, 65, 77, and 74 before taking the final exam. What are the possible grades on the final that will earn Eric a B grade? (Assume a maximum of 100 points are possible on the final.)

25. A metallurgist needs 2000 pounds of an alloy that is 80% copper. In stock, he has only 83% copper and 68% copper. How many pounds of each must be used?

CUMULATIVE REVIEW (7)

Determine whether or not the given number is a solution to the given equation in Exercises 1 and 2.

1. $3(2x - 5) = 5 - x; \quad x = 4$

2. $5x + 3 - (x + 2)(3x - 1) = -7; \quad x = 2$

Solve each equation for y in Exercises 3 and 4.

3. $6x + 2y = 9$

4. $3x - 4y + 7 = 0$

Factor completely each of the polynomials in Exercises 5–7.

5. $5x^2 + 28x - 12$

6. $8x^2 - 34x - 30$

7. $6xy + 8x + 15y + 20$

Perform the indicated operations in Exercises 8–10.

8. $\dfrac{x^2 + 3x - 4}{2x^2 - 7x + 5} \cdot \dfrac{2x - 5}{x + 7}$

9. $\dfrac{x}{3x + 2} + \dfrac{4}{x - 2}$

10. $\dfrac{x}{x^2 - 2x - 15} - \dfrac{x - 2}{x - 5}$

Solve each equation in Exercises 11–13.

11. $3(x - 1) - 2(4 - x) - 6x = 3$

12. $(x + 4)^2 = 1$

13. $\dfrac{7}{3x - 1} = \dfrac{6}{2x + 3}$

Solve each of the inequalities in Exercises 14 and 15; then graph the solution set.

14. $2x + 5 - (3 - 2x) \leq 6$

15. $-3(7 - 2x) \geq 2 + (3x - 10)$

Set up equations for each of Exercises 16–20. (You need not solve the equations.)

16. A manufacturer knows that one of his machines produces 4 defective parts for each 100 it produces. If on a particular day, the machine produced 875 parts, how many were defective?

17. Jimmy takes 5 days longer to complete a job than does Jerry. Working together, they can complete the work in 6 days. How long would it take Jimmy to complete the job?

18. The area of a rectangle is 480 square centimeters. The length is 2 centimeters less than twice the width. Find the length and width of the rectangle.

19. At 12:00 noon, a car leaves a town. At 1:00 P.M., another car leaves the same point and travels in the opposite direction. The speed of the second car is 2 mph faster than the first one. If the cars are 356 miles apart at 4:00 P.M., find the rate of each car.

20. How many liters each of a 30% salt solution and a 40% salt solution must be used to produce 40 liters of a 32% salt solution?

LINEAR EQUATIONS AND SYSTEMS OF EQUATIONS

◢ DID YOU KNOW ?

In Chapter 8, you will be introduced to the idea of a graph of an algebraic equation. A graph is simply a picture of an algebraic relationship. This topic is more formally called **analytic geometry.** It is a combination of algebra (the equation) and geometry (the picture).

The idea of combining algebra and geometry was not thought of until René Descartes wrote his famous *Discourse on the Method of Reasoning* in 1637. The third appendix in this book, "La Geometrie," made Descartes' system of analytic geometry known to the world. In fact, you will find that analytic geometry is sometimes called Cartesian geometry.

René Descartes is perhaps better known as a philosopher than as a mathematician; he is often referred to as the father of modern philosophy. His method of reasoning was to apply the same logical structure to philosophy that had been developed in mathematics and especially in geometry.

In the Middle Ages, the highest forms of knowledge were believed to be mathematics, philosophy, and theology. Many famous people in history who have reputations as poets, artists, philosophers, and theologians were also creative mathematicians. Almost every royal court had mathematicians whose work reflected glory on the royal sponsor who paid the mathematician for his research and his court presence.

Descartes, in fact, died in 1650 after accepting a position at the court of the young warrior-queen, Christina of Sweden. Apparently, the frail French philosopher-mathematician, who spent his mornings in bed doing mathematics, could not stand the climate of Sweden and the hardships imposed by Christina in her demand that Descartes tutor her in mathematics each morning at 5 o'clock in an unheated castle library.

Divide each problem that you examine into as many parts as you can and as you need to solve them more easily.

René Descartes (1596–1650)

◢ **CHAPTER OUTLINE**

T hanks to René Descartes (1596–1650), the ideas of algebra and geometry can be combined. He developed an entire theory connecting algebraic expressions and equations with points on geometric graphs. In particular in Chapter 8, you will see that equations with two variables can be related to graphs or "pictures" of straight lines in a plane. By studying the equations in special forms, you will be able to tell the position and basic properties of the corresponding lines before you start to draw the graphs. Also, you will see how the equations for word problems are easily set up and solved by considering two such equations (called a system) at the same time.

◢ **8.1 The Cartesian Coordinate System**

<table>
<tr><td>

OBJECTIVES

In this section, you will be learning to:

1. Name ordered pairs corresponding to points on graphs.

2. Graph ordered pairs in the Cartesian coordinate system.

3. Find ordered pairs that satisfy given equations.

</td></tr>
</table>

René Descartes (1596–1650), a famous French mathematician, developed a system for solving geometric problems using algebra. This system is called the **Cartesian coordinate system** in his honor. Descartes based his system on a relationship between points in a plane and **ordered pairs** of real numbers. We begin the discussion in this section by relating algebraic formulas with ordered pairs, and then we show how these ideas can be related to geometry.

Equations such as $d = 60t$, $I = 0.05P$, and $y = 2x + 3$ represent relationships between pairs of variables. For example, in the first equation, if $t = 3$, then $d = 60 \cdot 3 = 180$. With the understanding that t is first and d is second, we can write the pair (3,180) to represent $t = 3$ and $d = 180$. The pair (3,180) is called an **ordered pair.** Obviously, (180,3) is different from (3,180) if t is the first number and d is the second number.

We say that (3,180) **satisfies the equation** or **is a solution of the equation** $d = 60t$. Similarly, (100,5) satisfies $I = 0.05P$ where $P = 100$ and $I = 0.05(100) = 5$. Also, (2,7) satisfies $y = 2x + 3$, where $x = 2$ and $y = 2 \cdot 2 + 3 = 7$.

In an ordered pair such as (x,y), x is called the **first component** and y is called the **second component.** To find ordered pairs that satisfy an equation such as $y = 2x + 3$, we can choose **any** value for one variable and then find the corresponding value for the other variable by substituting into the equation. For example,

$y = 2x + 3$	**Ordered Pairs**
Choose $x = 1$; then $y = 2 \cdot 1 + 3 = 5$.	(1,5)
Choose $x = -2$; then $y = 2(-2) + 3 = -1$.	(−2,−1)
Choose $x = \dfrac{1}{2}$, then $y = 2\left(\dfrac{1}{2}\right) + 3 = 4$.	$\left(\dfrac{1}{2},4\right)$

The ordered pairs $(1,5)$, $(-2,-1)$, and $\left(\frac{1}{2},4\right)$ all satisfy the equation $y = 2x + 3$.

The variable assigned to the first component is also called the **independent variable,** and the variable assigned to the second component is called the **dependent variable.** Thus, we associate (x,y) with $y = 2x + 3$, and y "depends" on the values assigned to x.

We can also write the pairings in the form of tables. The choices for the values of the independent variable are arbitrary.

$d = 60t$		$I = 0.05P$		$y = 2x + 3$	
t	d	P	I	x	y
5	$60 \cdot 5 = 300$	100	$0.05(100) = 5$	-2	$2(-2) + 3 = -1$
10	$60 \cdot 10 = 600$	200	$0.05(200) = 10$	-1	$2(-1) + 3 = 1$
12	$60 \cdot 12 = 720$	500	$0.05(500) = 25$	0	$2(0) + 3 = 3$
15	$60 \cdot 15 = 900$	1000	$0.05(1000) = 50$	3	$2(3) + 3 = 9$

Now we are ready to show how the Cartesian coordinate system relates algebraic equations and ordered pairs to geometry. In this system, two number lines intersect at right angles and separate the plane into four **quadrants.** The **origin,** designated by the ordered pair $(0,0)$, is the point of intersection. The horizontal number line is called the **horizontal axis** or **x-axis.** The vertical number line is called the **vertical axis** or **y-axis.** Points that lie on either axis are not in any quadrant. They are simply on an axis (Figure 8.1).

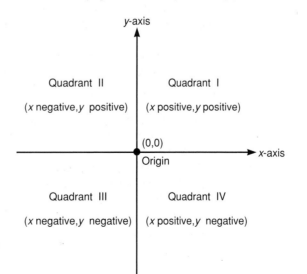

Figure 8.1

There is a one-to-one correspondence between points in a plane and ordered pairs of real numbers. In other words, for each point there is one and only one corresponding ordered pair of real numbers, and for each ordered pair of real numbers there is one and only one corresponding point. This important relationship is the cornerstone of the Cartesian coordinate system.

The graphs of the points $A(2,1)$, $B(-2,3)$, $C(-3,-2)$, $D(1,-2)$, and $E(3,0)$ are shown in Figure 8.2. [**Note:** An ordered pair of real numbers and the corresponding point on the graph are frequently used to refer to each other. Thus, the ordered pair $(2,1)$ and the point $(2,1)$ are interchangeable ideas.]

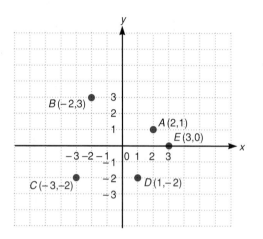

Figure 8.2

EXAMPLES   Graph the sets of ordered pairs in Examples 1 and 2.

1. $\{(-2,1), (0,2), (1,3), (2,-3)\}$

Solution

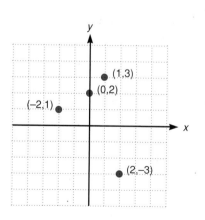

2. $\{(-1,3), (0,1), (1,-1), (2,-3), (3,-5)\}$
Solution

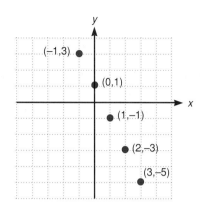

The points (or ordered pairs) in Example 2 can be shown to satisfy the equation $y = -2x + 1$. For example, using $x = -1$ in the equation,

$$y = -2(-1) + 1 = 2 + 1 = 3$$

and the ordered pair $(-1,3)$ satisfies the equation. Similarly, letting $y = 1$ gives

$$1 = -2x + 1$$
$$0 = -2x$$
$$0 = x$$

and the ordered pair $(0,1)$ satisfies the equation.

We can write all the ordered pairs in Example 2 in table form.

x	$-2x + 1 = y$
-1	$-2(-1) + 1 = 3$
0	$-2(0) + 1 = 1$
1	$-2(1) + 1 = -1$
2	$-2(2) + 1 = -3$
3	$-2(3) + 1 = -5$

EXAMPLES

3. Determine which, if any, of the ordered pairs $(0,-2)$, $\left(\frac{2}{3},0\right)$, and $(2,5)$ satisfy the equation $y = 3x - 2$.

    Solution   We will substitute 0, $\frac{2}{3}$, and 2 for x and see if the corresponding y-values match those in the given ordered pairs.

    $x = 0$: $y = 3(0) - 2 = -2$ so $(0,-2)$ satisfies the equation.

    $x = \frac{2}{3}$: $y = 3\left(\frac{2}{3}\right) - 2 = 0$ so $\left(\frac{2}{3},0\right)$ satisfies the equation.

    $x = 2$: $y = 3(2) - 2 = 4$ so $(2,4)$ satisfies the equation.

    The point $(2,5)$ does not satisfy the equation $y = 3x - 2$ since $y = 4$ when $x = 2$.

4. Determine the missing coordinate in each of the following ordered pairs so that the point will satisfy the equation $2x + 3y = 12$:
    $(0,\ \ )$, $(3,\ \ )$, $(\ \ ,0)$, $(\ \ ,-2)$.
    Solution

    For $(0,\ \ )$, let $x = 0$.

    $$2(0) + 3y = 12$$
    $$3y = 12$$
    $$y = 4$$

    The ordered pair is $(0,4)$.

    For $(3,\ \ )$, let $x = 3$.

    $$2(3) + 3y = 12$$
    $$6 + 3y = 12$$
    $$3y = 6$$
    $$y = 2$$

    The ordered pair is $(3,2)$.

    For $(\ \ ,0)$, let $y = 0$.

    $$2x + 3(0) = 12$$
    $$2x = 12$$
    $$x = 6$$

    The ordered pair is $(6,0)$.

    For $(\ \ ,-2)$, let $y = -2$.

    $$2x + 3(-2) = 12$$
    $$2x - 6 = 12$$
    $$2x = 18$$
    $$x = 9$$

    The ordered pair is $(9,-2)$.

5. Complete the table below so that each ordered pair will satisfy the equation $y = 1 - 2x$.

x	y
0	
	3
$\dfrac{1}{2}$	
5	

Solution Substituting each given value for x and y into the equation $y = 1 - 2x$ gives the following table of ordered pairs.

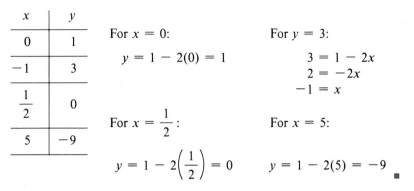

x	y
0	1
-1	3
$\dfrac{1}{2}$	0
5	-9

For $x = 0$:

$$y = 1 - 2(0) = 1$$

For $x = \dfrac{1}{2}$:

$$y = 1 - 2\left(\dfrac{1}{2}\right) = 0$$

For $y = 3$:

$$3 = 1 - 2x$$
$$2 = -2x$$
$$-1 = x$$

For $x = 5$:

$$y = 1 - 2(5) = -9$$

■

Practice Problems

1. Determine which ordered pairs satisfy the equation $3x + y = 14$.
   a. $(5, -1)$ **b.** $(4, 2)$ **c.** $(-1, 17)$
2. Given $3x + y = 5$, find the missing coordinate of each ordered pair so that it will satisfy the equation.
   a. $(0, \ )$ **b.** $\left(\dfrac{1}{3}, \ \right)$ **c.** $(\ , 2)$
3. Complete the table so that each ordered pair will satisfy the equation $y = \dfrac{2}{3}x + 1$.

x	y
0	
	-2
-3	
6	

Answers to Practice Problems **1.** All satisfy the equation.

2. a. $(0, 5)$ **b.** $\left(\dfrac{1}{3}, 4\right)$ **c.** $(1, 2)$

3.

x	y
0	1
$-\dfrac{9}{2}$	-2
-3	-1
6	5

EXERCISES 8.1

List the sets of ordered pairs corresponding to the graphs in Exercises 1–10. Assume that the grid lines are marked one unit apart.

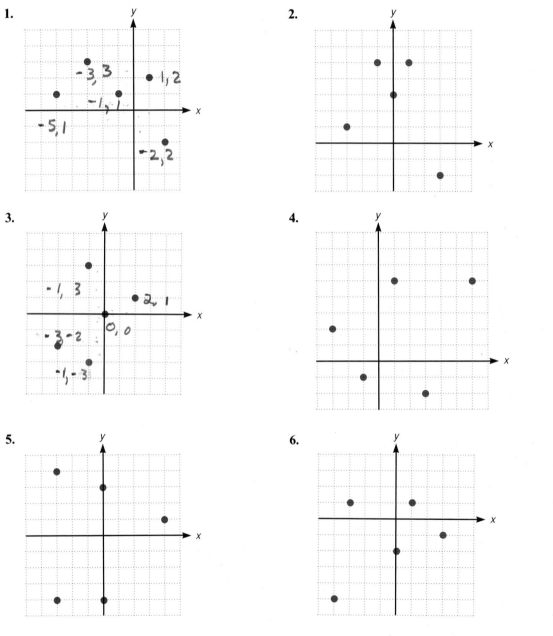

7.

8.

9.

10.

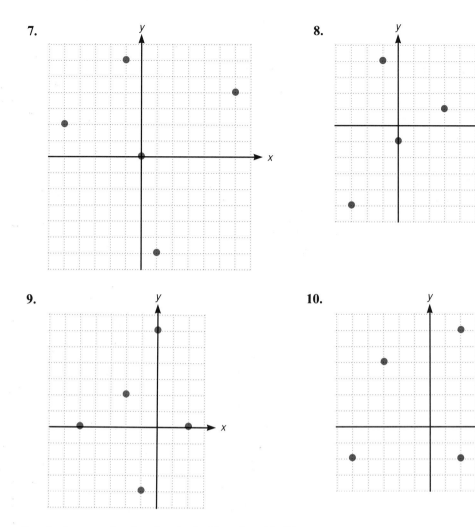

Graph the sets of ordered pairs in Exercises 11–24.

11. $\{(4,-1), (3,2), (0,5), (1,-1), (2,4)\}$

12. $\{(-1,-1), (-3,-2), (1,3), (0,0), (2,5)\}$

13. $\{(1,2), (0,2), (-1,2), (2,2), (-3,2)\}$

14. $\{(1,0), (3,0), (-2,1), (-1,1), (0,0)\}$

15. $\{(-1,4), (0,-3), (2,-1), (4,1)\}$

16. $\{(-1,-1), (0,1), (1,3), (2,5), (3,10)\}$

17. $\{(4,1), (0,-3), (1,-2), (2,-1)\}$

18. $\{(0,1), (1,0), (2,-1), (3,-2), (4,-3)\}$

19. $\{(1,4), (-1,-2), (0,1), (2,7), (-2,-5)\}$

20. $\left\{(1,-3), \left(-4,\frac{3}{4}\right), \left(2,-2\frac{1}{2}\right), \left(\frac{1}{2},4\right)\right\}$

21. $\left\{(0,0), \left(\frac{3}{2},2\right), \left(-1,\frac{7}{4}\right), \left(3,-\frac{1}{2}\right)\right\}$

22. $\left\{\left(\frac{3}{4},\frac{1}{2}\right), \left(2,-\frac{5}{4}\right), \left(\frac{1}{3},-2\right), \left(-\frac{5}{3},2\right)\right\}$

23. $\{(1.6,-2), (3,2.5), (-1,1.5), (0,-2.3)\}$

24. $\{(-2,2), (-3,1.6), (3,0.5), (1.4,0)\}$

Determine which of the given ordered pairs satisfy the equation in Exercises 25–30.

25. $2x - y = 4$
   a. $(1,1)$
   b. $(2,0)$
   c. $(1,-2)$
   d. $(3,2)$

26. $x + 2y = -1$
   a. $(1,-1)$
   b. $(1,0)$
   c. $(2,1)$
   d. $(3,-2)$

27. $4x + y = 5$
   a. $\left(\dfrac{3}{4},2\right)$
   b. $(4,0)$
   c. $(1,1)$
   d. $(0,3)$

28. $2x - 3y = 7$
   a. $(1,3)$
   b. $\left(\dfrac{1}{2},-2\right)$
   c. $\left(\dfrac{7}{2},0\right)$
   d. $(2,1)$

29. $2x + 5y = 8$
   a. $(4,0)$
   b. $(2,1)$
   c. $(1,1.2)$
   d. $(1.5,1)$

30. $3x + 4y = 10$
   a. $(-2,3)$
   b. $(0,2.5)$
   c. $(4,-2)$
   d. $(1.2,1.6)$

Determine the missing coordinate in each of the ordered pairs so that it will satisfy the equation given in Exercises 31–40.

31. $x - y = 4$
   $(0, \ ), (2, \ ), (\ ,0), (\ ,-3)$

32. $x + y = 7$
   $(0, \ ), (-1, \ ), (\ ,0), (\ ,3)$

33. $x + 2y = 6$
   $(0, \ ), (2, \ ), (\ ,0), (\ ,4)$

34. $3x + y = 9$
   $(0, \ ), (4, \ ), (\ ,0), (\ ,3)$

35. $4x - y = 8$
   $(0, \ ), (1, \ ), (\ ,0), (\ ,-4)$

36. $x - 2y = 2$
   $(0, \ ), (4, \ ), (\ ,0), (\ ,3)$

37. $2x + 3y = 6$
   $(0, \ ), (-1, \ ), (\ ,0), (\ ,-2)$

38. $5x + 3y = 15$
   $(0, \ ), (2, \ ), (\ ,0), (\ ,4)$

39. $3x - 4y = 7$
   $(0, \ ), (1, \ ), (\ ,0), \left(\ ,\dfrac{1}{2}\right)$

40. $2x + 5y = 6$
   $(0, \ ), \left(\dfrac{1}{2}, \ \right), (\ ,0), (\ ,2)$

Complete the tables in Exercises 41–50 so that each ordered pair will satisfy the given equation.

41. $y = 3x$

x	y
0	
	-3
-2	
	6

42. $y = -2x$

x	y
0	
	4
3	
	-2

43. $y = 2x - 3$

x	y
0	
	-1
-2	
$\dfrac{1}{2}$	

44. $y = 3x + 5$

x	y
0	
	-1
-2	
$\dfrac{1}{2}$	

45. $y = 7 - 3x$

x	y
0	
	0
-1	
$\frac{1}{3}$	

46. $y = 6 - 2x$

x	y
0	
	0
-2	
$\frac{1}{2}$	

47. $y = \dfrac{3}{4}x + 2$

x	y
0	
	5
-4	
	$\frac{5}{4}$

48. $y = \dfrac{3}{2}x - 1$

x	y
0	
	2
-2	
	$-\frac{5}{2}$

49. $3x - 5y = 9$

x	y
0	
	0
-2	
	-1

50. $4x + 3y = 6$

x	y
0	
	0
3	
	-1

◢ 8.2 Graphing Linear Equations

OBJECTIVES

In this section, you will be learning to:

1. Graph linear equations by choosing and plotting any two points on the lines.
2. Graph linear equations by calculating the x- and y-intercepts.
3. Graph horizontal lines.
4. Graph vertical lines.

In Section 8.1, we discussed ordered pairs and graphed a few points (ordered pairs) that satisfied particular equations. Now, suppose we want to graph **all** the points that satisfy the equation $y = 2x + 3$. The fact is, there are an infinite number of such points. In Figure 8.3, we have graphed five points to try to find a pattern.

x	$2x + 3 = y$
0	$2(0) + 3 = 3$
-1	$2(-1) + 3 = 1$
$\frac{1}{2}$	$2\left(\dfrac{1}{2}\right) + 3 = 4$
1	$2(1) + 3 = 5$
-2	$2(-2) + 3 = -1$

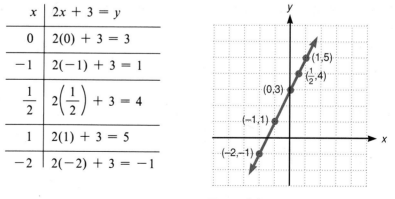

Figure 8.3

The five points in Figure 8.3 appear to lie on a straight line. They in fact do lie on a straight line, and any ordered pair that satisfies the equation $y = 2x + 3$ will also lie on that same line.

What determines whether or not the points that satisfy an equation will lie on a straight line? The points that satisfy any equation of the form

$$Ax + By = C$$

will lie on a straight line. The equation is called a **linear equation** and is considered the **standard form** for the equation of a line. We can write the equation $y = 2x + 3$ in the standard form $-2x + y = 3$.

Since we now know that the graph will be a straight line, only two points are necessary to determine the entire graph. (Two points determine a line.) The choice of the two points depends on the choice of any two values of x or any two values of y. A third point is sometimes chosen as insurance against a mistake and to help place the graph of the line in the right position.

EXAMPLES

1. Draw the graph of the linear equation $x + 3y = 6$.

Solution

$$
\begin{array}{ll}
x = 0 & x = 3 \\
0 + 3y = 6 & 3 + 3y = 6 \\
y = 2 & 3y = 3 \\
& y = 1
\end{array}
$$

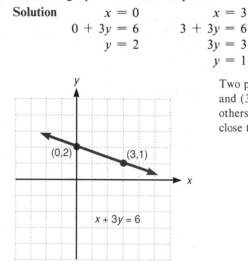

Two points on the graph are (0,2) and (3,1). You may have chosen two others. Avoid choosing two points close together.

2. Draw the graph of the linear equation $2x - 5y = 10$.

Solution

$$
\begin{array}{lll}
x = 5 & x = 0 & x = -5 \\
2 \cdot 5 - 5y = 10 & 2 \cdot 0 - 5y = 10 & 2(-5) - 5y = 10 \\
10 - 5y = 10 & 0 - 5y = 10 & -10 - 5y = 10 \\
-5y = 0 & y = -2 & -5y = 20 \\
y = 0 & & y = -4
\end{array}
$$

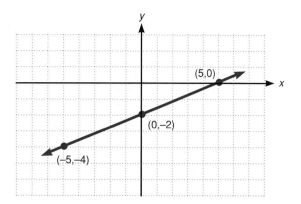

Graphing three points is a good idea, simply to be sure the graph is in the right position and no error has been made in the calculations for the other points.

While the choice of the values for x or y can be arbitrary, letting $x = 0$ will locate the point on the graph where the line crosses the y-axis. This point is called the **y-intercept**. The **x-intercept** is the point found by letting $y = 0$. These two points are generally easy to locate and are frequently used as the two points for drawing the graph of a linear equation.

EXAMPLES Graph the following linear equations by locating the x-intercepts and the y-intercepts.

3. $3y + 2x = 6$

   Solution

   $x = 0 \longrightarrow y = 2$
   $y = 0 \longrightarrow x = 3$

4. $3y - 2x = 6$

   Solution

   $x = 0 \longrightarrow y = 2$
   $y = 0 \longrightarrow x = -3$

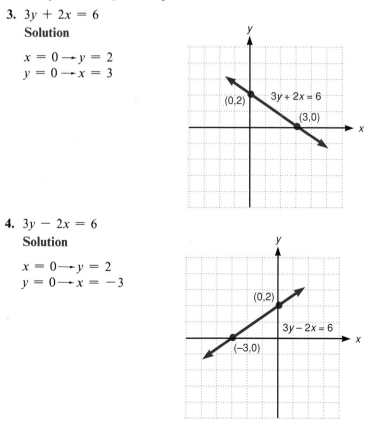

Since the *x*-coordinate of the *y*-intercept is always 0, we will agree to call the *y*-coordinate the *y*-intercept. In Examples 3 and 4 just given, we will say the *y*-intercept is 2 rather than giving the coordinates (0,2). Similarly, in Example 3, the *x*-intercept is 3 and in Example 4, the *x*-intercept is −3.

If the line goes through the origin, both the *x*-intercept and the *y*-intercept will be 0. In this case, some other point must be used.

EXAMPLE

5. Graph the linear equation $y - 3x = 0$.

   Solution   Locate two points on the graph.

$$x = 0 \longrightarrow y = 0$$
$$x = 2 \longrightarrow y = 6$$

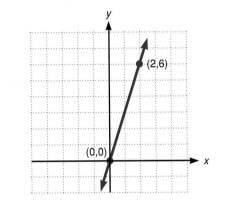

Now, consider the equation in standard form

$$0x + y = 3$$

or just

$$y = 3$$

Regardless of the value chosen for *x*, the corresponding *y*-value will be 3. Thus,

$$(-2,3), \ (0,3), \text{ and } (5,3)$$

all satisfy the equation $y = 3$. The graph of the equation $y = 3$ is a **horizontal line** and each *y*-coordinate is 3. (See Figure 8.4.)

Next, consider the equation in standard form

$$x + 0y = 2$$

or just

$$x = 2$$

Regardless of the value chosen for *y*, the corresponding *x*-value will be 2. Thus,

$$(2,4), \ (2,0), \text{ and } (2,-3)$$

all satisfy the equation $x = 2$. The graph of the equation $x = 2$ is a **vertical line** and each *x*-coordinate is 2. (See Figure 8.5.)

We can make the following two statements:

  1. The graph of any equation of the form $y = b$ is a horizontal line.
  2. The graph of any equation of the form $x = a$ is a vertical line.

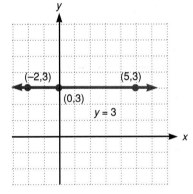

Figure 8.4

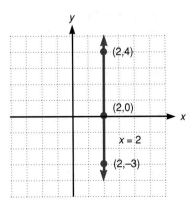

Figure 8.5

EXAMPLES Graph each of the following linear equations.

6. $2y = 5$

Solution $y = \dfrac{5}{2}$

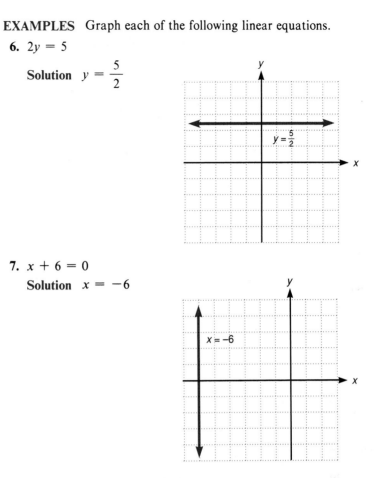

7. $x + 6 = 0$

Solution $x = -6$

◢ **EXERCISES 8.2**

Graph the linear equations in Exercises 1–35.

1. $y = 2x$

2. $y = 3x$

3. $y = -x$

4. $y = -5x$

5. $y = x - 4$

6. $y = x + 3$

7. $y = x + 2$

8. $y = x - 6$

9. $y = 4 - x$

10. $y = 8 - x$

11. $y = 2x - 1$

12. $y = 5 - 2x$

13. $3y = 12$

14. $10 - 2y = 0$

15. $2x - 8 = 0$

16. $9 - 4x = 0$

17. $x - 2y = 4$

18. $x + 3y = 5$

19. $2x - y = 1$

20. $2y = x$

21. $5y = 0$

22. $2y - 3 = 0$

23. $4x = 0$

24. $-\dfrac{3}{4}x - 1 = 0$

25. $2x + 3y = 7$

26. $4x + 3y = 11$

27. $3y = 2x - 4$

28. $3x - 2y = 6$

29. $5x + 2y = 9$

30. $2x - 7y = -14$

31. $4x + 2y = -10$

32. $y = \dfrac{1}{2}x + 1$

33. $y = \dfrac{1}{3}x - 3$

34. $\dfrac{2}{3}x + y = 4$

35. $\dfrac{1}{2}y = 2x - 1$

Graph the linear equations in Exercises 36–50 by locating the *x*-intercept and the *y*-intercept.

36. $x + y = 4$ **37.** $x - 2y = 6$ **38.** $3x - 2y = 6$ **39.** $3x - 4y = 12$

40. $5x + 2y = 10$ **41.** $3x + 7y = -21$ **42.** $2x - y = 9$ **43.** $4x + y = 7$

44. $x + 3y = 5$ **45.** $x - 6y = 3$ **46.** $\dfrac{1}{2}x - y = 4$ **47.** $\dfrac{2}{3}x - 3y = 4$

48. $\dfrac{1}{2}x - \dfrac{3}{4}y = 6$ **49.** $5x + 3y = 7$ **50.** $2x + 3y = 5$

◢ *8.3 The Slope-intercept Form*   **[y = mx + b]**

OBJECTIVES

In this section, you will be learning to:

1. Find the slopes of lines given two points.
2. Write the equations of lines given the slopes and *y*-intercepts.
3. Graph lines given the slopes and *y*-intercepts.
4. Find the slopes and *y*-intercepts of lines.

A carpenter is given a set of house plans that call for a 5 : 12 roof. What does this mean to the carpenter? It means that he must construct the roof so that for every 5 inches of rise (vertical distance), there are 12 inches of run (horizontal distance). That is, the ratio of rise to run is $\dfrac{5}{12}$ (Figure 8.6).

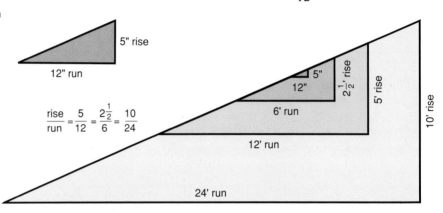

Figure 8.6

What if another roof is to be 7 : 12? Would the carpenter then construct the roof so that for every 7 inches of rise, there would be 12 inches of run? Of course. The ratio of rise to run would be $\dfrac{7}{12}$ (Figure 8.7).

Other examples involving the ratio of rise to run are the **slope** of a road and the **slope** of the side of a ditch.

For a straight line, the ratio of rise to run is called the **slope** of the line.

$$\text{slope} = \frac{\text{rise}}{\text{run}}$$

The graph of the linear equation $y = 3x - 1$ is given in Figure 8.8. What do you think is the slope of the line?

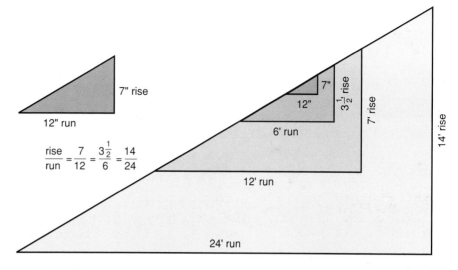

$$\frac{\text{rise}}{\text{run}} = \frac{7}{12} = \frac{3\frac{1}{2}}{6} = \frac{14}{24}$$

Figure 8.7

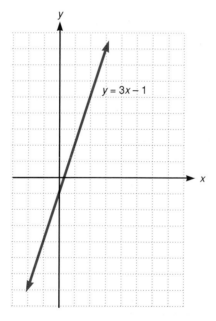

Figure 8.8

To calculate the slope, find **any two** points on the line; then find the rise and the run using those two points [Figure 8.9(a)]. (The ratio of rise to run is the same for any two points on a line.) For example,

$$
\begin{array}{ll}
x = -1 & x = 3 \\
y = 3(-1) - 1 & y = 3(3) - 1 \\
\quad = -3 - 1 & \quad = 9 - 1 \\
\quad = -4 & \quad = 8
\end{array}
$$

Now, using $P_1(-1,-4)$ and $P_2(3,8)$, the coordinates of P_3 are $(3,-4)$, as shown in Figure 8.9(b). P_3 has the same x-coordinate as P_2 and the same y-coordinate as P_1.

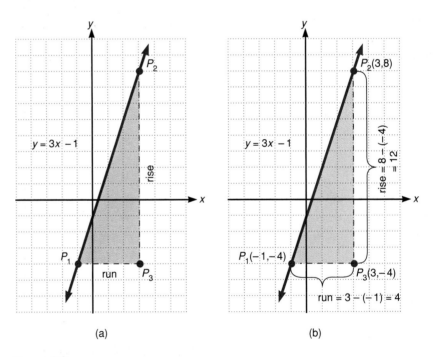

(a)   (b)

Figure 8.9

(**Note:** The notation P_1 is read "P sub 1," and the 1 is called a **subscript.** Similarly, P_2 is read "P sub 2" and P_3 is read "P sub 3.")

For the line $y = 3x - 1$,

$$
\text{slope} = \frac{\text{rise}}{\text{run}} = \frac{8 - (-4)}{3 - (-1)} = \frac{12}{4} = 3
$$

Now find the slope of the line $y = 3x + 2$. Locate two points on the line and calculate the ratio, $\dfrac{\text{rise}}{\text{run}}$. For example,

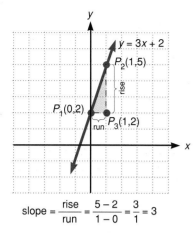

$$\text{slope} = \frac{\text{rise}}{\text{run}} = \frac{5-2}{1-0} = \frac{3}{1} = 3$$

Figure 8.10

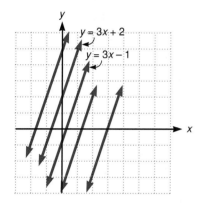

Figure 8.11

$$
\begin{array}{ll}
x = 0 & x = 1 \\
y = 3 \cdot 0 + 2 & y = 3 \cdot 1 + 2 \\
\quad = 0 + 2 & \quad = 3 + 2 \\
\quad = 2 & \quad = 5
\end{array}
$$

Thus, $P_1(0,2)$ and $P_2(1,5)$ are as shown in Figure 8.10. The point $P_3(1,2)$ is shown to help illustrate the rise and the run.

Notice that, as we have just illustrated, the two lines $y = 3x - 1$ and $y = 3x + 2$ have the same slope, 3 (Figure 8.11). This means that the lines are **parallel.** All lines with the same slope are parallel. All the lines in Figure 8.11 have slope 3 and therefore are parallel.

By using subscript notation, we can develop a formula for the slope of any line. Let $P_1(x_1,y_1)$ and $P_2(x_2,y_2)$ be two points on a line. Then $P_3(x_2,y_1)$ is at the right angle shown in Figure 8.12, and the slope can be calculated.

$$\text{slope} = \frac{\text{rise}}{\text{run}} = \frac{y_2 - y_1}{x_2 - x_1}$$

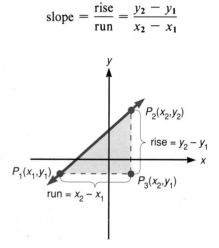

Figure 8.12

EXAMPLES

1. Using the formula for slope, find the slope of the line $2x + 3y = 6$.
   Solution

$$
\begin{array}{llll}
\text{Let} & x_1 = 0 & \text{Let} & x_2 = -3 \\
& 2 \cdot 0 + 3y_1 = 6 & & 2(-3) + 3y_2 = 6 \\
& 3y_1 = 6 & & -6 + 3y_2 = 6 \\
& y_1 = 2 & & 3y_2 = 12 \\
& & & y_2 = 4
\end{array}
$$

$$(x_1,y_1) = (0,2) \quad \text{and} \quad (x_2,y_2) = (-3,4)$$

$$\text{slope} = \frac{y_2 - y_1}{x_2 - x_1} = \frac{4 - 2}{-3 - 0} = \frac{2}{-3} = -\frac{2}{3}$$

2. Suppose that the order of the points in Example 1 is changed. That is, $(x_1, y_1) = (-3, 4)$ and $(x_2, y_2) = (0, 2)$. Will this make a difference in the slope?

Solution slope $= \dfrac{y_2 - y_1}{x_2 - x_1} = \dfrac{2 - 4}{0 - (-3)} = \dfrac{-2}{3} = -\dfrac{2}{3}$ ■

As demonstrated in Example 2, changing the order of the points does not make a difference in the value of the slope. Both the numerator and the denominator change sign, so the fraction has the same value. In Example 1, $\dfrac{2}{-3} = -\dfrac{2}{3}$, and in Example 2, $\dfrac{-2}{3} = -\dfrac{2}{3}$. The important procedure is that **the coordinates must be subtracted in the same order in both the numerator and the denominator.**

In general,

$$\text{slope} = \frac{y_2 - y_1}{x_2 - x_1} = \frac{y_1 - y_2}{x_1 - x_2}$$

The negative slope for the line $2x + 3y = 6$ means that the line slants (or slopes) downward to the right (Figure 8.13). **All lines with negative slopes slant downward to the right, and all lines with positive slopes slant upward to the right.**

In the beginning of this section, we graphed the line $y = 3x - 1$ and then found its slope to be 3. In Example 1, the line $2x + 3y = 6$ had a slope of $-\dfrac{2}{3}$. If we solve this equation for y, we get

$$2x + 3y = 6$$
$$3y = -2x + 6$$
$$y = \frac{-2x}{3} + \frac{6}{3}$$
$$y = -\frac{2}{3}x + 2$$

In both cases, when the equation is solved for y, the coefficient of x is the slope. This result is not an accident. The following discussion proves that this coefficient will always be the slope.

Given $y = mx + b$, then m is the **slope.**

Proof Suppose that the equation is solved for y and $y = mx + b$. Let (x_1, y_1) and (x_2, y_2) be two points on the line where $x_1 \neq x_2$. (More will be said about this condition in Section 8.4.) Then,

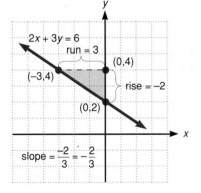

Figure 8.13

$$y_1 = mx_1 + b \qquad \text{and} \qquad y_2 = mx_2 + b$$

$$\text{slope} = \frac{y_2 - y_1}{x_2 - x_1} = \frac{(mx_2 + b) - (mx_1 + b)}{x_2 - x_1}$$

$$= \frac{mx_2 + b - mx_1 - b}{x_2 - x_1} = \frac{mx_2 - mx_1}{x_2 - x_1}$$

$$= \frac{m(x_2 - x_1)}{x_2 - x_1} = m$$

The slope is *m*, the coefficient of *x*, in the form $y = mx + b$.

For the line $y = mx + b$, the point where $x = 0$ is the point where the line will cross the *y*-axis. This point is called the **y-intercept.** By letting $x = 0$, we get

$$y = mx + b$$
$$y = m \cdot 0 + b$$
$$y = b$$

The point $(0,b)$ is the *y*-intercept. Generally, as discussed in Section 8.2, we say that *b* is the *y*-intercept. Thus, we have the following definition.

Slope-intercept Form	$y = mx + b$ is called the **slope-intercept form** for the equation of a line.
	m is the **slope** and *b* is the **y-intercept.**

EXAMPLES   Find the slope, *m*, and the *y*-intercept, *b*, of each of the following lines by rewriting the equation in the slope-intercept form. Then graph the line.

3. $2x + 3y = 3$
   Solution

$$3y = -2x + 3$$
$$y = \frac{-2x + 3}{3}$$
$$y = -\frac{2}{3}x + 1$$
$$m = -\frac{2}{3}$$
y-intercept $= 1$

We have used the slope and y-intercept to help draw the graph. Since $m = -\dfrac{2}{3} = \dfrac{-2}{3}$, move 3 units to the right and 2 units down from the y-intercept $(0,1)$ to locate another point on the line. Draw the line through this point and the y-intercept. We could also have moved 6 units right and 4 units down (or 3 units left and 2 units up) as long as the ratio remains $\dfrac{\text{rise}}{\text{run}} = -\dfrac{2}{3} = \dfrac{-2}{3} = \dfrac{2}{-3}$.

4. $x - 2y = 6$

Solution

$$-2y = -x + 6$$

$$y = \frac{-x + 6}{-2}$$

$$y = \frac{1}{2}x - 3$$

$$m = \frac{1}{2}$$

$$b = -3$$

Since $m = \dfrac{1}{2}$, moving 2 units right and 1 unit up from the y-intercept locates another point on the graph. Of course, another point could be located by substituting some value for x in the equation. For example, if $x = 4$, then

$$x - 2y = 6$$
$$4 - 2y = 6$$
$$-2y = 6 - 4$$
$$-2y = 2$$
$$\frac{-2y}{-2} = \frac{2}{-2}$$
$$y = -1$$

The point $(4,-1)$ is also on the graph.

5. $-4x + 2y = 7$

Solution

$$2y = 4x + 7$$

$$y = \frac{4x + 7}{2}$$

$$y = 2x + \frac{7}{2}$$

$$m = 2$$

$$b = \frac{7}{2}$$

$-4x + 2y = 7$

$(0, \frac{7}{2})$ 2

1

6. Find the equation of the line with y-intercept $(0, -2)$ and slope $m = \frac{3}{4}$.

Solution   In this case, we can substitute directly into the slope-intercept form $y = mx + b$ with $b = -2$ and $m = \frac{3}{4}$.

$$y = \frac{3}{4}x + (-2) \qquad \text{or} \qquad y = \frac{3}{4}x - 2$$

In standard form,

$$3x - 4y = 8$$

Both forms are correct, and either form is an acceptable answer.   ∎

Practice Problems

1. Find the slope of the line determined by the points $(1,5)$ and $(2,-2)$.

2. Find the equation of the line through the point $(0,3)$ with slope $-\frac{1}{2}$.

3. Find the slope, m, and y-intercept, b, for the equation $2x + 3y = 6$.

$$Y = mx + b$$
$$\frac{3y}{3} = \frac{-2x + 6}{3}$$
$$Y = 2x + 6$$
$$Y = -\frac{2}{3}x + 2$$

Answers to Practice Problems   1. $m = -7$ **2.** $y = -\frac{1}{2}x + 3$

3. $m = -\frac{2}{3}$, $b = 2$

EXERCISES 8.3

Find the slope of the line determined by each pair of points in Exercises 1–10.

1. $(1,4), (2,-1)$ $\frac{y_2-y_1}{}$ **2.** $(3,1), (5,0)$ **3.** $(4,7), (-3,-1)$ **4.** $(-6,2), (1,3)$

5. $(-3,8), (5,9)$ $\frac{}{x_2-x_1}$ **6.** $(0,0), (-6,-4)$ **7.** $(4,2), \left(-1,\frac{1}{2}\right)$ **8.** $\left(\frac{3}{4},2\right), \left(1,\frac{3}{2}\right)$

9. $\left(\frac{7}{2},3\right), \left(\frac{1}{2},-\frac{3}{4}\right)$ **10.** $\left(-2,\frac{4}{5}\right), \left(\frac{3}{2},\frac{1}{10}\right)$

Find the equation and draw the graph of the line passing through the given y-intercept with the given slope in Exercises 11–25.

11. $(0,0), m = \dfrac{2}{3}$ **12.** $(0,1), m = \dfrac{1}{5}$ **13.** $(0,-3), m = -\dfrac{3}{4}$ **14.** $(0,-2), m = \dfrac{4}{3}$

15. $(0,3), m = -\dfrac{5}{3}$ **16.** $(0,2), m = 4$ **17.** $(0,-1), m = 2$ **18.** $(0,4), m = -\dfrac{3}{5}$

19. $(0,-5), m = -\dfrac{1}{4}$ **20.** $(0,5), m = -3$ **21.** $(0,-4), m = \dfrac{3}{2}$ **22.** $(0,0), m = -\dfrac{1}{3}$

23. $(0,5), m = 1$ **24.** $(0,6), m = -1$ **25.** $(0,-6), m = \dfrac{2}{5}$

Find the slope, m, and the y-intercept, b, for each of the equations in Exercises 26–45. Then graph the line.

26. $y = 2x - 3$ **27.** $y = 3x + 4$ **28.** $y = 2 - x$ **29.** $y = -\dfrac{2}{3}x + 1$

30. $x + y = 5$ **31.** $2x - y = 3$ **32.** $3x - y = 4$ **33.** $4x + y = 0$

34. $x - 3y = 9$ **35.** $x + 4y = -8$ **36.** $3x + 2y = 6$ **37.** $2x - 5y = 10$

38. $4x - 3y = -3$ **39.** $7x + 2y = 4$ **40.** $6x + 5y = -15$ **41.** $3x + 8y = -16$

42. $x + 4y = 6$ **43.** $2x + 3y = 8$ **44.** $-3x + 6y = 4$ **45.** $5x - 2y = 3$

8.4 The Point-slope Form $[y - y_1 = m(x - x_1)]$

OBJECTIVES

In this section, you will be learning to:
1. Graph lines given the slope and one point on each line.
2. Write the equations of lines given the slope and one point on each line.
3. Write the equations of lines given two points on each line.
4. Write linear equations from information given in applied problems.
5. Find the slopes of horizontal and vertical lines.

Lines represented by equations in the **standard form** $Ax + By = C$ and in the **slope-intercept form** $y = mx + b$ have been discussed in Sections 8.2 and 8.3.

Now, suppose you are given the slope of a line and a point on the line. Can you draw the graph of the line? Probably. The technique is the same as that discussed in the examples in Section 8.3, but the given point need not be the y-intercept. Consider the following example.

EXAMPLE

1. Graph the line with slope $m = \dfrac{3}{4}$ which passes through the point $(4,5)$.

Solution   Start from the point (4,5) and locate another point on the line using the slope as $\dfrac{\text{rise}}{\text{run}} = \dfrac{3}{4}$.

Moving 4 units to the right (run) and 3 units up (rise) from (4,5) will give another point on the line.

From (4,5) you might have moved 4 units left and 3 units down, or 8 units right and then 6 units up. Just move so that the ratio of rise to run is 3 to 4, and you will locate a second point on the graph. (For a negative slope, move either to the right and then down or to the left and then up.)

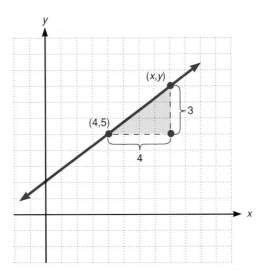

In Example 1, the graph is drawn given the slope $m = \dfrac{3}{4}$ and a point (4,5) on the line. How would you find the equation that corresponds to this line? If the slope is $\dfrac{3}{4}$ and (x,y) is to be a point on the line along with (4,5), then the following discussion will help.

$$\text{slope} = \frac{y - 5}{x - 4} \qquad \text{and} \qquad \text{slope} = \frac{3}{4}$$

Setting these equal to each other gives

$$\frac{y - 5}{x - 4} = \frac{3}{4}$$

This is the equation we want. To obtain a more common form, we multiply both sides by $x - 4$ and get

$$y - 5 = \frac{3}{4}(x - 4)$$

With this approach, we can say that a line with slope m that passes through a given fixed point (x_1, y_1) can be represented in the form

$$y - y_1 = m(x - x_1)$$

Point-slope Form	$y - y_1 = m(x - x_1)$ is called the **point-slope form** for the equation of a line.
	m is the slope, and (x_1, y_1) is a given point on the line.

EXAMPLES

2. Find the equation of the line that has slope $\dfrac{2}{3}$ and passes through the point $(1, 5)$.

   Solution Substitute into the point-slope form $y - y_1 = m(x - x_1)$.

   $$y - 5 = \frac{2}{3}(x - 1)$$

   This same equation can be written in the slope-intercept form and in standard form.

   $$y - 5 = \frac{2}{3}x - \frac{2}{3}$$

   $$y = \frac{2}{3}x - \frac{2}{3} + 5 \qquad -\frac{2}{3} + 5 = -\frac{2}{3} + \frac{15}{3} = \frac{13}{3}$$

   $$y = \frac{2}{3}x + \frac{13}{3} \qquad \text{Slope-intercept form}$$

   Or multiplying each term by 3 gives

   $$3y = 2x + 13$$
   $$-2x + 3y = 13 \qquad \text{Standard form}$$

3. Find three forms for the equation of the line with slope $-\dfrac{5}{2}$ that contains the point $(6, -1)$.

   Solution Use $y - y_1 = m(x - x_1)$.

   $$y - (-1) = -\frac{5}{2}(x - 6)$$

   $$y + 1 = -\frac{5}{2}(x - 6) \qquad \text{Point-slope form}$$

   $$y + 1 = -\frac{5}{2}x + 15$$

$$y = -\frac{5}{2}x + 14 \qquad \text{Slope-intercept form}$$

$$2y = -5x + 28$$

$$5x + 2y = 28 \qquad \text{Standard form} \qquad \blacksquare$$

You must understand that all three forms for the equation of a line are acceptable and correct. Your answer may be in a different **form** from that in the back of the text, but it should be algebraically equivalent.

If two points are given,

1. Use the formula $m = \dfrac{y_2 - y_1}{x_2 - x_1}$ to find the slope.

2. Use this slope m and either point in the point-slope formula $y - y_1 = m(x - x_1)$.

The following examples illustrate this common situation.

EXAMPLE

4. Find the equation of the line passing through the two points $(-1, -3)$ and $(5, -2)$.

Solution $m = \dfrac{y_2 - y_1}{x_2 - x_1} = \dfrac{-2 - (-3)}{5 - (-1)}$ Find m using the formula for slope.

$$= \frac{-2 + 3}{5 + 1} = \frac{1}{6}$$

Using the point-slope form [either point $(-1, -3)$ or point $(5, -2)$ will give the same result] yields

$$y - (-2) = \frac{1}{6}(x - 5) \qquad \text{Using } (5, -2)$$

$$y + 2 = \frac{1}{6}(x - 5) \qquad \text{Point-slope form}$$

or $\qquad\qquad y + 2 = \frac{1}{6}x - \frac{5}{6}$

$$y = \frac{1}{6}x - \frac{17}{6} \qquad \begin{array}{l}\text{Slope-intercept form} \\ -\dfrac{5}{6} - 2 = -\dfrac{5}{6} - \dfrac{12}{6} \\ \qquad = -\dfrac{17}{6}\end{array}$$

or $\qquad\qquad 6y = x - 17$

$$17 = x - 6y \qquad \text{Standard form } Ax + By = C$$

5. Find the equation of the line that contains the points $(-2,3)$ and $(5,1)$.

Solution $m = \dfrac{y_2 - y_1}{x_2 - x_1} = \dfrac{1 - 3}{5 - (-2)} = \dfrac{-2}{7} = -\dfrac{2}{7}$

Using $(-2,3)$ in the point-slope form gives

$$y - 3 = -\frac{2}{7}[x - (-2)]$$

$$y - 3 = -\frac{2}{7}(x + 2)$$

$$y - 3 = -\frac{2}{7}x - \frac{4}{7}$$

$$y = -\frac{2}{7}x - \frac{4}{7} + 3$$

$$y = -\frac{2}{7}x + \frac{17}{7}$$

Using $(5,1)$ in the point-slope form gives

$$y - 1 = -\frac{2}{7}(x - 5)$$

$$y - 1 = -\frac{2}{7}x + \frac{10}{7}$$

$$y = -\frac{2}{7}x + \frac{10}{7} + 1$$

$$y = -\frac{2}{7}x + \frac{17}{7}$$

Thus, we see that using either point yields the same equation. In standard form, $2x + 7y = 17$.   ∎

Slopes of Horizontal and Vertical Lines

In Section 8.2, we discussed the graphs of horizontal lines ($y = b$) and vertical lines ($x = a$), but the slopes of lines of these types were not discussed.

Consider the problem of determining the slope of the horizontal line $y = 3$. Two points on this line are $(-2,3)$ and $(5,3)$. Using these points in the formula for slope, we get

$$\text{slope} = m = \frac{y_2 - y_1}{x_2 - x_1} = \frac{3 - 3}{5 - (-2)} = \frac{0}{7} = 0$$

In fact, the numerator will always be 0 because the y-values will all be 3 regardless of the x-values.

We could write the equation

$$y = 3$$

in the slope-intercept form

$$y = 0x + 3$$

which effectively verifies that the slope is 0.

For a vertical line such as $x = 4$, two points on the line are $(4,1)$ and $(4,6)$. The slope formula gives

$$\text{slope} = \frac{6 - 1}{4 - 4} = \frac{5}{0}, \text{ which is } \textbf{undefined.}$$

The slope is undefined because 0 cannot be a denominator.

We can make the following two general statements:

1. **For horizontal lines ($y = b$ or $y = 0x + b$), the slope is 0.**
2. **For vertical lines ($x = a$), the slope is undefined.**

EXAMPLES

6. Find the equation and slope of the vertical line through the point $(5,2)$.

   Solution The equation is $x = 5$ and the slope is undefined.

7. Find the equation and slope of the horizontal line through the point $(-2,6)$.

   Solution The equation is $y = 6$ and the slope is 0. ∎

Practice Problems

1. Write an equation in standard form for the line passing through the two points $(6,2)$ and $(-3,1)$.
2. Write an equation for the horizontal line through the point $(-2,3)$.
3. Write an equation for the line through the point $(-5,2)$ and parallel to the line $3x + y = 8$.
4. Write the equation of the vertical line through the point $(-1,-2)$.

◢ EXERCISES 8.4

In Exercises 1–12, write an equation (in slope-intercept form or standard form) for the line passing through the given point with the given slope.

1. $(0,3)$, $m = 2$
2. $(1,4)$, $m = -1$
3. $(-1,3)$, $m = -\dfrac{2}{5}$

4. $(-2,-5)$, $m = \dfrac{7}{2}$
5. $(-3,2)$, $m = \dfrac{3}{4}$
6. $(6,-1)$, $m = \dfrac{5}{3}$

7. $(3,-1)$, $m = 0$
8. $(4,6)$, $m = 0$
9. $(2,-4)$, undefined slope

10. $(-3,2)$, undefined slope
11. $\left(1,\dfrac{2}{3}\right)$, $m = \dfrac{3}{2}$
12. $\left(\dfrac{1}{2},-2\right)$, $m = \dfrac{4}{3}$

In Exercises 13–24, write an equation (in slope-intercept form or standard form) for the line passing through the given points.

13. $(-2,3)$, $(1,2)$
14. $(-5,1)$, $(2,0)$
15. $(3,4)$, $(6,2)$
16. $(-4,-4)$, $(3,1)$

17. $(8,2)$, $(-1,2)$
18. $\left(\dfrac{1}{2},-1\right)$, $(3,-1)$
19. $(-5,2)$, $(1,4)$
20. $(-2,6)$, $(3,1)$

21. $(-2,4)$, $(-2,1)$
22. $(3,0)$, $(3,-3)$
23. $(-4,2)$, $\left(1,\dfrac{1}{2}\right)$
24. $(0,2)$, $\left(1,\dfrac{3}{4}\right)$

Answers to Practice Problems 1. $x - 9y = -12$ 2. $y = 3$ 3. $y = -3x - 13$
4. $x = -1$

25. Write an equation for the horizontal line through point $(-2,5)$.

26. Write an equation for the line parallel to the x-axis through point $(6,-2)$.

27. Write an equation for the line parallel to the y-axis through point $(-1,-3)$.

28. Write an equation for the vertical line through point $(-1,-1)$.

29. Write an equation for the line parallel to the line $3x - y = 4$ through the origin.

30. Write an equation for the line parallel to the line $2x = 5$ through point $(0,3)$.

31. Write an equation for the line through point $\left(\dfrac{3}{2},1\right)$ and parallel to the line $4y - 6 = 0$.

32. Write an equation for the horizontal line through point $(-1,-6)$.

33. Find an equation for the line parallel to $2x - y = 7$ having the same y-intercept as $x - 3y = 6$.

34. Find an equation for the line parallel to $3x - 2y = 4$ having the same y-intercept as $5x + 4y = 12$.

35. The rental rate schedule for a car rental is described as "$30 per day plus 15¢ per mile driven." Write a linear equation for the daily cost of renting a car for x miles $(C = mx + b)$.

36. For each job, a print shop charges $5.00 plus $0.05 per page printed. Write a linear equation for the cost of printing x pages $(C = mx + b)$.

37. A salesperson's weekly salary depends on the amount of her sales. Her weekly salary is $200 plus 9% of her weekly sales. Write a linear equation for her weekly salary if her sales are x dollars $(S = mx + b)$.

38. A manufacturer has determined that the cost, C, of producing x units is given by a linear equation $(C = mx + b)$. If it costs $752 to produce 12 units and $800 to produce 15 units, find a linear equation for the cost of producing x units.

39. A local amusement park found that if the admission was $7, they had about 1000 customers per day. When the price was dropped to $6, they had about 1200 customers per day. Write a linear equation for the price in terms of attendance x $(P = mx + b)$.

40. A department store manager has determined that if the price for a necktie is $15, the store will sell 100 neckties per month. However, only 50 neckties per month are sold when the price is raised to $20. Write a linear equation for the price in terms of the number sold, x $(P = mx + b)$.

8.5 Systems of Equations—Solutions by Graphing

OBJECTIVES

In this section, you will be learning to:

1. Determine if given points lie on specified lines.

2. Determine if given points lie on both lines in specified systems of equations.

3. Determine if systems of linear equations are consistent, inconsistent, or dependent by graphing.

4. Estimate the coordinates of the intersections of consistent systems of linear equations by graphing.

In the first part of this chapter, we analyzed straight lines and their characteristics in terms of linear equations. Many applications, as we shall see in Sections 8.6–8.8, involve solving pairs of linear equations. Such pairs are said to form a **system of linear equations.** In this section, we will solve systems by graphing so that we can have a "picture" in mind for the various possible relationships between two linear equations.

Visualize two straight lines on the same graph. What two lines did you see? No matter what specific lines you envisioned, there are only three basic positions for the lines **relative to each other.** What do you think they are? The following discussion analyzes these three basic possibilities.

Do the lines $y = -x + 4$ and $y = 2x + 1$ intersect (cross each other)? If so, where do they **intersect?** If not, why not? We can answer these questions by graphing both equations on the same graph, as shown in Figure 8.14.

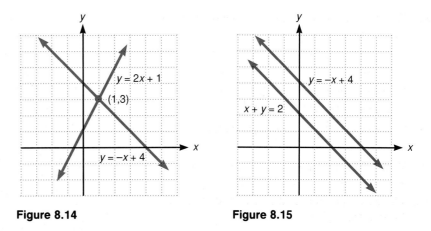

Figure 8.14   **Figure 8.15**

The lines appear to intersect at the point (1,3), or where $x = 1$ and $y = 3$. We can check this intersection by substituting $x = 1$ and $y = 3$ into **both** equations:

$$y = 2x + 1 \qquad\qquad y = -x + 4$$
$$3 \overset{?}{=} 2 \cdot 1 + 1 \qquad\qquad 3 \overset{?}{=} -1 + 4$$
$$3 = 3 \qquad\qquad\qquad 3 = 3$$

The lines do intersect at (1,3).

Do the lines $y = -x + 4$ and $x + y = 2$ intersect? If so, where do they intersect? If not, why not? The answers are in the graphs in Figure 8.15.

The lines do not intersect because they are **parallel.** In this case, we could have anticipated the result by writing both equations in the slope-intercept form and noting that both lines have the same slope, -1, and different y-intercepts.

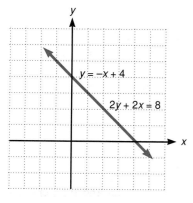

Figure 8.16

$$y = -x + 2 \qquad\qquad m = -1$$
$$y = -x + 4 \qquad\qquad m = -1$$

Do the lines $y = -x + 4$ and $2y + 2x = 8$ intersect? If so, where do they intersect? If not, why not? The graphs of both lines are shown in Figure 8.16. Why is there just one line?

There is just one line because both equations represent the same line. The lines not only intersect; they are the same line. That is, they **coincide.** Any point that satisfies one equation will also satisfy the other. We can easily see this by putting both equations in the slope-intercept form.

$$y = -x + 4 \qquad\qquad 2y + 2x = 8$$
$$2y = -2x + 8$$
$$\frac{2y}{2} = \frac{-2x + 8}{2}$$
$$y = -x + 4$$

Both equations are identical when written in the slope-intercept form.

These three examples constitute all three possible situations involving the graphs of two linear equations. When two linear equations are considered together, they are called a **system of linear equations,** or a **set of simultaneous equations.** The term **simultaneous** is frequently used to emphasize the idea that the solution of a system is the point that satisfies both equations at the same time, or simultaneously.

If a system has a unique solution (one point of intersection), the system is **consistent.** If a system has no solution (the lines are parallel with no point of intersection), the system is **inconsistent.** If a system has an infinite number of solutions (the lines coincide), the system is **dependent.**

The following table summarizes the basic ideas and terminology.

System	Graph	Intersection	Terms
$\begin{cases} y = -x + 4 \\ y = 2x + 1 \end{cases}$	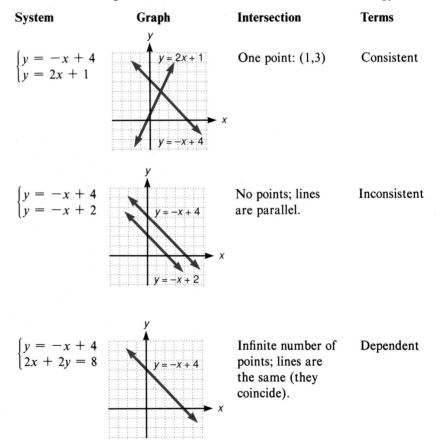	One point: (1,3)	Consistent
$\begin{cases} y = -x + 4 \\ y = -x + 2 \end{cases}$		No points; lines are parallel.	Inconsistent
$\begin{cases} y = -x + 4 \\ 2x + 2y = 8 \end{cases}$		Infinite number of points; lines are the same (they coincide).	Dependent

EXAMPLES Determine graphically whether the following systems are (a) consistent, (b) inconsistent, or (c) dependent. If the system is consistent, find (or estimate) the point of intersection.

1. $\begin{cases} x - y = 0 \\ 2x + y = 3 \end{cases}$

Solution The system is consistent. The point of intersection is $(1,1)$.

Check $x = 1, y = 1$

$$1 - 1 = 0$$
$$2 \cdot 1 + 1 = 3$$

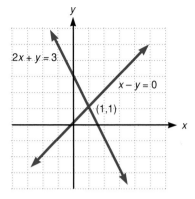

2. $\begin{cases} x - 2y = 1 \\ x + 3y = 0 \end{cases}$

Solution The system is consistent, but the point of intersection can only be estimated. This example points out the main weakness in solving a system graphically. In such cases, any reasonable estimate will be acceptable. For example, if you estimated $\left(\dfrac{1}{2}, -\dfrac{1}{4}\right)$, or $\left(\dfrac{3}{4}, -\dfrac{1}{3}\right)$, or some such point, your answer is acceptable. The actual point of intersection is $\left(\dfrac{3}{5}, -\dfrac{1}{5}\right)$. We will be able to locate this point precisely using the techniques of the next section.

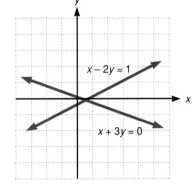

3. $\begin{cases} y = 3x \\ y - 3x = -4 \end{cases}$

Solution The system is inconsistent. The lines are parallel with the same slope, 3, and there are no points of intersection.

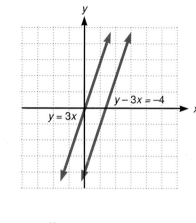

4. $\begin{cases} x + 2y = 6 \\ \qquad y = -\dfrac{1}{2}x + 3 \end{cases}$

Solution The system is dependent. All points that lie on one line also lie on the other line. For example, $(4,1)$ is a point on the line $x + 2y = 6$ since $4 + 2(1) = 6$. The point $(4,1)$ is also on the

line $y = -\dfrac{1}{2}x + 3$ since

$1 = -\dfrac{1}{2}(4) + 3.$

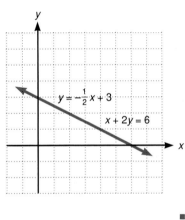

EXERCISES 8.5

Determine which of the given points lie on both of the lines determined by the systems of equations in Exercises 1–4.

1. $\begin{cases} x - y = 6 \\ 2x + y = 0 \end{cases}$
   a. $(1, -2)$
   b. $(4, -2)$
   c. $(2, -4)$
   d. $(-1, 2)$

2. $\begin{cases} x + 3y = 5 \\ \qquad 3y = 4 - x \end{cases}$
   a. $(2, 1)$
   b. $(2, -2)$
   c. $(-1, 2)$
   d. $(4, 0)$

3. $\begin{cases} 2x + 4y - 6 = 0 \\ 3x + 6y - 9 = 0 \end{cases}$

  a. $(1,1)$
  b. $(2,0)$
  c. $\left(0, \dfrac{3}{2}\right)$
  d. $(-1,3)$

4. $\begin{cases} 5x - 2y - 5 = 0 \\ \qquad 5x = -3y \end{cases}$

  a. $(1,0)$
  b. $\left(\dfrac{3}{5}, -1\right)$
  c. $(0,0)$
  d. $(1,4)$

Graph each of the systems in Exercises 5–30 and determine whether it is (a) consistent, (b) inconsistent, or (c) dependent. Estimate the coordinates of the intersection if the system is consistent.

5. $\begin{cases} 2x - y = 3 \\ x + 3y = 5 \end{cases}$

6. $\begin{cases} x + y - 5 = 0 \\ x - 4y = 5 \end{cases}$

7. $\begin{cases} 3x - y = 6 \\ \quad y = 3x \end{cases}$

8. $\begin{cases} x - y = 5 \\ \quad x = -3 \end{cases}$

9. $\begin{cases} x + 2y = 7 \\ 2x - y = -1 \end{cases}$

10. $\begin{cases} 5x - 4y = 5 \\ \quad 8y = 10x - 10 \end{cases}$

11. $\begin{cases} 4x - 2y = 10 \\ -6x + 3y = -15 \end{cases}$

12. $\begin{cases} \quad y = 2x + 5 \\ 4x - 2y = 7 \end{cases}$

13. $\begin{cases} \dfrac{1}{2}x + 2y = 7 \\ 2x = 4 - 8y \end{cases}$

14. $\begin{cases} 4x + 3y + 7 = 0 \\ 5x - 2y + 3 = 0 \end{cases}$

15. $\begin{cases} 2x + 3y = 4 \\ 4x - y = 1 \end{cases}$

16. $\begin{cases} 7x - 2y = 1 \\ \quad y = 3 \end{cases}$

17. $\begin{cases} 2x - 5y = 6 \\ \quad y = \dfrac{2}{5}x + 1 \end{cases}$

18. $\begin{cases} y = 4x - 3 \\ x = 2y - 8 \end{cases}$

19. $\begin{cases} \quad y = \dfrac{1}{2}x + 2 \\ x - 2y + 4 = 0 \end{cases}$

20. $\begin{cases} 2x + 3y = 5 \\ 3x - 2y = 1 \end{cases}$

21. $\begin{cases} \dfrac{2}{3}x + y = 2 \\ x - 4y = 3 \end{cases}$

22. $\begin{cases} x + y = 8 \\ \quad 5y = 2x + 5 \end{cases}$

23. $\begin{cases} \quad y = 2x \\ 2x + y = 4 \end{cases}$

24. $\begin{cases} x - y = 4 \\ \quad 2y = 2x - 4 \end{cases}$

25. $\begin{cases} y = x + 1 \\ y + x = -5 \end{cases}$

26. $\begin{cases} x + y = 4 \\ 2x - 3y = 3 \end{cases}$

27. $\begin{cases} 2x + y + 1 = 0 \\ 3x + 4y - 1 = 0 \end{cases}$

28. $\begin{cases} 4x + y = 6 \\ 2x + \dfrac{1}{2}y = 3 \end{cases}$

29. $\begin{cases} \dfrac{1}{2}x + \dfrac{1}{3}y = \dfrac{1}{6} \\ \dfrac{1}{4}x + \dfrac{1}{4}y = 0 \end{cases}$

30. $\begin{cases} \dfrac{1}{4}x - y = \dfrac{13}{4} \\ \dfrac{1}{3}x + \dfrac{1}{6}y = -\dfrac{1}{6} \end{cases}$

8.6 Systems of Equations—Solutions by Substitution

Solving a system of linear equations graphically can be time-consuming and does not always yield accurate results. When the system has one equation that can easily be solved for one of the variables, the algebraic technique of **substitution** is convenient. We simply substitute an expression for a variable from one equation into the other. For example, consider the system

$$\begin{cases} y = -2x + 5 \\ x + 2y = 1 \end{cases}$$

How would you substitute? The first equation is already solved for y. Would you put $-2x + 5$ for y in the second equation? Try this and see what happens.

$$y = -2x + 5$$
$$x + 2y = 1$$
$$x + 2(-2x + 5) = 1$$

We now have one equation in only one variable, x. We have reduced the problem of solving two equations in two variables to solving one equation in one variable. Solve the equation for x. Then find the corresponding y-value from **either of the two original equations.**

$$x + 2(-2x + 5) = 1$$
$$x - 4x + 10 = 1$$
$$-3x = -9$$
$$x = 3$$

Substituting $x = 3$ into $y = -2x + 5$ gives

$$y = -2 \cdot 3 + 5$$
$$= -6 + 5 = -1$$

The solution is $x = 3$ and $y = -1$, or the point $(3, -1)$.

Substitution is not the only algebraic technique for solving a system of linear equations. It does work in all cases but is generally used only when one equation is easily solved for one variable.

Solve the following system using the technique of substitution:

$$\begin{cases} 3x + y = 1 \\ 6x + 2y = 3 \end{cases}$$

Solving for y in the first equation, we have

$$3x + y = 1$$
$$y = -3x + 1$$

Substituting for y in the second equation yields

$$6x + 2y = 3$$
$$6x + 2(-3x + 1) = 3$$
$$6x - 6x + 2 = 3$$
$$2 = 3$$

This last equation, $2 = 3$, is never true; and this tells us that the system is inconsistent. The lines are parallel and there is no point of intersection. The system has no solution.

Solve the following system using the technique of substitution:

$$\begin{cases} x - 2y = 1 \\ 3x - 6y = 3 \end{cases}$$

Solving for x, we have

$$x - 2y = 1$$
$$x = 2y + 1$$

Substituting yields

$$3x - 6y = 3$$
$$3(2y + 1) - 6y = 3$$
$$6y + 3 - 6y = 3$$
$$3 = 3$$

This last equation, $3 = 3$, is always true; and this tells us that the system is dependent. The two lines are the same and there are an infinite number of points of intersection. Therefore, the solution consists of all the points that satisfy either of the original equations, $x - 2y = 1$ or $3x - 6y = 3$.

EXAMPLES   Solve the following systems of linear equations using the technique of substitution.

1. $$\begin{cases} y = \dfrac{5}{6}x + 1 \\ 2x + 6y = 7 \end{cases}$$

Solution $y = \dfrac{5}{6}x + 1$ is already solved for y. Substituting, we find

$$2x + 6y = 7$$
$$2x + 6\left(\frac{5}{6}x + 1\right) = 7$$
$$2x + 5x + 6 = 7$$
$$7x = 1$$
$$x = \frac{1}{7} \qquad y = \frac{5}{6} \cdot \frac{1}{7} + 1 = \frac{5}{42} + \frac{42}{42} = \frac{47}{42}$$

The solution is $\left(\dfrac{1}{7}, \dfrac{47}{42}\right)$, or $x = \dfrac{1}{7}$ and $y = \dfrac{47}{42}$.

2. $$\begin{cases} x + 3y = 0 \\ -4x + 2y = 7 \end{cases}$$

Solution   Solving for x, we have

$$x + 3y = 0$$
$$x = -3y$$

Substituting yields

$$-4x + 2y = 7$$
$$-4(-3y) + 2y = 7$$
$$12y + 2y = 7$$
$$14y = 7$$
$$y = \frac{1}{2} \qquad x = -3\left(\frac{1}{2}\right) = -\frac{3}{2}$$

The solution is $\left(-\frac{3}{2}, \frac{1}{2}\right)$, or $x = -\frac{3}{2}$ and $y = \frac{1}{2}$.

3. $\begin{cases} x = -5 \\ y = 2x + 9 \end{cases}$

Solution $x = -5$ is already solved for x. The equation represents a vertical line. Substituting, we have

$$y = 2x + 9$$
$$y = 2(-5) + 9 = -10 + 9 = -1$$

The solution is $x = -5$ and $y = -1$, or $(-5, -1)$.

4. $\begin{cases} x + y = 5 \\ 0.2x + 0.3y = 0.9 \end{cases}$

Solution   Solving the first equation for y, we have

$$x + y = 5$$
$$y = 5 - x$$

Substituting yields

$$0.2x + 0.3y = 0.9$$
$$0.2x + 0.3(5 - x) = 0.9$$
$$0.2x + 1.5 - 0.3x = 0.9$$
$$-0.1x = -0.6$$
$$\frac{-0.1x}{-0.1} = \frac{-0.6}{-0.1}$$
$$x = 6 \qquad y = 5 - x = 5 - 6 = -1$$

You could multiply each term by 10 to clear decimals at any step. The answer will be the same.

The solution is $x = 6$ and $y = -1$, or $(6, -1)$.   ■

We have solved a variety of word problems in several previous sections. Each of these problems was solved using a single equation with one variable. The following examples illustrate how some word problems can be solved using two variables. Of course, this also means we will need two equations (that is, a system of equations).

EXAMPLES

5. The sum of two numbers is 80, and their difference is 10. What are the two numbers?

Solution   Let x = one number
and y = the other number

Then we have the system

$$\begin{cases} x + y = 80 & \text{The sum is 80.} \\ x - y = 10 & \text{The difference is 10.} \end{cases}$$

Solving the first equation for y, then substituting gives

$$y = 80 - x$$
$$x - (80 - x) = 10$$
$$x - 80 + x = 10$$
$$2x = 90$$
$$x = 45 \qquad y = 80 - 45 = 35$$

The two numbers are 45 and 35.

Check $45 + 35 = 80$ and $45 - 35 = 10$

6. James has two investment accounts, one pays 6% interest and the other pays 10% interest. He has $1000 more in the 10% account than he has in the 6% account. If his total interest for one year is $740, how much does he have in each account?

Solution   Let x = amount at 6%
and y = amount at 10%

Then the system of equations is

$$\begin{cases} y = x + 1000 & \text{y is larger than x by \$1000.} \\ 0.06x + 0.10y = 740 & \text{The total interest is \$740.} \end{cases}$$

Substituting for y in the second equation,

$$0.06x + 0.10(x + 1000) = 740$$
$$6x + 10(x + 1000) = 74{,}000 \qquad \text{Multiply by 100.}$$
$$6x + 10x + 10{,}000 = 74{,}000$$
$$16x = 64{,}000$$
$$x = 4000 \qquad y = 4000 + 1000 = 5000$$

James has $4000 invested at 6% and $5000 invested at 10%.   ∎

Practice Problems

Solve the following systems using the technique of substitution.

1. $x + y = 3$ **2.** $y = 3x - 1$ **3.** $x + 2y = -1$
  $y = 2x$ $2x + y = 4$ $x - 4y = -4$

Answers to Practice Problems **1.** $x = 1, y = 2$ **2.** $x = 1, y = 2$ **3.** $x = -2, y = \dfrac{1}{2}$

EXERCISES 8.6

Solve the following systems in Exercises 1–35 using the technique of substitution. If the system is inconsistent or dependent, say so in your answer.

1. $\begin{cases} x + y = 6 \\ \quad y = 2x \end{cases}$

2. $\begin{cases} 5x + 2y = 21 \\ \quad x = y \end{cases}$

3. $\begin{cases} x - 7 = 3y \\ \quad y = 2x - 4 \end{cases}$

4. $\begin{cases} y = 3x + 4 \\ 2y = 3x + 5 \end{cases}$

5. $\begin{cases} x = 3y \\ 3y - 2x = 6 \end{cases}$

6. $\begin{cases} 4x = y \\ 4x - y = 7 \end{cases}$

7. $\begin{cases} x - 5y + 1 = 0 \\ \quad x = 7 - 3y \end{cases}$

8. $\begin{cases} 2x + 5y = 15 \\ \quad x = y - 3 \end{cases}$

9. $\begin{cases} 7x + y = 9 \\ \quad y = 4 - 7x \end{cases}$

10. $\begin{cases} 3y + 5x = 5 \\ \quad y = 3 - 2x \end{cases}$

11. $\begin{cases} 3x - y = 7 \\ x + y = 5 \end{cases}$

12. $\begin{cases} 4x - 2y = 5 \\ \quad y = 2x + 3 \end{cases}$

13. $\begin{cases} 6x + 3y = 9 \\ \quad y = 3 - 2x \end{cases}$

14. $\begin{cases} x - y = 5 \\ 2x + 3y = 0 \end{cases}$

15. $\begin{cases} 4x = 8 \\ 3x + y = 8 \end{cases}$

16. $\begin{cases} x + y = 8 \\ 3x + 2y = 8 \end{cases}$

17. $\begin{cases} y = 2x - 5 \\ 2x + y = -3 \end{cases}$

18. $\begin{cases} 2x + 3y = 5 \\ x - 6y = 0 \end{cases}$

19. $\begin{cases} 2y = 5 \\ 3x - 4y = -4 \end{cases}$

20. $\begin{cases} x + 5y = 1 \\ x - 3y = 5 \end{cases}$

21. $\begin{cases} 3x + 8y = -2 \\ x + 2y = -1 \end{cases}$

22. $\begin{cases} 4x - 5y = 9 \\ 3x + y = 2 \end{cases}$

23. $\begin{cases} 5x + 2y = -10 \\ \quad 7x = 4 - y \end{cases}$

24. $\begin{cases} x - 2y = -4 \\ 3x + y = -5 \end{cases}$

25. $\begin{cases} x + 4y = 3 \\ 3x - 4y = -23 \end{cases}$

26. $\begin{cases} 3x - y = -1 \\ 7x - 4y = 0 \end{cases}$

27. $\begin{cases} x + 5y = -1 \\ 2x + 7y = 1 \end{cases}$

28. $\begin{cases} x + 3y = 5 \\ 3x + 2y = 7 \end{cases}$

29. $\begin{cases} 3x - 4y - 39 = 0 \\ 2x - y - 13 = 0 \end{cases}$

30. $\begin{cases} \dfrac{x}{3} + \dfrac{y}{5} = 1 \\ x + 6y = 12 \end{cases}$

31. $\begin{cases} \dfrac{x}{5} + \dfrac{y}{4} - 3 = 0 \\ \dfrac{x}{10} - \dfrac{y}{2} + 1 = 0 \end{cases}$

32. $\begin{cases} 6x - y = 15 \\ 0.2x + 0.5y = 2.1 \end{cases}$

33. $\begin{cases} x + 2y = 3 \\ 0.4x + y = 0.6 \end{cases}$

34. $\begin{cases} 0.2x - 0.1y = 0 \\ \quad y = x + 10 \end{cases}$

35. $\begin{cases} 0.1x - 0.2y = 1.4 \\ 3x + y = 14 \end{cases}$

Solve each of the problems in Exercises 36–45 by setting up a system of two equations in two unknowns and solving the system.

36. The sum of two numbers is 56. Their difference is 10. Find the numbers.

37. The sum of two numbers is 40. The sum of twice the larger and 4 times the smaller is 108. Find the numbers.

38. The sum of two numbers is 36. Three times the smaller plus twice the larger is 87. Find the two numbers.

39. Carmen invested $9000, part in a 6% passbook account and the rest in a 10% certificate account. If her annual interest was $680, how much did she invest at each rate?

40. Mrs. Brown has $12,000 invested. Part is invested at 6% and the remainder at 8%. If the interest from the 6% investment exceeds the interest from the 8% investment by $230, how much is invested at each rate?

41. Ten thousand dollars are invested, part at 5.5% and part at 6%. The interest from the 5.5% investment exceeds the interest from the 6% investment by $251. How much is invested at each rate?

42. On two investments totaling $9500, Bill lost 3% on one and earned 6% on the other. If his net annual receipts were $282, how much was each investment?

43. The area of a rectangular field is 198 square meters. If it takes 58 meters of fencing to enclose the field, what are the dimensions of the field? (**Hint:** The length plus the width is 29 meters.)

44. The perimeter of a rectangle is 66 centimeters. The area of the rectangle is 270 square centimeters. Find the length and width. (**Hint:** After substituting and simplifying, you will have a quadratic equation.)

45. A farmer has 160 meters of fencing to build a rectangular corral. If he uses the barn for one of the longer sides, what will be the dimensions of the corral if the area is 3200 square meters? (**Hint:** After substituting and simplifying, you will have a quadratic equation.)

8.7 Systems of Equations—Solutions by Addition

OBJECTIVES

In this section, you will be learning to:

1. Solve systems of linear equations by addition.
2. Write equations of lines given two points by using the formula $y = mx + b$ and systems of linear equations.
3. Solve word problems by setting up and solving systems of linear equations.

In some cases, the technique of substitution for solving a system can lead to rather complicated algebraic steps. Just solving for x or y can involve several steps and fractions. Consider the system

$$\begin{cases} 4x - 3y = 1 \\ 3x - 2y = 4 \end{cases}$$

Solving the first equation for x,

$$4x - 3y = 1$$
$$4x = 1 + 3y$$
$$x = \frac{1 + 3y}{4}$$

Then, substitution gives

$$3\left(\frac{1 + 3y}{4}\right) - 2y = 4$$

We can finish this process and find the solution; but the point here is that the technique involves somewhat difficult algebraic manipulations.

Another method, called the **method of addition,** can simplify the process considerably. We begin the discussion with a relatively simple system:

$$\begin{cases} y = 4 - x \\ x = y + 6 \end{cases}$$

Put both equations in the standard form $Ax + By = C$ and set one equation under the other so that like terms are aligned. Then **add like terms.**

$$\begin{array}{r} x + y = \ \ 4 \\ \underline{x - y = \ \ 6} \\ 2x \ \ \ \ \ \ = 10 \end{array}$$

Since $+y$ and $-y$ have opposite coefficients, the y-terms are eliminated, and the resulting equation, $2x = 10$, has only one variable. Just as with the technique of substitution, the solution of the system is reduced to solving one equation in one variable.

$$\begin{array}{r} x + y = \ \ 4 \\ \underline{x - y = \ \ 6} \\ 2x \ \ \ \ \ \ = 10 \\ x \ \ \ \ \ \ = \ \ 5 \end{array}$$

Substitute $x = 5$ into one of the original equations.

$$\begin{array}{r} x + y = 4 \\ 5 + y = 4 \\ y = -1 \end{array}$$

The solution is $x = 5$ and $y = -1$, or $(5, -1)$.

If the two coefficients of one variable in the system are not opposites, then we multiply each equation by some nonzero constant so that either the two x-coefficients are opposites or the two y-coefficients are opposites. For example, consider the system

$$\begin{cases} 4x - 3y = 1 \\ 3x - 2y = 4 \end{cases}$$

Multiplying each term of the first equation by 2 and each term of the second equation by -3 will result in the y-coefficients being opposites. Or multiplying the terms of the first equation by 3 and the second equation by -4 will give opposite coefficients for the x-terms. Both cases yield the same solution and are illustrated below. The number used to multiply the terms of each equation is in brackets.

Method 1: Eliminate y-terms.

$$\begin{cases} [2] \quad 4x - 3y = 1 \\ [-3] \quad 3x - 2y = 4 \end{cases}$$

$$\begin{array}{r} 8x - 6y = \ \ \ \ 2 \\ \underline{-9x + 6y = -12} \\ -x \ \ \ \ \ \ = -10 \\ x = 10 \end{array}$$

Substitute $x = 10$ into one of the original equations.

$$\begin{array}{r} 4x - 3y = 1 \\ 4 \cdot 10 - 3y = 1 \\ 40 - 3y = 1 \\ -3y = -39 \\ y = 13 \end{array}$$

The solution is $x = 10$ and $y = 13$, or $(10, 13)$.

Method 2: Eliminate x-terms.

$$\begin{cases} [3] & 4x - 3y = 1 \\ [-4] & 3x - 2y = 4 \end{cases}$$

$$\begin{aligned} 12x - 9y &= 3 \\ -12x + 8y &= -16 \\ \hline -y &= -13 \\ y &= 13 \end{aligned}$$

Substitute $y = 13$ into one of the original equations.

$$\begin{aligned} 4x - 3y &= 1 \\ 4x - 3 \cdot 13 &= 1 \\ 4x - 39 &= 1 \\ 4x &= 40 \\ x &= 10 \end{aligned}$$

The solution is $x = 10$ and $y = 13$, or $(10,13)$.

As illustrated in Method 1 and Method 2, eliminating either the y-terms or the x-terms yields the same results. What if both the x- and y-terms are eliminated? In this situation, the system is either inconsistent or dependent. Just as with the technique of substitution, the resulting equation involving only constants tells which case is under consideration.

For the system

$$\begin{cases} 2x - y = 6 \\ 4x - 2y = 1 \end{cases}$$

multiplying the terms in the first equation by -2 and then adding gives

$$\begin{cases} [-2] & 2x - y = 6 \\ & 4x - 2y = 1 \end{cases}$$

$$\begin{aligned} -4x + 2y &= -12 \\ 4x - 2y &= 1 \\ \hline 0 &= -11 \end{aligned}$$

Since the equation $0 = -11$ is not true, the system is inconsistent. There is no solution.

For the system

$$\begin{cases} 2x - 2y = 1 \\ 3x - 3y = \dfrac{3}{2} \end{cases}$$

multiplying the terms in the first equation by 3 and the terms in the second equation by -2 and then adding gives

$$\begin{cases} [3] & 2x - 2y = 1 \\ [-2] & 3x - 3y = \dfrac{3}{2} \end{cases}$$

$$\begin{aligned} 6x - 6y &= 3 \\ -6x + 6y &= -3 \\ \hline 0 &= 0 \end{aligned}$$

Since the equation $0 = 0$ is always true, the system is dependent. The solution consists of all points that satisfy the equation $2x - 2y = 1$.

In solving a system by addition, find the constant for multiplying the terms of each equation by trying to get coefficients of like terms to be opposites. There are many possible choices. One approach is to find the least common multiple of the two coefficients already there, and then multiply so that one coefficient will be the LCM and the other coefficient, its opposite.

Thus, for the system

$$\begin{cases} 4x - 3y = 1 \\ 3x - 2y = 4 \end{cases}$$

in Method 1, the coefficients for the y-terms ended up being -6 and $+6$. The LCM for 3 and 2 is 6. In Method 2, the coefficients for the x-terms ended up being $+12$ and -12. The LCM for 4 and 3 is 12.

EXAMPLES   Solve the following systems of equations using the technique of addition.

1. $\begin{cases} 5x + 3y = -3 \\ 2x - 7y = 7 \end{cases}$

Solution

$$\begin{cases} [-2] & 5x + 3y = -3 \\ [5] & 2x - 7y = 7 \end{cases} \qquad \begin{aligned} -10x - 6y &= 6 \\ \underline{10x - 35y} &= 35 \\ -41y &= 41 \\ y &= -1 \end{aligned}$$

$$\begin{aligned} 5x + 3y &= -3 \\ 5x + 3(-1) &= -3 \\ 5x - 3 &= -3 \\ 5x &= 0 \\ x &= 0 \end{aligned}$$

The solution is $x = 0$ and $y = -1$, or $(0, -1)$.

2. $\begin{cases} y = -2 + 4x \\ 8x - 2y = 4 \end{cases}$

Solution   Rearranging gives $\begin{cases} -4x + y = -2 \\ 8x - 2y = 4 \end{cases}$

$$\begin{cases} & -4x + y = -2 \\ [\tfrac{1}{2}] & 8x - 2y = 4 \end{cases} \qquad \begin{aligned} -4x + y &= -2 \\ \underline{4x - y} &= 2 \\ 0 &= 0 \end{aligned}$$

The system is dependent. The solution is the set of all points that satisfy the equation $y = -2 + 4x$ (or the equation $8x - 2y = 4$).

3. $\begin{cases} x + 0.4y = 3.08 \\ 0.1x - y = 0.1 \end{cases}$

Solution

$$\begin{cases} & x + 0.4y = 3.08 \\ [-10] & 0.1x - y = 0.1 \end{cases}$$

$$\begin{aligned} 1.0x + 0.4y &= 3.08 \\ \underline{-1.0x + 10.0y} &= -1.0 \\ 10.4y &= 2.08 \\ y &= 0.2 \end{aligned}$$

$$\begin{aligned} x + 0.4y &= 3.08 \\ x + 0.4(0.2) &= 3.08 \\ x + 0.08 &= 3.08 \\ x &= 3 \end{aligned}$$

The solution is $x = 3$ and $y = 0.2$, or $(3, 0.2)$.

4. Using the formula $y = mx + b$, find the equation of the line determined by the two points $(3,5)$ and $(-6,2)$.

   Solution   Write two equations in m and b by substituting the coordinates of the points for x and y.

$$\begin{cases} 5 = 3m + b \\ [-1] \quad 2 = -6m + b \end{cases}$$

$$\begin{array}{l} 5 = 3m + b \\ \underline{-2 = 6m - b} \\ 3 = 9m \\ \dfrac{1}{3} = m \end{array}$$

$$\begin{array}{l} 5 = 3m + b \\ 5 = 3 \cdot \dfrac{1}{3} + b \\ 5 = 1 + b \\ 4 = b \end{array}$$

   The equation is $y = \dfrac{1}{3}x + 4$.

5. A small plane flew 300 miles in 2 hours. Then on the return trip, flying against the wind, it traveled only 200 miles in 2 hours. What were the wind velocity and the speed of the plane?

   Solution   Let s = speed of plane

   v = wind velocity

	rate	· time =	distance
With the wind	$s + v$	2	$2(s + v)$
Against the wind	$s - v$	2	$2(s - v)$

$$\begin{cases} 2(s + v) = 300 \\ 2(s - v) = 200 \end{cases}$$

$$\begin{array}{l} 2s + 2v = 300 \\ \underline{2s - 2v = 200} \\ 4s = 500 \\ s = 125 \end{array}$$

$$\begin{array}{l} 2(125 + v) = 300 \\ 125 + v = 150 \\ v = 25 \end{array}$$

   The speed of the plane was 125 mph, and the wind velocity was 25 mph.

6. How many ounces each of a 10% salt solution and a 15% salt solution must be used to produce 50 ounces of a 12% salt solution?

   Solution   Let x = amount 10% solution

   y = amount of 15% solution

	amount of solution	·	percent of salt	=	amount of salt
10% solution	x		0.10		$0.10x$
15% solution	y		0.15		$0.15y$
12% solution	50		0.12		$0.12(50)$

$$\begin{cases} \\ [100] \end{cases} \qquad \begin{array}{l} x + y = 50 \\ 0.10x + 0.15y = 0.12(50) \end{array} \qquad \begin{array}{l} x + y = 50 \\ 10x + 15y = 12(50) \end{array}$$

$$\begin{cases} [-10] \quad x + y = 50 \qquad -10x - 10y = -500 \\ \quad \underline{10x + 15y = 600} \qquad \underline{10x + 15y = 600} \end{cases}$$

$$\begin{array}{ll} & 5y = 100 \\ & y = 20 \qquad x + 20 = 50 \\ & x = 30 \end{array}$$

Use 30 ounces of the 10% solution and 20 ounces of the 15% solution. ∎

Practice Problems   Solve the following systems using the technique of addition.

1. $2x + 3y = 4$
$\ x - y = -3$

2. $3x + 4y = 12$
$\ \dfrac{1}{3}x - 8y = -5$

◢ EXERCISES 8.7

Solve the systems in Exercises 1–34 using the technique of addition. If the system is inconsistent or dependent, say so in your answer.

1. $\begin{cases} 2x - y = 7 \\ x + y = 2 \end{cases}$

2. $\begin{cases} x + 3y = 9 \\ x - 7y = -1 \end{cases}$

3. $\begin{cases} 3x + 2y = 0 \\ 5x - 2y = 8 \end{cases}$

4. $\begin{cases} 4x - y = 7 \\ 4x + y = -3 \end{cases}$

5. $\begin{cases} 2x + 2y = 5 \\ x + y = 3 \end{cases}$

6. $\begin{cases} y = 2x + 14 \\ x = 14 - 3y \end{cases}$

7. $\begin{cases} x = 11 + 2y \\ 2x - 3y = 17 \end{cases}$

8. $\begin{cases} 6x - 3y = 6 \\ y = 2x - 2 \end{cases}$

9. $\begin{cases} x - 2y = 4 \\ y = \dfrac{1}{2}x - 2 \end{cases}$

10. $\begin{cases} x = 3y + 4 \\ y = 6 - 2x \end{cases}$

11. $\begin{cases} 8x - y = 29 \\ 2x + y = 11 \end{cases}$

12. $\begin{cases} 7x - y = 16 \\ 2y = 2 - 3x \end{cases}$

13. $\begin{cases} 3x + y = -10 \\ 2y - 1 = x \end{cases}$

14. $\begin{cases} 3x + 3y = 18 \\ 4x + 4y = 32 \end{cases}$

15. $\begin{cases} 3x + 2y = 4 \\ x + 5y = -3 \end{cases}$

16. $\begin{cases} x + 2y = 0 \\ \quad\ 2x = 4y \end{cases}$

17. $\begin{cases} \dfrac{1}{2}x + y = -4 \\ 3x - 4y = 6 \end{cases}$

18. $\begin{cases} x + y = 1 \\ x - \dfrac{1}{3}y = \dfrac{11}{3} \end{cases}$

19. $\begin{cases} 4x + 3y = 2 \\ 3x + 2y = 3 \end{cases}$

20. $\begin{cases} 5x - 2y = 17 \\ 2x - 3y = 9 \end{cases}$

21. $\begin{cases} \dfrac{1}{2}x + 2y = 9 \\ 2x - 3y = 14 \end{cases}$

22. $\begin{cases} 3x + 2y = 14 \\ 7x + 3y = 26 \end{cases}$

23. $\begin{cases} 4x + 3y = 28 \\ 5x + 2y = 35 \end{cases}$

24. $\begin{cases} 2x + 7y = 2 \\ 5x + 3y = -24 \end{cases}$

25. $\begin{cases} 7x - 6y = -1 \\ 5x + 2y = 37 \end{cases}$

26. $\begin{cases} 9x + 2y = -42 \\ 5x - 6y = -2 \end{cases}$

27. $\begin{cases} 7x + 4y = 7 \\ 6x + 7y = 31 \end{cases}$

28. $\begin{cases} 6x - 5y = -40 \\ 8x - 7y = -54 \end{cases}$

29. $\begin{cases} \dfrac{3}{4}x - \dfrac{1}{2}y = -2 \\ \dfrac{1}{3}x - \dfrac{7}{6}y = 1 \end{cases}$

30. $\begin{cases} x + y = 12 \\ 0.05x + 0.25y = 1.6 \end{cases}$

31. $\begin{cases} x + 0.5y = 8 \\ 0.1x + 0.01y = 0.64 \end{cases}$

32. $\begin{cases} 0.5x - 0.3y = 7 \\ 0.3x - 0.4y = 2 \end{cases}$

33. $\begin{cases} 0.6x + 0.5y = 5.9 \\ 0.8x + 0.4y = 6 \end{cases}$

34. $\begin{cases} 2.5x + 1.8y = 7 \\ 3.5x - 2.7y = 4 \end{cases}$

In Exercises 35–40, write an equation for the line determined by the two given points using the formula $y = mx + b$ to set up a system of equations with m and b as the unknowns.

35. (2,3), (1,−2)

36. (4,7), (−3,2)

37. (−4,1), (5,2)

38. (0,6), (−3,−3)

39. (3,−4), (7,7)

40. (1,−3), (5,−3)

Solve each of the problems in Exercises 41–50 by setting up a system of two equations in two unknowns and solving the system.

41. A metallurgist has an alloy containing 20% copper and some containing 70% copper. How many pounds of each alloy must he use to make 50 pounds of an alloy containing 50% copper?

42. A manufacturer has received an order for 24 tons of a 60% copper alloy. His stock contains only 80% copper alloy and 50% copper alloy. How much of each will he need to fill the order?

43. A tobacco shop wants 50 ounces of tobacco that is 24% rare Turkish blend. How much each of a 30% Turkish blend and a 20% Turkish blend will be needed?

44. How many liters each of a 40% acid solution and a 55% acid solution must be used to produce 60 liters of a 45% acid solution?

45. Ken makes a 4-mile motorboat trip downstream in 20 minutes $\left(\dfrac{1}{3}\text{ hr}\right)$. The return trip takes 30 minutes $\left(\dfrac{1}{2}\text{ hr}\right)$. Find the rate of the boat in still water and the rate of the current.

46. Mr. McKelvey finds that flying with the wind he can travel 1188 miles in 6 hours. However, when flying against the wind, he travels only $\dfrac{2}{3}$ of the distance in the same amount of time. Find the speed of the plane in still air and the wind speed.

47. Randy made a business trip of 190 miles. He averaged 52 mph for the first part of the trip and 56 mph for the second part. If the total trip took $3\dfrac{1}{2}$ hours, how long did he travel at each rate?

48. Marian drove to a resort 335 miles from her home. She averaged 60 mph for the first part of her trip and 55 mph for the second part. If her total driving time was $5\frac{3}{4}$ hours, how long did she travel at each rate?

49. Mr. Green traveled to a city 200 miles from his home to attend a meeting. Due to car trouble, his average speed returning was 10 mph less than his speed going. If the total driving time for the round trip was 9 hours, at what rate of speed did he travel to the city?

50. A man and his son can paint their cabin in 3 hours. Working alone, it would take the son 8 hours longer than it would the father. How long would it take the father to paint the cabin alone?

Calculator Problems

Use a calculator to solve the systems of equations in Exercises 51–56.

51. $\begin{cases} 0.9x + 1.3y = 1.4 \\ 1.2x - 0.7y = 4.3 \end{cases}$

52. $\begin{cases} 1.8x + 2.0y = 4.4 \\ 4.2x + 1.2y = -3.6 \end{cases}$

53. $\begin{cases} 1.4x + 3.5y = 7.28 \\ 2.4x - 2.1y = -0.48 \end{cases}$

54. $\begin{cases} 2.2x + 1.5y = 7.69 \\ 4.0x - 0.8y = 7.28 \end{cases}$

55. $\begin{cases} 1.3x + 4.1y = 9.294 \\ 0.7x + 1.6y = 4.494 \end{cases}$

56. $\begin{cases} 0.09x + 0.17y = 0.6198 \\ 2.10x - 0.90y = 1.6140 \end{cases}$

8.8 *More Applications*

Systems of equations do occur in practical situations related to supply and demand in economics and marketing as well as in "fun type" reasoning problems involving such topics as coins and people's ages. The following problems will acquaint you with some of these applications and provide more practice in solving systems of equations.

EXAMPLES

1. Mike has $1.05 worth of change in nickels and quarters. If he has twice as many nickels as quarters, how many of each type of coin does he have?

   Solution Let n = number of nickels
   q = number of quarters

   $$\begin{cases} n = 2q \qquad \text{This equation relates the number of coins.} \\ 0.05n + 0.25q = 1.05 \qquad \text{This equation relates the values of the coins} \end{cases}$$

   By substitution, we get

   $$0.05(2q) + 0.25q = 1.05$$
   $$0.10q + 0.25q = 1.05$$
   $$10q + 25q = 105$$
   $$35q = 105$$
   $$q = 3 \text{ quarters}$$
   $$n = 2q = 6 \text{ nickels}$$

2. Pat is 6 years older than her sister Sue. In 3 years, she will be twice as old as Sue. How old is each girl now?

Solution   Let P = Pat's age now
S = Sue's age now

$$\begin{cases} P - S = 6 \\ P + 3 = 2(S + 3) \end{cases}$$ The difference in their ages is 6.
Each age is increased by 3.

Simplify the second equation and solve by addition.

$$P + 3 = 2(S + 3)$$
$$P + 3 = 2S + 6$$
$$P - 2S = 3$$

$$\begin{cases} & P - S = 6 \\ [-1] & P - 2S = 3 \end{cases}$$

$$\begin{aligned} P - \ \ S &= \ \ \ 6 \\ -P + 2S &= -3 \\ \hline S &= \ \ \ 3 \end{aligned}$$

$$P - 3 = 6$$
$$P = 9$$

Pat is 9 years old; Sue is 3 years old.

3. Three hot dogs and two orders of French fries cost \$5.80. Four hot dogs and four orders of fries cost \$8.60. What is the cost of a hot dog? What is the cost of an order of fries?

Solution   Let x = cost of one hot dog
y = cost of one order of fries

$$3x + 2y = 5.80$$
$$4x + 4y = 8.60$$

Solve using the addition method.

$$\begin{cases} [-2] & 3x + 2y = 5.80 \\ & 4x + 4y = 8.60 \end{cases}$$

$$\begin{aligned} -6x - 4y &= -11.60 \\ 4x + 4y &= \ \ \ \ 8.60 \\ \hline -2x \ \ \ \ \ &= -3.00 \\ x \ \ \ \ \ &= \ \ \ \ 1.50 \end{aligned}$$

$$3(1.50) + 2y = 5.80$$
$$4.50 + 2y = 5.80$$
$$2y = 1.30$$
$$y = 0.65$$

One hot dog costs \$1.50 and one order of fries costs \$0.65.   ■

◢ EXERCISES 8.8

Solve each problem in Exercises 1–26 by setting up a system of equations in two unknowns.

1. Sonja has some nickels and dimes. If she has 30 coins worth a total of \$2.00, how many of each type of coin does she have?

2. Louis has 27 coins consisting of quarters and dimes. The total value of the coins is \$5.40. How many of each type of coin does he have?

3. Jill is 8 years older than her brother Curt. Four years from now, Jill will be twice as old as Curt. How old is each at the present time?

4. Two years ago Anna was half as old as Beth. Eight years from now she will be two-thirds as old as Beth. How old are they now?

5. Raggs & Associates sells one style of trousers for $35 per pair and another style for $28 per pair. How many pairs of each style were sold if a total of 16 pairs yielded a revenue of $483?

6. Admission to the baseball game is $2.00 for general admission and $3.50 for reserved seats. The receipts were $36,250 for 12,500 paid admissions. How many of each ticket, general and reserved, were sold?

7. A men's clothing store sells two styles of sports jackets, one selling for $95 and one selling for $120. Last month, the store sold 40 jackets, with receipts totaling $4250. How many of each style did the store sell?

8. Seventy children and 160 adults attended the movie theater. The total receipts were $620. One adult ticket and 2 children's tickets cost $7. Find the price of each type of ticket.

9. Morton took some old newspapers and aluminum cans to the recycling center. Their total weight was 180 pounds. He received 1.5¢ per pound for the newspapers and 30¢ per pound for the cans. The total received was $14.10. How many pounds of each did Morton have?

10. A meat market has ground beef that is 40% fat and extra lean ground beef that is only 15% fat. How many pounds of each will be needed to obtain 50 pounds of lean ground beef that is 25% fat?

11. A dairyman wants to mix a 35% protein supplement and a standard 15% protein ration to make 1800 pounds of a high-grade 20% protein ration. How many pounds of each should he use?

12. An insect spray can be purchased in an 80% strength solution. How many gallons must be used with water to fill a 300-gallon tank on a spray truck with a 2% spray solution?

13. Frank bought 2 shirts and 1 pair of slacks for a total of $55. If he had bought 1 shirt and 2 pairs of slacks, he would have paid $68. What was the price of each shirt and each pair of slacks?

14. Four hamburgers and three orders of French fries cost $5.15. Three hamburgers and five orders of fries cost $5.10. What would one hamburger and one order of fries cost?

15. A small manufacturer produces two kinds of radios, model X and model Y. Model X takes 4 hours to produce and costs $8 each to make. Model Y takes 3 hours to produce and costs $7 each to make. If the manufacturer decides to allot a total of 58 hours and $126 each week, how many of each model will be produced?

16. A large cattle feed lot mixes two rations, I and II, to obtain the ration to be used. Ration I contains 20% crude protein and costs 45 cents per pound. Ration II contains 16% crude protein and costs 60 cents per pound. The daily ration contains a total of 300 pounds of protein and costs $885. How many pounds of each ration is used?

17. A furniture shop refinishes chairs. Employees use two methods to refinish a chair. Method I takes 1 hour and the material costs $3. Method II takes $1\frac{1}{2}$ hours and the material costs $1.50. Last week, they took 36 hours and spent $60 refinishing chairs. How many did they refinish with each method?

18. The Chopping Block sells two grades of hamburger. Grade A is 80% lean beef, while Grade B is 50% lean beef. Grade A sells for $2.40 per pound and Grade B sells for $2.00 per pound. How many pounds of each are sold if 20 pounds of lean beef is used and the sales amounted to $64.00?

19. Larry sells sound systems. Brand X takes 2 hours to install and $\frac{1}{2}$ hour to adjust. Brand Y takes $1\frac{1}{2}$ hours to install and $\frac{3}{4}$ hour to adjust. If in one week, he spent 24 hours for installations and 6 hours for adjustments, how many of each brand did Larry sell?

In economics, an important application is the law of supply and demand. Let S be the price at which the manufacturer is willing to supply x units of a product and D be the price at which the consumer is willing to buy x units of the product. The equilibrium price is the price when supply equals demand.

20. Determine the equilibrium price for a product if the supply equation is $S = \frac{1}{2}x + 10$ dollars and the demand equation is $D = 40 - x$ dollars.

21. Determine the equilibrium price of a product if the supply equation is $S = 10 + \frac{1}{4}x$ dollars and the demand equation is $D = 16 - \frac{1}{2}x$ dollars.

22. A wholesaler will supply x hair dryers at a price of $S = 14 + \frac{1}{6}x$ dollars each. The local consumers will buy x hair dryers if the price is $D = 32 - \frac{1}{3}x$ dollars. Find the equilibrium price for the hair dryer.

In economics, the break-even point is the number of items sold so that the total revenue received, R, is equal to the total costs incurred, C.

23. Mr. Catz has invented a new mouse trap. He plans to sell it in a small shop where the rent is $400 per month. The mouse traps cost him 15 cents each to produce and he plans to sell them for 95 cents each.
  a. Write equations for the cost, C, and revenue, R, if he sells x mouse traps. (**Hint:** The cost is the sum of the rent and $0.15x$. The revenue is $0.95x$.)
  b. Find the break-even point.

24. A company is considering a new product for production. It estimates that setting up for production will cost $9000. The material and labor are predicted to cost $2.40 per item. The company expects to be able to sell each item for $4.20.
  a. Write equations for the cost, C, and the revenue, R.
  b. Find the break-even point.

25. The local women's club plans to set up a hot dog stand at the county fair. The rental charge for the booth is $520. The wholesale cost of the hot dogs is 40 cents each and the women plan to sell them at 80 cents each.
  a. Write equations for the cost, C, and the revenue, R.
  b. Find the break-even point.

26. A furniture manufacturer sells lamp tables for $80 each. His cost includes a fixed overhead of $8000 plus a production cost of $30 per table.
  a. Write equations for the cost, C, and the revenue, R.
  b. Find the break-even point.

CHAPTER 8 SUMMARY

Key Terms and Formulas

In an **ordered pair** (x,y), x is called the **first component** and y is called the **second component**. The variable assigned to the first component is also called the **independent variable,** and the variable assigned to the second component is called the **dependent variable.** [8.1]

In the **Cartesian coordinate system,** two number lines intersect at right angles and separate the plane into four **quadrants**. The point of intersection $(0,0)$ is called the **origin.** The horizontal number line is called the **horizontal axis** or **x-axis.** The vertical number line is called the **vertical axis** or **y-axis.** [8.1]

The discussion of straight lines can be summarized as follows:

1. $Ax + By = C$ **Standard form**   [8.2]

2. $m = \dfrac{y_2 - y_1}{x_2 - x_1} = \dfrac{y_1 - y_2}{x_1 - x_2}$ **Slope of a line**   [8.3]

3. $y = mx + b$ **Slope-intercept form**   [8.3]

4. $y - y_1 = m(x - x_1)$ **Point-slope form**   [8.4]

5. $y = b$ **Horizontal line, slope 0**   [8.4]

6. $x = a$ **Vertical line, slope undefined**   [8.4]

When two linear equations are considered together, they are called a **system of linear equations** or a **set of simultaneous equations.**   [8.5]

If a system has a unique solution (one point of intersection), the system is **consistent.** If a system has no solution (the lines are parallel with no point of intersection), the system is **inconsistent.** If a system has an infinite number of solutions (the lines coincide), the system is **dependent.**   [8.5]

Procedures

To Solve a System by Substitution   [8.6]
Substitute the expression equal to one variable found in one of the equations for that variable in the other equation.

To Solve a System by Addition   [8.7]
Write each equation in standard form. Then multiply each equation by some nonzero constant so that either the two x-coefficients or the two y-coefficients are opposites. Finally, **add like terms** so that one equation in one variable is formed.

CHAPTER 8 REVIEW

List the set of ordered pairs corresponding to the graphs in Exercises 1–5.   [8.1]

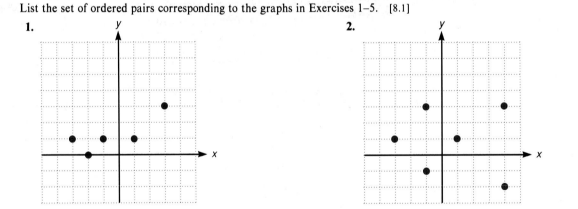

1.

2.

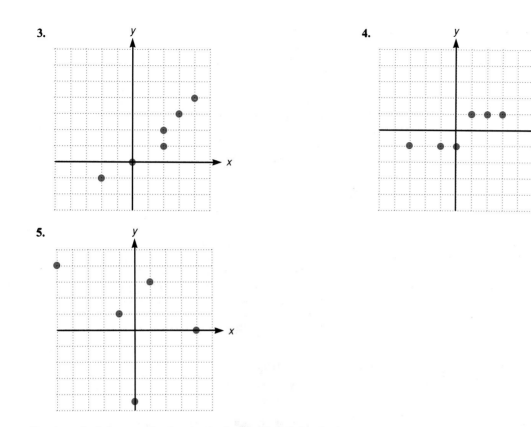

3.

4.

5.

Graph each of the sets of ordered pairs in Exercises 6–8. [8.1]

6. $\{(0,4), (2,2), (3,1), (-2,6)\}$

7. $\left\{\left(2,\dfrac{1}{2}\right), (-1,3), \left(\dfrac{2}{3},-2\right), (1,1)\right\}$

8. $\{(0,1.4), (2.5,6), (3,-1.75), (-2.25,0)\}$

Determine which of the given points lie on the line determined by the equations in Exercises 9 and 10. [8.1]

9. $4x - y = 7$
   a. $(2,1)$
   b. $(3,5)$
   c. $(1,-3)$
   d. $(0,7)$

10. $2x + 5y = 6$
   a. $\left(\dfrac{1}{2},1\right)$
   b. $(3,0)$
   c. $\left(0,\dfrac{6}{5}\right)$
   d. $(-2,2)$

Graph the linear equations in Exercises 11–14. [8.2]

11. $y = -4x$ **12.** $x + 2y = 4$ **13.** $x + 4 = 0$ **14.** $x - 3y = -2$

Graph the linear equations in Exercises 15–18 by locating the x-intercept and the y-intercept. [8.2]

15. $2x + y = 4$ **16.** $5x - y = 1$ **17.** $3y - 9 = 0$ **18.** $3x - 4y = 6$

For Exercises 19–22, determine the slope, m, and the y-intercept, b. Then graph the line. [8.3]

19. $y - 2x = 3$ **20.** $2x + 5y = 10$ **21.** $4x - 8 = 0$ **22.** $4x - 3y = 7$

Find the slope of the line determined by each pair of points in Exercises 23–26.   [8.3]

23. $(2,4)$, $(-1,6)$ **24.** $(6,-2)$, $(3,2)$ **25.** $(5,-4)$, $\left(\dfrac{3}{2},3\right)$ **26.** $\left(\dfrac{3}{4},5\right)$, $(2,3)$

In Exercises 27–32, graph the line and then write an equation (in slope-intercept form or standard form) for it.   [8.4]

27. $(3,7)$, $(5,2)$ **28.** $(-4,2)$, $(3,2)$ **29.** $(6,-1)$, $m = 0$

30. $\left(3,\dfrac{5}{2}\right)$, $m = -\dfrac{1}{4}$ **31.** $(-3,4)$, slope undefined   **32.** $(4,-5)$, $m = \dfrac{3}{2}$

33. Write an equation for the line parallel to the y-axis passing through $(-1,5)$.

34. Write an equation for the line parallel to $2x + y = 5$ passing through $(1,-2)$.

35. Write an equation for the line parallel to $x - 2y = 5$ having the same y-intercept as $5x + 3y = 9$.

36. In a biology experiment, Oraib observed that the bacteria count in a culture was approximately 1000 at the end of 1 hour. At the end of 3 hours, the bacteria count was about 2000. Write a linear equation describing the bacteria count, N, in terms of the time, t.

37. A television repairman charges $20 for the house call plus $17 per hour. Write a linear equation for the total charges, C, in terms of time, t.

Which of the given points in Exercises 38 and 39 lie on both of the lines determined by the given equations?   [8.5]

38. $\begin{cases} x - 2y = 7 \\ 2x - 3y = 5 \end{cases}$

    a. $(0,3)$
    b. $(7,0)$
    c. $(1,-1)$
    d. $(-11,-9)$

39. $\begin{cases} x - 2y = 6 \\ y = \dfrac{1}{2}x - 3 \end{cases}$

    a. $(0,6)$
    b. $(6,0)$
    c. $(2,-2)$
    d. $\left(3,-\dfrac{3}{2}\right)$

Graph each of the systems of equations in Exercises 40–43 and determine if the system is (a) consistent, (b) inconsistent, or (c) dependent.   [8.5]

40. $\begin{cases} 3x + y = 3 \\ x - 3y = -9 \end{cases}$ **41.** $\begin{cases} y = 4x - 6 \\ 8x - 2y = -4 \end{cases}$ **42.** $\begin{cases} \dfrac{1}{2}x + \dfrac{1}{3}y = 2 \\ \\ x = -\dfrac{2}{3}y + 4 \end{cases}$

43. $\begin{cases} x + y = 4 \\ 2x + 7y = -2 \end{cases}$

Solve the systems in Exercises 44–49 using substitution.   [8.6]

44. $\begin{cases} x + y = -4 \\ 2x + 7y = 2 \end{cases}$ **45.** $\begin{cases} x = 2y \\ \\ y = \dfrac{1}{2}x + 9 \end{cases}$ **46.** $\begin{cases} 4x + 3y = 8 \\ \\ x + \dfrac{3}{4}y = 2 \end{cases}$

47. $\begin{cases} 2x + y = 0 \\ 7x + 6y = -10 \end{cases}$ **48.** $\begin{cases} 2x + 3y = 0 \\ 2x - y = 0 \end{cases}$ **49.** $\begin{cases} x = -\dfrac{1}{3}y \\ \\ x + \dfrac{2}{3}y = -\dfrac{1}{3} \end{cases}$

Solve the systems in Exercises 50–55 using the technique of addition.   [8.7]

50. $\begin{cases} 2x + y = 7 \\ 2x - y = 1 \end{cases}$

51. $\begin{cases} 3x - 2y = 9 \\ x - 2y = 11 \end{cases}$

52. $\begin{cases} 2x + 4y = 9 \\ 3x + 6y = 8 \end{cases}$

53. $\begin{cases} x + 5y = 10 \\ \\ y = 2 - \dfrac{1}{5}x \end{cases}$

54. $\begin{cases} 2x - 5y = 1 \\ 2x + 3y = -7 \end{cases}$

55. $\begin{cases} 2x - \dfrac{5}{2}y = -\dfrac{1}{2} \\ \\ 3x - 2y = 1 \end{cases}$

Solve the systems in Exercises 56–61.   [8.6, 8.7]

56. $\begin{cases} x + 3y = 7 \\ 5x - 2y = 1 \end{cases}$

57. $\begin{cases} 3x - 5y = 17 \\ x + 2y = 4 \end{cases}$

58. $\begin{cases} 9x - 3y = 15 \\ 6x = 2y + 10 \end{cases}$

59. $\begin{cases} 9x - 2y = -4 \\ 3x + 4y = 1 \end{cases}$

60. $\begin{cases} -3x + y = 4 \\ 2x - 4y = 5 \end{cases}$

61. $\begin{cases} \dfrac{1}{2}x + \dfrac{1}{3}y = 1 \\ \\ \dfrac{2}{3}x - \dfrac{1}{4}y = \dfrac{1}{12} \end{cases}$

Write an equation for the line determined by the two given points. Use the formula $y = mx + b$ to set up a system of equations with m and b as the unknowns in Exercises 62 and 63.   [8.7]

62. $(3, -1), (2, 6)$

63. $(-2, 5), (4, -3)$

Set up a system of two equations in two unknowns in Exercises 64–70. Then solve.

64. Two trains leave Kansas City at the same time. One train travels east and the other travels west. The speed of the west-bound train is 5 mph greater than the speed of the east-bound train. After 6 hours, they are 510 miles apart. Find the rate of each train. (Assume that the trains travel in a straight line in directly opposite directions.)   [8.7]

65. To meet the government's specifications, an alloy must be 65% aluminum. How many pounds each of a 70% aluminum alloy and a 54% aluminum alloy will be needed to produce 640 pounds of the 65% aluminum alloy?   [8.7]

66. The Ski Club is planning to charter a bus to a ski resort. The cost will be $900, and each member will share the cost equally. If the club had 15 more members, the cost per person would be $10 less. How many are in the club now? (**Hint:** Substitution will result in a single quadratic equation.)   [8.7]

67. Pete's boat can travel 48 miles upstream in 4 hours. The return trip takes 3 hours. Find the speed of the boat in still water and the speed of the current.   [8.7]

68. The perimeter of a rectangle is 50 meters. The area of the rectangle is 154 square meters. Find the dimensions of the rectangle. (**Hint:** After substituting and simplifying, you will have a quadratic equation.)   [8.6]

69. A company produces baseball gloves. Overhead costs are $2400. Material and labor costs are $10 per glove. The baseball gloves sell for $22 each. Write an equation for total cost and an equation for total revenue, and find the break-even point.   [8.8]

70. A company manufactures two kinds of dresses, Model A and Model B. Each Model A dress takes 4 hours to produce and each costs $18. Each Model B takes 2 hours to produce and each costs $7. If during a week, there were 52 hours of production and costs of $198, how many of each model were produced?   [8.8]

CHAPTER 8 TEST

1. Which of the following points lie on the line determined by the equation $x - 4y = 7$?
   a. $(2,-2)$
   b. $(-1,-2)$
   c. $\left(5,-\dfrac{1}{2}\right)$
   d. $(0,7)$

2. List the ordered pairs corresponding to the points on the graph.

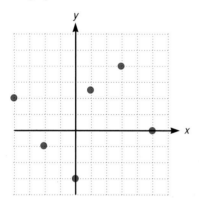

3. Graph the following set of ordered pairs:
   $$\left\{(0,2),\ (4,-1),\ (-3,2),\ (-1,-5),\ \left(2,\dfrac{3}{2}\right)\right\}$$

4. Graph $2x - y = 4$.

5. Graph the equation $3x + 4y = 9$ by locating the x-intercept and y-intercept.

6. What is the slope of the line determined by the points $(4,-3)$ and $\left(2,\dfrac{1}{2}\right)$?

7. Graph the line passing through the point $(1,-5)$ with the slope $m = \dfrac{5}{2}$.

8. For $3x + 5y = -15$, determine the slope, m, and the y-intercept, b. Then graph the line.

9. Write an equation for the line passing through $(-3,2)$ with the slope $m = \dfrac{1}{4}$.

10. Write an equation for the line passing through the points $(3,1)$ and $(-2,6)$.

11. Write an equation for the line parallel to the x-axis through the point $(3,-2)$.

12. Renting a carpet cleaning machine costs a flat rate of $9.00 plus $4.00 per hour. Write a linear equation representing the cost, C, in terms of the hours used, t.

13. Which of the following points lie on both of the lines determined by the equations $\begin{cases} 3x - 7y = 5 \\ 5x - 2y = -11 \end{cases}$
    a. $(1,8)$
    b. $(4,1)$
    c. $(-3,-2)$
    d. $(13,10)$

Graph each of the systems in Exercises 14–16 and determine if the system is (a) consistent, (b) inconsistent, or (c) dependent.

14. $\begin{cases} y = 2 - 5x \\ x - y = 6 \end{cases}$

15. $\begin{cases} x - y = 3 \\ 2x + 3y = 11 \end{cases}$

16. $\begin{cases} 3x + y = 10 \\ 6x + 2y = 5 \end{cases}$

Solve the systems in Exercises 17 and 18 using substitution.

17. $\begin{cases} 5x - 2y = 0 \\ y = 3x + 4 \end{cases}$

18. $\begin{cases} x = \dfrac{1}{3}y - 4 \\ 2x + \dfrac{3}{2}y = 5 \end{cases}$

Solve the systems in Exercises 19 and 20 using the method of addition.

19. $\begin{cases} x + 2y = 5 \\ 2x + 4y = 10 \end{cases}$

20. $\begin{cases} -2x + 3y = 6 \\ 4x + y = 1 \end{cases}$

Solve the systems in Exercises 21–23.

21. $\begin{cases} x + y = 2 \\ y = -2x - 1 \end{cases}$

22. $\begin{cases} 6x + 2y - 8 = 0 \\ y = -3x \end{cases}$

23. $\begin{cases} 7x + 5y = -9 \\ 6x + 2y = 6 \end{cases}$

24. Eight pencils and two pens cost $2.22. Three pens and four pencils cost $2.69. What is the price of each pen and each pencil?

25. The perimeter of a rectangle is 60 inches. The area of the rectangle is 221 square inches. Find the length and width of the rectangle. (**Hint:** After substituting and simplifying, you will have a quadratic equation.)

CUMULATIVE REVIEW (8)

Simplify the expressions in Exercises 1 and 2.

1. $\dfrac{x^5 \cdot x^2}{(x^2)^2}$

2. $\dfrac{y^3}{y^{-2}y^{-1}}$

Factor completely each of the expressions in Exercises 3 and 4.

3. $2x^2 + 15x - 8$

4. $12x^2 - 25x + 12$

Perform the indicated operations in Exercises 5 and 6.

5. $\dfrac{6x^2 - 7x - 3}{x^2 - 1} \div \dfrac{2x - 3}{x - 1}$

6. $\dfrac{x + 5}{(x - 5)^2} + \dfrac{x}{2x^2 - 50}$

Solve the equations in Exercises 7–9.

7. $3x(x - 1) = 3x^2 - 2(x - 4)$

8. $2x^2 + 9x - 18 = 0$

9. $\dfrac{x}{x + 3} + \dfrac{1}{x + 2} = 1$

10. Solve the formula $v = k + gt$ for t.

11. Graph the equation $x + 2y = 3$.

12. For $2x + 4y = 8$, determine the slope, m, and the y-intercept, b. Then graph the line.

13. Write an equation for the line parallel to the line $x - 3y = 1$ and passing through the point $(1,2)$.

Graph each of the systems in Exercises 14 and 15 and determine if the system is (a) consistent, (b) inconsistent, or (c) dependent.

14. $\begin{cases} 2x + 3y = 4 \\ 3x - y = 6 \end{cases}$

15. $\begin{cases} 5x + 2y = 3 \\ y = 4 \end{cases}$

Solve the systems in Exercises 16 and 17.

16. $\begin{cases} 3y - x = 7 \\ x + 2y = -2 \end{cases}$

17. $\begin{cases} x - \dfrac{2}{5}y = \dfrac{4}{5} \\ \dfrac{3}{4}x + \dfrac{3}{4}y = \dfrac{5}{4} \end{cases}$

18. A farmer and his son can plow a field with two tractors in 4 hours. If it would take the son 6 hours longer than the father to plow the field alone, how long would it take each working alone?

20. A boat can travel 24 miles downstream in 2 hours. The return trip takes 3 hours. Find the speed of the boat in still water and the speed of the current.

19. Bernice's secretary bought 20 stamps, some 17¢ and some 25¢. If he spent $4.52, how many of each kind did he buy?

9 REAL NUMBERS AND RADICALS

◢ DID YOU KNOW ?

An important method of reasoning related to mathematical proofs is proof by contradiction. See if you can follow the reasoning in the following "proof" that $\sqrt{2}$ is an irrational number.

We need the following two statements (which can be proven algebraically):

1. The square of an even integer is even if and only if the integer is even.
2. The square of an odd integer is odd if and only if the integer is odd.

Proof $\sqrt{2}$ is either an irrational number or a rational number. Suppose that $\sqrt{2}$ is a rational number and

$$\frac{a}{b} = \sqrt{2}$$

where a and b are integers and $\frac{a}{b}$ is reduced. (± 1 are the only factors common to a and b.)

$$\frac{a^2}{b^2} = 2 \qquad \text{Square both sides.}$$

$$a^2 = 2b^2 \qquad \text{This means } a^2 \text{ is an even integer.}$$

So, $a = 2n$ Since a^2 is even, a must be even.

$$a^2 = 4n^2 \qquad \text{Square both sides.}$$
$$a^2 = 4n^2 = 2b^2 \qquad \text{Substituting}$$
$$2n^2 = b^2 \qquad \text{This means } b^2 \text{ is an even integer.}$$

Therefore, b is an even integer.
But if a and b are both even, 2 is a common factor.

This contradicts the statement that $\frac{a}{b}$ is reduced.

Thus, our original supposition that $\sqrt{2}$ is rational is false, and $\sqrt{2}$ is an irrational number.

The number of grains of sand on the beach at Coney Island is much less than a googol—10,000,000,000,000,000,000,000,000,000,000,000,000,000, 000,000,000,000,000,000,000,000,000,000,000,000,000,000,000,000.

Edward Kasner

CHAPTER OUTLINE

*R*eal numbers and real number lines have been discussed and used throughout the text. In Chapter 9, the related terms real numbers, rational numbers, and irrational numbers are going to be presented in some depth. These are extremely important concepts since the real number system forms the foundation for most of the mathematical ideas we use today.

Calculators have helped a great deal in making operations with both rational numbers and irrational numbers seem easier. However, you should understand that, in many cases, calculators give only approximate values. To fully grasp this fact, as with much of mathematics, you need to understand some of the more abstract theoretical concepts such as those that will be introduced in Chapter 9.

9.1 Real Numbers and Simplifying Radicals

OBJECTIVES

In this section, you will be learning to:

1. Determine if real numbers are rational or irrational.

2. Simplify radicals including square roots and cube roots.

Real numbers were discussed in Chapter 7 when we graphed solutions to inequalities, such as $x < 3$, on real number lines. In this section, we are going to define and discuss real numbers in detail. Also, we are going to simplify expressions containing real numbers.

From previous discussions, we know about integers and rational numbers.

Integers: whole numbers and their opposites

$$. . . , -4, -3, -2, -1, 0, 1, 2, 3, 4, . . .$$

Rational numbers: any number that can be written in the form $\dfrac{a}{b}$

where a and b are integers and $b \neq 0$

In decimal form, all rational numbers can be written as **repeating decimals.** For example,

$$\frac{1}{3} = 0.33333 . . . \qquad\qquad = 0.\overline{3}$$

$$\frac{1}{4} = 0.2500000 . . . \qquad\qquad = 0.25\overline{0} \qquad \text{Repeating 0s.}$$

$$\frac{1}{7} = 0.142857142857142857 . . . = 0.\overline{142857}$$

$$\frac{5}{3} = 1.66666 . . . \qquad\qquad = 1.\overline{6}$$

The bar over the digits indicates that that pattern of digits is to be repeated without end. The pattern can be found by long division. As examples,

$$\frac{1}{3} = 1 \div 3 \qquad\qquad \frac{1}{7} = 1 \div 7$$

$$
\begin{array}{r}
.333\ldots = .\overline{3} \\
3)\overline{1.0000} \\
\underline{9} \\
10 \\
\underline{9} \\
10 \\
\underline{9}
\end{array}
\qquad\qquad
\begin{array}{r}
.142857\ldots = .\overline{142857} \\
7)\overline{1.000000} \\
\underline{7} \\
30 \\
\underline{28} \\
20 \\
\underline{14} \\
60 \\
\underline{56} \\
40 \\
\underline{35} \\
50 \\
\underline{49} \\
1
\end{array}
$$

If we square an integer, the result is called a **perfect square integer.** The perfect square integers are

0, 1, 4, 9, 16, 25, 36, 49, 64, 81, 100, 121, and so on.

Since $5^2 = 25$, the number 5 is called the **square root** of 25. We write $\sqrt{25} = 5$. Similarly,

$$\sqrt{0} = 0, \ \sqrt{1} = 1, \ \sqrt{4} = 2, \ \sqrt{9} = 3, \text{ and so on.}$$

The symbol $\sqrt{}$ is called a **radical sign.**

The number under the radical sign is called the **radicand.**

The complete expression, such as $\sqrt{25}$, is called a **radical.**

Because $(-9)^2 = 81$ and $9^2 = 81$, the number 81 has two square roots, one positive (called the **principal square root**) and one negative.

$$\sqrt{81} = 9$$
$$-\sqrt{81} = -9$$

Square Root	For real numbers a and b,
	a is a **square root** of b if $a^2 = b$.
	We write $a = \sqrt{b}$ if a is positive.

EXAMPLES

1. Since $6^2 = 36$, $\sqrt{36} = 6$

2. Since $\left(\dfrac{2}{3}\right)^2 = \dfrac{4}{9}$, $\sqrt{\dfrac{4}{9}} = \dfrac{2}{3}$

3. Since $8^2 = 64$, $\sqrt{64} = 8$

4. 49 has two square roots, one positive and one negative.

$$\sqrt{49} = 7$$

The $\sqrt{}$ sign is understood to represent the positive square root or the principal square root.

$$-\sqrt{49} = -7$$

The negative square root must have a $-$ sign in front of the radical sign.

5. $-\sqrt{\dfrac{25}{121}} = -\dfrac{5}{11}$

6. $\sqrt{100} = 10$ and $-\sqrt{100} = -10$ ∎

There are roots (or radicals) other than square roots. For example, $5^3 = 125$, so we say that 5 is the **cube root** of 125. Symbolically, $\sqrt[3]{125} = 5$.

Cube Root

For real numbers *a* and *b*,

$$a \text{ is the } \textbf{cube root} \text{ of } b \text{ if } a^3 = b.$$

We write $a = \sqrt[3]{b}$.

EXAMPLES

7. Since $2^3 = 8$, $\sqrt[3]{8} = 2$

8. Since $4^3 = 64$, $\sqrt[3]{64} = 4$

9. Since $(-6)^3 = -216$, $\sqrt[3]{-216} = -6$ ∎

Not all square roots and cube roots are integers or rational numbers, as illustrated in Examples 1–9. Numbers such as

$$\sqrt{5}, \sqrt{7}, \sqrt{39}, -\sqrt{10}, \sqrt[3]{2}, \sqrt[3]{4}, \text{ and } \sqrt[3]{36}$$

are called **irrational numbers.** In decimal form, irrational numbers can be written as **nonrepeating decimals.** (Or, in another sense, they cannot be written as repeating decimals.)

EXAMPLES The following numbers are all irrational numbers. Your calculator will show them as rounded off to seven or eight decimal places; however, the digits do continue without end and there is no pattern to the digits.

10. $\sqrt{2} = 1.4142136\ldots$

See "Did You Know?" on page 310 of this chapter for a formal proof of the fact that $\sqrt{2}$ is irrational.

11. $\sqrt[3]{4} = 1.5874011\ldots$

12. $\pi = 3.14159265358979\ldots$

π is the ratio of the circumference to the diameter of any circle.

13. $e = 2.7182818284590\ldots$

e is the base of natural logarithms and will be studied in detail in later courses.   ∎

Irrational numbers are just as important as rational numbers and just as useful in solving equations. Number lines have points corresponding to irrational numbers as well as rational numbers. Remember, in decimal form,

$$-\frac{1}{3} = -0.33333 \ldots \qquad \text{Repeating but never ending}$$

and $\sqrt{2} = 1.4142136 \ldots$ Nonrepeating and never ending

However, both numbers correspond to just one point on a line (Figure 9.1).

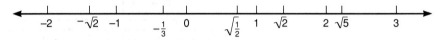

Figure 9.1

The following statement forms the basis for understanding number lines:

Every point on a number line is represented by one real number, and every real number is represented by one point on a number line.

To further emphasize the idea that irrational numbers correspond to points on a number line, consider a circle with a diameter of 1 unit rolling on a line. If the circle touches the line at the point 0, at what point on the line will the same point on the circle again touch the line?

The point will be at π on the number line because π is the circumference of the circle (Figure 9.2).

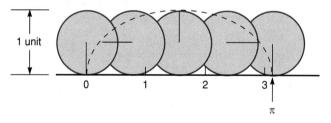

Figure 9.2

Real numbers are numbers that are either rational or irrational. That is, every rational number and every irrational number is also a real number, just as every integer is also a rational number. (Of course, every integer is a real number, too.)

The relationships between the various types of numbers we have discussed are shown in the following diagram.

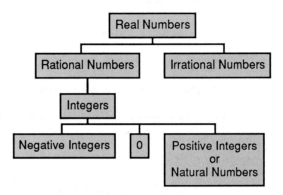

Cube roots of negative numbers are defined to be negative real numbers because the cube of a negative number is negative. But, **square roots of negative numbers are not classified as real numbers** because the square of a real number is nonnegative. For example,

$$\sqrt{-4}, \; \sqrt{-2}, \text{ and } \sqrt{-10} \text{ are not real numbers.}$$

These numbers are called **imaginary numbers** or **complex numbers** and will be studied in the next course in algebra.

Various roots, rational and irrational, can be related to solutions of equations, and we want such numbers to be in a **simplified form** for easier calculations and algebraic manipulations. We need the two properties of radicals stated here for square roots. (Similar properties are true for other radicals.)

If a and b are **positive** real numbers, then

1. $\sqrt{ab} = \sqrt{a}\sqrt{b}$ and

2. $\sqrt{\dfrac{a}{b}} = \dfrac{\sqrt{a}}{\sqrt{b}}$

Thus,

$$\sqrt{144} = \sqrt{36} \cdot \sqrt{4} = 6 \cdot 2 = 12$$

and
$$\sqrt{\frac{9}{25}} = \frac{\sqrt{9}}{\sqrt{25}} = \frac{3}{5}$$

To simplify $\sqrt{450}$, we can write

$$\sqrt{450} = \sqrt{25 \cdot 18} = \sqrt{25}\sqrt{18} = 5\sqrt{18}$$

Is $5\sqrt{18}$ the simplest form of $\sqrt{450}$? The answer is no, because $\sqrt{18}$ has a square number factor, 9, and $\sqrt{18} = \sqrt{9}\sqrt{2} = 3\sqrt{2}$.

We can write

$$\sqrt{450} = \sqrt{25 \cdot 18} = \sqrt{25} \cdot \sqrt{9} \cdot \sqrt{2} = 5 \cdot 3 \cdot \sqrt{2} = 15\sqrt{2}$$

or $\sqrt{450} = \sqrt{225 \cdot 2} = \sqrt{225} \cdot \sqrt{2} = 15\sqrt{2}$

In simplifying a square root, try to find the largest square factor of the radicand. **A square root is considered to be in simplest form when the radicand has no square number factor.**

EXAMPLES Simplify the following radicals.

14. $\sqrt{24}$

Solution Factor 24 so that one factor is a square number.

$$\sqrt{24} = \sqrt{4 \cdot 6} = \sqrt{4} \cdot \sqrt{6}$$
$$= 2\sqrt{6}$$

15. $\sqrt{72}$

Solution Find the largest square factor you can before simplifying.

$$\sqrt{72} = \sqrt{36 \cdot 2} = \sqrt{36} \cdot \sqrt{2}$$
$$= 6\sqrt{2}$$

Or, if you did not notice 36 as a factor, you could write

$$\sqrt{72} = \sqrt{9 \cdot 8} = \sqrt{9} \cdot \sqrt{8} = 3\sqrt{4 \cdot 2}$$
$$= 3 \cdot \sqrt{4} \cdot \sqrt{2} = 3 \cdot 2 \cdot \sqrt{2} = 6\sqrt{2}$$

16. $\sqrt{\dfrac{75}{4}}$

Solution $\sqrt{\dfrac{75}{4}} = \dfrac{\sqrt{75}}{\sqrt{4}} = \dfrac{\sqrt{25 \cdot 3}}{2} = \dfrac{\sqrt{25} \cdot \sqrt{3}}{2}$

$$= \dfrac{5\sqrt{3}}{2}$$

17. $\dfrac{3 + \sqrt{18}}{3}$

Solution $\dfrac{3 + \sqrt{18}}{3} = \dfrac{3 + \sqrt{9 \cdot 2}}{3} = \dfrac{3 + \sqrt{9} \cdot \sqrt{2}}{3} = \dfrac{3 + 3\sqrt{2}}{3}$

$$= \dfrac{3}{3} + \dfrac{3\sqrt{2}}{3} = 1 + \sqrt{2}$$

$$\left[\text{or, factoring,} \ \dfrac{3 + 3\sqrt{2}}{3} = \dfrac{\cancel{3}(1 + \sqrt{2})}{\cancel{3}} = 1 + \sqrt{2} \right]$$ ∎

Cube roots can be simplified if they have a perfect cube factor. Perfect cubes are:

$$0, 1, 8, 27, 64, 125, 216, \text{ and so on.}$$

$$0^3 \quad 1^3 \quad 2^3 \quad 3^3 \quad 4^3 \quad 5^3 \quad 6^3$$

A cube root is considered to be in simplest form when the root has no cube of a rational number as a factor.

EXAMPLES   Simplify the following cube roots.

18. $\sqrt[3]{54}$

    Solution   Factor 54 as $27 \cdot 2$ because 27 is a cube: $27 = 3^3$.

$$\sqrt[3]{54} = \sqrt[3]{27 \cdot 2} = \sqrt[3]{27} \cdot \sqrt[3]{2}$$
$$= 3\sqrt[3]{2}$$

19. $\sqrt[3]{40}$

    Solution   Factor 40 as $8 \cdot 5$ because $8 = 2^3$.

$$\sqrt[3]{40} = \sqrt[3]{8 \cdot 5} = \sqrt[3]{8} \cdot \sqrt[3]{5}$$
$$= 2\sqrt[3]{5}$$

20. $\sqrt[3]{\dfrac{250}{8}}$

    Solution $\sqrt[3]{\dfrac{250}{8}} = \dfrac{\sqrt[3]{250}}{\sqrt[3]{8}} = \dfrac{\sqrt[3]{125 \cdot 2}}{\sqrt[3]{8}} = \dfrac{\sqrt[3]{125} \cdot \sqrt[3]{2}}{\sqrt[3]{8}} = \dfrac{5\sqrt[3]{2}}{2}$ ■

Practice Problems

Simplify the following radical expressions.

1. $\sqrt{49}$ **2.** $\sqrt[3]{27}$ **3.** $-\sqrt{25}$ **4.** $\sqrt{\dfrac{3}{4}}$

5. $\sqrt{32}$ **6.** $\sqrt[3]{32}$ **7.** $\sqrt{\dfrac{54}{25}}$ **8.** $\dfrac{4 - \sqrt{12}}{2}$

For Calculators

A calculator can be used to find decimal approximations for radicals. If the root is a square root, then the $\boxed{\sqrt{x}}$ key can be used. For other roots, the $\boxed{y^x}$ key can be used.

Answers to Practice Problems   **1.** 7   **2.** 3   **3.** -5 **4.** $\dfrac{\sqrt{3}}{2}$ **5.** $4\sqrt{2}$ **6.** $2\sqrt[3]{4}$

7. $\dfrac{3\sqrt{6}}{5}$ **8.** $2 - \sqrt{3}$

EXAMPLES Follow the steps outlined here to find approximate values for the radicals. (**Note:** Not all calculators operate exactly as discussed here. You may need to make some adjustments for your calculator.)

21. $\sqrt{3}$

Solution

Step 1 Enter 3.

Step 2 Press the $\boxed{\sqrt{x}}$ key.

The result should read 1.732050808. (**Note:** On some calculators, you may have to press $\boxed{2nd}$, then $\boxed{x^2}$.)
Thus, $\sqrt{3} \approx 1.732050808$.

22. $\sqrt[4]{100}$

Solution (We will find in Chapter 11 that radicals can be written with fractional exponents. Thus, $\sqrt[4]{100} = 100^{1/4} = 100^{0.25}$.)

Step 1 Enter 100.

Step 2 Press the $\boxed{y^x}$ key.

Step 3 Enter 0.25.

Step 4 Press the $\boxed{=}$ key.

The result should read 3.16227766.
Thus, $\sqrt[4]{100} \approx 3.16227766$.
Note: To verify this result, press $\boxed{y^x}$, then enter 4, then press the $\boxed{=}$ key. This gives $(3.16227766)^4$, which should be 100 or very close to 100.

23. $\sqrt[3]{5}$

Solution (In Chapter 11, we will learn that $\sqrt[3]{5} = 5^{1/3}$. Here we use $\frac{1}{3} \approx 0.333333333$.)

Step 1 Enter 5.

Step 2 Press the $\boxed{y^x}$ key.

Step 3 Enter 0.333333333.

Step 4 Press the $\boxed{=}$ key.

The result should read 1.709975946.
Thus, $\sqrt[3]{5} \approx 1.709975946$.
Note: To check, press $\boxed{y^x}$, then enter 3, then press the $\boxed{=}$ key. This gives $(1.709975946)^3$, which should be 5 or very close to 5. ∎

◢ EXERCISES 9.1

Identify the rational numbers and irrational numbers in Exercises 1–20.

1. $\sqrt{64}$ **2.** $\sqrt{17}$ **3.** $-\sqrt{24}$ **4.** $\sqrt{81}$ **5.** $\sqrt{\dfrac{4}{9}}$ **6.** $\sqrt{\dfrac{7}{16}}$

7. $-\sqrt{\dfrac{12}{25}}$ **8.** $-\sqrt{\dfrac{25}{4}}$ **9.** $\sqrt{\dfrac{12}{75}}$ **10.** $\sqrt{\dfrac{18}{32}}$ **11.** $\sqrt[3]{27}$ **12.** $\sqrt[3]{125}$

13. $\sqrt[3]{-16}$ **14.** $\sqrt[3]{-64}$

15. 0.1010010001 . . . (nonrepeating)   **16.** 2.1632147985 . . . (nonrepeating)
17. 4.232323 . . . (repeating)   **18.** 1.9347193471 . . . (repeating)
19. −6.051051051 . . . (repeating)   **20.** −3.1629341251 . . . (nonrepeating)

Simplify each of the radicals in Exercises 21–70.

21. $\sqrt{36}$ **22.** $\sqrt{49}$ **23.** $\sqrt{\dfrac{1}{4}}$ **24.** $-\sqrt{121}$ **25.** $\sqrt[3]{-8}$ **26.** $\sqrt[3]{216}$

27. $\sqrt[3]{1}$ **28.** $\sqrt[3]{-125}$ **29.** $\sqrt{8}$ **30.** $\sqrt{18}$ **31.** $-\sqrt{56}$ **32.** $\sqrt{162}$

33. $\sqrt{\dfrac{5}{9}}$ **34.** $-\sqrt{\dfrac{7}{16}}$ **35.** $\sqrt{12}$ **36.** $-\sqrt{45}$ **37.** $\sqrt{288}$ **38.** $-\sqrt{63}$

39. $-\sqrt{72}$ **40.** $\sqrt{98}$ **41.** $-\sqrt{125}$ **42.** $\sqrt[3]{56}$ **43.** $\sqrt[3]{-128}$ **44.** $\sqrt[3]{-250}$

45. $\sqrt[3]{135}$ **46.** $\sqrt[3]{320}$ **47.** $\sqrt[3]{-72}$ **48.** $\sqrt{\dfrac{32}{49}}$ **49.** $-\sqrt{\dfrac{11}{64}}$ **50.** $-\sqrt{\dfrac{125}{100}}$

51. $\sqrt{\dfrac{28}{25}}$ **52.** $\sqrt{\dfrac{147}{100}}$ **53.** $\sqrt{\dfrac{32}{81}}$ **54.** $\sqrt{\dfrac{75}{121}}$ **55.** $\dfrac{\sqrt[3]{81}}{6}$ **56.** $\dfrac{\sqrt[3]{192}}{10}$

57. $\sqrt[3]{\dfrac{375}{8}}$ **58.** $\sqrt[3]{-\dfrac{48}{125}}$ **59.** $\dfrac{2-2\sqrt{3}}{4}$ **60.** $\dfrac{4+2\sqrt{6}}{2}$ **61.** $\dfrac{3+3\sqrt{6}}{6}$

62. $\dfrac{6-2\sqrt{3}}{8}$ **63.** $\dfrac{12+\sqrt{45}}{15}$ **64.** $\dfrac{7-\sqrt{98}}{14}$ **65.** $\dfrac{10-\sqrt{108}}{4}$ **66.** $\dfrac{4+\sqrt{288}}{20}$

67. $\dfrac{16-\sqrt{60}}{12}$ **68.** $\dfrac{12-\sqrt{18}}{21}$ **69.** $\dfrac{12-\sqrt{192}}{28}$ **70.** $\dfrac{14+\sqrt{147}}{35}$

▦ Calculator Problems

Use your calculator to estimate the indicated roots. Round to the nearest thousandth. (**Note:** To test your understanding, you might write your own estimate of the answers before you use your calculator.)

71. $\sqrt{65}$ **72.** $\sqrt{127}$ **73.** $\sqrt[3]{98}$ **74.** $\sqrt[3]{-146}$ **75.** $\sqrt[4]{82}$

76. $\sqrt[4]{236}$ **77.** $\sqrt[3]{37.9}$ **78.** $\sqrt[3]{92.7}$ **79.** $\sqrt[4]{86.9}$ **80.** $\sqrt[4]{153.2}$

9.2 *Addition and Multiplication with Radicals*

In this section, you will be learning to:

1. Add and/or subtract radical expressions.

2. Multiply radical expressions.

Sometimes, solving equations and simplifying algebraic expressions can involve operations with radicals (square roots, cube roots, fourth roots, and so on). The emphasis here will be on radicals that are square roots with some work on cube roots.

Like radicals are terms that either have the same radicals or can be simplified so that the radicals are the same. For example,

$7\sqrt{2}$ and $3\sqrt{2}$ are like radicals. The radical, $\sqrt{2}$, is the same in both terms.

$2\sqrt{5}$ and $2\sqrt{3}$ are **not** like radicals. The radicals, $\sqrt{5}$ and $\sqrt{3}$, are different.

Adding Like Radicals

To add (or subtract) like radicals, we proceed just as in combining like terms. We use the distributive property, then add the coefficients.

$$7x + 3x = (7 + 3)x = 10x$$

and
$$7\sqrt{2} + 3\sqrt{2} = (7 + 3)\sqrt{2} = 10\sqrt{2}$$

We may need to simplify before combining like radicals.

$$\sqrt{75} - \sqrt{27} = \sqrt{25 \cdot 3} - \sqrt{9 \cdot 3}$$
$$= 5\sqrt{3} - 3\sqrt{3} = 2\sqrt{3}$$

and
$$\sqrt{16x} + \sqrt{4x} = \sqrt{16 \cdot x} + \sqrt{4 \cdot x}$$
$$= 4\sqrt{x} + 2\sqrt{x} = 6\sqrt{x}$$

Combining cube roots is done in the same way.

$$\sqrt[3]{16x} + \sqrt[3]{54x} = \sqrt[3]{8 \cdot 2x} + \sqrt[3]{27 \cdot 2x}$$
$$= \sqrt[3]{8} \cdot \sqrt[3]{2x} + \sqrt[3]{27} \cdot \sqrt[3]{2x}$$
$$= 2\sqrt[3]{2x} + 3\sqrt[3]{2x} = 5\sqrt[3]{2x}$$

EXAMPLES   Simplify the following expressions.

1. $5\sqrt{2} - \sqrt{2}$

    Solution $5\sqrt{2} - \sqrt{2} = (5 - 1)\sqrt{2} = 4\sqrt{2}$

2. $3\sqrt{5} + 8\sqrt{5} - 14\sqrt{5}$

    Solution $3\sqrt{5} + 8\sqrt{5} - 14\sqrt{5} = (3 + 8 - 14)\sqrt{5} = -3\sqrt{5}$

3. $\sqrt{18} + \sqrt{8}$

    Solution $\sqrt{18} + \sqrt{8} = \sqrt{9}\sqrt{2} + \sqrt{4}\sqrt{2} = 3\sqrt{2} + 2\sqrt{2} = 5\sqrt{2}$

4. $\sqrt{5} - \sqrt{3} + \sqrt{12}$

    Solution $\sqrt{5} - \sqrt{3} + \sqrt{12} = \sqrt{5} - \sqrt{3} + \sqrt{4}\sqrt{3}$
$$= \sqrt{5} - \sqrt{3} + 2\sqrt{3} = \sqrt{5} + \sqrt{3}$$

5. $x\sqrt{x} + 7x\sqrt{x}$

    Solution $x\sqrt{x} + 7x\sqrt{x} = (x + 7x)\sqrt{x} = 8x\sqrt{x}$

6. $\sqrt[3]{125a} - \sqrt[3]{64a}$

    Solution $\sqrt[3]{125a} - \sqrt[3]{64a} = 5\sqrt[3]{a} - 4\sqrt[3]{a} = \sqrt[3]{a}$ ∎

Multiplying Radicals

To find the product of radicals, we proceed just as in multiplying polynomials, as the following examples illustrate.

$$5(x + y) = 5x + 5y$$
$$\sqrt{2}(\sqrt{7} + \sqrt{3}) = \sqrt{2}\sqrt{7} + \sqrt{2}\sqrt{3} = \sqrt{14} + \sqrt{6}$$

And, with binomials,

$$(x + 5)(x - 3) = x^2 - 3x + 5x - 15 = x^2 + 2x - 15$$
$$(\sqrt{2} + 5)(\sqrt{2} - 3) = (\sqrt{2})^2 - 3\sqrt{2} + 5\sqrt{2} - 15$$
$$= 2 + 2\sqrt{2} - 15 = -13 + 2\sqrt{2}$$

(**Note:** $\sqrt{2}\sqrt{2} = \sqrt{4} = 2$ or $(\sqrt{2})^2 = 2$.)

In general, $\sqrt{a}\sqrt{a} = a$ if a is positive. [Or $(\sqrt{a})^2 = a$.]

EXAMPLES   Find the following products and simplify.

7. $\sqrt{7}(\sqrt{7} - \sqrt{14})$

    Solution $\sqrt{7}(\sqrt{7} - \sqrt{14}) = \sqrt{7}\sqrt{7} - \sqrt{7}\sqrt{14}$
$$= 7 - \sqrt{98}$$
$$= 7 - \sqrt{49 \cdot 2} = 7 - 7\sqrt{2}$$

8. $(\sqrt{2} + 4)(\sqrt{2} - 4)$

    Solution $(\sqrt{2} + 4)(\sqrt{2} - 4) = (\sqrt{2})^2 - 4^2 = 2 - 16 = -14$

9. $(\sqrt{5} + \sqrt{3})(\sqrt{5} + \sqrt{3})$

    Solution $(\sqrt{5} + \sqrt{3})(\sqrt{5} + \sqrt{3}) = (\sqrt{5})^2 + 2\sqrt{5}\sqrt{3} + (\sqrt{3})^2$
$$= 5 + 2\sqrt{15} + 3 = 8 + 2\sqrt{15}$$

10. $(\sqrt{x} + 1)(\sqrt{x} - 1)$

    Solution $(\sqrt{x} + 1)(\sqrt{x} - 1) = (\sqrt{x})^2 - 1^2 = x - 1$ ∎

Practice Problems

1. $2\sqrt{3} - \sqrt{3}$ **2.** $-2(\sqrt{8} + \sqrt{2})$

3. $\sqrt{75} - \sqrt{27} + \sqrt{20}$ **4.** $(\sqrt{3} + \sqrt{8})(\sqrt{2} - \sqrt{3})$

5. $5\sqrt{x} - 4\sqrt{x} + 2\sqrt{x}$ **6.** $\sqrt[3]{8} + \sqrt[3]{27}$

7. $\sqrt[3]{16} + \sqrt[3]{250}$

Answers to Practice Problems   **1.** $\sqrt{3}$ **2.** $-6\sqrt{2}$ **3.** $2\sqrt{3} + 2\sqrt{5}$ **4.** $1 - \sqrt{6}$
5. $3\sqrt{x}$ **6.** 5   **7.** $7\sqrt[3]{2}$

EXERCISES 9.2

Simplify the radical expressions in Exercises 1–40.

1. $3\sqrt{2} + 5\sqrt{2}$
2. $7\sqrt{3} - 2\sqrt{3}$
3. $6\sqrt{5} + \sqrt{5}$
4. $4\sqrt{11} - 3\sqrt{11}$
5. $8\sqrt{10} - 11\sqrt{10}$
6. $6\sqrt{17} - 9\sqrt{17}$
7. $4\sqrt[3]{3} + 9\sqrt[3]{3}$
8. $11\sqrt[3]{14} - 6\sqrt[3]{14}$
9. $6\sqrt{11} - 5\sqrt{11} - 2\sqrt{11}$
10. $\sqrt{7} + 6\sqrt{7} - 2\sqrt{7}$
11. $\sqrt{a} + 4\sqrt{a} - 2\sqrt{a}$
12. $2\sqrt{x} - 3\sqrt{x} + 7\sqrt{x}$
13. $5\sqrt{x} + 3\sqrt{x} - \sqrt{x}$
14. $6\sqrt{xy} - 10\sqrt{xy} + \sqrt{xy}$
15. $3\sqrt{2} + 5\sqrt{3} - 2\sqrt{3} + \sqrt{2}$
16. $\sqrt{5} + \sqrt{4} - 2\sqrt{5} + 6$
17. $2\sqrt{a} + 7\sqrt{b} - 6\sqrt{a} + \sqrt{b}$
18. $4\sqrt{x} - 3\sqrt{x} + 2\sqrt{y} + 2\sqrt{x}$
19. $6\sqrt[3]{x} - 4\sqrt[3]{y} + 7\sqrt[3]{x} + 2\sqrt[3]{y}$
20. $5\sqrt[3]{x} + 9\sqrt[3]{y} - 10\sqrt[3]{y} + 4\sqrt[3]{x}$
21. $\sqrt{12} + \sqrt{27}$
22. $\sqrt{32} - \sqrt{18}$
23. $3\sqrt{5} - \sqrt{45}$
24. $2\sqrt{7} + 5\sqrt{28}$
25. $3\sqrt[3]{54} + 8\sqrt[3]{2}$
26. $2\sqrt[3]{128} + 5\sqrt[3]{-54}$
27. $\sqrt{50} + \sqrt{18} - 3\sqrt{12}$
28. $2\sqrt{48} - \sqrt{54} + \sqrt{27}$
29. $2\sqrt{20} - \sqrt{45} - \sqrt{36}$
30. $\sqrt{18} - 2\sqrt{12} + 5\sqrt{2}$
31. $\sqrt{8} - 2\sqrt{3} + \sqrt{27} - \sqrt{72}$
32. $\sqrt{80} + \sqrt{8} - \sqrt{45} + \sqrt{50}$
33. $5\sqrt[3]{16} - 4\sqrt[3]{24} + \sqrt[3]{-250}$
34. $\sqrt[3]{192} - 2\sqrt[3]{128} + \sqrt[3]{-81}$
35. $6\sqrt{2x} - \sqrt{8x}$
36. $5\sqrt{3x} + 2\sqrt{12x}$
37. $5y\sqrt{2y} - y\sqrt{18y}$
38. $9x\sqrt{xy} - x\sqrt{16xy}$
39. $4x\sqrt{3xy} - x\sqrt{12xy} - 2x\sqrt{27xy}$
40. $x\sqrt{32x} - x\sqrt{50x} + 2x\sqrt{18x}$

Multiply the expressions in Exercises 41–60.

41. $\sqrt{2}(3 - 4\sqrt{2})$
42. $2\sqrt{7}(\sqrt{7} + 3\sqrt{2})$
43. $3\sqrt{18} \cdot \sqrt{2}$
44. $2\sqrt{10} \cdot \sqrt{5}$
45. $-2\sqrt{6} \cdot \sqrt{8}$
46. $2\sqrt{15} \cdot 5\sqrt{6}$
47. $\sqrt{3}(\sqrt{2} + 2\sqrt{12})$
48. $\sqrt{2}(\sqrt{3} - \sqrt{6})$
49. $\sqrt{y}(\sqrt{x} + 2\sqrt{y})$
50. $\sqrt{x}(\sqrt{x} - 3\sqrt{y})$
51. $(5 + \sqrt{2})(3 - \sqrt{2})$
52. $(2\sqrt{3} + 1)(\sqrt{3} - 3)$
53. $(4\sqrt{3} + \sqrt{2})(\sqrt{3} - 2\sqrt{2})$
54. $(\sqrt{5} - \sqrt{3})(2\sqrt{5} + 3\sqrt{3})$
55. $(\sqrt{x} + 3)(\sqrt{x} - 3)$
56. $(\sqrt{a} + b)(\sqrt{a} + b)$
57. $(\sqrt{2} + \sqrt{7})(\sqrt{2} - \sqrt{7})$
58. $(\sqrt{x} + \sqrt{y})(\sqrt{x} - \sqrt{y})$
59. $(\sqrt{x} + 5)(\sqrt{x} - 3)$
60. $(\sqrt{3} + 7)(\sqrt{3} + 2\sqrt{7})$

9.3 Rationalizing Denominators

OBJECTIVE
In this section, you will be learning to rationalize the denominators of rational expressions containing radicals.

Each of the expressions

$$\frac{5}{\sqrt{3}}, \quad \frac{\sqrt{7}}{\sqrt{8}}, \quad \text{and} \quad \frac{2}{3 - \sqrt{2}}$$

contains a radical in the denominator that is an irrational number. Such expressions are not considered in simplest form because they are difficult to operate with algebraically. Calculations of sums and differences are much easier if the denominators are rational expressions. So, in simplifying, the objective is to find an equal fraction that has a rational number for a denominator.

That is, we want to simplify the expression by **rationalizing the denominator.** The following examples illustrate the method:

a. $\dfrac{5}{\sqrt{3}} = \dfrac{5 \cdot \sqrt{3}}{\sqrt{3} \cdot \sqrt{3}} = \dfrac{5\sqrt{3}}{3}$ Multiply the numerator and denominator by $\sqrt{3}$ because $\sqrt{3} \cdot \sqrt{3} = 3$, a rational number.

b. $\dfrac{4}{\sqrt{x}} = \dfrac{4 \cdot \sqrt{x}}{\sqrt{x} \cdot \sqrt{x}} = \dfrac{4\sqrt{x}}{x}$ Multiply the numerator and denominator by $\sqrt{x}$ because $\sqrt{x} \cdot \sqrt{x} = x$. We have no guarantee that x is rational, but the radical sign does not appear in the denominator of the expression. (Assume that $x > 0$.)

c. $\dfrac{3}{7\sqrt{2}} = \dfrac{3 \cdot \sqrt{2}}{7\sqrt{2} \cdot \sqrt{2}} = \dfrac{3\sqrt{2}}{7 \cdot 2} = \dfrac{3\sqrt{2}}{14}$ Multiply the numerator and denominator by $\sqrt{2}$ because $\sqrt{2} \cdot \sqrt{2} = 2$, a rational number.

d. $\dfrac{\sqrt{7}}{\sqrt{8}} = \dfrac{\sqrt{7} \cdot \sqrt{2}}{\sqrt{8} \cdot \sqrt{2}} = \dfrac{\sqrt{14}}{4}$ Multiply the numerator and denominator by $\sqrt{2}$ because $\sqrt{8} \cdot \sqrt{2} = \sqrt{16} = 4$, a rational number. (Note that $8 \cdot 2 = 16$ and 16 is a perfect square.)

If we had multiplied by $\sqrt{8}$, the results would have been the same, but the fraction would have to be reduced.

$$\frac{\sqrt{7}}{\sqrt{8}} = \frac{\sqrt{7} \cdot \sqrt{8}}{\sqrt{8} \cdot \sqrt{8}} = \frac{\sqrt{56}}{8} = \frac{\sqrt{4} \cdot \sqrt{14}}{8} = \frac{2\sqrt{14}}{8} = \frac{\sqrt{14}}{4}$$

Now we want to rationalize the denominator of a fraction such as

$$\frac{2}{3 - \sqrt{2}}$$

in which the denominator is of the form $a - b = 3 - \sqrt{2}$ where square roots are involved.

Recall that the product $(a - b)(a + b)$ results in the difference of two squares.

$$(a - b)(a + b) = a^2 - b^2$$

We consider two cases:

1. If the denominator is of the form $a - b$, we multiply both the numerator and denominator by $a + b$.
2. If the denominator is of the form $a + b$, we multiply both the numerator and denominator by $a - b$.

In either case, the denominator becomes

$$a^2 - b^2, \text{ the difference of two squares}$$

and we have a rational denominator. Thus,

$$\frac{2}{3 - \sqrt{2}} = \frac{2(3 + \sqrt{2})}{(3 - \sqrt{2})(3 + \sqrt{2})} \qquad \begin{array}{l} \text{If } a - b = 3 - \sqrt{2}, \text{ then} \\ a + b = 3 + \sqrt{2}. \end{array}$$

$$= \frac{2(3 + \sqrt{2})}{3^2 - (\sqrt{2})^2} \qquad \begin{array}{l} \text{The denominator is the difference of two} \\ \text{squares.} \end{array}$$

$$= \frac{2(3 + \sqrt{2})}{9 - 2} \qquad \text{The denominator is a rational number.}$$

$$= \frac{2(3 + \sqrt{2})}{7}$$

EXAMPLES Rationalize the denominator of each of the following expressions and simplify.

1. $\sqrt{\dfrac{1}{2}}$

   Solution $\sqrt{\dfrac{1}{2}} = \dfrac{\sqrt{1}}{\sqrt{2}} = \dfrac{1 \cdot \sqrt{2}}{\sqrt{2} \cdot \sqrt{2}} = \dfrac{\sqrt{2}}{2}$

2. $\dfrac{5}{\sqrt{5}}$

   Solution $\dfrac{5}{\sqrt{5}} = \dfrac{5 \cdot \sqrt{5}}{\sqrt{5} \cdot \sqrt{5}} = \dfrac{5\sqrt{5}}{5} = \sqrt{5}$

3. $\dfrac{6}{\sqrt{27}}$

   Solution $\dfrac{6}{\sqrt{27}} = \dfrac{6 \cdot \sqrt{3}}{\sqrt{27} \cdot \sqrt{3}} = \dfrac{6 \cdot \sqrt{3}}{\sqrt{81}} = \dfrac{6\sqrt{3}}{9} = \dfrac{2\sqrt{3}}{3}$

4. $\dfrac{-3}{\sqrt{xy}}$

   Solution $\dfrac{-3}{\sqrt{xy}} = \dfrac{-3 \cdot \sqrt{xy}}{\sqrt{xy} \cdot \sqrt{xy}} = \dfrac{-3\sqrt{xy}}{xy}$

5. $\dfrac{3}{\sqrt{5} + \sqrt{2}}$

   Solution $\dfrac{3}{\sqrt{5} + \sqrt{2}} = \dfrac{3(\sqrt{5} - \sqrt{2})}{(\sqrt{5} + \sqrt{2})(\sqrt{5} - \sqrt{2})} = \dfrac{3(\sqrt{5} - \sqrt{2})}{(\sqrt{5})^2 - (\sqrt{2})^2}$

   $$= \dfrac{3(\sqrt{5} - \sqrt{2})}{5 - 2} = \dfrac{3(\sqrt{5} - \sqrt{2})}{3} = \sqrt{5} - \sqrt{2}$$

6. $\dfrac{x}{\sqrt{x} - 3}$

Solution $\dfrac{x}{\sqrt{x} - 3} = \dfrac{x(\sqrt{x} + 3)}{(\sqrt{x} - 3)(\sqrt{x} + 3)}$

$= \dfrac{x(\sqrt{x} + 3)}{(\sqrt{x})^2 - (3)^2} = \dfrac{x(\sqrt{x} + 3)}{x - 9}$ ∎

Practice Problems   Rationalize each denominator and simplify.

1. $\sqrt{\dfrac{5}{2}}$ **2.** $\dfrac{\sqrt{7}}{\sqrt{18}}$

3. $\dfrac{4}{\sqrt{7} + \sqrt{3}}$ **4.** $\dfrac{5}{\sqrt{x} + 2}$

EXERCISES 9.3

Rationalize each denominator in Exercises 1–60 and simplify.

1. $\dfrac{5}{\sqrt{2}}$ **2.** $\dfrac{7}{\sqrt{5}}$ **3.** $\dfrac{-3}{\sqrt{7}}$ **4.** $\dfrac{-10}{\sqrt{2}}$ **5.** $\dfrac{6}{\sqrt{3}}$

6. $\dfrac{8}{\sqrt{2}}$ **7.** $\dfrac{\sqrt{18}}{\sqrt{2}}$ **8.** $\dfrac{\sqrt{25}}{\sqrt{3}}$ **9.** $\dfrac{\sqrt{27x}}{\sqrt{3x}}$ **10.** $\dfrac{\sqrt{45y}}{\sqrt{5y}}$

11. $\dfrac{\sqrt{ab}}{\sqrt{9ab}}$ **12.** $\dfrac{\sqrt{5}}{\sqrt{12}}$ **13.** $\dfrac{\sqrt{4}}{\sqrt{3}}$ **14.** $\sqrt{\dfrac{3}{8}}$ **15.** $\sqrt{\dfrac{9}{2}}$

16. $\sqrt{\dfrac{3}{5}}$ **17.** $\sqrt{\dfrac{1}{x}}$ **18.** $\sqrt{\dfrac{x}{y}}$ **19.** $\sqrt{\dfrac{2x}{y}}$ **20.** $\sqrt{\dfrac{x}{4y}}$

21. $\dfrac{2}{\sqrt{2y}}$ **22.** $\dfrac{-10}{3\sqrt{5}}$ **23.** $\dfrac{21}{5\sqrt{7}}$ **24.** $\dfrac{x}{5\sqrt{x}}$ **25.** $\dfrac{-2y}{5\sqrt{2y}}$

26. $\dfrac{3}{1 + \sqrt{2}}$ **27.** $\dfrac{2}{\sqrt{6} - 2}$ **28.** $\dfrac{-11}{\sqrt{3} - 4}$ **29.** $\dfrac{1}{\sqrt{5} - 3}$ **30.** $\dfrac{7}{3 - 2\sqrt{2}}$

31. $\dfrac{-6}{5 - 3\sqrt{2}}$ **32.** $\dfrac{11}{2\sqrt{3} + 1}$ **33.** $\dfrac{-\sqrt{3}}{\sqrt{2} + 5}$ **34.** $\dfrac{\sqrt{2}}{\sqrt{7} + 4}$ **35.** $\dfrac{\sqrt{7}}{1 - 3\sqrt{5}}$

36. $\dfrac{-3\sqrt{3}}{6 + \sqrt{3}}$ **37.** $\dfrac{1}{\sqrt{3} - \sqrt{5}}$ **38.** $\dfrac{-4}{\sqrt{7} - \sqrt{3}}$ **39.** $\dfrac{-5}{\sqrt{2} + \sqrt{3}}$ **40.** $\dfrac{7}{\sqrt{2} + \sqrt{5}}$

41. $\dfrac{4}{\sqrt{x} + 1}$ **42.** $\dfrac{-7}{\sqrt{x} - 3}$ **43.** $\dfrac{5}{6 + \sqrt{y}}$ **44.** $\dfrac{x}{\sqrt{x} + 2}$ **45.** $\dfrac{8}{2\sqrt{x} + 3}$

Answers to Practice Problems   **1.** $\dfrac{\sqrt{10}}{2}$ **2.** $\dfrac{\sqrt{14}}{6}$ **3.** $\sqrt{7} - \sqrt{3}$ **4.** $\dfrac{5(\sqrt{x} - 2)}{x - 4}$

46. $\dfrac{3\sqrt{x}}{\sqrt{2x} - 5}$ **47.** $\dfrac{\sqrt{4y}}{\sqrt{5y} - \sqrt{3}}$ **48.** $\dfrac{\sqrt{3x}}{\sqrt{2} + \sqrt{3x}}$ **49.** $\dfrac{3}{\sqrt{x} - \sqrt{y}}$ **50.** $\dfrac{4}{2\sqrt{x} + \sqrt{y}}$

51. $\dfrac{x}{\sqrt{x} + 2\sqrt{y}}$ **52.** $\dfrac{y}{\sqrt{x} - \sqrt{3y}}$ **53.** $\dfrac{\sqrt{3} + 1}{\sqrt{3} - 2}$ **54.** $\dfrac{\sqrt{2} + 4}{5 - \sqrt{2}}$ **55.** $\dfrac{\sqrt{5} - 2}{\sqrt{5} + 3}$

56. $\dfrac{1 + \sqrt{3}}{3 - \sqrt{3}}$ **57.** $\dfrac{\sqrt{x} + 1}{\sqrt{x} - 1}$ **58.** $\dfrac{\sqrt{x} - 4}{\sqrt{x} + 3}$ **59.** $\dfrac{\sqrt{x} + 2}{\sqrt{3x} + y}$ **60.** $\dfrac{3 - \sqrt{x}}{2\sqrt{x} + y}$

9.4 Solving Radical Equations

OBJECTIVE

In this section, you will be learning to solve equations with radical expressions by squaring both sides of the equations.

If an equation involves radicals with radicands that include variables, then the equation is called a **radical equation.** For example,

$$\sqrt{x} = \underline{5}, \qquad \sqrt{2x + 1} = 3, \qquad \text{and} \qquad x + 1 = \sqrt{3x - 1}$$

are all radical equations.

To solve such equations, we need the following property of real numbers.

> For real numbers a and b, if $a = b$, then $a^2 = b^2$.

We use this property by **squaring both sides of a radical equation that contains square roots.** However, because squaring positive numbers and squaring negative numbers can sometimes give the same result, we must be sure to check all potential solutions in the original equation. For example, consider the equation

$$x = \underline{-2} \qquad \text{an equation with one solution}$$

Squaring both sides, $(x)^2 = (-2)^2$ gives

$$x^2 = 4 \qquad \text{An equation with two solutions}$$
$$x^2 - 4 = 0$$
$$(x + 2)(x - 2) = 0$$
$$x + 2 = 0 \qquad \text{or} \qquad x - 2 = 0$$
$$x = -2 \qquad\qquad\qquad x = 2$$

Thus, squaring both sides created an equation with more solutions than the original equation.

Example 1 illustrates how to solve a radical equation and how to check each potential solution in the original equation.

EXAMPLE

1. Solve the radical equation $\sqrt{18 - x} = x - 6$.

Solution

$$\sqrt{18 - x} = x - 6$$

$$(\sqrt{18 - x})^2 = (x - 6)^2 \qquad \text{Square both sides.}$$

$$18 - x = x^2 - 12x + 36$$

$$0 = x^2 - 11x + 18 \qquad \begin{array}{l}\text{One side must be 0}\\ \text{to solve by factoring.}\end{array}$$

$$0 = (x - 2)(x - 9) \qquad \text{Factor.}$$

$$x - 2 = 0 \quad \text{or} \quad x - 9 = 0$$

$$x = 2 \qquad\qquad x = 9$$

There are two **potential** solutions, 2 and 9.

Check $x = 2$:   *Check* $x = 9$:

$$\sqrt{18 - 2} \stackrel{?}{=} 2 - 6 \qquad\qquad \sqrt{18 - 9} \stackrel{?}{=} 9 - 6$$

$$\sqrt{16} \stackrel{?}{=} -4 \qquad\qquad\qquad \sqrt{9} \stackrel{?}{=} 3$$

$$4 \neq -4 \qquad\qquad\qquad\qquad 3 = 3$$

So 2 is **not** a solution.   So 9 is a solution.

The only solution to the original equation is 9.   ∎

The process of squaring both sides will give a new equation. This equation will have all the solutions of the original equation and possibly, as illustrated in Example 1, some solutions that are not solutions of the original equation. These "extra" solutions, if they exist, are called **extraneous solutions.**

In Example 1, $x = 2$ is an extraneous solution. It is a solution of the quadratic equation

$$0 = x^2 - 11x + 18$$

but it is not a solution of the original radical equation

$$\sqrt{18 - x} = x - 6$$

The checking shows that substituting $x = 2$ in the radical equation gives

$$4 \neq -4$$

We can see that squaring both sides does give a new equation that is true.

$$(4)^2 = (-4)^2$$

Solving radical equations by squaring does not always yield extraneous solutions, but we emphasize that all solutions must be checked in the original equation.

To Solve an Equation with a Square Root Radical Expression	1. Isolate the radical expression on one side of the equation.
	2. Square both sides of the equation.
	3. Solve the new equation.
	4. Check each solution of the new equation in the original equation and eliminate any extraneous solutions.

The following examples illustrate a variety of possible results in solving radical equations.

EXAMPLES Solve the following radical equations.

2. $\sqrt{2x - 3} = 5$

   Solution

   $$\sqrt{2x - 3} = 5$$
   $$(\sqrt{2x - 3})^2 = (5)^2 \qquad \text{Square both sides.}$$
   $$2x - 3 = 25$$
   $$2x = 28$$
   $$x = 14$$

   Check $\sqrt{2 \cdot 14 - 3} \overset{?}{=} 5$

   $$\sqrt{28 - 3} \overset{?}{=} 5$$
   $$\sqrt{25} \overset{?}{=} 5$$
   $$5 = 5$$

   The solution is 14.

3. $\sqrt{x + 1} = -3$

   Solution   We can stop right here. There is **no real solution** to this equation because the radical on the left is nonnegative and cannot possibly equal -3, a negative number.

   Suppose we did not notice this relationship. Then, proceeding as usual,

   $$\sqrt{x + 1} = -3$$
   $$(\sqrt{x + 1})^2 = (-3)^2 \qquad \text{Square both sides.}$$
   $$x + 1 = 9$$
   $$x = 8$$

   Check $\sqrt{8 + 1} \overset{?}{=} -3$

   $$\sqrt{9} \overset{?}{=} -3$$
   $$3 \neq -3$$

   So 8 does **not** check, and there are no solutions.

4. $a - 3 = \sqrt{3a - 9}$

    Solution $a - 3 = \sqrt{3a - 9}$

$$(a - 3)^2 = (\sqrt{3a - 9})^2 \qquad \text{Square both sides.}$$

$$a^2 - 6a + 9 = 3a - 9$$

$$a^2 - 9a + 18 = 0$$

$$(a - 6)(a - 3) = 0 \qquad\qquad \text{Factor.}$$

$$a - 6 = 0 \quad \text{or} \quad a - 3 = 0$$

$$a = 6 \qquad\qquad a = 3$$

    Check $a = 6$:   *Check* $a = 3$:

$$6 - 3 \stackrel{?}{=} \sqrt{3(6) - 9} \qquad\qquad 3 - 3 \stackrel{?}{=} \sqrt{3(3) - 9}$$

$$3 \stackrel{?}{=} \sqrt{18 - 9} \qquad\qquad 0 \stackrel{?}{=} \sqrt{9 - 9}$$

$$3 \stackrel{?}{=} \sqrt{9} \qquad\qquad 0 \stackrel{?}{=} \sqrt{0}$$

$$3 = 3 \qquad\qquad 0 = 0$$

    There are two solutions, 6 and 3.

5. $\sqrt{2x + 5} + 3 = 18$

    Solution $\sqrt{2x + 5} + 3 = 18$

$$\sqrt{2x + 5} = 15 \qquad \text{Isolate the radical on one side.}$$

$$(\sqrt{2x + 5})^2 = 15^2 \qquad \text{Square both sides.}$$

$$2x + 5 = 225$$

$$2x = 220$$

$$x = 110$$

    Check $\sqrt{2 \cdot 110 + 5} + 3 \stackrel{?}{=} 18$

$$\sqrt{225} + 3 \stackrel{?}{=} 18$$

$$15 + 3 \stackrel{?}{=} 18$$

$$18 = 18$$

    The solution is 110.   ∎

◢ EXERCISES 9.4

Solve the equations in Exercises 1–40. Check all solutions in the original equation.

1. $\sqrt{x + 3} = 6$ **2.** $\sqrt{x - 1} = 4$ **3.** $\sqrt{2x - 1} = 3$ **4.** $\sqrt{3x + 1} = 5$

5. $\sqrt{3x + 4} = -5$ **6.** $\sqrt{2x - 5} = -1$ **7.** $\sqrt{5x + 4} = 7$ **8.** $\sqrt{3x - 2} = 4$

9. $\sqrt{6 - x} = 3$ **10.** $\sqrt{11 - x} = 5$ **11.** $\sqrt{x - 4} + 6 = 2$ **12.** $\sqrt{2x - 7} + 5 = 3$

13. $\sqrt{4x + 1} - 2 = 3$ **14.** $\sqrt{6x + 4} - 3 = 5$ **15.** $\sqrt{2x + 11} = 3$ **16.** $\sqrt{3x + 7} = 1$

17. $\sqrt{5 - 2x} = 7$ **18.** $\sqrt{4 - 3x} = 5$ **19.** $\sqrt{5x - 4} = 14$ **20.** $\sqrt{7x - 5} = 4$

21. $\sqrt{2x - 1} = \sqrt{x + 1}$ **22.** $\sqrt{3x + 2} = \sqrt{x + 4}$ **23.** $\sqrt{x + 2} = \sqrt{2x - 5}$

24. $\sqrt{2x - 5} = \sqrt{3x - 9}$ **25.** $\sqrt{4x - 3} = \sqrt{2x + 5}$ **26.** $\sqrt{4x - 6} = \sqrt{3x - 1}$

27. $\sqrt{x + 6} = x + 4$ **28.** $\sqrt{x + 4} = x - 8$ **29.** $\sqrt{x - 2} = x - 2$ **30.** $\sqrt{x + 8} = x - 4$

31. $x + 1 = \sqrt{x + 7}$ **32.** $x + 2 = \sqrt{x + 4}$ **33.** $\sqrt{4x + 1} = x - 5$ **34.** $\sqrt{3x + 1} = x - 3$

35. $\sqrt{2 - x} = x + 4$ **36.** $x + 2 = \sqrt{x + 14}$ **37.** $x - 3 = \sqrt{27 - 3x}$

38. $x + 1 = 2\sqrt{x + 4}$ **39.** $2x + 3 = \sqrt{3x + 7}$ **40.** $3x + 1 = \sqrt{x + 5}$

CHAPTER 9 SUMMARY

Key Terms and Formulas

A **rational number** is any number that can be written in the form $\dfrac{a}{b}$ where a and b are integers and $b \neq 0$.   [9.1]

In decimal form, all **rational numbers** can be written as **repeating decimals.**   [9.1]

In decimal form, **irrational numbers** can be written as **nonrepeating decimals.**   [9.1]

Real numbers are numbers that are either rational or irrational.   [9.1]

The symbol $\sqrt{}$ is called a **radical sign,** and the number under the radical sign is called the **radicand.** The complete expression $\sqrt{a}$ is called a **radical.**   [9.1]

For real numbers a and b, a is a **square root** of b if $a^2 = b$. We write $a = \sqrt{b}$ if a is positive.   [9.1]

A positive square root is called the **principal square root.**   [9.1]

For real numbers a and b, a is the **cube root** of b if $a^3 = b$. We write $a = \sqrt[3]{b}$.   [9.1]

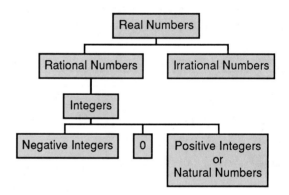

Properties and Rules

If a and b are **positive** real numbers, then

1. $\sqrt{ab} = \sqrt{a}\sqrt{b}$ and   **2.** $\sqrt{\dfrac{a}{b}} = \dfrac{\sqrt{a}}{\sqrt{b}}$ [9.1]

Procedures

To Find the Sum of Radicals   [9.2]
  1. Simplify each radical expression.
  2. Use the distributive property to combine any like radicals.

To Rationalize the Denominator of a Fraction   [9.3]
Multiply the numerator and denominator by a number that will give a denominator that is a rational number.

To Solve an Equation with a Square Root Radical Expression   [9.4]
  1. Isolate the radical expression on one side of the equation.
  2. Square both sides of the equation.
  3. Solve the new equation.
  4. Check each solution of the new equation in the original equation and eliminate any extraneous solutions.

CHAPTER 9 REVIEW

Identify the rational numbers and irrational numbers in Exercises 1–6. [9.1]

1. $-\sqrt{81}$ **2.** $\sqrt{12}$ **3.** $-\sqrt{\dfrac{18}{50}}$ **4.** $\sqrt{\dfrac{5}{16}}$ **5.** $\sqrt{\dfrac{4}{25}}$ **6.** $\sqrt[3]{49}$

Simplify the radical expressions in Exercises 9–20. [9.1]

7. $\sqrt{169}$ **8.** $\sqrt{196}$ **9.** $\sqrt{\dfrac{81}{144}}$ **10.** $\sqrt{\dfrac{225}{49}}$ **11.** $-\sqrt{48}$ **12.** $\sqrt{54}$

13. $\sqrt[3]{40}$ **14.** $\sqrt[3]{-250}$ **15.** $\sqrt{\dfrac{7}{196}}$ **16.** $-\sqrt{\dfrac{5}{121}}$ **17.** $-\sqrt{\dfrac{15}{48}}$ **18.** $\sqrt{\dfrac{75}{64}}$

19. $\dfrac{\sqrt[3]{24}}{10}$ **20.** $\sqrt[3]{\dfrac{135}{64}}$

Simplify the radical expressions in Exercises 21–35. [9.1, 9.2]

21. $\dfrac{8 + \sqrt{12}}{2}$ **22.** $\dfrac{20 - \sqrt{48}}{8}$ **23.** $\dfrac{18 + \sqrt{45}}{30}$ **24.** $\dfrac{18 + \sqrt{72}}{12}$

25. $\dfrac{\sqrt{12} - \sqrt{20}}{4}$ **26.** $\dfrac{\sqrt{27} + \sqrt{18}}{3}$ **27.** $11\sqrt{2} + \sqrt{50}$ **28.** $4\sqrt{20} - \sqrt{45}$

29. $\sqrt[3]{-54} + 5\sqrt[3]{2}$ **30.** $\sqrt{50x} - 2\sqrt{8x}$ **31.** $\sqrt{12} - 5\sqrt{3} + \sqrt{48}$

32. $4\sqrt{3} - \sqrt{8} + \sqrt{75}$ **33.** $\sqrt{49} - \sqrt{56} + 2\sqrt{25}$ **34.** $\sqrt[3]{81} + \sqrt[3]{-24} + \sqrt[3]{250}$

35. $\sqrt{12y} - 3\sqrt{48y} + \sqrt{75y}$

Multiply each expression in Exercises 36–45 and simplify if possible. [9.2]

36. $\sqrt{24} \cdot \sqrt{6}$ **37.** $2\sqrt{30} \cdot \sqrt{20}$ **38.** $2\sqrt{48} \cdot \sqrt{2}$

39. $\sqrt{14} \cdot \sqrt{21}$ **40.** $(\sqrt{3} - 5)(\sqrt{3} + 5)$ **41.** $(\sqrt{2} + 3)(\sqrt{2} + 5)$

42. $(\sqrt{6} + 2\sqrt{3})(3\sqrt{6} - \sqrt{3})$ **43.** $(\sqrt{x} + \sqrt{2})(\sqrt{x} - \sqrt{2})$ **44.** $(5 - \sqrt{3})(7 + \sqrt{3})$

45. $(2\sqrt{x} + 3)(\sqrt{x} - 8)$

Rationalize each denominator in Exercises 46–60 and simplify if possible. [9.3]

46. $\dfrac{9}{\sqrt{3}}$ **47.** $\sqrt{\dfrac{3}{20}}$ **48.** $\sqrt{\dfrac{5}{27}}$ **49.** $\dfrac{\sqrt{2}}{\sqrt{5}}$ **50.** $\dfrac{14}{\sqrt{7x}}$

51. $-\sqrt{\dfrac{9}{2y}}$ **52.** $\dfrac{2}{\sqrt{5y}}$ **53.** $\dfrac{2}{\sqrt{3} - 5}$ **54.** $\dfrac{1}{\sqrt{2} + 3}$ **55.** $\dfrac{6}{\sqrt{6} - 3}$

56. $\dfrac{4}{\sqrt{5} - \sqrt{13}}$ **57.** $\dfrac{\sqrt{3}}{\sqrt{x} - 7}$ **58.** $\dfrac{2}{\sqrt{2} + \sqrt{5}}$ **59.** $\dfrac{12}{\sqrt{7} + \sqrt{3}}$ **60.** $\dfrac{\sqrt{y} - 1}{\sqrt{y} + 3}$

Solve the equations in Exercises 61–70. Check all solutions in the original equation. [9.4]

61. $\sqrt{7x + 2} = 3$ **62.** $\sqrt{4x - 3} = 5$ **63.** $\sqrt{2x - 9} = 7$

64. $\sqrt{5x + 6} = 14$ **65.** $\sqrt{5x - 3} = \sqrt{2x + 3}$ **66.** $\sqrt{9 - x} = \sqrt{3x - 7}$

67. $x + 1 = \sqrt{x + 21}$ **68.** $x - 7 = \sqrt{19 - x}$ **69.** $2x - 1 = \sqrt{x + 7}$

70. $3x + 2 = \sqrt{9x + 10}$

CHAPTER 9 TEST

Identify the rational numbers and the irrational numbers in Exercises 1 and 2.

1. $\sqrt{48}$

2. $\dfrac{\sqrt[3]{-64}}{7}$

Simplify the radical expressions in Exercises 3–8.

3. $\sqrt{125}$ **4.** $\sqrt{144}$ **5.** $\sqrt[3]{72}$ **6.** $-\sqrt{\dfrac{16}{64}}$ **7.** $-\sqrt{\dfrac{24}{36}}$ **8.** $\sqrt[3]{\dfrac{48}{125}}$

Simplify the radical expressions in Exercises 9–14.

9. $\dfrac{\sqrt{96} + 12}{24}$

10. $\dfrac{21 - \sqrt{98}}{14}$

11. $5\sqrt{8} + 3\sqrt{2}$

12. $2\sqrt{27} + 5\sqrt{48} - 4\sqrt{3}$

13. $3\sqrt{2x} - 4\sqrt{50x} + 2\sqrt{72x}$

14. $\sqrt[3]{135} + \sqrt[3]{-40} + \sqrt[3]{108}$

Multiply the expressions in Exercises 15–18 and simplify if possible.

15. $3\sqrt{5} \cdot 5\sqrt{5}$

16. $\sqrt{15} \cdot (\sqrt{3} + \sqrt{5})$

17. $(5 + \sqrt{11})(5 - \sqrt{11})$

18. $(6 + \sqrt{2})(5 - \sqrt{2})$

Rationalize each denominator in Exercises 19–22 and simplify if possible.

19. $\dfrac{3}{\sqrt{18}}$

20. $\sqrt{\dfrac{4}{75}}$

21. $\dfrac{2}{\sqrt{3} - 1}$

22. $\dfrac{\sqrt{2}}{\sqrt{3} + \sqrt{2}}$

Solve the equations in Exercises 23–25. Check all solutions in the original equation.

23. $\sqrt{5x - 1} = 13$

24. $\sqrt{4x + 2} = \sqrt{12 - x}$

25. $2x - 5 = \sqrt{13 - 6x}$

CUMULATIVE REVIEW (9)

Complete the square by adding the correct term in Exercises 1 and 2. Then factor as indicated.

1. $x^2 + 8x + \underline{\hspace{1cm}} = (\quad)^2$

2. $x^2 - \underline{\hspace{1cm}} + 49 = (\quad)^2$

Perform the indicated operations in Exercises 3 and 4.

3. $\dfrac{2x + 10}{x^2 - x - 6} \cdot \dfrac{x^2 + 6x + 8}{x + 5}$

4. $\dfrac{x}{x + 6} - \dfrac{2x - 5}{x^2 + 4x - 12}$

Graph the equations given in Exercises 5 and 6.

5. $y = \dfrac{3}{4}x - 2$

6. $2x + 3y = 12$

Simplify the radical expressions in Exercises 7–10.

7. $-\sqrt{63}$

8. $\sqrt[3]{\dfrac{81}{64}}$

9. $2\sqrt{50} - \sqrt{98} + 3\sqrt{8}$

10. $\sqrt{\dfrac{5}{12x}}$

Solve the equations in Exercises 11–14.

11. $x(x + 4) - (x + 4)(x - 3) = 0$

12. $(2x + 1)(x - 4) = 5$

13. $\dfrac{x - 1}{2x + 1} + \dfrac{1}{x + 2} = \dfrac{1}{2}$

14. $x + 3 = \sqrt{x + 15}$

Solve the systems of equations in Exercises 15 and 16.

15. $\begin{cases} 2x - y = 7 \\ 3x + 2y = 0 \end{cases}$

16. $\begin{cases} 4x + 3y = 15 \\ 2x - 5y = 1 \end{cases}$

Set up equations and solve the problems in Exercises 17–20.

17. Two automobiles start from the same place at the same time and travel in opposite directions. One travels 12 mph faster than the other. At the end of $3\dfrac{1}{2}$ hours, they are 350 miles apart. Find the rate of each automobile.

18. The owner of a candy store had 28 pounds of candy that sold for $1.20 per pound. How many pounds of candy selling for $1.90 per pound did he use to obtain a mixture which sold for $1.40 per pound?

19. A man and his daughter can paint their cabin in 3 hours. Working alone, it would take the daughter 8 hours longer than it would the father. How long would it take each working alone?

20. For lunch, Jason had one burrito and 2 tacos. Matt had 2 burritos and 3 tacos. If Jason spent $3.35 and Matt spent $5.90, find the price of one burrito and the price of one taco.

DID YOU KNOW ?

The quadratic formula is a general method for solving second-degree equations of the form $ax^2 + bx + c = 0$, where a, b, and c can be any real numbers. The quadratic formula is a very old formula; it was known to Babylonian mathematicians around 2000 B.C. However, Babylonian, and later Greek, mathematicians always discarded negative solutions of quadratic equations because they felt that these solutions had no physical meaning. Greek mathematicians always tried to interpret their algebraic problems from a geometrical viewpoint and hence the development of the geometric method of "completing the square."

Consider the following equation: $x^2 + 6x = 7$. A geometric figure is constructed having areas x^2, $3x$ and $3x$.

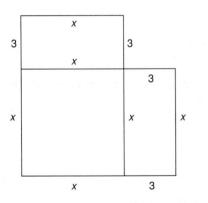

Note that to make the figure a square, one must add a 3-by-3 section (area = 9). Thus, 9 must be added to both sides of the equation to restore

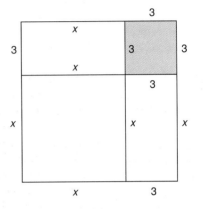

equality. Therefore,

$$x^2 + 6x = 7 \quad \text{Original equation}$$

and
$$x^2 + 6x + 9 = 7 + 9 \quad \text{Adding 9 to}$$
$$x^2 + 6x + 9 = 16 \quad \text{complete the square}$$

So, the square with side $x + 3$ now has an area of 16 square units. Therefore, the sides must be of length 4, which means $x + 3 = 4$. Hence, $x = 1$.

Note that actually the solution set of the original equation is $\{1, -7\}$, since $(-7)^2 + 6(-7) = 49 - 42 = 7$. Thus the Greek mathematicians "lost" the negative solution because of their strictly geometric interpretation of quadratic equations. There were, therefore, many quadratic equations that the Greek mathematicians could not solve because both solutions were negative numbers or complex numbers. Negative solutions to equations were almost completely ignored until the early 1500s during the Renaissance.

What is the difference between method and device? A method is a device you can use twice.

George Pólya in *How to Solve It*

CHAPTER OUTLINE

You have solved quadratic equations by factoring in Chapter 5. However, in many "real-life" situations, factoring is simply not practical because the coefficients do not "cooperate" and the solutions to the equations are irrational numbers (involving square roots) or are nonreal numbers.

In this chapter, you will study three new related techniques for solving quadratic equations culminating in the infamous **quadratic formula.** You will study and use this formula in almost every mathematics course you ever take after this one.

10.1 The Square Root Method

Quadratic equations were introduced in Chapter 5, and we have since used quadratic equations in a variety of applications. Factoring has been the only technique we have used for solving these equations. While factoring is a good technique and we want to use it much of the time, there are quadratic expressions that are not easily factored and some that are not factorable at all using real numbers. In this chapter, we will develop and discuss methods other than factoring for solving quadratic equations.

Consider the equation $x^2 = 5$. We can solve by factoring in a rather unusual way, as follows:

$$x^2 = 5$$

$x^2 - 5 = 0$	Get 0 on one side.
$x^2 - (\sqrt{5})^2 = 0$	Difference of two squares with $5 = (\sqrt{5})^2$
$(x + \sqrt{5})(x - \sqrt{5}) = 0$	Factor.

$$x + \sqrt{5} = 0 \quad \text{or} \quad x - \sqrt{5} = 0 \qquad \text{There are two irrational}$$
$$x = -\sqrt{5} \qquad\qquad\qquad x = \sqrt{5} \qquad \text{solutions.}$$

We can solve this equation more directly by **taking square roots of both sides,** as the following statement indicates.

For an equation in the form

$$x^2 = c \quad \text{where } c \text{ is positive}$$

then $x = \sqrt{c}$ or $x = -\sqrt{c}$. We can write $x = \pm\sqrt{c}$.

EXAMPLES Solve the following quadratic equations.

1. $3x^2 = 51$

Solution $3x^2 = 51$

$x^2 = 17$	Divide both sides by 3 so that the coefficient of x^2 is 1.
$x = \pm\sqrt{17}$	Keep in mind that the expression $x = \pm\sqrt{17}$ represents two equations, $x = \sqrt{17}$ and $x = -\sqrt{17}$.

2. $(x + 4)^2 = 21$

Solution $(x + 4)^2 = 21$

$x + 4 = \pm\sqrt{21}$

$x = -4 \pm \sqrt{21}$ There are two solutions,
$-4 + \sqrt{21}$ and $-4 - \sqrt{21}$.

3. $(x - 2)^2 = 50$

Solution $(x - 2)^2 = 50$

$x - 2 = \pm\sqrt{50}$

$x = 2 \pm 5\sqrt{2}$ Simplify the radical.
$(\sqrt{50} = \sqrt{25}\sqrt{2} = 5\sqrt{2})$

4. $(2x + 4)^2 = 72$

Solution $(2x + 4)^2 = 72$

$2x + 4 = \pm\sqrt{72}$

$2x = -4 \pm 6\sqrt{2}$ Simplify the radical.
$(\sqrt{72} = \sqrt{36}\sqrt{2} = 6\sqrt{2})$

$x = \dfrac{-4 \pm 6\sqrt{2}}{2}$ Factor 2 and simplify.

$= \dfrac{\cancel{2}(-2 \pm 3\sqrt{2})}{\cancel{2}} = -2 \pm 3\sqrt{2}$

5. $(x + 7)^2 + 10 = 8$

Solution $(x + 7)^2 + 10 = 8$

$(x + 7)^2 = -2$

There is no real solution. The square of a real number cannot be negative. ∎

The Pythagorean Theorem

Among the most interesting and useful applications of squares and square roots are those dealing with **right triangles** and **the Pythagorean Theorem**. (Pythagoras was a famous Greek mathematician.)

   A **right triangle** is a triangle in which one angle is a right angle (measures 90°). Two of the sides are perpendicular and are called **legs**. The longest side is called the **hypotenuse** and is opposite the 90° angle.

The Pythagorean Theorem   In a right triangle, the square of the hypotenuse is equal to the sum of the squares of the two legs.

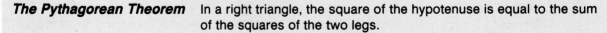

$c^2 = a^2 + b^2$

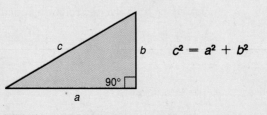

a²+b² = c²
100+24 = 576
C² = a²+b²
26² = 10+24
24 = 100-
576 = 100 + 12²

C² = a²+b²
c² = 10²+b²
26² = 100+b²
576

676
676 = 100+b²

676-100 ≤ b²
576 = b²

12

a = 10
c = 26

EXAMPLES

6. What is the length of the hypotenuse of a right triangle if one leg is 8 cm long and the other leg is 6 cm long?

Solution   By the Pythagorean Theorem,

$a^2+b^2 = c^2$
$10^2+b^2 = 26^2$
$100+b^2 = 26^2$

$$c^2 = 8^2 + 6^2$$
$$c^2 = 64 + 36$$
$$c^2 = 100$$
$$c = \sqrt{100} = 10 \text{ cm}$$

The hypotenuse is 10 cm long. (Note that only the positive square root is used because the negative square root does not make sense as a solution to the problem.)

7. If the hypotenuse of a right triangle is 15 meters long and one leg is 10 meters long, what is the length of the other leg?

Solution   By the Pythagorean Theorem,

$$x^2 + 10^2 = 15^2$$
$$x^2 + 100 = 225$$
$$x^2 = 125$$
$$x = \sqrt{125} = \sqrt{25} \cdot \sqrt{3} = 5\sqrt{3} \text{ m} \ (\approx 8.66 \text{ m})$$

$a^2+b^2 = c^2$
$a^2 - c^2 = -b^2$
$100 - 676 = b^2$
$576 = -b^2$

The other leg is $5\sqrt{3}$ m long.   ▪

EXERCISES 10.1

Solve the quadratic equations in Exercises 1–30.

1. $x^2 = 121$	**2.** $x^2 = 81$	**3.** $3x^2 = 108$	**4.** $5x^2 = 245$
5. $x^2 = 35$	**6.** $x^2 = 42$	**7.** $x^2 - 62 = 0$	**8.** $x^2 - 75 = 0$
9. $x^2 - 45 = 0$	**10.** $x^2 - 98 = 0$	**11.** $3x^2 = 54$	**12.** $5x^2 = 60$
13. $9x^2 = 4$	**14.** $4x^2 = 25$	**15.** $(x - 1)^2 = 4$	**16.** $(x + 3)^2 = 9$
17. $(x + 2)^2 = -25$	**18.** $(x - 5)^2 = 36$	**19.** $(x + 1)^2 = \dfrac{1}{4}$	**20.** $(x - 9)^2 = -\dfrac{9}{25}$
21. $(x - 3)^2 = \dfrac{4}{9}$	**22.** $(x - 2)^2 = \dfrac{1}{16}$	**23.** $(x - 6)^2 = 18$	**24.** $(x + 8)^2 = 75$
25. $2(x - 7)^2 = 24$	**26.** $3(x + 11)^2 = 60$	**27.** $(3x + 4)^2 = 27$	**28.** $(2x + 1)^2 = 48$
29. $(5x - 2)^2 = 63$	**30.** $(4x - 3)^2 = 125$		

Use the Pythagorean Theorem to decide if each of the triangles determined by the three sides given in Exercises 31–34 is a right triangle.

31. 8 cm, 15 cm, 17 cm

32. 4 cm, 5 cm, 6 cm

33. $\sqrt{2}$ in., $\sqrt{2}$ in., 2 in.

34. 3 ft, 6 ft, $3\sqrt{5}$ ft

Determine the length of the missing side in each of the Exercises 35–38. (See figure below.)

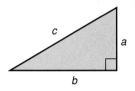

35. $a = 9, b = 12, c = ?$

36. $a = 10, b = 24, c = ?$

37. $a = 6, c = 12, b = ?$

38. $b = 10, c = 30, a = ?$

39. The hypotenuse of a right triangle is twice the length of one of the legs. The length of the other leg is $4\sqrt{3}$ feet. Find the length of the leg and the hypotenuse.

41. The two legs of a right triangle are the same length. The hypotenuse is 6 centimeters long. Find the length of the legs.

40. One leg of a right triangle is three times the other. The length of the hypotenuse is 20 centimeters. Find the lengths of the legs.

42. The two legs of a right triangle are the same length. The hypotenuse is $4\sqrt{2}$ meters long. Find the length of the legs.

Calculator Problems

Use a calculator to solve the equations in Exercises 43–50. Round off your answers to the nearest hundredth.

43. $x^2 = 647$

44. $x^2 = 378$

45. $19x^2 = 523$

46. $14x^2 = 795$

47. $6x^2 = 17.32$

48. $15x^2 = 229.63$

49. $2.1x^2 = 35.82$

50. $4.7x^2 = 118.34$

10.2 Completing the Square

OBJECTIVES

In this section, you will be learning to:

1. Determine the constant terms that will make incomplete trinomials perfect square trinomials.

2. Solve quadratic equations by completing the square.

3. Recognize when quadratic equations can be solved by factoring and when they must be solved by completing the square.

In Section 5.2, we discussed factoring perfect square trinomials and we also added terms to binomials so that the resulting trinomial was a perfect square. The process was called **completing the square.** Now we will show how the procedure of completing the square can be used to solve quadratic equations.

As review, we complete the square for $x^2 + 12x$ and $a^2 - 20a$.

$$x^2 + 12x + \underline{} = ()^2$$

The number 36 will complete the square since

$$\frac{1}{2}(12) = 6 \quad \text{and} \quad 6^2 = 36.$$

So,

$$x^2 + 12x + \underline{36} = (x + 6)^2$$

Similarly, for

$$a^2 - 20a + \underline{} = ()^2$$

we have

$$\frac{1}{2}(-20) = -10 \quad \text{and} \quad (-10)^2 = 100$$
$$a^2 - 20a + \underline{\ 100\ } = (a - 10)^2$$

If the coefficient of x^2 is **not 1,** we proceed as follows:

$$2x^2 + 20x + \underline{} = 2()^2$$
$$2(x^2 + 10x + \underline{}) = 2()^2 \qquad \text{Factor 2.}$$
$$2(x^2 + 10x + \underline{\ 25\ }) = 2(x + 5)^2 \qquad \text{Complete the square of the}$$

expression inside the parentheses.

$$2x^2 + 20x + \underline{\ 50\ } = 2(x + 5)^2 \qquad$$ The number to be added to the
original expression is $2 \cdot 25 = 50$.

The following procedure is used in the examples to solve quadratic equations by completing the square.

To Solve Quadratic Equations by Completing the Square	1. Arrange terms with variables on one side and constants on the other.
	2. Divide each term by the coefficient of x^2. (We want the coefficient of x^2 to be 1.)
	3. Find the number that completes the square of the quadratic expression and add this number to **both** sides of the equation.
	4. Find the positive and negative square roots of both sides.
	5. Solve for x. Remember, there will usually be two solutions.

EXAMPLES   Solve the following quadratic equations by completing the squares.

1. $x^2 - 6x + 4 = 0$

   Solution $x^2 - 6x + 4 = 0$

   $$x^2 - 6x = -4 \qquad$$ Add -4 to both sides of the equation.

   $$x^2 - 6x + 9 = -4 + 9 \qquad$$ Add 9 to both sides of the equation. The left side is now a **perfect square trinomial.**

   $$\frac{1}{2}(-6) = -3 \text{ and } (-3)^2 = 9$$

   $$(x - 3)^2 = 5 \qquad$$ Simplify.

   $$x - 3 = \pm\sqrt{5} \qquad$$ Find square roots.

   $$x = 3 \pm \sqrt{5} \qquad$$ Solve for x.

2. $x^2 + 5x = 7$

Solution

$$x^2 + 5x = 7$$

$$x^2 + 5x + \frac{25}{4} = 7 + \frac{25}{4}$$

Complete the square on the left. $\frac{1}{2} \cdot 5 = \frac{5}{2}$ and $\left(\frac{5}{2}\right)^2 = \frac{25}{4}$

$$\left(x + \frac{5}{2}\right)^2 = \frac{53}{4}$$

Simplify:
$\left(7 + \frac{25}{4} = \frac{28}{4} + \frac{25}{4} = \frac{53}{4}\right)$

$$x + \frac{5}{2} = \pm\sqrt{\frac{53}{4}}$$

Find square roots.

$$x = -\frac{5}{2} \pm \sqrt{\frac{53}{4}}$$

Solve for x.

$$x = -\frac{5}{2} \pm \frac{\sqrt{53}}{2}$$

Special property of square roots

$\sqrt{\frac{a}{b}} = \frac{\sqrt{a}}{\sqrt{b}}$ for $a > 0$ and $b > 0$

or $\quad x = \frac{-5 \pm \sqrt{53}}{2}$

Combine the fractions.

3. $6x^2 + 12x - 9 = 0$

Solution

$$6x^2 + 12x - 9 = 0$$

$$6x^2 + 12x = 9$$

Add 9 to both sides of the equation.

$$\frac{6x^2}{6} + \frac{12x}{6} = \frac{9}{6}$$

Divide each term by 6 **so that the leading coefficient will be 1.**

$$x^2 + 2x = \frac{3}{2}$$

The leading coefficient is 1.

$$x^2 + 2x + 1 = \frac{3}{2} + 1$$

Complete the square,
$\frac{1}{2} \cdot 2 = 1$ and $1^2 = 1$.

$$(x + 1)^2 = \frac{5}{2} \qquad \text{Simplify.}$$

$$x + 1 = \pm \sqrt{\frac{5}{2}} \qquad \text{Find square roots.}$$

$$x = -1 \pm \sqrt{\frac{5}{2}} \qquad \text{Solve for } x.$$

$$x = -1 \pm \frac{\sqrt{5}}{\sqrt{2}} \cdot \frac{\sqrt{2}}{\sqrt{2}} \qquad \text{Rationalize the denominator.}$$

$$x = -1 \pm \frac{\sqrt{10}}{2} \qquad \text{Simplify.}$$

$$\text{or} \qquad x = \frac{-2 \pm \sqrt{10}}{2} \qquad \text{Combine the fractions.}$$

4. $2x^2 + 5x - 8 = 0$
Solution

$$2x^2 + 5x - 8 = 0$$

$$2x^2 + 5x = 8 \qquad \text{Add 8 to both sides of the equation.}$$

$$\frac{2x^2}{2} + \frac{5x}{2} = \frac{8}{2} \qquad \text{Divide each term by 2 \textbf{so that the} \textbf{leading coefficient will be 1.}}$$

$$x^2 + \frac{5}{2}x = 4 \qquad \text{Simplify.}$$

$$x^2 + \frac{5}{2}x + \frac{25}{16} = 4 + \frac{25}{16} \qquad \text{Complete the square,} \quad \frac{1}{2} \cdot \frac{5}{2} = \frac{5}{4}$$
$$\text{and } \left(\frac{5}{4}\right)^2 = \frac{25}{16}.$$

$$\left(x + \frac{5}{4}\right)^2 = \frac{89}{16} \qquad \text{Simplify.} \quad \left(4 + \frac{25}{16} = \frac{64}{16} + \frac{25}{16} = \frac{89}{16}\right)$$

$$x + \frac{5}{4} = \pm \sqrt{\frac{89}{16}} \qquad \text{Find the square roots.}$$

$$x = -\frac{5}{4} \pm \frac{\sqrt{89}}{4} \qquad \text{Solve for } x; \text{ also,}$$
$$\sqrt{\frac{89}{16}} = \frac{\sqrt{89}}{\sqrt{16}} = \frac{\sqrt{89}}{4}$$

$$\text{or} \qquad x = \frac{-5 \pm \sqrt{89}}{4} \qquad \text{Combine the fractions.}$$

5. $x^2 + 8x = -4$

Solution

$$x^2 + 8x = -4$$
$$x^2 + 8x + 16 = -4 + 16 \qquad \text{Complete the square.}$$
$$(x + 4)^2 = 12 \qquad \text{Simplify.}$$
$$x + 4 = \pm\sqrt{12} \qquad \text{Find square roots.}$$
$$x = -4 \pm \sqrt{12} \qquad \text{Solve for } x.$$
$$x = -4 \pm 2\sqrt{3} \qquad \sqrt{12} = \sqrt{4} \cdot \sqrt{3} = 2\sqrt{3}$$
$$\text{by special property}$$
$$\sqrt{ab} = \sqrt{a} \cdot \sqrt{b} \text{ for } a > 0$$
$$\text{and } b > 0 \qquad \blacksquare$$

Solving quadratic equations by completing the square is a good technique, and it provides practice with several algebraic operations. However, we do not want to forget that, in general, factoring is the preferred technique whenever the factors are relatively easy to find. Some of the following exercises allow the option of solving by factoring or completing the square.

 **EXERCISES 10.2**

Add the correct constants in Exercises 1–10 in order to make the trinomial factorable as indicated. (Constant factors of the x^2 term must be dealt with in Exercises 3, 4, 9, and 10.)

1. $x^2 + 12x + \underline{\hspace{0.4cm}} = (\hspace{0.5cm})^2$

2. $x^2 - 14x + \underline{\hspace{0.4cm}} = (\hspace{0.5cm})^2$

3. $2x^2 - 16x + \underline{\hspace{0.4cm}} = 2(\hspace{0.5cm})^2$

4. $5x^2 + 10x + \underline{\hspace{0.4cm}} = 5(\hspace{0.5cm})^2$

5. $x^2 - 3x + \underline{\hspace{0.4cm}} = (\hspace{0.5cm})^2$

6. $x^2 + 5x + \underline{\hspace{0.4cm}} = (\hspace{0.5cm})^2$

7. $x^2 + x + \underline{\hspace{0.4cm}} = (\hspace{0.5cm})^2$

8. $x^2 - 7x + \underline{\hspace{0.4cm}} = (\hspace{0.5cm})^2$

9. $2x^2 + 4x + \underline{\hspace{0.4cm}} = 2(\hspace{0.5cm})^2$

10. $3x^2 + 18x + \underline{\hspace{0.4cm}} = 3(\hspace{0.5cm})^2$

Solve the quadratic equations in Exercises 11–30 by completing the squares.

11. $x^2 + 6x - 7 = 0$

12. $x^2 + 8x + 12 = 0$

13. $x^2 - 4x - 45 = 0$

14. $x^2 - 10x + 21 = 0$

15. $x^2 - 3x - 40 = 0$

16. $x^2 + x - 42 = 0$

17. $3x^2 + x - 4 = 0$

18. $2x^2 + x - 6 = 0$

19. $4x^2 - 4x - 3 = 0$

20. $3x^2 + 11x + 10 = 0$

21. $x^2 + 6x + 3 = 0$

22. $x^2 - 4x - 3 = 0$

23. $x^2 + 2x - 5 = 0$

24. $x^2 + 8x + 2 = 0$

25. $3x^2 - 2x - 1 = 0$

26. $2x^2 + 5x + 2 = 0$

27. $x^2 + x - 3 = 0$

28. $x^2 + 5x + 1 = 0$

29. $2x^2 + 3x - 1 = 0$

30. $3x^2 - 4x - 2 = 0$

Solve the quadratic equations in Exercises 31–50 by factoring or completing the square.

31. $x^2 - 9x + 2 = 0$

32. $x^2 - 8x - 20 = 0$

33. $x^2 + 7x - 14 = 0$

34. $x^2 + 5x + 3 = 0$

35. $x^2 - 11x - 26 = 0$

36. $x^2 + 6x - 4 = 0$

37. $2x^2 + 5x + 2 = 0$

38. $6x^2 - 2x - 2 = 0$

39. $3x^2 - 6x + 3 = 0$

40. $5x^2 + 15x - 5 = 0$

41. $4x^2 + 20x - 8 = 0$

42. $2x^2 - 7x + 4 = 0$

43. $6x^2 - 8x + 1 = 0$

44. $5x^2 - 10x + 3 = 0$

45. $2x^2 + 7x + 4 = 0$

46. $3x^2 - 5x - 3 = 0$

47. $4x^2 - 2x - 3 = 0$

48. $2x^2 - 9x + 7 = 0$

49. $3x^2 + 8x + 5 = 0$

50. $5x^2 + 11x - 1 = 0$

10.3 The Quadratic Formula: $x = \dfrac{-b \pm \sqrt{b^2 - 4ac}}{2a}$

OBJECTIVES

In this section, you will be learning to:

1. Write quadratic equations in standard form.

2. Identify the coefficients of quadratic equations in standard form.

3. Solve quadratic equations by using the quadratic formula.

4. Choose the most expedient method for solving quadratic equations.

Now we are interested in developing a formula that will be useful in solving quadratic equations of any form. **This formula will always work,** but you should not forget the factoring and completing the square techniques because they can be easier to apply than the formula.

The **general quadratic equation** is

$$ax^2 + bx + c = 0, \qquad a \neq 0$$

We want to solve the general quadratic equation for x in terms of the coefficients a, b, and c. The technique is to **complete the square** (Section 10.2), treating a, b, and c as constants.

Development of the Quadratic Formula

$ax^2 + bx + c = 0$	Begin with the general quadratic equation.
$ax^2 + bx = -c$	Add $-c$ to both sides.
$\dfrac{ax^2}{a} + \dfrac{bx}{a} = \dfrac{-c}{a}$	Divide each term by a.
$x^2 + \dfrac{b}{a}x = \dfrac{-c}{a}$	Rewrite: $\dfrac{bx}{a} = \dfrac{b}{a}x.$
$x^2 + \dfrac{b}{a}x + \left(\dfrac{b}{2a}\right)^2 = \left(\dfrac{b}{2a}\right)^2 + \dfrac{-c}{a}$	Complete the square, $\dfrac{1}{2}\left(\dfrac{b}{a}\right) = \dfrac{b}{2a}.$
$\left(x + \dfrac{b}{2a}\right)^2 = \dfrac{b^2}{4a^2} + \dfrac{-c}{a}$	Simplify.
$\left(x + \dfrac{b}{2a}\right)^2 = \dfrac{b^2}{4a^2} + \dfrac{-c \cdot 4a}{a \cdot 4a}$	Common denominator is $4a^2$.
$\left(x + \dfrac{b}{2a}\right)^2 = \dfrac{b^2 - 4ac}{4a^2}$	Simplify.
$x + \dfrac{b}{2a} = \pm\sqrt{\dfrac{b^2 - 4ac}{4a^2}}$	Find the square roots.
$x + \dfrac{b}{2a} = \dfrac{\pm\sqrt{b^2 - 4ac}}{\pm\sqrt{4a^2}}$	Use the relationship $\sqrt{\dfrac{a}{b}} = \dfrac{\sqrt{a}}{\sqrt{b}}$ if a, $b > 0$.

$$x + \frac{b}{2a} = \pm\frac{\sqrt{b^2 - 4ac}}{2a} \qquad \text{Simplify.}$$

$$x = \frac{-b}{2a} \pm \frac{\sqrt{b^2 - 4ac}}{2a} \qquad \text{Solve for } x.$$

$$x = \frac{-b \pm \sqrt{b^2 - 4ac}}{2a} \qquad \text{THE QUADRATIC FORMULA}$$

Special Note: The expression $b^2 - 4ac$ is called the **discriminant.** If $b^2 - 4ac < 0$, then there are no real number solutions because the square root of a negative number is not a real number. A discussion of negative discriminants is given in later courses in algebra.

Now the solutions to quadratic equations can be found by going directly to the formula.

EXAMPLES   Solve the following quadratic equations using the quadratic formula.

1. $2x^2 + x - 2 = 0$

Solution $a = 2, b = 1,$ and $c = -2$

$$x = \frac{-1 \pm \sqrt{1^2 - 4(2)(-2)}}{2 \cdot 2} = \frac{-1 \pm \sqrt{1 + 16}}{4} = \frac{-1 \pm \sqrt{17}}{4}$$

2. $3x^2 - 5x + 1 = 0$

Solution $a = 3, b = -5,$ and $c = 1$

$$x = \frac{-(-5) \pm \sqrt{(-5)^2 - 4(3)(1)}}{2 \cdot 3} = \frac{5 \pm \sqrt{25 - 12}}{6} = \frac{5 \pm \sqrt{13}}{6}$$

3. $\frac{1}{6}x^2 - x + \frac{1}{2} = 0$

Solution

$$6 \cdot \frac{1}{6}x^2 - 6 \cdot x + 6 \cdot \frac{1}{2} = 6 \cdot 0 \qquad \begin{array}{l}\text{Multiply each term by 6, the} \\ \text{least common denominator.} \\ \text{Integer coefficients are much} \\ \text{easier to use in the formula.}\end{array}$$

$$x^2 - 6x + 3 = 0$$

$a = 1, b = -6,$ and $c = 3$

$$x = \frac{-(-6) \pm \sqrt{(-6)^2 - 4(1)(3)}}{2 \cdot 1} = \frac{6 \pm \sqrt{36 - 12}}{2}$$

$$= \frac{6 \pm \sqrt{24}}{2} = \frac{6 \pm 2\sqrt{6}}{2} = \frac{\cancel{2}(3 \pm 2\sqrt{6})}{\cancel{2}}$$

$$= 3 \pm \sqrt{6}$$

$$\left(\text{or } \frac{6 \pm \sqrt{24}}{2} = \frac{6 \pm \sqrt{4} \cdot \sqrt{6}}{2} = \frac{6 \pm 2\sqrt{6}}{2} = \frac{6}{2} \pm \frac{2\sqrt{6}}{2} = 3 \pm \sqrt{6} \right)$$

4. $2x^2 - 25 = 0$

Solution We could add 25 to both sides, divide by 2, and then take the square roots. We can arrive at the same result using the quadratic formula, but note that $b = 0$.

$2x^2 - 25 = 0$ (or $2x^2 + 0x - 25 = 0$)
$a = 2$, $b = 0$, and $c = -25$

$$x = \frac{-(0) \pm \sqrt{0^2 - 4(2)(-25)}}{2 \cdot 2} = \frac{\pm\sqrt{200}}{4} = \frac{\pm 10\sqrt{2}}{4}$$

$$= \frac{\pm 5\sqrt{2}}{2}$$

5. $-5x^2 + 3x = -2$

Solution $-5x^2 + 3x + 2 = 0$ One side must be 0.

$a = -5$, $b = 3$, and $c = 2$

$$x = \frac{-3 \pm \sqrt{3^2 - 4(-5)(2)}}{2(-5)} = \frac{-3 \pm \sqrt{9 + 40}}{-10}$$

$$= \frac{-3 \pm \sqrt{49}}{-10} = \frac{-3 \pm 7}{-10}$$

$$x = \frac{-3 + 7}{-10} = \frac{4}{-10} = -\frac{2}{5} \quad \text{or} \quad x = \frac{-3 - 7}{-10} = \frac{-10}{-10} = 1$$

This example could also be solved by factoring:

$-5x^2 + 3x + 2 = 0$
$5x^2 - 3x - 2 = 0$ Multiply both sides of the equation by -1.
$(5x + 2)(x - 1) = 0$

$5x + 2 = 0 \qquad \text{or} \qquad x - 1 = 0$
$5x = -2 \qquad\qquad x = 1$
$x = -\dfrac{2}{5}$

6. $4x^2 + 12x + 9 = 0$

Solution $a = 4$, $b = 12$, $c = 9$

$$x = \frac{-12 \pm \sqrt{12^2 - 4(4)(9)}}{2 \cdot 4} = \frac{-12 \pm \sqrt{144 - 144}}{8}$$

$$= \frac{-12 \pm \sqrt{0}}{8} \qquad (\sqrt{0} = 0)$$

$$= -\frac{12}{8} = -\frac{3}{2}$$

Note that when the discriminant is 0, there is only one solution. This equation could also be solved by factoring in the following manner:

$$4x^2 + 12x + 9 = 0$$
$$(2x + 3)^2 = 0$$
$$2x + 3 = 0 \qquad \text{The factor } 2x + 3 \text{ is repeated.}$$
$$2x = -3$$
$$x = -\frac{3}{2} \qquad -\frac{3}{2} \text{ is called a \textbf{double solution}}$$

or a **double root.**   ■

Equations such as those in Examples 5 and 6 should be solved by factoring because factoring is generally easier than applying the quadratic formula. However, if you do not "see" the factors readily, the quadratic formula will give the solutions, as illustrated in Examples 5 and 6.

EXAMPLE

7. $x^2 + x + 1 = 0$

Solution $a = 1, b = 1, c = 1$

$$x = \frac{-1 \pm \sqrt{1^2 - 4(1)(1)}}{2 \cdot 1} = \frac{-1 \pm \sqrt{1 - 4}}{2}$$
$$= \frac{-1 \pm \sqrt{-3}}{2} \qquad \text{(Not a real number)}$$

There is no real solution. This example illustrates the fact that not every equation has real solutions. None of the exercises in this text has this kind of answer. Such solutions are called **complex numbers** and will be discussed in detail in the next course in algebra.   ■

Practice Problems   Solve the equations using the quadratic formula.

1. $x^2 + 5x + 1 = 0$
2. $4x^2 - x = 1$
3. $x^2 - 4x - 4 = 0$
4. $-2x^2 + 3 = 0$ **(Note: Here $b = 0$.)**
5. $5x^2 - x = 0$ **(Note: Here $c = 0$.)**

Answers to Practice Problems   1. $x = \dfrac{-5 \pm \sqrt{21}}{2}$ **2.** $x = \dfrac{1 \pm \sqrt{17}}{8}$
3. $x = 2 \pm 2\sqrt{2}$ **4.** $x = \dfrac{0 \pm 2\sqrt{6}}{-4}$ or $x = \dfrac{\pm \sqrt{6}}{2}$
5. $x = \dfrac{1 \pm \sqrt{1}}{10}$ or $x = 0, x = \dfrac{1}{5}$ (This problem could be solved easily by factoring.)

EXERCISES 10.3

Write each of the quadratic equations in Exercises 1–12 in the form $ax^2 + bx + c = 0$ with $a > 0$ and a, b, c as integers; then identify the constants a, b, and c.

1. $x^2 - 3x = 2$

2. $x^2 + 2 = 5x$

3. $x = 2x^2 + 6$

4. $5x^2 = 3x - 1$

5. $4x + 3 = 7x^2$

6. $x = 4 - 3x^2$

7. $4 = 3x^2 - 9x$

8. $6x + 4 = 3x^2$

9. $x^2 + 5x = 3 - x^2$

10. $x^2 + 4x - 1 = 2x + 3x^2$

11. $\dfrac{2}{3}x^2 + x - 1 = 0$

12. $\dfrac{x^2}{4} + x - \dfrac{1}{2} = 0$

Solve the quadratic equations in Exercises 13–32 using the quadratic formula.

13. $x^2 + 4x - 1 = 0$

14. $x^2 - 3x + 1 = 0$

15. $x^2 - 3x = 4$

16. $x^2 + 5x = 2$

17. $-2x^2 + x = -1$

18. $-3x^2 + x = -1$

19. $5x^2 + 3x - 2 = 0$

20. $-2x^2 + 5x - 1 = 0$

21. $6x^2 - 3x - 1 = 0$

22. $4x^2 - 7x + 3 = 0$

23. $x^2 - 7 = 0$

24. $2x^2 + 5x - 3 = 0$

25. $x^2 + 4x = x - 2x^2$

26. $x^2 - 2x + 1 = 2 - 3x^2$

27. $3x^2 + 4x = 0$

28. $4x^2 - 10 = 0$

29. $\dfrac{2}{5}x^2 + x - 1 = 0$

30. $2x^2 + 3x - \dfrac{3}{4} = 0$

31. $-\dfrac{x^2}{2} - 2x + \dfrac{1}{3} = 0$

32. $\dfrac{x^2}{3} - x - \dfrac{1}{5} = 0$

Solve the quadratic equations in Exercises 33–50 using any method (factoring, completing the square, or quadratic formula).

33. $2x^2 + 7x + 3 = 0$

34. $5x^2 - x - 4 = 0$

35. $9x^2 - 6x - 1 = 0$

36. $3x^2 - 7x + 1 = 0$

37. $2x^2 - 2x - 1 = 0$

38. $10x^2 = x + 24$

39. $9x^2 + 12x = -2$

40. $3x^2 - 11x = 4$

41. $5x^2 = 7x + 5$

42. $-4x^2 + 11x - 5 = 0$

43. $-4x^2 + 12x - 9 = 0$

44. $10x^2 + 35x + 30 = 0$

45. $6x^2 + 2x = 20$

46. $25x^2 + 4 = 20$

47. $3x^2 - 4x + \dfrac{1}{3} = 0$

48. $\dfrac{3}{4}x^2 - 2x + \dfrac{1}{8} = 0$

49. $\dfrac{11}{2}x + 1 = 3x^2$

50. $\dfrac{3}{7}x^2 = \dfrac{1}{2}x + 1$

10.4 Applications

OBJECTIVE

In this section, you will be learning to solve applied problems using quadratic equations.

Application problems are designed to teach you to read carefully, to think clearly, and to translate from English to algebraic expressions and equations. The problems do not tell you directly to add, subtract, multiply, divide, or square. You must decide on a method of attack from the wording of the problem, combined with your previous experience and knowledge. Study the following examples and read the explanations carefully. They are similar to some, but not all, of the problems in the exercises. Example 1 makes use of the Pythagorean Theorem that was discussed in Section 10.1.

EXAMPLE 1: The Pythagorean Theorem

The length of a rectangle is 7 feet longer than the width. If one diagonal measures 13 feet, what are the dimensions of the rectangle?

> **Solution**   Draw a diagram for problems involving geometric figures whenever possible.

Let x = width of the rectangle
        $x + 7$ = length of the rectangle

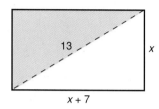

$$(x + 7)^2 + x^2 = 13^2 \qquad \text{Use the Pythagorean Theorem.}$$
$$x^2 + 14x + 49 + x^2 = 169$$
$$2x^2 + 14x + 49 - 169 = 0$$
$$2x^2 + 14x - 120 = 0$$
$$2(x^2 + 7x - 60) = 0$$
$$2(x - 5)(x + 12) = 0$$

$x - 5 = 0$ or $x + 12 = 0$ A negative number does not
$ x = 5$ $\cancel{x = -12}$ fit the conditions of the
                                         problem.

$$x = 5 \text{ ft width}$$
$$x + 7 = 12 \text{ ft length}$$

Check $5^2 + 12^2 \overset{?}{=} 13^2$

$ 25 + 144 \overset{?}{=} 169$

$ 169 = 169$

The width is 5 ft and the length is 12 ft.

EXAMPLE 2: Work

Working for a janitorial service, a woman and her daughter clean a building in 5 hours. If the daughter were to do the job by herself, she would take 24 hours longer than her mother would take. How long would it take her mother to clean the building without the daughter's help?

> **Solution**   This problem is similar to the work problems discussed in Section 7.4.

Let x = time for mother alone
         $x + 24$ = time for daughter alone

	Hours	Part in 1 hour
Mother	x	$\dfrac{1}{x}$
Daughter	$x + 24$	$\dfrac{1}{x + 24}$
Together	5	$\dfrac{1}{5}$

$$\underbrace{\begin{matrix}\text{part done} \\ \text{by mother} \\ \text{in 1 hour}\end{matrix}}_{\dfrac{1}{x}} + \underbrace{\begin{matrix}\text{part done by} \\ \text{daughter in} \\ \text{1 hour}\end{matrix}}_{\dfrac{1}{x + 24}} = \underbrace{\begin{matrix}\text{part done working} \\ \text{together in} \\ \text{1 hour}\end{matrix}}_{\dfrac{1}{5}}$$

$$\frac{1}{x}(5x)(x + 24) + \frac{1}{x + 24}(5x)(x + 24) = \frac{1}{5}(5x)(x + 24)$$

Multiply each term by the LCM of the denominators.

$$5(x + 24) + 5x = x(x + 24)$$
$$5x + 120 + 5x = x^2 + 24x$$
$$0 = x^2 + 24x - 10x - 120$$
$$0 = x^2 + 14x - 120$$
$$0 = (x - 6)(x + 20)$$

$$x - 6 = 0 \quad \text{or} \quad x + 20 = 0$$
$$x = 6 \qquad \qquad \cancel{x = -20}$$

The mother could do the job alone in 6 hours.

EXAMPLE 3: Distance, Rate, Time

An airplane travels at a speed of 200 mph in still air. The plane, flying with a tailwind, is clocked over a distance of 960 miles. The same plane, flying against a headwind, takes 2 hours more time to complete the return trip. What was the wind velocity?

Solution   The basic formula is $d = rt$ (distance = rate $\times$ time). Also,
$$t = \frac{d}{r} \text{ and } r = \frac{d}{t}.$$

If we know or can represent any two of the three quantities, the formula using these two should be used.

Let $x =$ wind velocity
   $200 + x =$ speed going with the wind
   $200 - x =$ returning against the wind
    $960 =$ distance each way

We know distance and can represent rate (or speed), so the formula $t = \dfrac{d}{r}$ is used for representing time.

	Rate	Time	Distance
Going	$200 + x$		960
Returning	$200 - x$		960

	Rate	Time	Distance
Going	$200 + x$	$\dfrac{960}{200 + x}$	960
Returning	$200 - x$	$\dfrac{960}{200 - x}$	960

$$\underbrace{\text{time}}_{\text{returning}} - \underbrace{\text{time}}_{\text{going}} = \underbrace{\text{difference}}_{\text{in time}}$$

$$\frac{960}{200 - x} - \frac{960}{200 + x} = 2$$

$$\frac{960}{200 - x}(200 - x)(200 + x) - \frac{960}{200 + x}(200 - x)(200 + x) = 2(200 - x)(200 + x)$$

$$960(200 + x) - 960(200 - x) = 2(40{,}000 - x^2)$$
$$192{,}000 + 960x - 192{,}000 + 960x = 80{,}000 - 2x^2$$
$$2x^2 + 1920x - 80{,}000 = 0$$
$$x^2 + 960x - 40{,}000 = 0$$
$$(x - 40)(x + 1000) = 0$$

$$x - 40 = 0 \quad \text{or} \quad x + 1000 = 0$$
$$x = 40 \qquad\qquad \cancel{x = -1000}$$

The wind velocity was 40 mph.

$$\textit{Check} \quad \frac{960}{200 - 40} - \frac{960}{200 + 40} \overset{?}{=} 2$$

$$\frac{960}{160} - \frac{960}{240} \overset{?}{=} 2$$

$$6 - 4 \overset{?}{=} 2$$

$$2 = 2$$

EXAMPLE 4: Geometry

A square piece of cardboard has a small square, 2 in. by 2 in., cut from each corner. The edges are then folded up to form a box with a volume of 5000 cu in. What are the dimensions of the box? (**Hint:** The volume is the product of the length, width, and height: $V = \ell w h$.)

Solution   Draw a diagram illustrating the information.

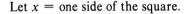
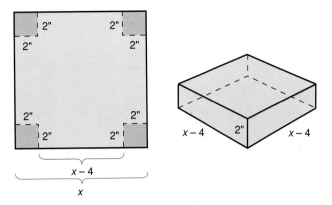

Let x = one side of the square.

$$2(x - 4)(x - 4) = 5000$$
$$2(x^2 - 8x + 16) = 5000$$
$$x^2 - 8x + 16 = 2500$$
$$x^2 - 8x - 2484 = 0$$
$$(x - 54)(x + 46) = 0$$

$$x - 54 = 0 \quad \text{or} \quad x + 46 = 0$$
$$x = 54 \qquad \qquad \cancel{x = -46}$$
$$x - 4 = 50$$

The dimensions of the box are 50 in. by 50 in. by 2 in.   ■

◢ EXERCISES 10.4

Determine a quadratic equation for each of the following problems. Then solve the equation.

1. The sum of a positive number and its square is 132. Find the number.

2. The dimensions of a rectangle can be represented by two consecutive even integers. The area of the rectangle is 528 square centimeters. Find the width and the length of the rectangle. (**Hint:** Area = length · width.)

3. A rectangle has a length 5 meters less than twice its width. If the area is 63 square meters, find the dimensions of the rectangle.
(**Hint:** Area = length · width.)

4. The sum of a positive number and its square is 992. Find the number.

5. The area of a rectangular field is 198 square meters. If it takes 58 meters of fencing to enclose the field, what are the dimensions of the field? (**Hint:** The length plus the width is 29 meters.)

6. The length of a rectangle exceeds the width by 5 cm. If both were increased by 3 cm, the area would be increased by 96 sq cm. Find the dimensions of the original rectangle. [**Hint:** The original rectangle has area $w(w + 5)$.]

7. The Willsons have a rectangular swimming pool that is 10 ft longer than it is wide. The pool is completely surrounded by a concrete deck that is 6 ft wide. The total area of the pool and the deck is 1344 sq ft. Find the dimensions of the pool.

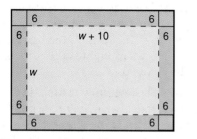

8. The product of two positive consecutive odd integers exceeds their sum by 287. Find the numbers.

9. The sum of the squares of two positive consecutive integers is 221. Find the positive integers.

10. The difference between two positive numbers is 9. If the small number is added to the square of the larger number, the result is 147. Find the numbers.

11. The difference between a positive number and 3 is four times the reciprocal of the number. Find the number. $\left(\textbf{Hint:}\text{ The reciprocal of } x \text{ is } \dfrac{1}{x}.\right)$

12. The sum of a positive number and 5 is fourteen times the reciprocal of the number. Find the number. $\left(\textbf{Hint:}\text{ The reciprocal of } x \text{ is } \dfrac{1}{x}.\right)$

13. Each side of a square is increased by 10 cm. The area of the resulting square is 9 times the area of the original square. Find the length of the sides of the original square.

14. If 5 meters are added to each side of a square, the area of the resulting square is four times the area of the original square. Find the length of the sides of the original square.

15. The diagonal of a rectangle is 13 meters. The length is 2 meters more than twice the width. Find the dimensions of the rectangle.

16. The length of a rectangle is 4 meters more than its width. If the diagonal is 20 meters, what are the dimensions of the rectangle?

17. A right triangle has two equal sides. If the hypotenuse is 12 cm, what is the length of the equal sides?

18. Mr. Prince owns a 15-unit apartment complex. The rent for each apartment is presently $200 per month and all units are rented. Each time the rent is increased $20, he will lose 1 tenant. What is the rental rate if he receives $3120 monthly in rent? (**Hint:** Let x = number of empty units.)

19. The Ski Club is planning to charter a bus to a ski resort. The cost will be $900, and each member will share the cost equally. If the club had 15 more members, the cost per person would be $10 less. How many are in the club now? $\left(\textbf{Hint:}\text{ If } x = \text{ number in club now,}\right.$

$\left.\dfrac{900}{x} = \text{ cost per person.}\right)$

20. A sporting goods store owner estimates that if he sells a certain model of basketball shoes for x dollars a pair, he will be able to sell $40 - x$ pairs. Find the price if his sales were $375. Is there more than one possible answer?

21. Mr. Green traveled to a city 200 miles from his home to attend a meeting. Due to car trouble, his average speed returning was 10 mph less than his speed going. If the total driving time for the round trip was 9 hours, at what rate of speed did he travel to the city?

	Rate	Time	Distance
Going	x		200
Returning	$x - 10$		200

22. A motorboat takes a total of 2 hours to travel 8 miles downstream and 4 miles back on a river that is flowing at a rate of 2 mph. Find the rate of the boat in still water.

23. A small motorboat travels 12 mph in still water. It takes 2 hours longer to travel 45 miles going upstream than it does going downstream. Find the rate of the current. (**Hint:** $12 + c$ = rate going downstream and $12 - c$ = rate going upstream.)

24. Recently Mr. and Mrs. Roberts spent their vacation in San Francisco, which is 540 miles from their home. Being a little reluctant to return home, the Roberts took 2 hours longer on their return trip and their average speed was 9 mi/hr slower than when they were going. What was their average rate of speed as they traveled from home to San Francisco?

25. The Blumin Garden Club planned to give their president a gift of appreciation costing $120 and to divide the cost evenly. In the meantime, 5 members dropped out of the club. If it now costs each of the remaining members $2 more than originally planned, how many members initially participated in the gift buying?

$$\left(\textbf{Hint: } \text{If } x = \text{number in club initially,} \right.$$

$$\left. \frac{120}{x} = \text{cost per member.} \right)$$

26. A rectangular sheet of metal is 6 in. longer than it is wide. A box is to be made by cutting out 3-in. squares at each corner and folding up the sides. If the box has a volume of 336 cu in., what were the dimensions of the sheet metal? (See Example 4.)

27. A box is to be made out of a square piece of cardboard by cutting out 2-in. squares at each corner and folding up the sides. If the box has a volume of 162 cu in., how big was the piece of cardboard? (See Example 4.)

28. A man and his son can paint their cabin in 3 hours. Working alone, it would take the son 8 hours longer than it would the father. How long would it take the father to paint the cabin alone?

29. Two pipes can fill a tank in 8 minutes if both are turned on. If only one is used, it would take 30 minutes longer for the smaller pipe to fill the tank than the larger pipe. How long will it take this small pipe to fill the tank?

30. A farmer and his son can plow a field with two tractors in 4 hours. If it would take the son 6 hours longer than the father to plow the field alone, how long would it take each if they worked alone?

CHAPTER 10 SUMMARY

Key Terms and Formulas

The **general quadratic equation** is

$$ax^2 + bx + c = 0, \quad a \neq 0 \quad [10.3]$$

The **quadratic formula** is

$$x = \frac{-b \pm \sqrt{b^2 - 4ac}}{2a} \quad [10.3]$$

The expression $b^2 - 4ac$ is called the **discriminant.** [10.3]

Properties and Rules

For an equation in the form

$$x^2 = c \quad \text{where } c \text{ is positive}$$

then $x = \sqrt{c}$ or $x = -\sqrt{c}$. We can write $x = \pm\sqrt{c}$. [10.1]

The Pythagorean Theorem [10.1]
In a right triangle, the square of the hypotenuse is equal to the sum of the squares of the two legs.

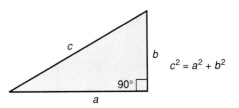

Procedures

To Solve Quadratic Equations by Completing the Square [10.2]

1. Arrange terms with variables on one side and constants on the other.
2. Divide each term by the coefficient of x^2. (We want the coefficient of x^2 to be 1.)

3. Find the number that completes the square of the quadratic expression and add this number to **both** sides of the equation.
4. Find the positive and negative square roots of both sides.
5. Solve for x. Remember, there will usually be two solutions.

CHAPTER 10   REVIEW

Solve the quadratic equation in Exercises 1–8.   [10.1]

1. $x^2 = 49$

2. $x^2 = 25$

3. $x^2 - \dfrac{1}{4} = 0$

4. $9x^2 - 162 = 0$

5. $(x + 2)^2 = 4$

6. $(x - 1)^2 = 9$

7. $(x - 5)^2 = 7$

8. $3(x - 4)^2 = 24$

In Exercises 9 and 10, use the Pythagorean Theorem to decide whether each of the triangles determined by the sides a, b, and c is a right triangle.   [10.1]

9. $a = 4$ cm, $b = 7$ cm, $c = \sqrt{65}$ cm

10. $a = 2\sqrt{3}$ in., $b = 2$ in., $c = 4$ in.

In Exercises 11 and 12, solve for the missing side. (See figure below.)   [10.1]

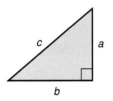

11. $a = 7$ ft, $b = 6\sqrt{2}$ ft, $c = ?$

12. $a = 8$ cm, $b = 10$ cm, $c = ?$

Add the correct constant in Exercises 13–18 to make the trinomial factorable as indicated.   [10.2]

13. $x^2 - 4x +$ _____ $= ($ $)^2$

14. $3x^2 + 18x +$ _____ $= 3($ $)^2$

15. $x^2 - \dfrac{2}{3}x +$ _____ $= ($ $)^2$

16. $x^2 - x +$ _____ $= ($ $)^2$

17. $x^2 + 9x +$ _____ $= ($ $)^2$

18. $5x^2 - 10x +$ _____ $= 5($ $)^2$

Solve the quadratic equations in Exercises 19–24 by completing the square.   [10.2]

19. $x^2 - 6x = 7$

20. $x^2 + 4x = -2$

21. $x^2 + 2x - 4 = 0$

22. $2x^2 + 8x - 1 = 0$

23. $2x^2 + 2x - 3 = 0$

24. $3x^2 - 2x - 2 = 0$

Solve the quadratic equations in Exercises 25–32 by factoring or completing the square.   [10.2]

25. $x^2 + 6x - 16 = 0$

26. $x^2 + 8x - 48 = 0$

27. $2x^2 + 7x - 4 = 0$

28. $5x^2 - 6x = 0$

29. $5x^2 - 10x = -3$

30. $4x^2 + 2x = 3$

31. $3x^2 + 6x - 5 = 0$

32. $8x^2 + 24x = 0$

Write each of the quadratic equations in Exercises 33–38 in the form $ax^2 + bx + c = 0$ where a, b, and c are integers. Identify the constants a, b, and c. Then solve using the quadratic formula. [10.3]

33. $2x^2 + 3x - 2 = 0$ **34.** $3x^2 + x - 4 = 0$ **35.** $2x^2 = 2x + 1$

36. $2x^2 + 5x = 6$ **37.** $5x^2 = 1 - x$ **38.** $\dfrac{3}{4}x^2 = 2x - \dfrac{1}{8}$

Solve the quadratic equations in Exercises 39–46 using any method. [10.3]

39. $2x^2 = 72$ **40.** $(x - 2)^2 = 17$ **41.** $x^2 - 6x + 4 = 0$ **42.** $5x^2 - 2x - 7 = 0$

43. $x^2 - 7x = 60$ **44.** $4x^2 - 1 = 3x + 3$ **45.** $3x^2 - 9x - 12 = 0$ **46.** $3x^2 - 4x - 2 = 0$

Determine an equation for each of the Exercises 47–54, then solve. [10.4]

47. The product of two consecutive positive even integers is 168. Find the integers.

48. The length of a rectangle is three centimeters greater than twice the width. If the area of the rectangle is 275 square centimeters, find the length and width.

49. The diagonal of a rectangle is 15 meters. The length of the rectangle is 6 meters less than twice the width. Find the length and width.

50. A square garden has a 3-foot walk surrounding it. If the walk is removed and the space is included in the garden, the area will then be four times the original area. Find the length of one side of the original square.

51. A boat travels 12 miles downstream and then returns. The current in the stream is moving at 2 mph. If the time for the total round trip is $4\dfrac{1}{2}$ hours, find the speed of the boat in still water.

52. A tank can be filled by two pipes in 4 hours. The larger pipe alone will fill the tank in 6 hours less time than the smaller one. Find the time required by each pipe to fill the tank alone.

53. A store manager ordered $300 worth of shirts. If each shirt had cost $3 more, he would have obtained 5 fewer shirts for the same amount of money. How much did each shirt cost and how many did he purchase?

54. A rectangular sheet of metal is three times as long as it is wide. A box is to be made by cutting out 2-inch squares at each corner and folding up the sides. If the volume of the box is 312 cubic inches, what are the dimensions of the sheet of metal?

CHAPTER 10 TEST

Solve the equations in Exercises 1 and 2.

1. $(x - 5)^2 = 49$ **2.** $8x^2 = 96$

3. Find the missing side in the right triangle shown here.

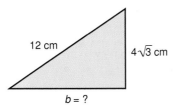

12 cm $4\sqrt{3}$ cm $b = ?$

Add the correct constant in Exercises 4–6 in order to make the trinomial factorable as indicated.

4. $x^2 - 24x + \underline{\hspace{1cm}} = (\hspace{1cm})^2$ **5.** $x^2 + 9x + \underline{\hspace{1cm}} = (\hspace{1cm})^2$

6. $3x^2 + 9x + \underline{\hspace{1cm}} = 3(\hspace{1cm})^2$

Solve the quadratic equations in Exercises 7–10 by factoring.

7. $x^2 - 3x + 2 = 0$ **8.** $5x^2 + 12x = 0$ **9.** $3x^2 - x - 10 = 0$ **10.** $2x^2 - 11x + 12 = 0$

Solve the quadratic equations in Exercises 11–14 by completing the square.

11. $x^2 + 6x + 8 = 0$ **12.** $x^2 - 2x - 5 = 0$ **13.** $2x^2 + 3x - 3 = 0$ **14.** $3x^2 + x - 1 = 0$

Solve the quadratic equations in Exercises 15–18 using the quadratic formula.

15. $3x^2 + 8x + 2 = 0$ **16.** $2x^2 = 4x + 3$ **17.** $\frac{1}{2}x^2 - x = 2$ **18.** $\frac{1}{3}x^2 + 2x = 1$

Solve the quadratic equations in Exercises 19–22 using any method.

19. $x^2 + 6x = 2$ **20.** $3x^2 - 7x + 2 = 0$ **21.** $2x^2 - 3x = 1$ **22.** $5x^2 - 2x - 3 = 0$

Determine an equation for each of the Exercises 23–25, then solve.

23. The diagonal of a square is 36 inches. Find the length of each side.

24. A small boat travels 8 mph in still water. It takes 10 hours longer to travel 60 miles upstream than it takes to travel 60 miles downstream. Find the rate of the current.

25. A rectangular sheet of metal is 3 inches longer than it is wide. A box is to be made by cutting out 4-inch squares at each corner and folding up the sides. If the box has a volume of 720 cubic inches, what are the dimensions of the sheet of metal?

◢◢ CUMULATIVE REVIEW (10)

Simplify each expression in Exercises 1 and 2.

1. $(-3x^2y)^2$

2. $\dfrac{(2x^2)(-6x^3)}{3x^{-1}}$

Solve the inequalities in Exercises 3 and 4 and graph the solution.

3. $2(x - 5) - 4 \le x - 3(x - 1)$

4. $x - (2x + 5) \ge 9 - (4 - x)$

Perform the indicated operations in Exercises 5 and 6.

5. $\dfrac{x^2 - x - 20}{2x^2 - x - 1} \cdot \dfrac{2x + 1}{x^2 - 25}$

6. $\dfrac{x}{x + 3} + \dfrac{5}{x^2 + 6x + 9}$

Graph the linear equations in Exercises 7 and 8.

7. $y = -\dfrac{5}{2}x + 3$

8. $4x + 3y = 6$

Simplify the radical expressions in Exercises 9 and 10.

9. $\sqrt{147} + \sqrt{72} - \sqrt{48}$

10. $(2\sqrt{3} + 1)(\sqrt{3} - 4)$

Solve the equations in Exercises 11–15.

11. $5(2x - 3) + 6 = 2(3x + 4) + 1$

12. $\dfrac{4}{x} + \dfrac{6}{x - 3} = 1$

13. $x - 1 = \sqrt{2 - 2x}$

14. $6x^2 - 19x + 15 = 0$

15. $2x^2 = 5x - 2$

16. Solve the system of equations.
$$\begin{cases} 2x - 5y = 11 \\ 5x - 3y = -1 \end{cases}$$

17. The length of a rectangle is five more than twice the width. The area is 52 sq cm. Find the width and length.

18. At a small college, the ratio of women to men is 9:7. If there is a total enrollment of 6400 students, how many women and how many men are enrolled?

19. Louise has $2700 invested. Part of the money is invested at 5.5% and the remainder is invested at 7.2%. Her total annual interest is $168.90. How much is invested at each rate?

20. Glen traveled 6 miles downstream and returned. His speed downstream was 5 mph faster than his speed upstream. If the total trip took 1 hour, find his speed in each direction.

11 ADDITIONAL TOPICS

DID YOU KNOW ?

Most people recognize that mathematics has many applications in science and the social sciences. But what about the arts? Surely mathematics has had no impact on the fine arts—or has it?

As previously mentioned, the Pythagoreans investigated the relationship between rational numbers and harmony. For a long time after the Greek period, music was considered to be applied mathematics. The basic law of musical harmony is a mathematical equation stating that a musical tone is composed of a fundamental note and overtones whose frequencies are all integral multiples of the fundamental tone. Modern investigations into the physics of sound use mathematics heavily.

In painting, when styles demanded more realistic presentations of subjects, the science of projective geometry was developed by artist-mathematicians to add perspective to their paintings. Albrecht Dürer, the German artist, traveled to Italy in 1506 to learn the secret art of perspective which was being used by Italian painters in the early Renaissance period. In modern art, the Dutch artist Mauritis Escher has done a great deal of work investigating symmetry patterns in art. Symmetry patterns have algebraic interpretations and have also been investigated by mathematicians. Surrealism may have mathematical qualities because of the time-space transformations that distort size, shape, volume, and time. Optical art and computer-generated art also have a mathematical basis.

Stories, novels, and poems have been written using mathematical ideas for inspiration. William Wordsworth (1770–1850) wrote a poem entitled "Geometry." Science fiction writers often use mathematical ideas such as five-dimensional worlds, computers gone wild, probability, and geometry as themes for their short stories or novels. Arthur C. Clark, the author of *2001: A Space Odyssey,* has written several short stories involving computer technology. You will find that mathematics is a discipline with applications in all fields, arts as well as science.

A scientist worthy of the name, above all a mathematician, experiences in his work the same impression as an artist, his pleasure is as great and of the same nature.

Jules Henri Poincaré (1854–1912)

 **CHAPTER OUTLINE**

*I*n this final chapter of the text, you will find several topics somewhat unrelated. Each of these topics is presented in the next course in algebra, and the intention here is, if time permits, to give you a running start at your next course. We hope you have had a successful experience and are looking forward to expanding your knowledge of the marvelous science of mathematics.

11.1 Fractional Exponents

OBJECTIVES

In this section, you will be learning to:

1. Write radical expressions with fractional exponents in radical form.

2. Simplify expressions with fractional exponents.

In Chapter 9, we discussed radicals involving square roots and cube roots. For example,

$$\sqrt{36} = 6 \text{ because } 6^2 = 36$$

and

$$\sqrt[3]{125} = 5 \text{ because } 5^3 = 125.$$

The 3 in $\sqrt[3]{125}$ is called the **index.** In $\sqrt{36}$, the index is understood to be 2. That is, we could write $\sqrt[2]{36}$ for the square root. Examples of other roots are

$$\sqrt[4]{81} = 3 \text{ because } 3^4 = 81,$$
$$\sqrt[5]{-32} = -2 \text{ because } (-2)^5 = -32,$$

and

$$\sqrt[5]{\frac{1}{32}} = \frac{1}{2} \text{ because } \left(\frac{1}{2}\right)^5 = \frac{1}{32}.$$

The symbol $\sqrt[n]{b}$ is read "the nth root of b." **If $a = \sqrt[n]{b}$, then $a^n = b$.** We need to know the conditions on a and b that will guarantee that $\sqrt[n]{b}$ is a real number.

As we discussed in Chapter 9, $\sqrt{-4}$ is not classified as a real number because the square of a real number is positive or zero. Such numbers are called **complex numbers** and will be discussed in the next course in algebra. In the general case,

$$\sqrt[n]{b} \text{ is not a real number if } n \text{ is even and } b \text{ is negative.}$$

In other words, even roots of negative numbers are **not** real numbers.

EXAMPLES

1. $\sqrt[3]{-27} = -3$ because $(-3)^3 = -27$

2. $\sqrt[4]{0.0016} = 0.2$ because $(0.2)^4 = 0.0016$

3. $\sqrt[5]{0.00001} = 0.1$ because $(0.1)^5 = 0.00001$

4. $\sqrt[6]{-64}$ is not a real number.   ∎

The following definition shows how radicals can be expressed using fractional exponents. We will find that algebraic operations with radicals are generally easier to perform when fractional exponents are used.

$\sqrt[n]{b}$ | If n is a positive integer and $\sqrt[n]{b}$ is a real number, then
$$\sqrt[n]{b} = b^{1/n}.$$

Thus, by the definition,

$$\sqrt{25} = (25)^{1/2} = 5$$

and $$\sqrt[3]{8} = (8)^{1/3} = 2.$$

EXAMPLES Simplify the following expressions.

5. $(49)^{1/2}$

   Solution $(49)^{1/2} = \sqrt{49} = 7$

6. $(-216)^{1/3}$

   Solution $(-216)^{1/3} = \sqrt[3]{-216} = -6$

7. $25^{1/2}$

   Solution $25^{1/2} = \sqrt{25} = 5$

8. $\left(\dfrac{16}{81}\right)^{1/4}$

   Solution $\left(\dfrac{16}{81}\right)^{1/4} = \sqrt[4]{\dfrac{16}{81}} = \dfrac{2}{3}$ ∎

Now consider the problem of evaluating the expression $8^{2/3}$. We need the following definition for using rational exponents.

$(\sqrt[n]{b})^m$ | If n is a positive integer and m is any integer and $\sqrt[n]{b}$ is a real number, then
$$(\sqrt[n]{b})^m = (b^{1/n})^m = b^{m/n}.$$

Using this definition, we have

$$8^{2/3} = (8^{1/3})^2 = (2)^2 = 4$$

and $$32^{3/5} = (32^{1/5})^3 = (2)^3 = 8.$$

Now that we have $b^{m/n}$ defined for rational exponents, we can consider simplifying expressions such as

$$x^{2/3} \cdot x^{1/2} \quad \text{and} \quad 36^{-1/2} \quad \text{and} \quad 2^{3/4} \cdot 2^{1/4}.$$

All the previous properties of exponents apply to rational exponents.

EXAMPLES   Using the properties of exponents, simplify each expression. Assume that all variables are positive.

9. $x^{2/3} \cdot x^{1/6}$

Solution $x^{2/3} \cdot x^{1/6} = x^{2/3 + 1/6} = x^{4/6 + 1/6} = x^{5/6}$ Add the exponents.

10. $\dfrac{a^{3/4}}{a^{1/2}}$

Solution $\dfrac{a^{3/4}}{a^{1/2}} = a^{3/4 - 1/2} = a^{3/4 - 2/4} = a^{1/4}$ Subtract the exponents.

11. $36^{-1/2}$

Solution $36^{-1/2} = \dfrac{1}{36^{1/2}} = \dfrac{1}{6}$

12. $(2y^{1/4})^3 \cdot (3y^{1/8})^2$

Solution $(2y^{1/4})^3 \cdot (3y^{1/8})^2 = 2^3 y^{3/4} \cdot 3^2 y^{1/4}$ Multiply the exponents.
$= 8 \cdot 9 y^{3/4 + 1/4}$ Simplify.
$= 72y$ Add the exponents.

13. $2^{3/4} \cdot 2^{1/4}$

Solution $2^{3/4} \cdot 2^{1/4} = 2^{3/4 + 1/4}$ Keep the base 2.
$= 2^1$ Add the exponents.
$= 2$ ∎

◢ EXERCISES 11.1

Write each of the expressions in Exercises 1–10 as radical expressions, then simplify.

1. $16^{1/2}$ 2. $49^{1/2}$ 3. $-8^{1/3}$ 4. $216^{1/3}$ 5. $81^{1/4}$
6. $16^{1/4}$ 7. $(-32)^{1/5}$ 8. $243^{1/5}$ 9. $0.0004^{1/2}$ 10. $0.09^{1/2}$

Simplify the expressions in Exercises 10–30.

11. $\left(\dfrac{4}{9}\right)^{1/2}$ 12. $\left(\dfrac{1}{27}\right)^{1/3}$ 13. $\left(\dfrac{8}{125}\right)^{1/3}$ 14. $\left(\dfrac{16}{9}\right)^{1/2}$ 15. $36^{-1/2}$

16. $64^{-1/3}$ 17. $8^{-1/3}$ 18. $100^{-1/2}$ 19. $81^{-1/4}$ 20. $16^{-1/4}$

21. $4^{3/2}$ 22. $64^{2/3}$ 23. $(-27)^{2/3}$ 24. $36^{3/2}$ 25. $9^{5/2}$

26. $8^{5/3}$ 27. $16^{-3/4}$ 28. $25^{-3/2}$ 29. $(-8)^{-2/3}$ 30. $(-125)^{-2/3}$

Using the properties of exponents, simplify each expression in Exercises 31–60. Assume that all variables are positive.

31. $x^{1/3} \cdot x^{1/3}$ **32.** $x^{1/4} \cdot x^{1/4}$ **33.** $5^{1/2} \cdot 5^{-3/2}$ **34.** $2^{-2/3} \cdot 2^{1/3}$ **35.** $x^{1/5} \cdot x^{2/5}$

36. $x^{1/8} \cdot x^{5/8}$ **37.** $\dfrac{x^{3/4}}{x^{1/4}}$ **38.** $\dfrac{x^{5/7}}{x^{3/7}}$ **39.** $\dfrac{x^{4/5}}{x^{2/5}}$ **40.** $\dfrac{x^{2/3}}{x^{2/3}}$

41. $\dfrac{2^{2/3}}{2^{-1/3}}$ **42.** $\dfrac{a^{5/4}}{a^{-1/4}}$ **43.** $a^{2/3} \cdot a^{1/2}$ **44.** $y^{3/4} \cdot y^{1/2}$ **45.** $x^{3/4} \cdot x^{-1/8}$

46. $y^{1/2} \cdot y^{-5/6}$ **47.** $\dfrac{x^{2/5}}{x^{1/2}}$ **48.** $\dfrac{x^{1/2}}{x^{2/3}}$ **49.** $\dfrac{6^2}{6^{1/2}}$ **50.** $\dfrac{10^3}{10^{4/3}}$

51. $\dfrac{x^{3/4}}{x^{-1/2}}$ **52.** $\dfrac{a^{2/5}}{a^{-1/10}}$ **53.** $\dfrac{x^{5/3}}{x^{-1/3}}$ **54.** $\dfrac{y^{1/2}}{y^{-2/3}}$

55. $(7x^{1/3})^2 \cdot (4x^{1/2})$ **56.** $(2x^{1/2})^3 \cdot (3x^{1/3})^2$ **57.** $(9x^2)^{1/2} \cdot (8x^3)^{1/3}$ **58.** $(x^{2/3})^3 \cdot (25x^4)^{1/2}$

59. $(16x^8)^{1/4} \cdot (9x^2)^{-1/2}$ **60.** $(2x^{3/4})^2 \cdot (3x^{3/2})^{-2}$

11.2 Distance Between Two Points and Midpoint Formula

OBJECTIVES

In this section, you will be learning to:

1. Find the distance between two points.
2. Determine if triangles are right triangles given the coordinates of the vertices by using the distance formula and the Pythagorean Theorem.
3. Show that specified geometric relationships are true by using the distance formula.
4. Find the perimeters of triangles given the coordinates of the vertices by using the distance formula.
5. Find the coordinates of the midpoints of line segments.

Distance Between Two Points

A formula particularly useful in fields of study where geometry is involved, such as surveying, machining, and engineering, is that for finding the distance between two points. The formula is related to the Cartesian coordinate system discussed in Chapter 8.

We first discuss the distance between two points that are on a horizontal line or on a vertical line as indicated by the following two questions.

What is the distance between the two points $P_1(2,3)$ and $P_2(6,3)$?
What is the distance between the points $P_3(-1,-4)$ and $P_4(-1,1)$?

See Figure 11.1.

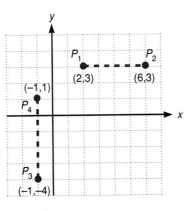

Figure 11.1

Did you find

$$\text{distance } (P_1 \text{ to } P_2) = 6 - 2 = 4?$$

This is correct since P_1 and P_2 lie on a horizontal line and have the same y-coordinate. Did you find

$$\text{distance } (P_3 \text{ to } P_4) = 1 - (-4) = 5?$$

This is correct since P_3 and P_4 lie on a vertical line and have the same x-coordinate.

For the distance from P_1 to P_2, why not take

$$\text{distance } (P_1 \text{ to } P_2) = 2 - 6 = -4?$$

The reason is that the term **distance** is taken to mean a nonnegative number (positive or 0). How can a nonnegative number be guaranteed after we subtract the coordinates? The answer is to take the absolute value of the difference of the coordinates.

The distance between two points, as indicated in the following formulas, is represented by d.

For $P_1(x_1, y_1)$ and $P_2(x_2, y_1)$ on a horizontal line,

$$d = |x_2 - x_1| \quad (\text{or } d = |x_1 - x_2|)$$

For $P_1(x_1, y_1)$ and $P_2(x_1, y_2)$ on a vertical line,

$$d = |y_2 - y_1| \quad (\text{or } d = |y_1 - y_2|)$$

EXAMPLES

1. Find the distance, d, between the two points $(5,7)$ and $(-3,7)$.

   Solution   Since the points are on a horizontal line (they have the same y-coordinate),

   $$d = |-3 - 5| = |-8| = 8$$

   or $\quad\quad d = |5 - (-3)| = |8| = 8$

2. Find the distance, d, between the two points $(2,8)$ and $\left(2, -\frac{1}{2}\right)$.

   Solution   Since the points are on a vertical line (they have the same x-coordinate),

   $$d = \left| 8 - \left(-\frac{1}{2}\right)\right| = \left|8\frac{1}{2}\right| = 8\frac{1}{2}$$

   or $\quad\quad d = \left|-\frac{1}{2} - 8\right| = \left|-8\frac{1}{2}\right| = 8\frac{1}{2}$

■

What if the points do not lie on a vertical line or a horizontal line? For example, what is the distance between the two points $P_1(-1,2)$ and $P_2(5,4)$? We apply the Pythagorean Theorem and calculate the length of the hypotenuse as shown in Figure 11.2.

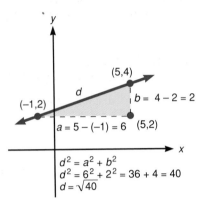

Figure 11.2

We could write directly $d = \sqrt{a^2 + b^2}$. Carrying this idea one step further, we can write a formula for d involving the coordinates of two general points $P_1(x_1,y_1)$ and $P_2(x_2,y_2)$.

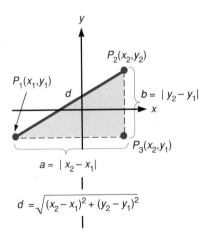

Formula for the Distance Between Two Points

The formula for the distance between two points $P_1(x_1,y_1)$ and $P_2(x_2,y_2)$ is

$$d = \sqrt{(x_2 - x_1)^2 + (y_2 - y_1)^2}$$

Do we need $|x_2 - x_1|$ and $|y_2 - y_1|$ in the formula? Evidently not, but why not? The answer is that even though $x_2 - x_1$ and $y_2 - y_1$ might be negative, both are being squared, so neither of the terms $(x_2 - x_1)^2$ or $(y_2 - y_1)^2$ will ever be negative. Also, the order of subtraction may be reversed because the squares will give the same value. **In the actual calculation of d, be sure to add the squares before taking the square root.**

EXAMPLES

3. Find the distance between the two points $(3,4)$ and $(-2,7)$.

Solution $\begin{aligned} d &= \sqrt{(3 - (-2))^2 + (4 - 7)^2} \\ &= \sqrt{5^2 + (-3)^2} \\ &= \sqrt{25 + 9} \\ &= \sqrt{34} \end{aligned}$

4. Find the distance between the two points $\left(\dfrac{1}{2}, \dfrac{2}{3}\right)$ and $\left(\dfrac{3}{4}, \dfrac{5}{3}\right)$.

Solution $\begin{aligned} d &= \sqrt{\left(\dfrac{1}{2} - \dfrac{3}{4}\right)^2 + \left(\dfrac{2}{3} - \dfrac{5}{3}\right)^2} \\ &= \sqrt{\left(-\dfrac{1}{4}\right)^2 + \left(-\dfrac{3}{3}\right)^2} \\ &= \sqrt{\dfrac{1}{16} + 1} \\ &= \sqrt{\dfrac{1}{16} + \dfrac{16}{16}} \\ &= \sqrt{\dfrac{17}{16}} = \dfrac{\sqrt{17}}{4} \end{aligned}$

5. Show that the triangle determined by the points $A(-2,1)$, $B(3,4)$, and $C(1,-4)$ is an isosceles triangle. An isosceles triangle has two equal sides.

Solution   The length of the line segment AB is the distance between the points A and B. We denote this distance by $|AB|$. Thus, to show that the triangle ABC is isosceles, we need to show that $|AB| = |AC|$ or that $|AB| = |BC|$ or $|AC| = |BC|$. If none of these relationships is true, then the triangle does not have two equal sides and is not isosceles.

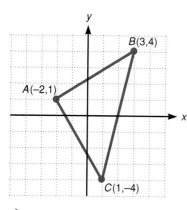

$$|AB| = \sqrt{(-2 - 3)^2 + (1 - 4)^2} = \sqrt{(-5)^2 + (-3)^2}$$
$$= \sqrt{25 + 9} = \sqrt{34}$$
$$|AC| = \sqrt{(-2 - 1)^2 + [1 - (-4)]^2} = \sqrt{(-3)^2 + (5)^2}$$
$$= \sqrt{9 + 25} = \sqrt{34}$$

Since $|AB| = |AC|$, the triangle is isosceles.   ∎

The Midpoint Formula

Another formula that involves the coordinates of points in a plane is that for finding the midpoint of the segment joining the two given points. This point is found by averaging the corresponding x- and y-coordinates.

Formula for the Midpoint Between Two Points	The formula for the midpoint between two points $P_1(x_1, y_1)$ and $P_2(x_2, y_2)$ is $$\left(\frac{x_1 + x_2}{2}, \frac{y_1 + y_2}{2}\right)$$

EXAMPLE

6. Find the coordinates of the midpoint of the line segment joining the two points $P_1(-1, 2)$ and $P_2(4, 6)$.

Solution   The midpoint is

$$\left(\frac{-1 + 4}{2}, \frac{2 + 6}{2}\right) = \left(\frac{3}{2}, 4\right)$$

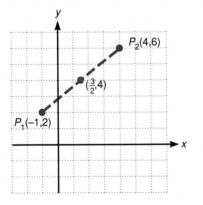

EXERCISES 11.2

For Exercises 1–24, (a) find the distance between the two given points and (b) find the coordinates of the midpoint of the line segment joining the two given points.

1. $(-3, 6), (-3, 2)$ **2.** $(5, 7), (-1, 7)$ **3.** $(4, -3), (7, -3)$ **4.** $(-1, 2), (5, 2)$

5. $(3, 1), \left(-\dfrac{1}{2}, 1\right)$ **6.** $\left(\dfrac{4}{3}, 7\right), \left(\dfrac{4}{3}, -\dfrac{2}{3}\right)$ **7.** $(3, 1), (2, 0)$ **8.** $(4, 6), (5, -2)$

9. $(1, 5), (-1, 2)$ **10.** $(0, 0), (-3, 4)$ **11.** $(2, -7), (-3, 5)$ **12.** $(5, -3), (7, -3)$

13. $\left(\dfrac{3}{7}, \dfrac{4}{7}\right), (0, 0)$ **14.** $(-5, 2), (1, 1)$ **15.** $(4, 1), (7, 5)$ **16.** $(10, 7), (1, 7)$

17. $(-10,3), (2,-2)$ **18.** $\left(\frac{7}{3},2\right), \left(-\frac{2}{3},1\right)$ **19.** $(-3,2), (3,-6)$ **20.** $\left(\frac{3}{4},6\right), \left(\frac{3}{4},-2\right)$

21. $(4,0), (0,-3)$ **22.** $(0,-2), (4,-3)$ **23.** $\left(\frac{4}{5},\frac{2}{7}\right), \left(-\frac{6}{5},\frac{2}{7}\right)$ **24.** $(6,8), (2,5)$

25. Use the distance formula and the Pythagorean Theorem to decide if the triangle determined by the points $A(1,-2)$, $B(7,1)$, and $C(5,5)$ is a right triangle.

26. Using the distance formula and the Pythagorean Theorem, decide if the triangle determined by the points $A(-5,-1)$, $B(2,1)$, and $C(-1,6)$ is a right triangle.

In Exercises 27 and 28, show that the triangle determined by the given points is an isosceles triangle (has two equal sides).

27. $A(1,1)$, $B(5,9)$, $C(9,5)$ **28.** $A(1,-4)$, $B(3,2)$, $C(9,4)$

In Exercises 29 and 30, show that the triangle determined by the given points is an equilateral triangle (all sides equal). [**Hint:** $(\sqrt{a})^2 = a$.]

29. $A(1,0)$, $B(3,\sqrt{12})$, $C(5,0)$ **30.** $A(0,5)$, $B(0,-3)$, $C(\sqrt{48},1)$

In Exercises 31 and 32, show that the diagonals of the rectangle $ABCD$ are equal.

31. $A(2,-2)$, $B(2,3)$, $C(8,3)$, $D(8,-2)$ **32.** $A(-1,1)$, $B(-1,4)$, $C(4,4)$, $D(4,1)$

In Exercises 33–35, find the perimeter of the triangle determined by the given points.

33. $A(-5,0)$, $B(3,4)$, $C(0,0)$ **34.** $A(-6,-1)$, $B(-3,3)$, $C(6,4)$ **35.** $A(-2,5)$, $B(3,1)$, $C(2,-2)$

36. Suppose that $\left(\frac{3}{4},-5\right)$ is the midpoint of the line segment AB where $A(2,y)$ and $B(x,-3)$. Find the values of x and y.

37. Suppose that $(2,7)$ is the midpoint of the line segment AB where $A(-1,1)$ and $B(x,y)$. Find the values of x and y.

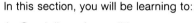 11.3 Linear Inequalities

OBJECTIVES	We have solved and graphed first-degree inequalities on a number line. (See Section 7.2.) As an example,

In this section, you will be learning to:

1. Graph linear inequalities.
2. Graph the points that satisfy given pairs of linear inequalities.

for $x < 5$
the graph is

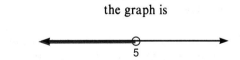

The points in the graph are on a line (one-dimensional) and lie to one side of the point represented by 5.

In Chapter 8, we discussed linear equations in two variables and their graphs. Now we are interested in graphing linear inequalities.

A **linear inequality** is an inequality of the form

$$Ax + By < C$$

($\leq$, $>$, $\geq$ can also be used.)

The graph of a linear inequality is based on the idea that a straight line separates a plane into two **half-planes.** The points on one side of the line are in one half-plane and the points on the other side are in the other half-plane. The line is called the **boundary** of each half-plane.

The graphs of linear inequalities are half-planes. If the boundary line is not included, the half-plane is said to be **open** [Figure 11.3(a)]. If the boundary line is included, the half-plane is said to be **closed** [Figure 11.3(b)].

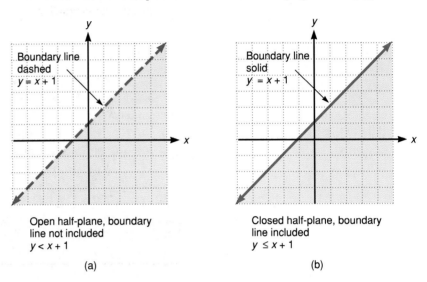

Open half-plane, boundary
line not included
$y < x + 1$

(a)

Closed half-plane, boundary
line included
$y \leq x + 1$

(b)

Figure 11.3

The boundary line is graphed dotted if the inequality is $<$ or $>$. It is graphed solid if the inequality is $\leq$ or $\geq$.

The following steps show how to graph the open half-plane represented by the inequality $y + 4x > 3$.

Step 1 Graph the line $y + 4x = 3$
as a dashed line.

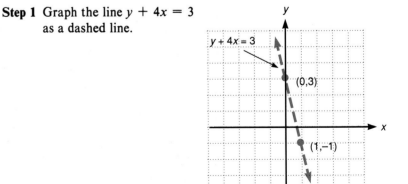

Step 2 Test one point not on the line in the inequality $y + 4x > 3$.

Test point: $(0,0)$

$0 + 4 \cdot 0 > 3$

$\qquad 0 > 3 \quad$ false

The test point does not satisfy the inequality.

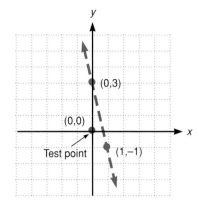

Step 3 Shade in the half-plane on the opposite side of the test point since the point did not satisfy the inequality.

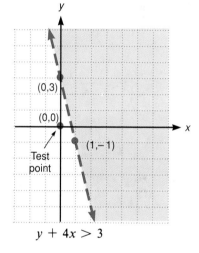

$$y + 4x > 3$$

Steps for Graphing a Linear Inequality	**Step 1** Graph the corresponding straight line (dashed for $<$ or $>$, solid for $\leq$ or $\geq$).
	Step 2 Choose any test point not on the line and test it in the inequality.
	Step 3 Shade the half-plane (a) on the same side as the test point if it satisfies the inequality or (b) on the opposite side of the test point if it does not satisfy the inequality.

EXAMPLES

1. Graph the points (open half-plane) that satisfy the linear inequality $2x + 3y < 6$.

   Solution

   Step 1 Graph the line
   $$2x + 3y = 6$$
   as a dashed line.

   Step 2 Test point (0,0):

   $$2 \cdot 0 + 3 \cdot 0 < 6$$
   $$0 < 6 \quad \text{true}$$

   Step 3 Shade the half-plane on the same side as (0,0).

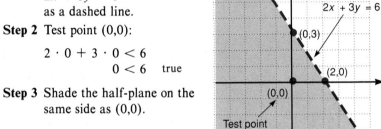

$$2x + 3y < 6$$

2. Graph the points (closed half-plane) that satisfy the linear inequality $5x - y \leq 4$.

   Solution The test point (3,0) gives

   $$5 \cdot 3 - 0 \leq 4$$
   $$15 \leq 4 \quad \text{false}$$

   The half-plane is on the opposite side of this point.

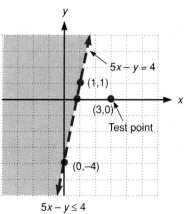

$$5x - y \leq 4$$

3. Graph the points that satisfy both inequalities $x \leq 2$ and $y \geq -x + 1$.

Solution We must find a region where the points satisfy **both** inequalities. The test point $(0,3)$ satisfies both inequalities:

$$0 \leq 2 \text{ and } 3 \geq -0 + 1$$

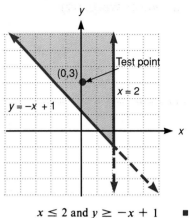

$$x \leq 2 \text{ and } y \geq -x + 1 \quad \blacksquare$$

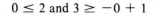

EXERCISES 11.3

Graph the half-planes that satisfy the linear inequalities in Exercises 1–30.

1. $y > 3x$	**2.** $x - y > 0$	**3.** $y \leq 2x - 1$	**4.** $y < 4 - x$
5. $y > -7$	**6.** $2y - x \geq 2$	**7.** $2x - 3 \leq 0$	**8.** $4x + y < 2$
9. $5x - 2y \geq 4$	**10.** $3y - 8 \leq 2$	**11.** $y < -\dfrac{x}{4}$	**12.** $2y - 3x \leq 4$
13. $2x - y < 1$	**14.** $4x + 3y \leq 6$	**15.** $2x - 3y \geq -3$	**16.** $2y + 5x > 0$
17. $2x + 5y \geq 10$	**18.** $4x + 2y \leq 10$	**19.** $3x < 4 - y$	**20.** $x < 2y + 3$
21. $2x - 3y < 9$	**22.** $x \leq 4 - 3y$	**23.** $5x - 2y \geq 4$	**24.** $3x + 6 \geq 2y$
25. $4y \geq 3x - 8$	**26.** $\dfrac{1}{2}x + y < 3$	**27.** $2x + \dfrac{1}{3}y < 2$	**28.** $2x - \dfrac{1}{2}y > -1$

29. $\dfrac{1}{2}x + \dfrac{2}{3}y \geq \dfrac{5}{6}$ **30.** $\dfrac{1}{4}x - \dfrac{1}{2}y \leq \dfrac{3}{4}$

Graph the points that satisfy both inequalities in Exercises 31–40.

31. $y < 2$ and $x \geq -3$ **32.** $2x + 5 < 0$ and $y \geq 2$ **33.** $x \geq -4$ and $x < 3$

34. $y \leq 5$ and $y > -2$ **35.** $x \leq 3$ and $2x + y > 7$ **36.** $2x - y > 4$ and $y < -1$

37. $x - 3y \leq 3$ and $x < 5$ **38.** $3x - 2y \geq 8$ and $y \geq 0$ **39.** $x - y \geq 0$ and $3x - 2y \geq 4$

40. $y \geq -x - 2$ and $x + y \geq -2$

11.4 Functions and Function Notation

In Chapter 8, we discussed sets of ordered pairs and their graphs, particularly straight lines. In this section, we will discuss the use of ordered pairs under the condition that they follow specific rules to be classified as **functions.** The concept of function forms the basis for applying mathematical concepts to real-life problems. The related function notation and terminology is used throughout the world of mathematics.

We begin with the definition of a relation. For the purposes of this text, we will discuss only relations that involve real numbers. However, the student should be aware that relations can be defined that involve pairs other than real numbers. As examples, parents and children can be paired, cities and states can be paired, governors and states can be paired, and so on.

Relation	A **relation** is a set of ordered pairs of real numbers. A **domain** of a relation is the set of all first components in the relation. The **range** of a relation is the set of all second components in the relation.

EXAMPLES Find the domain and range of each relation.

1. $r = \{(-2,3), (0,0), (1,2), (1,3), (\sqrt{2},\sqrt{3})\}$

  Solution Domain $= \{-2, 0, 1, \sqrt{2}\}$ All first components (**Note:** Even though 1 appears as a first component more than once, it is listed only once in the domain.)

  Range $= \{3, 0, 2, \sqrt{3}\}$ All second components

2. $f = \{(-1,1), (1,5), (2,3), (\sqrt{5}, 1)\}$

  Solution Domain $= \{-1, 1, 2, \sqrt{5}\}$

  Range $= \{1, 5, 3\}$ ■

Functions

The relation in Example 2 has the special property that each first component appears only once. This means that the relation f is called a **function.** In the relation r of Example 1, the first component 1 appears more than once and has two different corresponding second components, namely 2 and 3. The relation r is **not** a function.

The following three definitions of a function are all equivalent.

Function	A **function** is a relation in which each domain element has only one corresponding range element. OR A **function** is a relation in which each first component appears only once. OR A **function** is a relation in which no two ordered pairs have the same first component.

EXAMPLES   Find the domain and range of each relation and determine whether or not it is a function.

3. $s = \{(0,0), (1,1), (2,4), (3,0)\}$

Solution   Domain $= \{0, 1, 2, 3\}$

Range $= \{0, 1, 4\}$

s is a function. Each first component appears only once. The fact that 0 appears twice as a **second** component has no relationship to the function concept.

4. $t = \{(1,5), (3,5), (\sqrt{2}, 5), (-1,5)\}$

Solution   Domain $= \{1, 3, \sqrt{2}, -1\}$

Range $= \{5\}$

t is a function. Each first component has only one corresponding second component. The fact that all the second components are the same has no effect on the definition of a function.   ∎

Vertical Line Test

If one point in the graph of a relation is directly above or below another point in the graph, then these points have the same first coordinate (or x-coordinate). Such a relation would **not** be a function. Therefore, we can tell whether or not a graph represents a function by using the **vertical line test.** If **any** vertical line intersects a graph of a relation in more than one point, then the relation graphed is **not** a function (Figure 11.4).

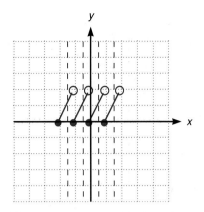

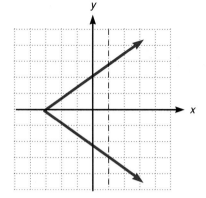

This graph represents a function. Several vertical lines are drawn to illustrate that vertical lines do not intersect the graph in more than one point. Each *x*-value has only one corresponding *y*-value.

This graph is not a function because the vertical line drawn intersects the graph in more than one point. Thus for the *x*-value there is more than one corresponding *y*-value.

(a)

(b)

Figure 11.4

EXAMPLES   Use the vertical line test to determine whether or not each graph represents a function.

5.

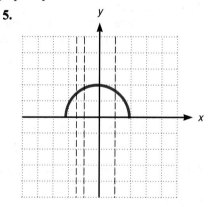

Solution   A function. No vertical line will intersect the graph in more than one point. Three vertical lines are drawn to illustrate this.

6.

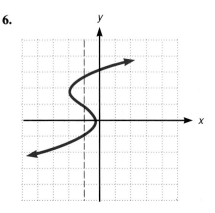

Solution   Not a function. Some vertical line will intersect the graph in more than one point.   ∎

Function Notation f(x)

Ordered pairs can be associated with equations using two variables. Thus, (x,y) is associated with the equations

$$y = 2x + 3, \qquad y = \sqrt{x - 1}, \qquad \text{and} \qquad y = x^2 + 2x + 3.$$

In each of these equations, y is said to "depend" on the value of x. We say that y **is a function of** x and y can be represented using a special notation called **function notation.**

In function notation, the symbol $f(x)$, read "f of x," is used to represent the y-value or the variable dependent on x. For example,

$$\text{if } y = 2x + 3$$

we can write $f(x) = 2x + 3$; then

$$
\begin{aligned}
f(1) &= 2 \cdot 1 + 3 = 5 & &f(1) \text{ means replace } x \text{ with } 1. \\
f(3) &= 2 \cdot 3 + 3 = 9 & &f(3) \text{ means replace } x \text{ with } 3. \\
f(-4) &= 2(-4) + 3 = -5 & &f(-4) \text{ means replace } x \text{ with } -4.
\end{aligned}
$$

(**Note:** $f(x)$ does **not** mean to multiply f by x. $f(x)$ is a notation unto itself.)

EXAMPLES

7. If $f(x) = -2x + 1$, find $f(-1)$, $f(0)$, and $f(3)$.

   Solution $f(-1) = -2(-1) + 1 = 2 + 1 = 3$
   $\qquad\qquad\quad f(0) = -2(0) + 1 = 0 + 1 = 1$
   $\qquad\qquad\quad f(3) = -2(3) + 1 = -6 + 1 = -5$

8. If $g(x) = x^2 - x + 1$, find $g(0)$, $g(3)$, and $g(-2)$. (Letters other than f can be used in function notation.)

   Solution $\quad g(0) = 0^2 - 0 + 1 = 1$
   $\qquad\qquad\quad g(3) = 3^2 - 3 + 1 = 9 - 3 + 1 = 7$
   $\qquad\quad g(-2) = (-2)^2 - (-2) + 1 = 4 + 2 + 1 = 7$

9. Suppose $h(x) = 5$. Find $h(-1)$, $h(6)$, and $h(\sqrt{2})$.

Solution $h(x) = 5$ is called a **constant function.** Regardless of the value for x, the corresponding functional value [y-value or $h(x)$ value] is 5.

$$h(-1) = 5$$
$$h(6) = 5$$
$$h(\sqrt{2}) = 5$$ ■

EXERCISES 11.4

Find the domain and range of each relation in Exercises 1–12. Then determine whether or not it is a function.

1. $\{(0,0), (3,1), (-2,4), (1,1), (2,0)\}$

2. $\{(1,5), (-1,3), (-2,1), (2,7), (0,5)\}$

3. $\{(0,2), (-1,1), (2,4), (3,5), (2,-1)\}$

4. $\{(0,0), (-1,-2), (1,2), (3,6), (2,2)\}$

5. $\{(0,1), (-1,3), (3,4), (2,-5), (1,2)\}$

6. $\{(5,3), (2,1), (0,-3), (2,4), (1,0)\}$

7. $\{(0,-1), (1,0), (-1,-2), (\sqrt{3},2), (-2,-3)\}$

8. $\{(-1,-3), (0,0), (1,3), (2,\sqrt{6}), (-2,-6)\}$

9. $\{(-3,4), (-1,4), (0,4), (\sqrt{2},4), (3,4)\}$

10. $\{(2,-3), (2,1), (2,0), (2,\sqrt{2}), (2,3)\}$

11. $\{(1,\sqrt{3}), (2,7), (4,-1), (0,0), (0,\sqrt{2})\}$

12. $\{(2,4), (\sqrt{3},4), (1,-4), (0,-4), (\sqrt{5},-4)\}$

Use the vertical line test to determine whether or not each of the graphs in Exercises 13–24 represents a function.

13.

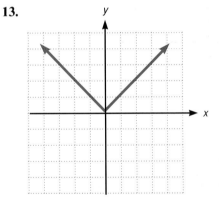

14.

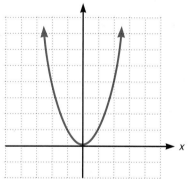

15.

16.

17.

18.

19.

20.

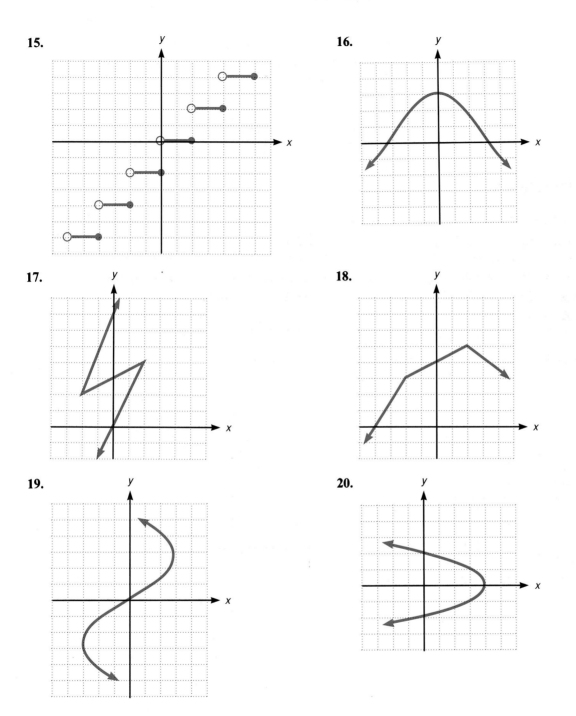

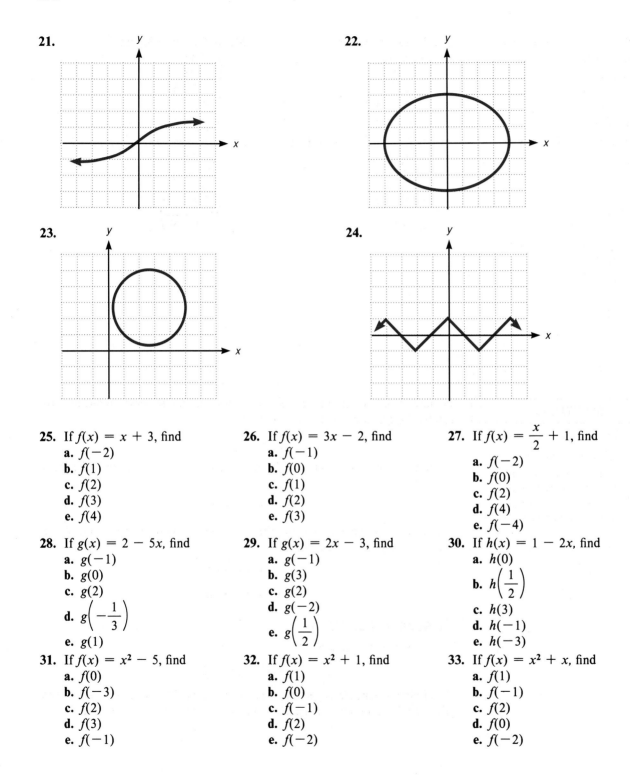

21.

22.

23.

24.

25. If $f(x) = x + 3$, find
  a. $f(-2)$
  b. $f(1)$
  c. $f(2)$
  d. $f(3)$
  e. $f(4)$

26. If $f(x) = 3x - 2$, find
  a. $f(-1)$
  b. $f(0)$
  c. $f(1)$
  d. $f(2)$
  e. $f(3)$

27. If $f(x) = \dfrac{x}{2} + 1$, find
  a. $f(-2)$
  b. $f(0)$
  c. $f(2)$
  d. $f(4)$
  e. $f(-4)$

28. If $g(x) = 2 - 5x$, find
  a. $g(-1)$
  b. $g(0)$
  c. $g(2)$
  d. $g\left(-\dfrac{1}{3}\right)$
  e. $g(1)$

29. If $g(x) = 2x - 3$, find
  a. $g(-1)$
  b. $g(3)$
  c. $g(2)$
  d. $g(-2)$
  e. $g\left(\dfrac{1}{2}\right)$

30. If $h(x) = 1 - 2x$, find
  a. $h(0)$
  b. $h\left(\dfrac{1}{2}\right)$
  c. $h(3)$
  d. $h(-1)$
  e. $h(-3)$

31. If $f(x) = x^2 - 5$, find
  a. $f(0)$
  b. $f(-3)$
  c. $f(2)$
  d. $f(3)$
  e. $f(-1)$

32. If $f(x) = x^2 + 1$, find
  a. $f(1)$
  b. $f(0)$
  c. $f(-1)$
  d. $f(2)$
  e. $f(-2)$

33. If $f(x) = x^2 + x$, find
  a. $f(1)$
  b. $f(-1)$
  c. $f(2)$
  d. $f(0)$
  e. $f(-2)$

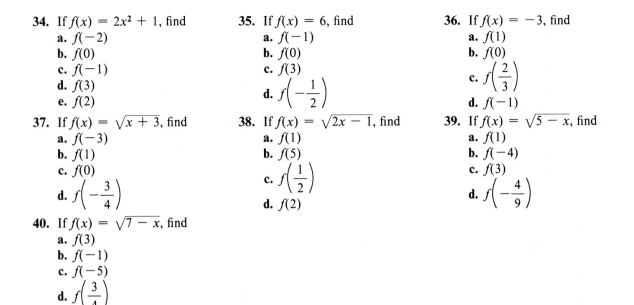

34. If $f(x) = 2x^2 + 1$, find
 a. $f(-2)$
 b. $f(0)$
 c. $f(-1)$
 d. $f(3)$
 e. $f(2)$

35. If $f(x) = 6$, find
 a. $f(-1)$
 b. $f(0)$
 c. $f(3)$
 d. $f\left(-\dfrac{1}{2}\right)$

36. If $f(x) = -3$, find
 a. $f(1)$
 b. $f(0)$
 c. $f\left(\dfrac{2}{3}\right)$
 d. $f(-1)$

37. If $f(x) = \sqrt{x + 3}$, find
 a. $f(-3)$
 b. $f(1)$
 c. $f(0)$
 d. $f\left(-\dfrac{3}{4}\right)$

38. If $f(x) = \sqrt{2x - 1}$, find
 a. $f(1)$
 b. $f(5)$
 c. $f\left(\dfrac{1}{2}\right)$
 d. $f(2)$

39. If $f(x) = \sqrt{5 - x}$, find
 a. $f(1)$
 b. $f(-4)$
 c. $f(3)$
 d. $f\left(-\dfrac{4}{9}\right)$

40. If $f(x) = \sqrt{7 - x}$, find
 a. $f(3)$
 b. $f(-1)$
 c. $f(-5)$
 d. $f\left(\dfrac{3}{4}\right)$

◣ 11.5 *Graphing Parabolas* [y = ax² + bx + c]

We have discussed linear equations and their graphs in some detail. For example, the graph of the linear equation

$$y = 2x + 5$$

has slope 2 and y-intercept (0,5). The equation can also be written in the function form

$$f(x) = 2x + 5.$$

In fact, any linear equation **not** in the form $x = a$ can be thought of as a **linear function**. In general, we can write

$$y = mx + b$$

as

$$f(x) = mx + b$$

Quadratic Functions

Now consider the function

$$y = x^2 - 2x - 3.$$

This function is **not** linear so its graph is **not** a straight line. The nature of the graph can be seen by plotting several points, as shown in Figure 11.5.

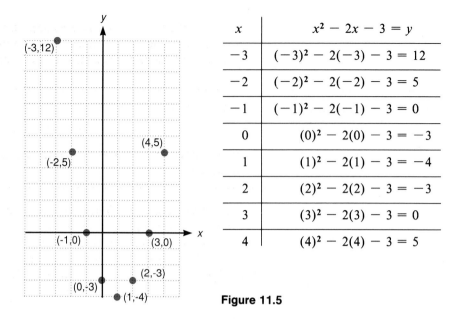

x	$x^2 - 2x - 3 = y$
-3	$(-3)^2 - 2(-3) - 3 = 12$
-2	$(-2)^2 - 2(-2) - 3 = 5$
-1	$(-1)^2 - 2(-1) - 3 = 0$
0	$(0)^2 - 2(0) - 3 = -3$
1	$(1)^2 - 2(1) - 3 = -4$
2	$(2)^2 - 2(2) - 3 = -3$
3	$(3)^2 - 2(3) - 3 = 0$
4	$(4)^2 - 2(4) - 3 = 5$

Figure 11.5

As we knew, the points do not lie on a straight line. The graph can be seen by drawing a smooth curve called a **parabola** through the points, as shown in Figure 11.6. The point $(1, -4)$ is the "turning point" of the curve and is called the **vertex** of the parabola. The line $x = 1$ is the **line of symmetry** or **axis of symmetry** for the parabola. The curve is a "mirror image" of itself on either side of the line $x = 1$.

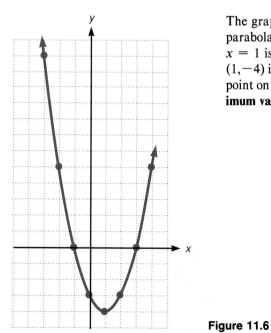

The graph of $y = x^2 - 2x - 3$ is a parabola.

$x = 1$ is the line of symmetry.

$(1, -4)$ is the vertex. This is the lowest point on the curve and -4 is the **minimum value** for y.

Figure 11.6

Parabolas (or parabolic arcs) occur frequently in describing information in real life. For example, the paths of projectiles (thrown balls, artillery shells, arrows) and revenue functions in business can be illustrated using parabolas.

Quadratic Function

Any function of the form

$$y = ax^2 + bx + c$$

where *a, b,* and *c* are real constants and *a* ≠ 0 is called a **quadratic function.**

The graph of every quadratic function is a parabola. The position of the parabola, its shape, and whether it "opens up" or "opens down" can be determined by investigating the function and its coefficients.

The following information about quadratic functions will be used here without proof. A thorough development is part of the next course in algebra.

For the quadratic function $y = ax^2 + bx + c$:

1. If $a > 0$, the parabola "opens upward."
2. If $a < 0$, the parabola "opens downward."
3. $x = -\dfrac{b}{2a}$ is the line of symmetry.
4. $\left(-\dfrac{b}{2a}, \dfrac{4ac - b^2}{4a}\right)$ is the vertex. This point is the lowest point on the curve if the parabola opens upward or is the highest point if the parabola opens downward.

Several graphs related to the basic form

$$y = ax^2 \quad \text{where } b = 0 \quad \text{and} \quad c = 0$$

are shown in Figure 11.7.

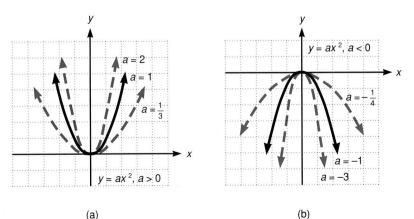

(a)   (b)

Figure 11.7

Since $b = 0$ and $c = 0$,

$$\left(-\frac{b}{2a}, \frac{4ac - b^2}{4a}\right) = (0,0)$$

is the vertex in each case and the axis of symmetry is the y-axis ($x = 0$).

EXAMPLES   For each quadratic function, find (a) its vertex and (b) its line of symmetry. Then plot a few specific points and graph the parabola.

1. $y = x^2 - 1$

   Solution $a = 1, b = 0$, and $c = -1$

   a. The vertex is at $\left(-\dfrac{0}{2 \cdot 1}, \dfrac{4 \cdot 1(-1) - 0^2}{4 \cdot 1}\right) = (0, -1)$.

   b. The line of symmetry is $x = -\dfrac{0}{2 \cdot 1} = 0$.

x	$x^2 - 1 = y$
-2	$(-2)^2 - 1 = 3$
-1	$(-1)^2 - 1 = 0$
0	$(0)^2 - 1 = -1$
1	$(1)^2 - 1 = 0$
2	$(2)^2 - 1 = 3$

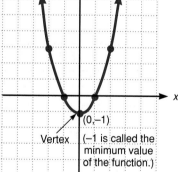

(0,-1)

Vertex   (-1 is called the minimum value of the function.)

2. $f(x) = x^2 - 6x + 1$

Solution $a = 1, b = -6, c = 1$

a. The vertex is at $\left(-\dfrac{-6}{2 \cdot 1}, \dfrac{4 \cdot 1 \cdot 1 - (-6)^2}{4 \cdot 1} \right) = (3, -8)$.

b. The line of symmetry is $x = -\dfrac{-6}{2 \cdot 1} = 3$.

x	$x^2 - 6x + 1 = y$
1	$(1)^2 - 6(1) + 1 = -4$
3	$(3)^2 - 6(3) + 1 = -8$
5	$(5)^2 - 6(5) + 1 = -4$

Vertex   (3,–8)

(–8 is the minimum value of the function.)

3. $y = -2x^2 + 4x$

Solution $a = -2, b = 4, c = 0$

a. The vertex is at $\left(-\dfrac{4}{2(-2)}, \dfrac{4(-2)(0) - 4^2}{4(-2)} \right) = (1, 2)$.

b. The line of symmetry is $x = -\dfrac{4}{2(-2)} = 1$.

x	$-2x^2 + 4x = y$
0	$-2(0)^2 + 4(0) = 0$
1	$-2(1)^2 + 4(1) = 2$
2	$-2(2)^2 + 4(2) = 0$

(2 is called the maximum value of the function.)

(1,2)   Vertex

4. Find the dimensions of the rectangle with maximum area if the perimeter is 40 meters.

Solution Let A = area
ℓ = length
w = width

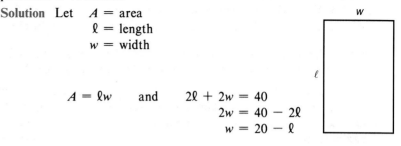

$$A = \ell w \qquad \text{and} \qquad 2\ell + 2w = 40$$
$$2w = 40 - 2\ell$$
$$w = 20 - \ell$$

So, substituting for w,

$$A = \ell(20 - \ell)$$
$$A = 20\ell - \ell^2$$
$$A = -\ell^2 + 20\ell$$

This is a quadratic function with $a = -1$, $b = 20$, and $c = 0$. The maximum area occurs at the vertex of the corresponding parabola where

$$\ell = -\frac{b}{2a} = -\frac{20}{2(-1)} = 10$$

and
$$w = 20 - \ell = 20 - 10 = 10$$

The maximum occurs when $\ell = 10$ m and $w = 10$ m. (The rectangle is in fact a square and the area = 100 m².) ∎

◢ **EXERCISES 11.5**

For each quadratic function in Exercises 1–20, find (a) its vertex and (b) its line of symmetry. Then plot a few specific points and graph the equation.

1. $y = x^2 + 4$ **2.** $y = x^2 - 6$ **3.** $y = 8 - x^2$ **4.** $y = -2 - x^2$

5. $y = x^2 - 2x - 3$ **6.** $y = x^2 - 4x + 5$ **7.** $y = x^2 + 6x$ **8.** $y = x^2 - 8x$

9. $y = -x^2 - 4x + 2$ **10.** $y = -x^2 - 2x - 2$ **11.** $y = 2x^2 - 10x + 3$ **12.** $y = 2x^2 - 12x - 5$

13. $y = x^2 - 3x - 1$ **14.** $y = x^2 + 5x + 4$ **15.** $y = -x^2 + x - 3$ **16.** $y = -x^2 - 7x + 3$

17. $y = 2x^2 + 7x - 4$ **18.** $y = 2x^2 - x - 3$ **19.** $y = 3x^2 + 5x + 2$ **20.** $y = 3x^2 - 9x + 5$

21. The perimeter of a rectangle is 40 meters. What are the dimensions of the rectangle with maximum area?

22. The perimeter of a rectangle is 56 feet. What are the dimensions of the rectangle with maximum area?

In Exercises 23–26 use the formula $h = -16t^2 + v_o t + h_o$ where h is the height of the object after time, t; v_o is the initial velocity; and h_o is the initial height.

23. A ball is thrown vertically upward from the ground with an initial velocity of 112 ft per sec. (a) When will the ball reach its maximum height? (b) What will be the maximum height?

24. A ball is thrown vertically upward from the ground with an initial velocity of 104 ft per sec. (a) When will the ball reach its maximum height? (b) What will be the maximum height?

25. A stone is projected vertically upward from a platform that is 32 ft high at a rate of 128 ft per sec. (a) When will the stone reach its maximum height? (b) What will be the maximum height?

26. A stone is projected vertically upward from a platform that is 20 ft high at a rate of 160 ft per sec. (a) When will the stone reach its maximum height? (b) What will be the maximum height?

In business, the term **revenue** represents income. The revenue (income) is found by multiplying the number of units sold times the price per unit. Revenue = (price) · (units sold).

27. A store owner estimates that by charging x dollars each for a certain lamp, he can sell $40 - x$ lamps each week. What price will yield maximum revenue?

28. When fishing reels are priced at p dollars each, local consumers will buy $36 - p$ fishing reels. What price will yield maximum revenue?

29. A manufacturer produces calculators. She estimates that by selling them for x dollars each, she will be able to sell $80 - 2x$ calculators each week. (a) What price will yield a maximum revenue? (b) What will be the maximum revenue?

30. A manufacturer produces radios. He estimates that by selling them for x dollars each, he will be able to sell $100 - x$ radios each month. (a) What price will yield a maximum revenue? (b) What will be the maximum revenue?

CHAPTER 11 SUMMARY

Key Terms and Formulas

The formula for the **distance between two points** $P_1(x_1, y_1)$ and $P_2(x_2, y_2)$ is

$$d = \sqrt{(x_2 - x_1)^2 + (y_2 - y_1)^2} \quad [11.2]$$

The formula for the **midpoint between two points** $P_1(x_1, y_1)$ and $P_2(x_2, y_2)$ is

$$\left(\frac{x_1 + x_2}{2}, \frac{y_1 + y_2}{2} \right) \quad [11.2]$$

A **linear inequality** is an inequality of the form

$$Ax + By < C$$

$(\leq, >, \geq$ can also be used.)   [11.3]

A **relation** is a set of ordered pairs of real numbers.   [11.4]

A **domain** of a relation is the set of all first components in the relation.   [11.4]

The **range** of a relation is the set of all second components in the relation.   [11.4]

A **function** is a relation in which each domain element has only one corresponding range element.   [11.4]
OR
A **function** is a relation in which each first component appears only once.
OR
A **function** is a relation in which no two ordered pairs have the same first component.

Vertical line test   [11.4]
If **any** vertical line intersects a graph in more than one point, then the relation graphed is **not** a function.

The function notation $f(x)$ is read "f of x."   [11.4]

Any function of the form

$$y = ax^2 + bx + c$$

where a, b, and c are real constants and $a \neq 0$ is called a **quadratic function.**   [11.5]

The graph of every quadratic function is a **parabola.**   [11.5]

Properties and Rules

If $a = \sqrt[n]{b}$, then $a^n = b$.   [11.1]

$\sqrt[n]{b}$ is not a real number if n is even and b is negative.   [11.1]

If n is a positive integer and $\sqrt[n]{b}$ is a real number, then

$$\sqrt[n]{b} = b^{1/n}.   [11.1]$$

If n is a positive integer and m is any integer and $\sqrt[n]{b}$ is a real number, then

$$(\sqrt[n]{b})^m = (b^{1/n})^m = b^{m/n}.   [11.1]$$

All the properties of exponents apply to rational exponents.   [11.1]

For the quadratic function $y = ax^2 + bx + c$:   [11.5]
1. If $a > 0$, the parabola "opens upward."
2. If $a < 0$, the parabola "opens downward."
3. $x = -\dfrac{b}{2a}$ is the line of symmetry.
4. $\left(-\dfrac{b}{2a}, \dfrac{4ac - b^2}{4a}\right)$ is the vertex.

Procedures

Steps for Graphing a Linear Inequality   [11.3]
Step 1 Graph the corresponding straight line (dashed for $<$ or $>$, solid for $\leq$ or $\geq$).
Step 2 Choose any test point not on the line and test it in the inequality.

Step 3 Shade the half-plane (a) on the same side as the test point if it satisfies the inequality or (b) on the opposite side of the test point if it does not satisfy the inequality.

CHAPTER 11   REVIEW

Write each of the expressions in Exercises 1–4 as a radical expression, then simplify.   [11.1]

1. $9^{1/2}$

2. $(-0.027)^{1/3}$

3. $\left(\dfrac{16}{81}\right)^{3/4}$

4. $36^{-3/2}$

Simplify the expressions in Exercises 5–16. Assume that all variables are positive.   [11.1]

5. $x^{2/5} \cdot x^{1/5}$

6. $x \cdot x^{1/2}$

7. $y^2 \cdot y^{-2/3}$

8. $y^{1/4} \cdot y^0$

9. $a^{1/3} \cdot a^{-1/2}$

10. $a^{5/7} \cdot a^{2/7}$

11. $\dfrac{x^{4/5}}{x^{1/2}}$

12. $\dfrac{x^{2/3}}{x^{-1/2}}$

13. $(6x^{1/3})(5x^{1/2})$

14. $(3x^{2/3})^2(2x^{1/4})$

15. $(x^{3/4})^2(16x)^{1/2}$

16. $(8x^4)^{1/3}(16x^3)^{-1/2}$

For Exercises 17–24, (a) find the distance between the two points and (b) find the coordinates of the midpoint of the line segment joining the two points.   [11.2]

17. $(4,7), (-4,5)$

18. $(5,5), (3,1)$

19. $(-4,-2), (-5,1)$

20. $\left(\dfrac{1}{2},2\right), \left(-\dfrac{3}{2},4\right)$

21. $\left(-\dfrac{1}{2},3\right), \left(-\dfrac{1}{2},5\right)$

22. $\left(\dfrac{4}{3},2\right), \left(\dfrac{1}{3},2\right)$

23. $(-4,-2), (-3,1)$

24. $(-6,2), (-1,8)$

Graph the linear inequalities in Exercises 25–30.   [11.3]

25. $y < 5x$

26. $y > 2 - 3x$

27. $4x - y \leq 4$

28. $3x - 6 \geq 2y$

29. $\dfrac{2}{3}x + \dfrac{1}{2}y \geq \dfrac{5}{6}$

30. $\dfrac{1}{2}x + \dfrac{1}{4}y \leq 1$

Graph the points that satisfy both inequalities in Exercises 31–34. [11.3]

31. $x \geq 2$ and $y \geq 3$

32. $x > -1$ and $y < 5$

33. $x - y \leq 0$ and $x \geq 0$

34. $2x + y > 4$ and $y \leq \frac{3}{2}x + 4$

Find the domain and range of each relation in Exercises 35–38. Then determine whether or not it is a function. [11.4]

35. $\{(0,3), (1,-4), (2,5), (6,6)\}$

36. $\{(1,-2), (3,-2), (-2,-2), (6,-2)\}$

37. $\{(4,2), (3,-5), (4,0), (0,0)\}$

38. $\{(1,2), (-3,4), (-1,3), (4,3)\}$

Use the vertical line test to determine whether or not each of the graphs in Exercises 39 and 40 represents a function. [11.4]

39.
40.

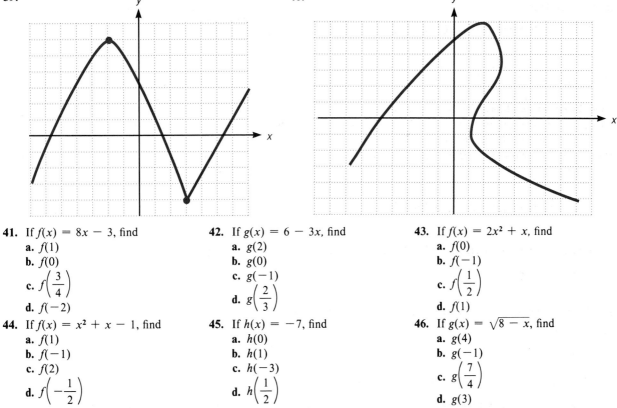

41. If $f(x) = 8x - 3$, find
  a. $f(1)$
  b. $f(0)$
  c. $f\left(\frac{3}{4}\right)$
  d. $f(-2)$

42. If $g(x) = 6 - 3x$, find
  a. $g(2)$
  b. $g(0)$
  c. $g(-1)$
  d. $g\left(\frac{2}{3}\right)$

43. If $f(x) = 2x^2 + x$, find
  a. $f(0)$
  b. $f(-1)$
  c. $f\left(\frac{1}{2}\right)$
  d. $f(1)$

44. If $f(x) = x^2 + x - 1$, find
  a. $f(1)$
  b. $f(-1)$
  c. $f(2)$
  d. $f\left(-\frac{1}{2}\right)$

45. If $h(x) = -7$, find
  a. $h(0)$
  b. $h(1)$
  c. $h(-3)$
  d. $h\left(\frac{1}{2}\right)$

46. If $g(x) = \sqrt{8 - x}$, find
  a. $g(4)$
  b. $g(-1)$
  c. $g\left(\frac{7}{4}\right)$
  d. $g(3)$

For each of the quadratic functions in Exercises 47–52, find (a) its vertex and (b) its line of symmetry. Then plot a few specific points and graph the equation. [11.5]

47. $y = x^2 + 4x - 2$

48. $y = x^2 - 8x$

49. $y = -2x^2 + 10x + 1$

50. $y = -x^2 - 6x + 2$

51. $y = x^2 - 3x + 7$

52. $y = 2x^2 - 5x + 1$

53. Show that the triangle determined by points $A(-5,1)$, $B(-2,4)$, and $C(1,1)$ is an isosceles triangle (two equal sides). [11.2]

54. Show that the triangle determined by points $A(-4,1)$, $B(4,9)$, and $C(2,3)$ is an isosceles triangle (two equal sides). [11.2]

55. Find the perimeter of the triangle determined by the points $A(2,0)$, $B(4,0)$, and $C(7,4)$. [11.2]

56. Find the perimeter of the triangle determined by the points $A(-3,-1)$, $B(1,2)$, and $C(0,4)$. [11.2]

57. Find the maximum height of a projectile by using the formula $h = -16t^2 + 144t$, where h is the height in feet and t is the time in seconds.   [11.5]

58. The manager of a jewelry department has determined that if he charges x dollars for a particular style of watch, he can sell $160 - 2x$ watches. Find the maximum revenue.   [11.5]

59. A baker has found that it costs him 10 cents to make each sweet roll. If he charges x cents each, he can sell $120 - 2x$ rolls daily. Find the selling price that will yield maximum profits. [**Hint:** Profit = Revenue − Cost.]   [11.5]

60. Find the dimensions of the rectangle with maximum area if the perimeter is 64 centimeters.   [11.5]

CHAPTER 11   TEST

Simplify the expressions in Exercises 1–6. Assume that all variables are positive.

1. $16^{3/4}$

2. $\left(\dfrac{4}{9}\right)^{-3/2}$

3. $a^{3/4} \cdot a^{-1/2}$

4. $\dfrac{y^{3/5}}{y^{-1/2}}$

5. $(16x^2)^{1/2}(27x^3)^{2/3}$

6. $(2x^{2/3})^2(x^{1/6})^4$

For Exercises 7 and 8, (a) find the distance between the two points and (b) find the coordinates of the midpoint of the line segment joining the two points.

7. $(-5,7), (3,-1)$

8. $(2,5), (-4,-6)$

Graph the linear inequalities in Exercises 9 and 10.

9. $2x - y \le 3$

10. $4x + 3y \le 6$

Graph the points that satisfy both inequalities in Exercises 11 and 12.

11. $x \ge 0$ and $y \ge 4$

12. $3x + y \le 5$ and $x \ge 1$

Find the domain and range of each relation in Exercises 13 and 14. Then determine whether or not it is a function.

13. $\{(1,7), (-2,5), (3,3), (1,1)\}$

14. $\left\{\left(3,\dfrac{1}{2}\right), (4,1), (1,4), \left(\dfrac{2}{3},6\right)\right\}$

15. Use the vertical line test to determine whether or not the graph represents a function.

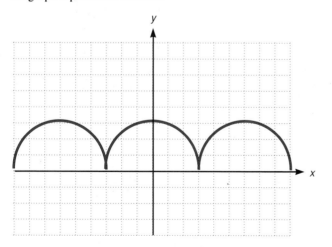

16. If $f(x) = 7 - 4x$, find
 a. $f(0)$
 b. $f\left(\dfrac{1}{2}\right)$
 c. $f(-1)$
 d. $f\left(\dfrac{3}{4}\right)$

17. If $g(x) = 11$, find
 a. $g(2)$
 b. $g(0)$
 c. $g\left(-\dfrac{1}{2}\right)$
 d. $g(-3)$

18. If $h(x) = x^2 + x - 2$, find
 a. $h(0)$
 b. $h(1)$
 c. $h(-2)$
 d. $h\left(\dfrac{1}{2}\right)$

19. If $f(x) = \sqrt{3x + 1}$, find
 a. $f(0)$
 b. $f(5)$
 c. $f\left(-\dfrac{1}{3}\right)$
 d. $f(2)$

For each of the quadratic functions in Exercises 20–22, find (a) its vertex and (b) its line of symmetry. Then plot a few specific points and graph the equation.

20. $y = -x^2 + 4$

21. $y = 2x^2 + 4x - 3$

22. $y = x^2 + 5x - 1$

23. Find the perimeter of the triangle determined by the points $A(-1,2)$, $B(2,6)$, and $C(-3,-6)$.

24. The height of a projectile is given by the formula $h = -16t^2 + 192t$, where h is the height in feet and t is the time in seconds. Find the time when the height is maximum.

25. The perimeter of a rectangular garden plot is 76 feet. Find the dimensions if the area is maximum.

CUMULATIVE REVIEW (11)

Simplify.

1. $4x + 3 - 2(x + 4) + (2 - x)$

2. $4x - [3x - 2(x + 5) + 7 - 5x]$

Solve.

3. $5(x - 8) - 3(x - 6) = 0$

4. $\dfrac{5}{8}x + 4 = \dfrac{3}{4}x + 7$

5. $x^2 - 8x + 15 = 0$

6. $x^2 + 5x - 36 = 0$

7. $\dfrac{1}{x} + \dfrac{2}{3x} = 1$

8. $\dfrac{3}{x - 1} + \dfrac{5}{x + 1} = 2$

9. $(x - 4)^2 = 9$

10. $(x + 3)^2 = 6$

11. $x^2 + 4x - 2 = 0$

12. $2x^2 - 3x - 4 = 0$

13. $x - 2 = \sqrt{x + 18}$

14. $x + 3 = 2\sqrt{2x + 6}$

Solve the inequality and graph the solution.

15. $15 - 3x + 1 \geq x + 4$

16. $4(6 - x) \geq -2(3x + 1)$

Solve the system of equations.

17. $\begin{cases} x + 5y = 9 \\ 2x - 3y = 1 \end{cases}$

18. $\begin{cases} 6x - y = 11 \\ 4x + 3y = -11 \end{cases}$

Solve each formula for the given letter.

19. $P = a + 2b$, for b

20. $A = \dfrac{h}{2}(a + b)$, for a

Simplify.

21. $\dfrac{4x^5 \cdot x^{-2}}{2x}$

22. $(4x^3)^2 \cdot (2x^2)^{-1}$

23. $\left(\dfrac{m^2 n^3}{m^3 n}\right)^2$

24. $\left(\dfrac{3x^2 y^{-1}}{7x^3 y}\right)^{-2}$

Factor.

25. $x^2 - 12x + 36$ **26.** $25x^2 - 49$ **27.** $3x^2 + 10x + 3$ **28.** $3x^2 + 6x - 72$

29. $15x^2 - 14x - 8$ **30.** $x^3 - 8x^2 + 4x - 32$

Perform the indicated operations.

31. $(5x^2 + 3x - 7) + (2x^2 - 3x - 7)$

32. $(2x^2 - 7x + 3) - (5x^2 - 3x + 4)$

33. $\dfrac{x^2}{x^2 - 4} \cdot \dfrac{x^2 - 3x + 2}{x^2 - x}$

34. $\dfrac{3x}{x^2 - 6x - 7} \div \dfrac{2x^2}{x^2 - 8x + 7}$

35. $\dfrac{2x}{x^2 - 25} - \dfrac{x + 1}{x + 5}$

36. $\dfrac{4x}{x^2 - 3x - 4} + \dfrac{x - 2}{x^2 + 3x + 2}$

Simplify.

37. $\sqrt{63}$ **38.** $\sqrt{243}$ **39.** $\sqrt[3]{250}$ **40.** $\sqrt{\dfrac{25}{12}}$

Graph.

41. $y = \dfrac{2}{3}x + 4$ **42.** $4x - 3y = 6$ **43.** $2x - 3y \leq 12$ **44.** $y = x^2 - 8x + 7$

45. Write the equation $5x + 2y = 6$ in slope-intercept form. Find the slope and the y-intercept.

46. Given the two points $(-5,2)$ and $(3,5)$, find:
    a. the slope of the line through the two points,
    b. the distance between the two points,
    c. the midpoint of the line segment joining the points, and
    d. an equation of the line through the two points.

47. Two cars leave the same town at the same time traveling in opposite directions. The speed of one car exceeds the speed of the other by 8 mph. At the end of $2\frac{1}{2}$ hours, the cars are 260 miles apart. Find the speed of each car.

48. Anna has money in two separate savings accounts; one pays 5% and one pays 8%. The amount invested at 5% exceeds the amount invested at 8% by $800. The total annual interest is $209. How much is invested at each rate?

49. Ben can clean the weeds from a vacant lot in 10 hours. Beth can do the same job in 8 hours. How long would it take if they worked together?

50. The speed of a boat in still water is 8 mph. It travels 6 miles upstream and returns. The total trip takes 1 hr 36 min. Find the speed of the current.

51. How many liters of a solution that is 40% insecticide must be mixed with 40 liters of a solution that is 25% insecticide to produce a solution that is 30% insecticide?

52. The length of a rectangle is one inch less than twice the width. The length of the diagonal is 17 inches. Find the width and length of the rectangle.

53. George bought 7 pairs of socks. Some cost $2.50 per pair and some cost $3.00 per pair. The total cost was $19.50. How many pairs of each type did he buy?

54. Edna has a pet pig. The pig's weekly diet must include 16 grams of supplement A and 23 grams of supplement B. These supplements can be found in two products, X and Y. Each cup of product X contains 2 grams of supplement A and 3 grams of supplement B. Each cup of product Y contains 3 grams of supplement A and 4 grams of supplement B. How many cups of each product must be included in the pig's diet?

CHAPTER 1

Exercises 1.1, Page 10

1. $3 + 7$ **3.** $4 \cdot 19$ **5.** $6 \cdot 5 + 6 \cdot 8$ **7.** $(2 \cdot 3) \cdot x$ **9.** $(3 + x) + 7$

11. Commutative property for multiplication, $6 \cdot 4 = 4 \cdot 6$
$$24 = 24$$

13. Associative property for addition, $8 + (4 + 4) = (8 + 4) + 4$
$$8 + 8 = 12 + 4$$
$$16 = 16$$

15. Distributive property, $5(4 + 18) = 5(22)$
$$5 \cdot 4 + 90 = 20 + 90$$
$$110 = 110$$

17. Associative property for multiplication, $(6 \cdot 4) \cdot 9 = 6 \cdot (4 \cdot 9)$
$$24 \cdot 9 = 6 \cdot 36$$
$$216 = 216$$

19. Commutative property for addition, $4 + 34 = 34 + 4$
$$38 = 38$$

21. $18 + 48 = 66$ **23.** $90 + 20 = 110$ **25.** 295   **27.** 286   **29.** 1215   **31.** 2300   **33.** 14
35. Undefined   **37.** 17   **39.** 25   **41.** 49   **43.** 125   **45.** 81   **47.** 2^5 **49.** $2^2 5^3 a^2$ **51.** $7^2 a^4 b^3$
53. $2 \cdot 3^3$ **55.** $2^3 \cdot 3 \cdot 7$ **57.** $2^2 \cdot 7^2$ **59.** 13   **61.** 33   **63.** 33   **65.** 11   **67.** 5   **69.** 14
71. 39   **73.** 5

Exercises 1.2, Page 20

1. 40   **3.** 0   **5.** $\dfrac{2}{5}$ **7.** $\dfrac{3}{7}$ **9.** (a) 1, 2, 3, 6   (b) 6, 12, 18, 24, 30, 36   **11.** (a) 1, 3, 5, 15

(b) 15, 30, 45, 60, 75, 90   **13.** 120   **15.** $40xy$ **17.** $210x^2y^2$ **19.** $\dfrac{1}{6}$ **21.** $\dfrac{1}{2}$ **23.** $\dfrac{7}{9}$ **25.** $\dfrac{5}{23}$

27. $\dfrac{143}{144}$ **29.** $\dfrac{4}{3}$ **31.** $\dfrac{2}{5x}$ **33.** $\dfrac{8b}{153}$ **35.** $\dfrac{2}{5ab}$ **37.** Undefined   **39.** $\dfrac{5}{9x}$ **41.** $\dfrac{3ax}{32}$ **43.** $\dfrac{83}{60}$

45. $\dfrac{5}{42}$ **47.** $\dfrac{33}{70}$ **49.** $\dfrac{3}{4x}$ **51.** $\dfrac{5}{24x}$ **53.** 0   **55.** $\dfrac{1}{24}$ **57.** $\dfrac{21}{100}$ **59.** $\dfrac{7}{4}$ **61.** $\dfrac{1}{8}$ **63.** $\dfrac{49}{40}$

65. $\dfrac{12}{5}$ **67.** $\dfrac{991}{336}$ **69.** 48   **71.** 344 yd

Exercises 1.3, Page 26

1. 91%   **3.** 137%   **5.** 0.69   **7.** 0.113   **9.** 0.375   **11.** 0.05   **13.** 317.23   **15.** 263.51   **17.** 34.04
19. 141.17   **21.** 108.72   **23.** 7.626   **25.** 15.1   **27.** 10.23   **29.** 55.86   **31.** 88.74   **33.** 793.8
35. 54.15   **37.** 20.655   **39.** 38   **41.** $\frac{63}{4}$ **43.** $\frac{42}{5}$ **45.** $36,464; $9116   **47.** 9720 ft   **49.** 10.5 in.
51. 9   **53.** $660   **55.** $1252.90   **57.** $2183.50   **59.** 19.6572   **61.** 218.2992   **63.** 0.05565
65. 10.1875 in.   **67.** 9 pieces

Exercises 1.4, Page 30

1. 9   **3.** 4   **5.** 1   **7.** 1   **9.** 1   **11.** 10   **13.** 33   **15.** 39   **17.** 15   **19.** 7   **21.** $4\frac{1}{2}, 4\frac{1}{3}$
23. $6, 5\frac{2}{3}$ **25.** $\frac{5}{4}, \frac{3}{4}$ **27.** 5.8, 3.5   **29.** 15, 9.25   **31.** 9.9, 0.7   **33.** 6 is a solution.
35. 3 is not a solution.   **37.** 4 is not a solution   **39.** 5 is a solution.   **41.** 1 is a solution.   **43.** 3 is a solution.
45. $\frac{1}{2}$ is a solution.   **47.** $\frac{1}{4}$ is a solution.   **49.** 0.2 is a solution.   **51.** 1.4 is not a solution.
53. 2.25 is a solution.   **55.** 2.36 is not a solution.   **57.** 2.08 is a solution.

Exercises 1.5, Page 37

1. e **3.** k **5.** c **7.** d **9.** g **11.** l **13.** m **15.** h **17.** 70 cm; 300 cm²   **19.** 54 mm; 126 mm²
21. 31.4 in.; 78.5 in.²   **23.** 1436.03 in.³   **25.** 729 m³   **27.** 56 in.   **29.** 63 cm²   **31.** 6154.4 ft³
33. 125 ft³   **35.** 272 cm²   **37.** 529.88 in.³   **39.** 28 cm; 48 cm²   **41.** 212.0428 m²   **43.** 130.0776 in.²
45. 6334233.209 cm³

Chapter 1 Review, Page 41

1. 36   **2.** 40   **3.** 9   **4.** 15   **5.** 126   **6.** 459   **7.** 39   **8.** 124   **9.** 840   **10.** 196   **11.** 3318
12. 24   **13.** 21 + 42 = 63   **14.** 72 + 56 = 128   **15.** 17   **16.** 26   **17.** 81   **18.** 121   **19.** 2^3a^2b
20. $3^25^2ab^3$ **21.** $2 \cdot 2 \cdot 3 \cdot 5 = 2^2 \cdot 3 \cdot 5$ **22.** $3 \cdot 3 \cdot 17 = 3^2 \cdot 17$ **23.** 108   **24.** $72xy^2$ **25.** 90
26. 12   **27.** $\frac{5}{11}$ **28.** $\frac{2}{3}$ **29.** 30   **30.** 52   **31.** Commutative property for addition
32. Associative property for addition   **33.** Distributive property   **34.** $\frac{43}{40}$ **35.** $\frac{1}{36}$ **36.** $\frac{22}{15y}$ **37.** $\frac{1}{4a}$
38. $\frac{15}{2}$ **39.** $\frac{5}{6}$ **40.** $\frac{7}{5}$ **41.** $\frac{2}{9}$ **42.** $\frac{263}{576}$ **43.** $\frac{3}{5}$ **44.** 0   **45.** $\frac{7}{16}$ **46.** 73%   **47.** 6.5%
48. 0.07   **49.** 0.125   **50.** 32.53   **51.** 4.69   **52.** 14.6   **53.** 58.19   **54.** 5.292   **55.** 10.8   **56.** 58.3
57. 43.7   **58.** 12   **59.** 35   **60.** $3\frac{1}{2}$ **61.** $\frac{1}{4}$ **62.** 1.7   **63.** 3.66   **64.** 2 is not a solution.
65. 3 is a solution.   **66.** $\frac{1}{2}$ is a solution.   **67.** 1.5 is a solution.   **68.** 76 ft   **69.** 254.34 m²   **70.** 693 in.³
71. $3\frac{1}{3}$ cups   **72.** $21   **73.** 20.8 in.   **74.** $365   **75.** $68.01

Chapter 1 Test, Page 43

1. 1794 **2.** 366 **3.** 23 **4.** Distributive property **5.** Associative property of multiplication

6. Commutative property of addition **7.** 53 **8.** $2 \cdot 3^2 xy^3$ **9.** $3 \cdot 3 \cdot 5 \cdot 7$ **10.** $90x^2y^2$ **11.** 88 **12.** $\dfrac{6}{13}$

13. 60 **14.** $\dfrac{3}{5}$ **15.** $\dfrac{3}{10}$ **16.** $\dfrac{29}{40}$ **17.** 21.06 **18.** 13.588 **19.** 42.5 **20.** 9 **21.** 4 is a solution.

22. 87.92 cm **23.** 33 in.² **24.** 102 ft² **25.** $32

CHAPTER 2

Exercises 2.1, Page 49

1. > **3.** > **5.** < **7.** = **9.** < **11.** < **13.** = **15.** < **17.** True **19.** True **21.** False
23. True **25.** False **27.** True **29.** True **31.** False **33.** True **35.** True

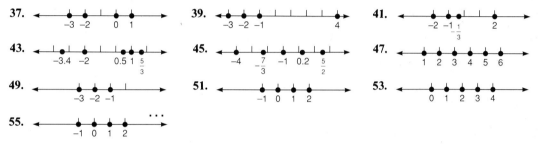

55.

Exercises 2.2, Page 52

11. True **13.** True **15.** True **17.** False **19.** True
21. $x = 4$ or $x = -4$ **23.** $x = 9$ or $x = -9$ **25.** $y = 0$ **27.** No solution **29.** $x = 4.7$ or $x = -4.7$

31. $y = 12$ or $y = -12$ **33.** $x = \dfrac{4}{7}$ or $x = -\dfrac{4}{7}$ **35.** $x = \dfrac{5}{4}$ or $x = -\dfrac{5}{4}$ **37.** No solution

39. No solution **41.** **43.**

45. **47.**

49. Sometimes **51.** Always **53.** Sometimes

Exercises 2.3, Page 56

1. 13 **3.** −4 **5.** 0 **7.** 5 **9.** −13 **11.** −2 **13.** −10 **15.** 0 **17.** 17 **19.** −19

21. $-9\dfrac{5}{6}$ **23.** $-3\dfrac{7}{16}$ **25.** 22.02 **27.** −55.52 **29.** −7 **31.** −16 **33.** −26 **35.** 0 **37.** −32

39. 5 **41.** −83 **43.** 12 **45.** −32 **47.** −2 is a solution. **49.** −4 is a solution. **51.** −7 is a solution.
53. 2 is not a solution. **55.** −1 is a solution. **57.** Sometimes **59.** Always **61.** 85.179 **63.** −97.714
65. 0.2633

Exercises 2.4, Page 61

1. −11 **3.** 6 **5.** −4.7 **7.** 0 **9.** $\dfrac{5}{16}$ **11.** 5 **13.** −10 **15.** 12 **17.** 3 **19.** $-6\dfrac{7}{8}$

21. 24.74 **23.** −15 **25.** 30 **27.** −57 **29.** 0 **31.** $-25\dfrac{7}{8}$ **33.** −31.04 **35.** −15 **37.** −1

39. −3 **41.** 7 **43.** $\dfrac{19}{30}$ **45.** −15.26 **47.** < **49.** < **51.** = **53.** > **55.** > **57.** <

59. = **61.** −8 is a solution. **63.** −2 is a solution. **65.** 4 is not a solution. **67.** 5 is not a solution.
69. 1 is a solution. **71.** 1.1 is a solution. **73.** 0.262 **75.** 20.405 **77.** <

Exercises 2.5, Page 66

1. −12 **3.** 56 **5.** 57 **7.** −224 **9.** −32.4 **11.** 2.73 **13.** $-\dfrac{3}{5}$ **15.** $\dfrac{3}{4}$ **17.** −288 **19.** 0

21. 4 **23.** −6 **25.** −3 **27.** 13 **29.** 0 **31.** Undefined **33.** −1.06 **35.** $\dfrac{4}{7}$ **37.** Negative

39. Negative **41.** Negative **43.** Zero **45.** Undefined **47.** True **49.** True **51.** False **53.** False
55. True **57.** −12 is a solution. **59.** −72 is a solution. **61.** −8 is not a solution. **63.** 5 is a solution.
65. −4 is a solution. **67.** −3.445911 **69.** −2.671

Exercises 2.6, Page 68

1. $(-23) + 13 + (-6) = -16$ lb **3.** $4° + 3° + (-9°) = -2°$ **5.** $\$47 + (-\$22) + (\$8) + (-\$45) = -\$12$
7. $14° + (-6°) + (11°) + (-15°) = 4°$ **9.** $\$187 + (-\$241) + (\$82) + (\$26) = \$54$ **11.** 4 yd

13. 8th floor **15.** \$40 **17.** −13 **19.** −10.8 **21.** $2\dfrac{1}{8}$ **23.** −15 **25.** $-4\dfrac{7}{16}$ **27.** \$220

29. −19° **31.** −\$8 **33.** −14,777 ft **35.** 52 yr **37.** 5 under par **39.** $1\dfrac{7}{8}$

Chapter 2 Review, Page 71

1. < **2.** = **3.** > **4.** > **5.** > **6.** > **7.**

8. **9.** **10.**

11.

12.

13.

14.

15.

16.

17. $x = 6$ or $x = -6$ **18.** $x = 10$ or $x = -10$ **19.** $y = \dfrac{3}{5}$ or $y = -\dfrac{3}{5}$

20. $y = 2.9$ or $y = -2.9$ **21.** No solution   **22.** $y = 1.724$ or $y = -1.724$ **23.**

24.   **25.**

26.

27. -8 **28.** -12 **29.** -15 **30.** 1.8 **31.** $-\dfrac{2}{15}$ **32.** $-\dfrac{23}{12}$ **33.** -1.18 **34.** -114 **35.** 253

36. 160 **37.** $\dfrac{6}{25}$ **38.** 5.85 **39.** 4 **40.** Undefined   **41.** -7 **42.** 0 **43.** -6.8 **44.** $\dfrac{10}{3}$

45. Sometimes   **46.** Never   **47.** Always   **48.** Sometimes   **49.** Sometimes   **50.** Never   **51.** Always
52. 6 is a solution.   **53.** -2 is a solution.   **54.** -5 is a solution.   **55.** -3 is not a solution.

56. -16 is a solution.   **57.** 16 is a solution.   **58.** $\dfrac{3}{4}$ is a solution.   **59.** $-\dfrac{2}{3}$ is a solution.

60. -2.6 is a solution.   **61.** 1.3 is not a solution.   **62.** \$10   **63.** \$88   **64.** 3   **65.** $1\dfrac{3}{16}$ **66.** 17.49

Chapter 2 Test, Page 72

1. (a) $<$ (b) $>$ **2.**   **3.**

4. $y = 7$ or $y = -7$ **5.** $x = \dfrac{7}{8}$ or $x = -\dfrac{7}{8}$ **6.** $-2, -1, 0, 1, 2$

7. $\ldots, -11, -10, -9, 9, 10, 11, \ldots$

8. 13   **9.** 6.68   **10.** $\dfrac{1}{8}$ **11.** 19   **12.** 11   **13.** 162

14. -4 **15.** -25.132 **16.** $-\dfrac{5}{36}$ **17.** $-\dfrac{9}{4}$ **18.** Never   **19.** Always   **20.** 5 is a solution.

21. -1.5 is a solution.   **22.** 15 is not a solution.   **23.** -60 **24.** $-\dfrac{5}{8}$ **25.** \$15.47

Cumulative Review (2), Page 73

1. Associative property of multiplication   **2.** Commutative property of multiplication   **3.** Distributive property

4. 8   **5.** 19   **6.** 18   **7.** $\dfrac{1}{6}$ **8.** $-\dfrac{7}{45}$ **9.** 9.88   **10.** -20 **11.** 1.3 is not a solution.

12. $\dfrac{1}{2}$ is a solution.   **13.** (a) 65 cm   (b) 259 cm^2 **14.** \$15.44   **15.** \$32.40   **16.** -6 **17.** $-2.3, 2.3$

18. $-5, -4, -3, -2, -1, 0, 1, 2, 3, 4, 5$ **19.** -8 is a solution.   **20.** -5

CHAPTER 3

Exercises 3.1, Page 78

1. $15x$ **3.** $3x$ **5.** $5y^2$ **7.** $12x^2$ **9.** $7x + 2$ **11.** $x - 3y$ **13.** $4x^2 + 3y$ **15.** $2x + 3$
17. $7x - 8y$ **19.** $8x + y$ **21.** $2x^2 - x$ **23.** $-2x^2 + 14x$ **25.** $3x^2 - xy + y^2$ **27.** $2x$ **29.** $-y$
31. 0 **33.** $10x$ **35.** x **37.** -12 **39.** -7 **41.** 0 **43.** 2 **45.** 0 **47.** $13x - 7; 45$
49. $9y + 2; -25$ **51.** $5x - 3y; 13$ **53.** $7x + 2y; -10$ **55.** $6y + 2xy; 42$ **57.** $5x; -30$ **59.** $6x; 18$
61. 0.6557 **63.** 7.0898 **65.** 10.62172

Exercises 3.2, Page 86

1. $y = 6$ **3.** $y = 0$ **5.** $y = 5$ **7.** $y = -\dfrac{5}{8}$ **9.** $y = -1.7$ **11.** $x = -6$ **13.** $x = -6$

15. $x = -\dfrac{1}{2}$ **17.** $x = 1.9$ **19.** $x = 7$ **21.** $x = 14$ **23.** $y = 5$ **25.** $x = -2$ **27.** $x = 3$

29. $x = -25$ **31.** $x = 10$ **33.** $y = 36$ **35.** $x = 12$ **37.** $x = -\dfrac{7}{3}$ **39.** $x = \dfrac{6}{5}$ **41.** $y = 2$

43. $y = -2$ **45.** $\$9013.50$ **47.** 29 cm **49.** $\$3.70$ **51.** 7 hr **53.** 3 yr **55.** 6 in. **57.** -50.753
59. $y = 26.087$ **61.** $y = -1036$

Exercises 3.3, Page 91

1. $x = -11$ **3.** $x = 2$ **5.** $x = 10$ **7.** $x = -2$ **9.** $x = 4$ **11.** $x = 2$ **13.** $x = \dfrac{1}{5}$ **15.** $y = \dfrac{5}{2}$

17. $y = 1.8$ **19.** $x = -0.9$ **21.** $x = 18$ **23.** $x = 5$ **25.** $x = 18$ **27.** $x = \dfrac{4}{3}$ **29.** $x = -\dfrac{10}{3}$

31. $x = 6$ **33.** $x = 3$ **35.** $x = -6$ **37.** $x = \dfrac{1}{2}$ **39.** $x = -\dfrac{2}{7}$ **41.** $x = -21$ **43.** $x = -2$

45. $x = -3$ **47.** $x = 10$ **49.** $x = -5$ **51.** $x = 7$ **53.** $x = 2$ **55.** $x = -\dfrac{7}{2}$ **57.** $x = -\dfrac{7}{3}$

59. $x = -\dfrac{1}{2}$ **61.** 24 ft **63.** 10 cm **65.** 18 months **67.** $x = -50.210169$ **69.** $x = 27.90$

71. $x = 4.59$

Exercises 3.4, Page 95

1. The product of 4 and a number **3.** 1 more than the product of 2 and a number
5. The difference between 7 times a number and 5.3 **7.** The product of -2 with the difference between a number and 8
9. 5 times the quantity found by taking the sum of twice a number and 3

11. $\dfrac{2}{3}$ added to the product of 6 with the difference between a number and 1

13. 7 added to the product of 3 with a number; the product of 3 with the sum of a number and 7
15. The difference between 7 times a number and 3; the product of 7 with the difference between a number and 3

17. $x + 6$ **19.** $x - 4$ **21.** $3x - 5$ **23.** $\dfrac{x - 3}{7}$ **25.** $3(x - 8)$ **27.** $3x - 5$ **29.** $3x + (8 - 2x)$

31. $8(x - 6) + 4$ **33.** $(3x - 9) - 5x$ **35.** $4(x + 5) - 2x$ **37.** $0.11x$ dollars   **39.** $(60h + 20)$ minutes

41. $0.20m + 20$ dollars   **43.** $0.09x + 250$ dollars   **45.** $2(2w - 3) + 2w$ cm

Exercises 3.5, Page 99

1. 36   **3.** 4   **5.** 3   **7.** -6 **9.** 13   **11.** -35 **13.** -7 **15.** -4 **17.** 12   **19.** 22, 23, 24

21. 29, 31   **23.** $-54, -52, -50$ **25.** 83, 88   **27.** 37, 77   **29.** 11, 6, 22   **31.** 27 ft   **33.** \$285

35. \$910   **37.** \$60   **39.** 11 yr; 33 yr   **41.** 8 ft; 10 ft   **43.** 23 won; 9 lost   **45.** \$280

47. \$27.45 for blouse; \$39.95 for skirt   **49.** 9 minutes

Exercises 3.6, Page 105

1. $b = P - a - c$ **3.** $m = \dfrac{f}{a}$ **5.** $w = \dfrac{A}{\ell}$ **7.** $n = \dfrac{R}{p}$ **9.** $P = A - I$ **11.** $m = 2A - n$ **13.** $s = \dfrac{P}{4}$

15. $t = \dfrac{d}{r}$ **17.** $t = \dfrac{I}{Pr}$ **19.** $b = \dfrac{P - a}{2}$ **21.** $\beta = 180 - \alpha - \gamma$ **23.** $x = \dfrac{y - b}{m}$ **25.** $r^2 = \dfrac{A}{4\pi}$

27. $M = \dfrac{(IQ)C}{100}$ **29.** $h = \dfrac{3V}{\pi r^2}$ **31.** $I = \dfrac{E}{R}$ **33.** $L = \dfrac{R}{2A}$ **35.** $t = \dfrac{v - k}{g}$ **37.** $h = \dfrac{S - 2\pi r^2}{2\pi r}$

39. $r = \dfrac{S - a}{S}$ **41.** $C = \dfrac{5}{9}(F - 32)$ **43.** $m = \dfrac{2gk}{v^2}$ **45.** $y = x - 3$ **47.** $y = 8 - 2x$ **49.** $x = 5 - 2y$

51. $y = \dfrac{4x - 9}{3}$ **53.** $x = \dfrac{4 - 8y}{3}$ **55.** $y = \dfrac{3 - 1.2x}{1.5}$ or $y = \dfrac{30 - 12x}{15}$

Exercises 3.7, Page 107

1. 48 ft per sec   **3.** 2 sec   **5.** 100 milligrams   **7.** 5 yr   **9.** 16 studs   **11.** \$1030   **13.** 230 calculators

15. 5%   **17.** $C = 2\pi r; r = 17$ cm   **19.** $p = b + 2a; a = 22$ **21.** $A = \pi r^2; A = 49\pi$ m²

23. $R = \dfrac{E}{I}; R = 6$ ohms   **25.** $A = \dfrac{1}{2}h(a + b); b = 10$ cm

Chapter 3 Review, Page 110

1. $7x + 1$ **2.** $-2x - 12$ **3.** 0   **4.** $4x$ **5.** $12x^2 + 29x$ **6.** 12   **7.** 15   **8.** 11   **9.** -22 **10.** 7

11. $x = -1.5$ **12.** $x = \dfrac{5}{4}$ **13.** $x = -\dfrac{3}{2}$ **14.** $x = -4.1$ **15.** $x = -3$ **16.** $x = 2$ **17.** $x = 6$

18. $x = -4$ **19.** $x = -3$ **20.** $x = -4$ **21.** $x = 3$ **22.** $x = \dfrac{25}{4}$ **23.** $x = \dfrac{32}{3}$ **24.** $x = -40$

25. $x = -4$ **26.** The difference between 3 times a number and 1

27. The difference between 4 and the product of a number and 7   **28.** The product of 5 with the sum of a number and 1

29. Twice the difference between 4 times a number and 1   **30.** 4 divided by the sum of a number and 7

31. $3x + 4$ **32.** $9 - 2x$ **33.** $11(x - 4)$ **34.** $24x + 5$ hr   **35.** $2x + 17$ points

36. $m = \dfrac{E}{c^2}$ **37.** $b = \dfrac{Fd^2}{ka}$ **38.** $h = \dfrac{3V}{\pi r^2}$ **39.** $y = \dfrac{8 - 2x}{7}$ **40.** $x = \dfrac{-2y - 3}{5}$

41. $240 = 2(85) + 2w$; $w = 35$ in.   **42.** $6400 = 150t + 1000$; $t = 36$ mo

43. $2x + 3 = 3x - 8$; $x = 11$ **44.** $3x = 2x + 10$; $x = 10$ **45.** $n + (n + 2) = 84$; 41, 43

46. $5 - 2n = (n + 1) + 19$; $-5, -4$ **47.** $x + (x + 9.35) = 16.25$; \$3.45 for ball; \$12.80 for bat   **48.** \$245

49. $V = 7200 - (0.7200)(0.08)(7)$ **50.** $147\pi = \dfrac{1}{3}\pi(7)^2h$; $h = 9$ in.
$V = \$3168$

Chapter 3 Test, Page 111

1. $8x^2 - 17x$ **2.** $5x - 10$ **3.** -3 **4.** 17 **5.** $x = -3$ **6.** $x = 5$ **7.** $x = 0$ **8.** $x = 14$

9. $x = 2$ **10.** $x = -2$ **11.** $x = 20$ **12.** The product of 4 with the difference between a number and 2

13. 7 subtracted from the product of 3 with the sum of a number and 4   **14.** $3(2x + 5)$ **15.** $4x + 3$ qt

16. $0.09x$ **17.** $h = \dfrac{S}{2\pi r}$ **18.** $m = \dfrac{N - p}{rt}$ **19.** $y = \dfrac{5x + 7}{3}$ **20.** $-9, -13$ **21.** 16, 17   **22.** 24 in.

23. \$24.50   **24.** \$0.95 for shake; \$1.85 for hamburger   **25.** (a) $A = \dfrac{1}{2}bh$ (b) 56 m²

Cumulative Review (3), Page 112

1. $24 + 42 = 66$ **2.** $6x + 2$ **3.** $-3x - 15$ **4.** 24 **5.** $\dfrac{65}{24}$ **6.** -13 **7.** 15 **8.** -7

9. -2 is a solution.   **10.** 21 **11.** $-\dfrac{5}{3}, \dfrac{5}{3}$ **12.** $4x + 7$ **13.** $x - 7$ **14.** $x^2 + 13x$ **15.** $x = -1$

16. $x = 3$ **17.** $x = -11$ **18.** $v = \dfrac{h + 16t^2}{t}$ **19.** $h = \dfrac{A - 2\pi r^2}{2\pi r}$ **20.** 16 in., 48 in.

CHAPTER 4

Exercises 4.1, Page 119

1. $3^3 = 27$ **3.** $8^3 = 512$ **5.** $\dfrac{1}{4^2} = \dfrac{1}{16}$ **7.** $\dfrac{1}{6^3} = \dfrac{1}{216}$ **9.** $(-4)^3 = -64$ **11.** 54 **13.** -54 **15.** $\dfrac{4}{9}$

17. $-\dfrac{5}{4}$ **19.** x^4 **21.** y^{11} **23.** $\dfrac{1}{y^2}$ **25.** $\dfrac{5}{y^4}$ **27.** $\dfrac{1}{x^2}$ **29.** y^3 **31.** $3^2 = 9$ **33.** $9^3 = 729$

35. $\dfrac{1}{10^3} = \dfrac{1}{1000}$ **37.** x^2 **39.** x^2 **41.** x^4 **43.** $\dfrac{1}{x^4}$ **45.** x^6 **47.** x^2 **49.** y^2 **51.** $3x^3$ **53.** $10x^5$

55. $36x^3$ **57.** $-14x^5$ **59.** $-12x^6$ **61.** $4y$ **63.** $3y^2$ **65.** $-2y^2$ **67.** $-7x^2$ **69.** $10^3 = 1000$

71. $\dfrac{1}{10^2} = \dfrac{1}{100}$ **73.** x^5 **75.** $y^0 = 1$ **77.** 25.6036 **79.** 0.003815

Exercises 4.2, Page 127

1. $36x^6$ **3.** $-27x^6$ **5.** x^3y^6 **7.** $64a^8b^2$ **9.** -1 **11.** $\dfrac{n^6}{16m^4}$ **13.** $-\dfrac{2y^6}{27x^{15}}$ **15.** $\dfrac{16x^2}{y^4}$ **17.** $\dfrac{9x^4}{y^6}$

19. $\dfrac{y^2}{x^2}$ **21.** $\dfrac{y^{10}}{4x^2}$ **23.** $\dfrac{1}{64a^6b^9}$ **25.** $25x^2y^4$ **27.** $16a^4b^4$ **29.** $\dfrac{1}{8a^3b^6}$ **31.** $\dfrac{y^8}{16x^8}$ **33.** $\dfrac{8a^6}{b^9}$ **35.** $\dfrac{49y^4}{x^6}$

37. $\dfrac{3y^4}{2x^3}$ **39.** $\dfrac{4}{9x^{10}}$ **41.** 8.6×10^4 **43.** 3.62×10^{-2} **45.** 1.83×10^7 **47.** 2.17×10^{-4} **49.** 4.5×10^6

51. 1.23×10^{10} **53.** 4.5×10^{-2} **55.** 7.5×10^{-9} **57.** 1.3×10^6 **59.** 2.5×10^{-3} **61.** 6.0×10^3

63. 3.9×10^0 **65.** 1.67×10^{-24} grams   **67.** 5.866×10^{12} miles

69. 1.8×10^{12} cm per min; 1.08×10^{14} cm per hr

Exercises 4.3, Page 132

1. $4x$; first degree   **3.** $x^3 + 3x^2 - 2x$; third degree   **5.** $-2x^2$; second degree   **7.** 0; no degree

9. $-x^5 - x^2$; fifth degree   **11.** $2x^3 + 4x$; third degree   **13.** 4; zero degree   **15.** $3x^2 + x + 1$; second degree.

17. $2x^2 + 2x + 2$ **19.** $4x^2 + x - 4$ **21.** $5x^2 + 6x - 10$ **23.** $4x^2 + 7x - 8$ **25.** $5x^2 + 14x - 2$

27. $2x^3 - 3x^2 - 3x - 6$ **29.** $3x^2 - 2x + 5$ **31.** $3x^3 + 2x^2 + 3x - 4$ **33.** $5x^3 + 11x^2 + 10x - 14$

35. x **37.** $2x^2 + 5x - 11$ **39.** $x^2 - x - 12$ **41.** $3x^2 - 4x - 8$ **43.** $2x^2 + 13x + 9$

45. $-7x^2 - x + 9$ **47.** $4x^2 + 3x + 11$ **49.** $6x^2 - 7x + 18$ **51.** $14x^2 - 15$ **53.** $2x^3 + 4x^2 + 3x - 10$

55. $2x^3 - 5x^2 + 11x - 11$ **57.** $-12x + 22$ **59.** $-x - 15$ **61.** $-9x + 26$ **63.** $10x^3 - 3x^2 + 7$

65. $-7x^2 + 6x - 2$ **67.** $5x - 7$ **69.** $2x^2 - 3x$

Exercises 4.4, Page 136

1. $6x^5$ **3.** $-20x^3$ **5.** $-3x^2 - 2x$ **7.** $-4x^3 - 4x$ **9.** $7x^4 + 14x^3 - 7x^2$ **11.** $-x^4 - 5x^2 + 4x$

13. $6x^2 - x - 2$ **15.** $9x^2 - 25$ **17.** $-10x^2 + 39x - 14$ **19.** $x^3 + 5x^2 + 8x + 4$ **21.** $x^2 + x - 12$

23. $x^2 - 2x - 48$ **25.** $x^2 - 3x + 2$ **27.** $3x^2 - 3x - 60$ **29.** $x^3 + 11x^2 + 24x$ **31.** $2x^2 - 7x - 4$

33. $6x^2 + 17x - 3$ **35.** $4x^2 - 9$ **37.** $16x^2 + 8x + 1$ **39.** $x^3 + 2x^2 + x + 12$ **41.** $3x^2 - 8x - 35$

43. $-16x^3 + 50x^2 + 25x - 14$ **45.** $12x^4 + 28x^3 + 47x^2 + 34x + 48$

47. $4x^2 + 17x - 42$
$$[2 + 6][4(2) - 7] = (8)(1)$$
$$= 8$$
$$4(2)^2 + 17(2) - 42 = 8$$

49. $5x^2 + 17x - 12$
$$[5(2) - 3][2 + 4] = (7)(6)$$
$$= 42$$
$$5(2)^2 + 17(2) - 12 = 42$$

51. $3x^2 + 2x - 16$
$$[2 - 2][3(2) + 8] = (0)(14)$$
$$= 0$$
$$3(2)^2 + 2(2) - 16 = 0$$

53. $6x^2 - x - 35$
$$[3(2) + 7][2(2) - 5] = (13)(-1)$$
$$= -13$$
$$6(2)^2 - (2) - 35 = -13$$

55. $25x^2 + 20x + 4$
$$[5(2) + 2][5(2) + 2] = (12)(12)$$
$$= 144$$
$$25(2)^2 + 20(2) + 4 = 144$$

57. $x^3 - 4x^2 + 2x - 8$
$$[(2)^2 + 2][2 - 4] = (6)(-2)$$
$$= -12$$
$$(2)^3 - 4(2)^2 + 2(2) - 8 = -12$$

59. $9x^2 - 16$
$$[3(2) - 4][3(2) + 4] = (2)(10)$$
$$= 20$$
$$9(2)^2 - 16 = 20$$

61. $x^3 - 8$
$$[2 - 2][(2)^2 + 2(2) + 4] = (0)(12)$$
$$= 0$$
$$(2)^3 - 8 = 0$$

63. $25x^2 - 60x + 36$
$$[5(2) - 6][5(2) - 6] = (4)(4)$$
$$= 16$$
$$25(2)^2 - 60(2) + 36 = 16$$
67. $x^3 - 6x^2 + 11x - 6$
$$(2-1)(2-2)(2-3) = (1)(0)(-1) = 0$$
$$2^3 - 6 \cdot 2^2 + 11 \cdot 2 - 6 = 8 - 24 + 22 - 6$$
$$= 0$$

65. $3x^3 - 2x^2 + 26x + 9$
$$[3(2) + 1][(2)^2 - (2) + 9] = (7)(11)$$
$$= 77$$
$$3(2)^3 - 2(2)^2 + 26(2) + 9 = 77$$
69. $x^4 + 2x^3 + x^2 - 4$
$$(2^2 + 2 + 2)(2^2 + 2 - 2) = (8)(4) = 32$$
$$2^4 + 2 \cdot 2^3 + 2^2 - 4 = 16 + 16 + 4 - 4$$
$$= 32$$

Exercises 4.5, Page 141

1. $x^2 - 9$; difference of squares   **3.** $x^2 - 10x + 25$; perfect square trinomial   **5.** $x^2 - 36$; difference of squares
7. $x^2 + 16x + 64$; perfect square trinomial   **9.** $2x^2 + x - 3$ **11.** $9x^2 - 24x + 16$; perfect square trinomial
13. $4x^2 - 1$; difference of squares   **15.** $9x^2 - 12x + 4$; perfect square trinomial
17. $9 + 6x + x^2$; perfect square trinomial   **19.** $25 - 10x + x^2$; perfect square trinomial
21. $25x^2 - 81$; difference of squares   **23.** $4x^2 + 28x + 49$; perfect square trinomial
25. $81x^2 - 4$; difference of squares   **27.** $10x^4 - 11x^2 - 6$ **29.** $1 + 14x + 49x^2$; perfect square trinomial
31. $5x^2 + 11x + 2$ **33.** $4x^2 + 13x - 12$ **35.** $3x^2 - 25x + 42$ **37.** $25 + 10x + x^2$ **39.** $x^4 - 1$
41. $x^4 + 6x^2 + 9$ **43.** $x^6 - 4x^3 + 4$ **45.** $x^4 + 3x^2 - 54$ **47.** $x^2 - \dfrac{4}{9}$ **49.** $x^2 - \dfrac{9}{16}$
51. $x^2 + \dfrac{6}{5}x + \dfrac{9}{25}$ **53.** $x^2 - \dfrac{5}{3}x + \dfrac{25}{36}$ **55.** $x^2 - \dfrac{1}{4}x - \dfrac{1}{8}$ **57.** $x^2 + \dfrac{5}{6}x + \dfrac{1}{6}$ **59.** $x^2 - 1.96$
61. $x^2 - 5x + 6.25$ **63.** $x^2 - 4.6225$ **65.** $x^2 + 2.48x + 1.5376$ **67.** $2.0164x^2 + 27.264x + 92.16$
69. $129.96x^2 - 12.25$ **71.** $93.24x^2 + 142.46x - 104.04$

Chapter 4 Review, Page 143

1. $6^7 = 279,936$ **2.** $(-3)^5 = -243$ **3.** $\dfrac{1}{7}$ **4.** $\dfrac{1}{5}$ **5.** y^5 **6.** y **7.** $2x^4$ **8.** $\dfrac{2y^2}{3}$ **9.** $\dfrac{1}{x^7}$

10. $2x$ **11.** $64x^6y^3$ **12.** $\dfrac{49x^{10}}{y^4}$ **13.** $\dfrac{36x^4}{y^{10}}$ **14.** $\dfrac{1}{x^6y^2}$ **15.** $\dfrac{a^6}{b^4}$ **16.** $\dfrac{x}{3y^4}$ **17.** $\dfrac{y^3}{8x^9}$ **18.** 1 **19.** x^4y^2

20. $\dfrac{x^8}{9y^6}$ **21.** 4.27×10^6 **22.** 2.3×10^{-4} **23.** 6.3 **24.** 4.0×10^6 **25.** 1.2×10^{-1} **26.** $5x^2 + 1$;
second degree   **27.** $4x^2 + 9x$; second degree   **28.** $x^2 + 4x$; second degree   **29.** $3x^2 + 8x$; second degree
30. $-x^3 + 3x^2 + x - 2$; third degree   **31.** $-x^3 - 6x^2 + 2x - 4$; third degree   **32.** $7x^2 + 10x - 5$
33. $2x^2 - 8x + 12$ **34.** $x^3 - 2x^2 - 6x - 7$ **35.** $-x^2 + x - 5$ **36.** $3x^3 + 4x^2 - 7x - 3$
37. $-x^3 + 7x^2 - 6$ **38.** $6x^2 - 7x + 1$ **39.** $-x - 17$ **40.** $-2x^2 - 3x$ **41.** $-10x + 4$ **42.** $5x - 11$
43. $2x^3 + 5x^2 + 3x$ **44.** $-18x + 26$ **45.** $7x - 38$ **46.** $-3x^3 + 12x$ **47.** $x^3 - x^5$
48. $x^2 - 36$; difference of squares   **49.** $x^2 + x - 12$ **50.** $9x^2 + 42x + 49$; perfect square trinomial
51. $4x^2 - 1$; difference of squares   **52.** $x^4 - 25$; difference of squares   **53.** $x^4 - 4x^2 + 4$; perfect square trinomial
54. $x^2 - \dfrac{4}{25}$; difference of squares   **55.** $x^2 + \dfrac{5}{4}x + \dfrac{25}{64}$; perfect square trinomial
56. $-x^2 + 5x - 2$
$$-1[(3)^2 - 5(3) + 2] = -1(-4)$$
$$= 4$$
$$-(3)^2 + 5(3) - 2 = 4$$
57. $3x^3 + 6x^2 - 3x$
$$3(3)[(3)^2 + 2(3) - 1] = (9)(14)$$
$$= 126$$
$$3(3)^3 + 6(3)^2 - 3(3) = 126$$

58. $12x^2 - 13x - 14$
$[3(3) + 2][4(3) - 7] = (11)(5)$
$= 55$
$12(3)^2 - 13(3) - 14 = 55$

60. $5x^2 - 27x - 18$
$[3 - 6][5(3) + 3] = (-3)(18)$
$= -54$
$5(3)^2 - 27(3) - 18 = -54$

62. $x^4 - 16$
$(3^2 + 4)(3^2 - 4) = (13)(5) = 65$
$3^4 - 16 = 81 - 16 = 65$

64. $4x^3 - 3x^2 - x$
$4(3) + 1(3)^2 - (3) = (13)(6)$
$= 78$
$4(3)^3 - 3(3)^2 - 3 = 78$

59. $2x^2 - x - 36$
$[2(3) - 9][3 + 4] = (-3)(7)$
$= -21$
$2(3)^2 - (3) - 36 = -21$

61. $6x^2 + x - 12$
$[3(3) - 4][2(3) + 3] = (5)(9)$
$= 45$
$6(3)^2 + 3 - 12 = 45$

63. $x^3 - 6x^2 + 8x - 48$
$(3 - 6)(3^2 + 8) = (-3)(17)$
$= -51$
$(3)^3 - 6(3)^2 + 8(3) - 48 = -51$

65. $5x^3 + 3x^2 - 12x + 4$
$5(3) - 2(3)^2 + (3) - 2 = (13)(10)$
$= 130$
$5(3)^3 + 3(3)^2 - 12(3) + 4 = 130$

Chapter 4 Test, Page 144

1. $-12x^6$ **2.** $2x^9$ **3.** $\dfrac{25y^6}{x^4}$ **4.** $\dfrac{x}{3y^2}$ **5.** $\dfrac{x^2}{4y^2}$ **6.** $4x^2y^4$ **7.** 1.15 **8.** 5.2×10^{-3} **9.** $8x^2 + 3x$;
second degree   **10.** $-x^3 + 3x^2 + 3x - 1$; third degree   **11.** $3x^2 + 2x - 7$ **12.** $2x^2 + 9x - 8$
13. $5x^3 - x^2 + 6x + 5$ **14.** $-2x^2 + x - 6$ **15.** $7x^2 + 14x + 4$ **16.** $3x^3 - 4x^2 - 5x + 8$ **17.** $-3x + 2$
18. $20x - 8$ **19.** $x - 9$ **20.** $-15x^4 + 45x^3$ **21.** $25x^2 - 16$ **22.** $16 - 24x + 9x^2$
23. $6x^3 - 27x^2 - 105x$ **24.** $x^4 - 81$ **25.** $x^2 - \dfrac{8}{5}x + \dfrac{16}{25}$

Cumulative Review (4), Page 144

1. $3(x + 15)$ **2.** $2(3x + 8)$ **3.** $1, 2, 3, 6, 9, 18, 27, 54$ **4.** $1, 2, 3, 4, 6, 8, 12, 24$ **5.** $16, 2$ **6.** $-4, -9$
7. $-4, 14$ **8.** $-12, 4$ **9.** $5x^2 + 16x$ **10.** $5x - 18$ **11.** $x = \dfrac{8}{5}$ **12.** $x = -\dfrac{1}{2}$ **13.** $4x^5$

14. $\dfrac{x^4y^2}{3}$ **15.** $14x^3 + 18x^2$ **16.** $9x^2 - 64$ **17.** $4x^2 - 36x + 81$ **18.** $8x^2 - 14x - 15$ **19.** $66, 68$
20. $16 \text{ cm} \times 25 \text{ cm}$

CHAPTER 5

Exercises 5.1, Page 152

1. x^2 **3.** x^4 **5.** $-4y$ **7.** $3x^3$ **9.** $2x^2y$ **11.** $11(x - 11)$ **13.** $4y(4y^2 + 3)$ **15.** $-8(a + 2b)$
17. $-3a(2x - 3y)$ **19.** $2x^2y(8x^2 - 7)$ **21.** $-14x^2y(y^2 - 1)$ **23.** $5(x^2 - 3x - 1)$ **25.** $4m(2x - 3y + z)$
27. $17x^3(2x^2 - 3x + 1)$ **29.** $x^4(15x^3 + 24x^2 - 32)$ **31.** $z(a^2x^2 - y^2)$ **33.** $7x^2y^4(x^2y^2 + 4)$
35. $5xy^2(3 - 4xy - 5x^4y^5)$ **37.** $-x^2y(16x^3 + 15x^2 - 3)$ **39.** $(x + 1)(b + c)$ **41.** $(x^2 + 6)(x + 3)$
43. $(x + 6y)(x - 4)$ **45.** $(y - 4)(5x + z)$ **47.** $(2x - 3y)(z - 8)$ **49.** $(x^3 + 7)(x - 3)$
51. $x(2x^2 - 7)(2x - 3)$ **53.** $x = 0, x = 7$ **55.** $x = 0, x = -6$ **57.** $x = 0, x = 4$ **59.** $x = 0, x = 2$
61. $x = 0, x = -\dfrac{2}{3}$ **63.** $x = 0, x = \dfrac{3}{4}$ **65.** 24 **67.** 0 or 8 **69.** 0 and 2 or 6 and 8

Exercises 5.2, Page 157

1. $(x + 1)(x - 1)$ **3.** $(x + 7)(x - 7)$ **5.** $(x + 2)^2$ **7.** $(x - 6)^2$ **9.** $(4x + 3)(4x - 3)$
11. $(3x + 1)(3x - 1)$ **13.** $(5 + 2x)(5 - 2x)$ **15.** $(x - 7)^2$ **17.** $(x + 3)^2$ **19.** $3(x + 3y)(x - 3y)$
21. $x(x + y)(x - y)$ **23.** $\left(x + \dfrac{1}{2}\right)\left(x - \dfrac{1}{2}\right)$ **25.** $\left(x + \dfrac{3}{4}\right)\left(x - \dfrac{3}{4}\right)$ **27.** $(x^2 + 1)(x + 1)(x - 1)$
29. $2(x - 8)^2$ **31.** $a(y + 1)^2$ **33.** $4y^2(x - 3)^2$ **35.** $x^2 - 6x + \underline{9} = (x - 3)^2$
37. $x^2 + 4x + \underline{4} = (x + 2)^2$ **39.** $x^2 + \underline{8x} + 16 = (x + 4)^2$ **41.** $x^2 - \underline{18x} + 81 = (x - 9)^2$
43. $x^2 + x + \dfrac{\underline{1}}{4} = \left(x + \dfrac{1}{2}\right)^2$ **45.** $x^2 - 9x + \dfrac{81}{4} = \left(x - \dfrac{9}{2}\right)^2$ **47.** $x^2 + \underline{5x} + \dfrac{25}{4} = \left(x + \dfrac{5}{2}\right)^2$
49. $x^2 - \underline{3x} + \dfrac{9}{4} = \left(x - \dfrac{3}{2}\right)^2$ **51.** $x = -10, x = 10$ **53.** $x = -5, x = 5$ **55.** $x = -4$ **57.** $x = 3$
59. $x = -\dfrac{7}{2}, x = \dfrac{7}{2}$ **61.** $x = 2$ **63.** $x = -\dfrac{3}{2}$ **65.** 9 cm **67.** 30 in., 6 in. **69.** $-5, 5$

Exercises 5.3, Page 166

1. $(x + 3)(x + 2)$ **3.** $(2x - 5)(x + 1)$ **5.** $(6x + 5)(x + 1)$ **7.** $(x - 1)(x - 2)$ **9.** $(x - 5)(x + 2)$
11. $(x - 14)(x + 1)$ **13.** Not factorable **15.** $(2x + 1)(x - 1)$ **17.** $(4t + 1)(t - 1)$ **19.** $(5a - 6)(a + 1)$
21. $(7x - 2)(x + 1)$ **23.** $(4x + 3)(x + 5)$ **25.** $(x + 8)(x - 2)$ **27.** $(3x - 2)(2x - 5)$
29. $(3x - 1)(x - 2)$ **31.** $(3x - 1)^2$ **33.** Not factorable **35.** $(4y + 3)(3y - 4)$ **37.** $5(x^2 + 9)$
39. $(4a - 3)(a - 2)$ **41.** Not factorable **43.** $(4x + y)(4x - y)$ **45.** $(8x - 3)^2$
47. $2(3x - 5)(x + 2)$ **49.** $5(2x + 3)(x + 2)$ **51.** $2(3x + 2y)(3x - 2y)$ **53.** $x^2(7x^2 - 5x + 3)$
55. $-2(6m + 1)(m - 2)$ **57.** $3x(2x - 1)(x + 2)$ **59.** $9xy(x^2y^2 + 1)$ **61.** $3x(2x - 9)^2$
63. $4xy(4y - 3)(3y - 4)$ **65.** $7y^2(3y - 2)(y - 4)$ **67.** $(a + 3)(x + y)$ **69.** $(5 + b)(x + y)(x - y)$

Exercises 5.4, Page 169

1. $x = 2, x = 3$ **3.** $x = -2, x = 7$ **5.** $x = 3$ **7.** $x = -5, x = 5$ **9.** $x = 0, x = 2$
11. $x = 0, x = 4, x = -1$ **13.** $x = 0, x = -3$ **15.** $x = -3, x = 4$ **17.** $x = 2, x = 4$
19. $x = -4, x = 3$ **21.** $x = -\dfrac{1}{2}, x = 3$ **23.** $x = -\dfrac{2}{3}, x = 2$ **25.** $x = -\dfrac{1}{2}, x = 4$
27. $x = -2, x = \dfrac{4}{3}$ **29.** $x = \dfrac{3}{2}$ **31.** $x = -1, x = 3$ **33.** $x = -8, x = -2$ **35.** $x = -5, x = 2$
37. $x = -5, x = 7$ **39.** $x = -6, x = 2$ **41.** $x = -1, x = \dfrac{2}{3}$ **43.** $x = -\dfrac{3}{2}, x = 4$ **45.** $x = 0, x = \dfrac{8}{5}$
47. $x = -2, x = 2$ **49.** $x = -1$ **51.** $x = 2$ **53.** $x = 3$ **55.** $x = -5, x = 10$ **57.** $x = -6, x = -2$
59. $x = \dfrac{1}{2}$ **61.** $x = 0, x = 2, x = 4$ **63.** $x = 0, x = -\dfrac{2}{3}, x = -\dfrac{1}{2}$

Exercises 5.5, Page 173

1. 0, 7 **3.** 4 **5.** 6, 13 or $-13, -6$ **7.** 6 **9.** 3, 11 **11.** $-9, 4$ or 4, 9 **13.** 8, 9 **15.** 6, 7 **17.** 8, 9
19. 10, 12 or $-12, -10$ **21.** 6 in., 12 in. **23.** 5 m, 17 m **25.** 13 ft, 9 ft **27.** 12 m, 7 m **29.** 14 ft
31. 13 cm, 8 cm **33.** 8 m by 12 m **35.** 9 rows **37.** 3 cm by 10 cm **39.** 10 ft by 30 ft or 15 ft by 20 ft

Exercises 5.6, Page 175

1. 2260.8 cu in.   **3.** 5 in.   **5.** 800 lb   **7.** 6 in.   **9.** 10 in.   **11.** 200 ft   **13.** 4 sec or 6 sec
15. 12 amps or 20 amps   **17.** \$16 or \$20   **19.** \$4

Exercises 5.7, Page 176

1. $(m + 6)(m + 1)$ **3.** $(x + 9)(x + 2)$ **5.** $(x + 10)(x - 10)$ **7.** $(m - 3)(m + 2)$ **9.** Not factorable
11. $(8x + 1)(8x - 1)$ **13.** $(x + 5)^2$ **15.** $(x + 12)(x - 3)$ **17.** $(x + 9)(x + 4)$ **19.** $5(x - 6)(x - 8)$
21. $-4(x - 10)(x + 5)$ **23.** $3(x + 7)(x - 7)$ **25.** $3x(x + 2)(x + 3)$ **27.** $4x(2x + 5)(2x - 5)$
29. $(3x - 2)(x - 5)$ **31.** $2(x - 3)(2x - 1)$ **33.** $(2x - 5)(6x - 1)$ **35.** $(3x - 7)(2x + 5)$
37. $(2x + 5)(4x - 7)$ **39.** $(5x + 6)(4x - 9)$ **41.** $3(4x^2 - 20x - 25)$ **43.** $(3 - x)(8 + 3x)$
45. $(2x - 3)(4x - 5)$ **47.** $(4x + 5)(5x - 4)$ **49.** $(6x - 1)(3x - 2)$ **51.** $7x(6 + 5x)(6 - 5x)$
53. $x(7x - 2)(3x - 1)$ **55.** $3x(3x - 2)(4x + 5)$ **57.** $2x(4x - 11)(2x - 1)$ **59.** $-5(4x + 3)(6x - 5)$
61. $(y - 4)(x + 3)$ **63.** $(x - 6)(x + 2y)$ **65.** $(x^2 - 5)(x - 8)$

Chapter 5 Review, Page 178

1. x^3 **2.** x^3 **3.** $2x$ **4.** $-8x^2y$ **5.** $4x^2y$ **6.** $5(x - 2)$ **7.** $-4x(3x + 4)$ **8.** $8xy(2x - 3)$
9. $5x^2(2x^2 - 5x + 1)$ **10.** $(2x + 1)(2x - 1)$ **11.** $(y - 10)^2$ **12.** $(9x + 2y)(9x - 2y)$ **13.** $(2x - 3)^2$
14. $a(c^2 + ab^2 + d)$ **15.** $3(x + 4y)(x - 4y)$ **16.** $(x - 9)(x + 2)$ **17.** $5(x + 4)^2$ **18.** $(5x + 2)(x + 3)$
19. $(x + 6)(x - 5)$ **20.** Not factorable   **21.** $(5x + 2)^2$ **22.** $(3x + 2)(x + 1)$ **23.** $2x(x - 5)^2$
24. $4x(x^2 + z^2)$ **25.** $(3x - 2)(2x + 1)$ **26.** $2x(4x + 3)(x - 2)$ **27.** $(x^2 + 7)(x + 2)(x - 2)$
28. $(x^2 + 2)(x + 2)(x - 2)$ **29.** $(y + 3)(x + 2)$ **30.** $(a + b)(x - 2)$ **31.** $x^2 - 4x + \underline{4} = (x - 2)^2$
32. $x^2 + 18x + \underline{81} = (x + 9)^2$ **33.** $x^2 - 8x + \underline{16} = (x - 4)^2$ **34.** $x^2 - \underline{10x} + 25 = (x - 5)^2$
35. $x^2 + \underline{5x} + \dfrac{25}{4} = \left(x + \dfrac{5}{2}\right)^2$ **36.** $x = 7, x = -1$ **37.** $x = 0, x = -\dfrac{5}{3}$ **38.** $x = 0, x = -5, x = 2$
39. $x = 0, x = 7$ **40.** $x = -6, x = -2$ **41.** $x = 7, x = -4$ **42.** $x = -3, x = 3$
43. $x = 0, x = -6, x = 1$ **44.** $x = 11, x = -5$ **45.** $x = -10, x = 6$ **46.** $x = 1, x = -5$
47. $x = 0, x = -7$ **48.** $x = 0, x = -2$ **49.** $x = 0, x = 4$ **50.** $x = -\dfrac{3}{5}, x = 2$ **51.** $x = -3$ or $x = \dfrac{3}{4}$
52. $x = -\dfrac{5}{4}, x = 1$ **53.** $x = -\dfrac{5}{2}, x = 5$ **54.** $x = -\dfrac{4}{3}, x = \dfrac{5}{2}$ **55.** 3, 12   **56.** 12 cm by 17 cm
57. 8, 9   **58.** 13 ft, 6 ft   **59.** 22 in., 17 in.   **60.** $2\dfrac{1}{2}$ sec or 3 sec

Chapter 5 Test, Page 179

1. $2x^3$ **2.** $-7y^2$ **3.** $x^3y(20y + 18 - 15x)$ **4.** $(x - 5)(x - 4)$ **5.** $(x + 7)^2$ **6.** $(6x + 1)(6x - 1)$
7. $(6x - 5)(x + 1)$ **8.** $(3x - 8)(x + 3)$ **9.** $(4x + 5y)(4x - 5y)$ **10.** $x(2x - 3)(x + 1)$
11. $(2x - 3)(3x - 2)$ **12.** $(y + 7)(2x - 3)$ **13.** $x^2 - 10x + \underline{25} = (x - 5)^2$
14. $x^2 + 16x + \underline{64} = (x + 8)^2$ **15.** $x^2 + \underline{24x} + 144 = (x + 12)^2$ **16.** $x = -2, x = \dfrac{5}{3}$
17. $x = 8, x = -1$ **18.** $x = 0, x = -6$ **19.** $x = 5, x = -3$ **20.** $x = \dfrac{3}{4}, x = -5$
21. $x = -\dfrac{5}{4}, x = \dfrac{3}{2}$ **22.** $x = \dfrac{3}{2}, x = 4$ **23.** 6, 20 or $-4, -30$ **24.** 18, 19   **25.** 15 cm, 11 cm

Cumulative Review (5), Page 180

1. 120   **2.** $168x^2y$ **3.** $\dfrac{55}{48}$ **4.** $\dfrac{19}{60a}$ **5.** $\dfrac{3}{10}$ **6.** $\dfrac{75x}{23}$ **7.** $-\dfrac{xy^4}{3}$ **8.** $7x^2y$ **9.** $13x + 1$

10. $x^2 + 12x + 3$ **11.** $3x^2 - 5x + 3$ **12.** $5x - 8$ **13.** $x^2 - 3x - 5$ **14.** $-4x^2 - 2x + 3$

15. $2x^2 + x - 28$ **16.** $-3x^2 - 17x + 6$ **17.** $4x(2x - 5)$ **18.** $6(x + 4)(x - 4)$ **19.** $(2x - 3)(x - 6)$

20. $(3x + 4)(2x - 3)$

CHAPTER 6

Exercises 6.1, Page 187

1. $x^2 + 2x + \dfrac{3}{4}$ **3.** $2x^2 - 3x - \dfrac{3}{5}$ **5.** $2x + 5$ **7.** $x + 6 - \dfrac{3}{x}$ **9.** $2x + 3 - \dfrac{3}{2x}$ **11.** $2x - 3 - \dfrac{1}{x}$

13. $x + 3y - \dfrac{11y}{7x}$ **15.** $\dfrac{3x}{4} - 2y - \dfrac{y^2}{x}$ **17.** $\dfrac{5x}{8} - 1 + \dfrac{2}{y}$ **19.** $\dfrac{8x^2}{9} - xy + \dfrac{5}{9y}$ **21.** $12\dfrac{2}{23}$ **23.** $7\dfrac{24}{59}$

25. $x - 8 + \dfrac{18}{x + 3}$ **27.** $a + \dfrac{-15}{a - 2}$ **29.** $y + 3 + \dfrac{-30}{y - 4}$ **31.** $5y - 11 + \dfrac{48}{y + 5}$ **33.** $4c - 5 + \dfrac{1}{2c + 3}$

35. $2m + 1 + \dfrac{-3}{5m - 3}$ **37.** $x - 2 + \dfrac{10}{x + 5}$ **39.** $y^2 - 7y + 12$ **41.** $3a^2 + 1 + \dfrac{2}{4a - 1}$

43. $x^2 + 7x + 35 + \dfrac{170}{x - 5}$ **45.** $x^2 - 3x + 9$ **47.** $2x + \dfrac{3x + 3}{x^2 - 2}$

Exercises 6.2, Page 193

1. $\dfrac{x - 2}{2x + 1}; x \ne -\dfrac{1}{2}$ **3.** $\dfrac{1}{x - 4}; x \ne 4, 0$ **5.** $7; x \ne -2$ **7.** $1; x \ne -\dfrac{5}{3}$ **9.** $-1; x \ne 1, -1$

11. $\dfrac{2}{x + 3}; x \ne -3$ **13.** $\dfrac{x + 2}{x - 5}; x \ne 5, -5$ **15.** $\dfrac{x - 6}{x + 3}; x \ne -3$ **17.** $\dfrac{-(x - 3)}{2x(2 + x)}$ or $\dfrac{3 - x}{2x(2 + x)}; x \ne 0, 2, -2$

19. $\dfrac{4x + 5}{x + 2}; x \ne -2, -\dfrac{5}{4}$ **21.** $\dfrac{2x}{x + 2}$ **23.** $\dfrac{2x(x - 3)}{x - 1}$ **25.** $3(x - 2)$ **27.** $\dfrac{2(x - 1)}{x + 1}$ **29.** $\dfrac{5x}{6}$

31. $\dfrac{x(x - 2)}{(x + 1)(x - 1)}$ **33.** $\dfrac{x + 3}{x + 4}$ **35.** $\dfrac{3x + 1}{x + 1}$ **37.** $\dfrac{-(x - 5)}{(x - 2)(x + 7)}$ or $\dfrac{5 - x}{(x - 2)(x + 7)}$ **39.** -1

41. $\dfrac{x(2x - 1)}{x + 4}$ **43.** $\dfrac{4x}{x - 3}$ **45.** $\dfrac{x - 3}{2x - 3}$ **47.** $\dfrac{3x - 2}{3x + 2}$ **49.** $\dfrac{(x + 3)(x - 4)}{(2x + 1)(x - 1)}$ **51.** $2x^2$

53. $\dfrac{x(x - 3)}{(x + 1)^2}$ **55.** $\dfrac{x(x + 5)}{2x + 1}$ **57.** $\dfrac{x - 1}{x + 1}$ **59.** $\dfrac{2x - 1}{(6x + 1)(2x + 1)}$

Exercises 6.3, Page 199

1. 4 **3.** 2 **5.** $\dfrac{x-3}{x+1}$ **7.** $\dfrac{1}{x-1}$ **9.** $\dfrac{x-2}{x+2}$ **11.** $\dfrac{x}{2x-1}$ **13.** $\dfrac{x-5}{x-4}$ **15.** $\dfrac{2}{x-3}$ **17.** $\dfrac{-2}{5(x-2)}$

19. $\dfrac{3x+8}{x(x+4)}$ **21.** $\dfrac{2x^2}{(x+4)(x-4)}$ **23.** $\dfrac{3}{x-2}$ **25.** $\dfrac{x+3}{x-5}$ **27.** $\dfrac{x^2+x+1}{(x-1)(x+2)}$ **29.** $\dfrac{-x^2-11x-4}{(x+4)(x-4)}$

31. $\dfrac{4x+24}{(x+2)(x-2)}$ **33.** $\dfrac{-9x-4}{12(x-2)}$ **35.** $\dfrac{-2x-17}{(x+5)(x-1)}$ **37.** $\dfrac{3x^2-20x}{(x+6)(x-6)}$ **39.** $\dfrac{2x^2+2x-1}{(x-1)(x-7)}$

41. $\dfrac{x^2+4x+2}{2(x+2)(x+3)}$ **43.** $\dfrac{-x^2-3x}{2(x+4)(x-4)}$ **45.** $\dfrac{-1}{(x+2)(x+1)(x-1)}$ **47.** $\dfrac{5x^2+18x}{(x+4)(x+3)(x-3)}$

49. $\dfrac{-(2x^2+8x+3)}{(x+2)(x+2)(x+1)}$ **51.** $\dfrac{-2x^2+13x}{(x+4)(x-4)(x+1)}$ **53.** $\dfrac{3x^2-5x-7}{(2x-1)(x+3)(x-4)}$

55. $\dfrac{4x^2+17x}{(x+5)(x+2)(x-2)}$ **57.** $\dfrac{10x^2-17x-3}{(x+4)(x-4)}$ **59.** $\dfrac{x+1}{(x-2)(x-1)}$

Exercises 6.4, Page 204

1. $\dfrac{8}{7}$ **3.** $\dfrac{10}{7}$ **5.** $\dfrac{1}{2}$ **7.** 3 **9.** $\dfrac{2x}{3y}$ **11.** $16xy$ **13.** $\dfrac{8}{7x^2y^3}$ **15.** $\dfrac{3x}{x-2}$ **17.** $\dfrac{2x(x+3)}{2x-1}$

19. $\dfrac{x+3}{x}$ **21.** $\dfrac{x+2}{4x}$ **23.** $\dfrac{2x}{3(x+6)}$ **25.** $\dfrac{x}{x-1}$ **27.** $\dfrac{x}{y+x}$ **29.** $\dfrac{2(x+1)}{x+2}$ **31.** $\dfrac{x+2}{x+3}$

33. $-\dfrac{5}{x+1}$ **35.** $\dfrac{29}{4(4x+5)}$ **37.** $\dfrac{x^2-3x-6}{x(x-1)}$ **39.** $\dfrac{x^2-4x-2}{(x+4)(x-4)}$

Exercises 6.5, Page 210

1. $x=\dfrac{3}{2}$ **3.** $x=7$ **5.** $x\neq 0;\, x=10$ **7.** $x\neq -4;\, x=20$ **9.** $x\neq 0;\, x=\dfrac{10}{3}$ **11.** $x\neq 0;\, x=\dfrac{3}{2}$

13. $x\neq 1;\, x=-3$ **15.** $x\neq 0,2;\, x=4$ **17.** $x\neq -7,4;\, x=-62$ **19.** $x\neq 6;\, x=5$

21. $x\neq 3;\, x=\dfrac{13}{5}$ **23.** $x\neq -3,0;\, x=6$ **25.** $x\neq -3,-2;\, x=-\dfrac{3}{2}$ **27.** $x\neq \dfrac{1}{2},4;\, x=-3$

29. $x\neq -1,\dfrac{1}{4};\, x=\dfrac{2}{3}$ **31.** $x\neq -5,4;\, x=-2$ **33.** $x\neq -1,1;\, x=-9, x=2$ **35.** $x\neq -1,3;\, x=-2$

37. $x\neq -5,-2;\, x=-\dfrac{22}{5}, x=-1$ **39.** \$8.50 **41.** 45 mi **43.** 144 bulbs **45.** 600 times at bat

47. 8.8 quarts **49.** 40 ft

Exercises 6.6, Page 216

1. 21, 27 **3.** 20 **5.** 400, 250 **7.** $\dfrac{7}{9}$ **9.** 45 mph, 60 mph **11.** 48 mph **13.** 500 mph, 520 mph

15. 480 mph **17.** 9 mph **19.** 14 mph **21.** $2\dfrac{2}{5}$ hr **23.** $11\dfrac{1}{4}$ min **25.** 45 min **27.** $1\dfrac{1}{2}$ hr, 3 hr

29. 10 days, 15 days

Exercises 6.7, Page 218

1. 384 rpm   **3.** 36 teeth   **5.** $139\frac{7}{32}$ sq ft   **7.** 192 lb per sq ft   **9.** 700 lb per sq ft   **11.** 1.5 ml

13. 9 sacks, 27 cu ft sand, 45 cu ft gravel   **15.** 900 lb   **17.** $1\frac{1}{3}$ ft from 960-lb weight   **19.** $5\frac{1}{3}$ ohms

Chapter 6 Review, Page 221

1. $2x^2 + x - \frac{5}{3}$ **2.** $4x - 7 + \frac{3}{x}$ **3.** $\frac{3x}{5} + 2y + \frac{y^2}{x}$ **4.** $1 - 3y + \frac{8y^2}{7}$ **5.** $2x^2 - \frac{3xy}{2} + y^2$ **6.** $10\frac{8}{37}$

7. $12\frac{60}{61}$ **8.** $x - 2$ **9.** $x - 15 - \frac{1}{x + 1}$ **10.** $x + 2 + \frac{4}{4x - 3}$ **11.** $2x^2 - x + 3 - \frac{2}{x + 3}$

12. $4x + 13 + \frac{35}{x - 2}$ **13.** $x - 3 + \frac{18}{x + 3}$ **14.** $2x^2 - 3x + 2 - \frac{3}{3x + 2}$ **15.** $x^2 - 4x + 16$

16. $\frac{1}{x + 1}$; $x \neq 0, -1$ **17.** $-\frac{1}{3}$; $x \neq 4$ **18.** $\frac{2}{3}$; $x \neq -3$ **19.** $\frac{x}{x + 4}$; $x \neq -3, -4$

20. $\frac{x + 5}{2(x - 3)}$; $x = 3$ **21.** $-\frac{x - 5}{4 + x}$; $x \neq 4, -4$ **22.** $\frac{2x + 1}{x + 1}$; $x \neq -1, \frac{3}{4}$ **23.** $-\frac{x}{2}$; $x \neq 0, \frac{2}{5}$

24. $x - y$ **25.** 2   **26.** 3   **27.** $\frac{3}{4}$ **28.** $\frac{7}{3}$ **29.** $\frac{16x - 28}{x(x - 4)}$ **30.** $\frac{4}{(y + 2)(y + 3)}$

31. $\frac{x^2 - 2x}{3(x + 2)(x + 1)}$ **32.** $\frac{x + 4}{3x}$ **33.** $\frac{3x(x + 2)(x + 2)}{x + 3}$ **34.** $\frac{2x + 1}{x - 1}$ **35.** $\frac{4x + 25}{(x + 5)(x - 5)}$

36. $\frac{12x^2 + 52x - 24}{(x + 6)(x + 3)(x - 2)}$ **37.** $\frac{2x^2 + 14x - 8}{(x + 3)(x - 2)(x - 1)}$ **38.** $\frac{-4}{(x - 1)(x + 4)}$ **39.** $\frac{x - 2}{x + 3}$

40. $\frac{x - 4}{(x + 2)(x - 2)}$ **41.** $\frac{7 - x}{7x^2}$ **42.** $\frac{5x^2 - x - 47}{(x - 5)(x + 2)(x + 3)}$ **43.** $\frac{4}{13}$ **44.** $\frac{19}{14}$ **45.** $2(x + 1)$

46. $\frac{x - 1}{x + 1}$ **47.** $\frac{3x}{2 - x}$ **48.** $x = 52$ **49.** $x = 14\frac{2}{5}$ **50.** $x = \frac{9}{2}$ **51.** $x = \frac{13}{2}$ **52.** $x = 7$

53. $x = -8, x = 3$ **54.** $4\frac{1}{2}$ in. by 6 in.   **55.** 161   **56.** 24, 60   **57.** 20 mph, 25 mph   **58.** 3 mph

59. $2\frac{2}{9}$ hr   **60.** $4\frac{1}{2}$ days, 9 days   **61.** 25,000 lb   **62.** 3 ohms

Chapter 6 Test, Page 223

1. $2x + \frac{3}{2} - \frac{3}{x}$ **2.** $\frac{5}{3} + 2y + \frac{y^2}{x}$ **3.** $x + 9$ **4.** $x - 6 - \frac{2}{2x + 3}$ **5.** $x^2 + 3x + 9$

6. $\frac{x + 3}{x}$; $x \neq 0, \frac{1}{2}$ **7.** $\frac{2x + 1}{x + 1}$ **8.** $-\frac{x^2}{2}$ **9.** $\frac{7(4x + 3)}{x(x + 4)}$ **10.** $\frac{1}{2(2x + 1)}$ **11.** $\frac{10x - 13}{(x + 1)(x - 1)(x - 2)}$

12. $\dfrac{8x^2 - 19x + 15}{(x + 5)(x - 5)(x - 2)}$ **13.** $\dfrac{3x^2 - 8x - 27}{(x - 1)(x - 4)(x + 3)}$ **14.** $\dfrac{x(x - 3)}{2(x + 1)}$ **15.** $\dfrac{1}{x + 1}$ **16.** $\dfrac{6(x + 3)}{x + 18}$

17. $x = \dfrac{15}{26}$ **18.** $x = -\dfrac{5}{11}$ **19.** $x = \dfrac{1}{3}$ **20.** $x = 2$ **21.** $3\dfrac{3}{5}$ ft   **22.** 40   **23.** 65, 91

24. 42 mph, 57 mph   **25.** 4 hr

Cumulative Review (6), Page 224

1. 3.57   **2.** 5   **3.** $3x^2 - 5x - 28$ **4.** $4x^2 - 20x + 25$ **5.** $t = \dfrac{A - p}{pr}$ **6.** $6x - 23$ **7.** $10x + 4$

8. $(2x + 3)(2x + 5)$ **9.** $2(x + 5)(x - 2)$ **10.** $8x^2$ **11.** $5x^3y^2$ **12.** $\dfrac{4x^2}{x - 3}$ **13.** $\dfrac{14(x - 2)}{5(x - 6)}$

14. $\dfrac{11x + 4}{(2x + 3)(x - 1)}$ **15.** $\dfrac{3x^2 - 19x + 28}{(x - 7)(x - 5)}$ **16.** $x = -1$ **17.** $x = -5, x = 6$ **18.** $x = -45$

19. 60 mph, 230 mph   **20.** $2\dfrac{2}{3}$ hr, 8 hr

CHAPTER 7

Exercises 7.1, Page 230

1. $x = -5$ **3.** $x = 5$ **5.** $x = 6$ **7.** $x = \dfrac{5}{2}$ **9.** $x = 2$ **11.** $x = 3$ **13.** $x = \dfrac{3}{2}$ **15.** $x = \dfrac{1}{2}$

17. $x = 3$ **19.** $x = 3$ **21.** $x = -7$ **23.** $x = -1$ **25.** $x = 21$ **27.** $x = -\dfrac{3}{2}$ **29.** $x = 20$

31. $x = -90$ **33.** $x = 17$ **35.** $x = \dfrac{13}{5}$ **37.** $x = -3$ **39.** $x = \dfrac{1}{9}$ **41.** $x = -13$ **43.** $x = \dfrac{10}{3}$

45. $x = \dfrac{1}{4}$ **47.** $x = -\dfrac{1}{27}$ **49.** $x = \dfrac{2}{3}$ **51.** $x = 0, x = 5$ **53.** $x = 0, x = -\dfrac{9}{5}$ **55.** $x = -1, x = 7$

57. $x = -5, x = 2$ **59.** $x = -1, x = \dfrac{6}{5}$ **61.** $x = -\dfrac{3}{5}, x = 2$ **63.** $x = \dfrac{1}{3}, x = 2$ **65.** $x = -2, x = \dfrac{5}{3}$

67. $x = 15$ **69.** $x = 0$ **71.** $x = -\dfrac{3}{2}$ **73.** $x = 1, x = 7$ **75.** $x = -3, x = \dfrac{2}{3}$ **77.** $x = 24$

79. $x \approx 12.41$

Exercises 7.2, Page 237

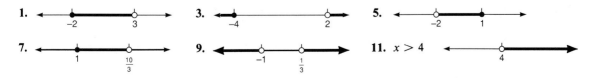

11. $x > 4$

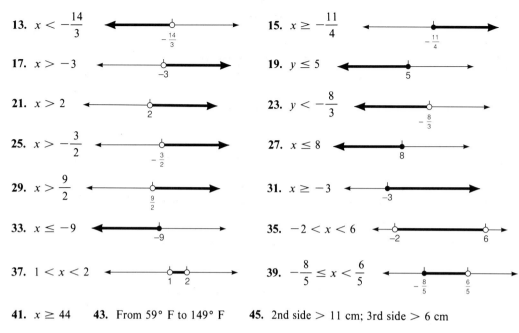

13. $x < -\dfrac{14}{3}$

15. $x \geq -\dfrac{11}{4}$

17. $x > -3$

19. $y \leq 5$

21. $x > 2$

23. $y < -\dfrac{8}{3}$

25. $x > -\dfrac{3}{2}$

27. $x \leq 8$

29. $x > \dfrac{9}{2}$

31. $x \geq -3$

33. $x \leq -9$

35. $-2 < x < 6$

37. $1 < x < 2$

39. $-\dfrac{8}{5} \leq x < \dfrac{6}{5}$

41. $x \geq 44$ **43.** From 59° F to 149° F **45.** 2nd side > 11 cm; 3rd side > 6 cm

47. At least 86 **49.** At least 171

Exercises 7.3, Page 241

1. 8 cm by 40 cm **3.** 140 ft by 220 ft **5.** 17 m by 35 m **7.** 10 m, 20 m, 21 m **9.** 9 m, 21 m, 19 m
11. 6 cm, 6 cm, 10 cm **13.** 40 mph, 45 mph **15.** June—25 mph; Sue—45 mph **17.** 3 hr **19.** 18 mi
21. 45 mi **23.** 6 hr **25.** 14 m, 8 m **27.** 3 mph **29.** 6 cm, 2 cm **31.** 30 mi

Exercises 7.4, Page 246

1. $1200 **3.** $1150 **5.** $800 @ 13%; $1500 @ 9% **7.** $900 @ 12%; $2100 @ 10%
9. $750 @ 7%; $2250 @ 8% **11.** 16 oz **13.** 84 oz **15.** 6.4 qt **17.** 12 liters **19.** 50 lb
21. $2500 @ 8%; $3500 @ 12% **23.** $3800 @ 15%; $4200 @ 12% **25.** 8 tons of 80%, 16 tons of 50%
27. 50 lb of 30%, 40 lb of 12%

Chapter 7 Review, Page 249

1. $x = 4$ **2.** $x = -6$ **3.** $x = -1$ **4.** $x = 6$ **5.** $x = -15$ **6.** $x = 4$ **7.** $x = 1$ **8.** $x = \dfrac{34}{3}$

9. $x = -31$ **10.** $x = \dfrac{100}{13}$ **11.** $x = -\dfrac{72}{7}$ **12.** $x = -\dfrac{3}{2}$ **13.** $x = \dfrac{5}{4}$ **14.** $x = 1$ **15.** $x = -9$

16. $x = 4$ **17.** $x = 0, x = -\dfrac{8}{3}$ **18.** $x = -\dfrac{3}{4}, x = 2$ **19.** $x = 18$ **20.** $x = -9, x = 2$

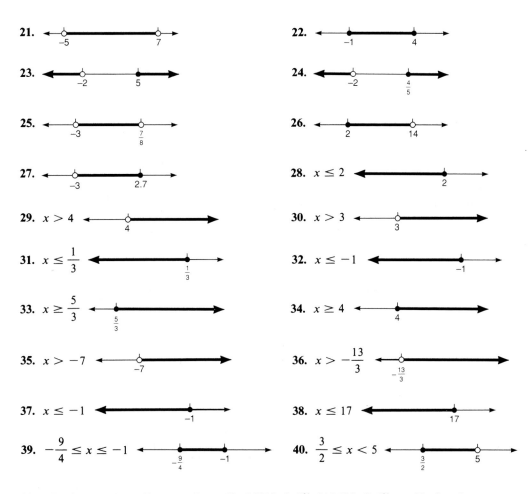

21. —5 ○——————● 7

22. ●————● −1 4

23. ←————○ −2 ● 5 →

24. ←————○ −2 ● 4/5 →

25. ○——————○ −3 7/8

26. ●——————○ 2 14

27. ○——————● −3 2.7

28. $x \leq 2$ ←——————● 2

29. $x > 4$ ○——————→ 4

30. $x > 3$ ○——————→ 3

31. $x \leq \dfrac{1}{3}$ ←——————● 1/3

32. $x \leq -1$ ←——————● −1

33. $x \geq \dfrac{5}{3}$ ●——————→ 5/3

34. $x \geq 4$ ●——————→ 4

35. $x > -7$ ○——————→ −7

36. $x > -\dfrac{13}{3}$ ○——————→ −13/3

37. $x \leq -1$ ←——————● −1

38. $x \leq 17$ ←——————● 17

39. $-\dfrac{9}{4} \leq x \leq -1$ ●————● −9/4 −1

40. $\dfrac{3}{2} \leq x < 5$ ●————○ 3/2 5

41. 15 m by 110 m **42.** $x < 12$ **43.** $6500 @ 4%; $19,500 @ 6% **44.** 5 mph
45. Between 12 and 17 yr, inclusive **46.** 30 lb **47.** $x < 12$ **48.** 19 cm; 6 cm
49. Soo—42 mph; Lisa—57 mph **50.** 20 liters @ 55%, 40 liters @ 40%

Chapter 7 Test, Page 250

1. $x = 1$ **2.** $x = 6$ **3.** $x = \dfrac{19}{10}$ **4.** $x = 20$ **5.** $x = 0, x = \dfrac{3}{2}$ **6.** $x = -\dfrac{3}{2}, x = 2$ **7.** $x = -7$

8. $x = \dfrac{8}{5}$ **9.** ←○————● −4 1.5 →

10. ←○————○ −1 3/2 →

11. ○————● −3/4 3

12. $x > \dfrac{17}{9}$ ←○——————→ 17/9

13. $-\dfrac{3}{5} < x < 1$ 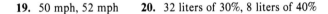 **14.** $x \le \dfrac{9}{2}$

15. $x > \dfrac{19}{10}$ **16.** $x \ge 7$

17. 8 ft by 25 ft **18.** 52 mph, 65 mph **19.** \$5000 @ 18%; 6200 @ 12% **20.** $3 \le x \le 12$
21. 200 gal **22.** \$960 @ 6%; \$1600 @ 8% **23.** 10 miles **24.** $x \ge 86$ **25.** 1600 lb of 83%, 400 lb of 68%

Cumulative Review (7), Page 251

1. $x = 4$ is not a solution. **2.** $x = 2$ is a solution. **3.** $y = \dfrac{9}{2} - 3x$ **4.** $y = \dfrac{3}{4}x + \dfrac{7}{4}$ **5.** $(5x - 2)(x + 6)$

6. $2(x - 5)(4x + 3)$ **7.** $(2x + 5)(3y + 4)$ **8.** $\dfrac{x + 4}{x + 7}$ **9.** $\dfrac{x^2 + 10x + 8}{(3x + 2)(x - 2)}$ **10.** $\dfrac{6 - x^2}{(x - 5)(x + 3)}$

11. $x = -14$ **12.** $x = -5, x = -3$ **13.** $x = \dfrac{27}{4}$ **14.** $x \le 1$

15. $x \ge \dfrac{13}{3}$ **16.** 35 parts **17.** 15 days **18.** 30 cm, 16 cm

19. 50 mph, 52 mph **20.** 32 liters of 30%, 8 liters of 40%

CHAPTER 8

Exercises 8.1, Page 260

1. $(-5,1), (-3,3), (-1,1), (1,2), (2,-2)$ **3.** $(-3,-2), (-1,3), (-1,-3), (0,0), (2,1)$
5. $(-3,-4), (-3,4), (0,3), (0,-4), (4,1)$ **7.** $(-5,2), (-1,6), (0,0), (1,-6), (6,4)$
9. $(-5,0), (-2,2), (-1,-4), (0,6), (2,0)$
11. **13.** **15.**

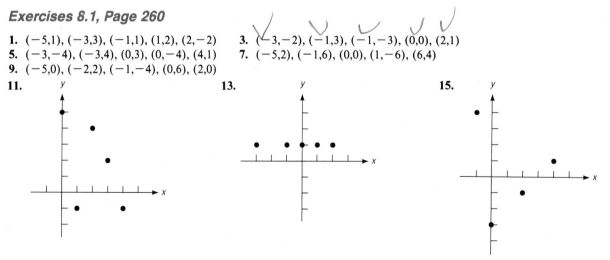

17.

19.

21.

23.

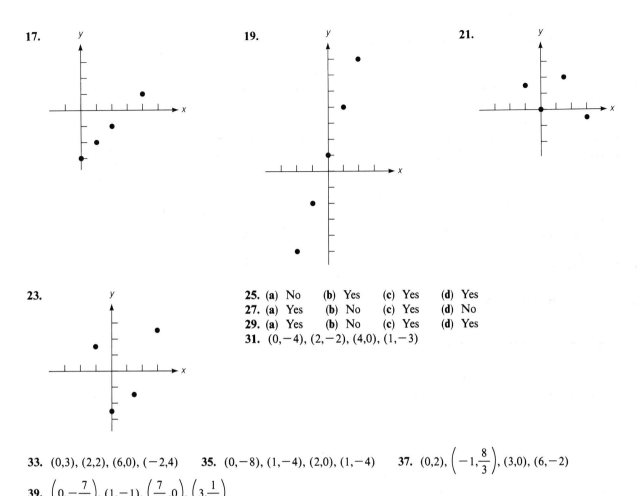

25. **(a)** No   **(b)** Yes   **(c)** Yes   **(d)** Yes
27. **(a)** Yes   **(b)** No   **(c)** Yes   **(d)** No
29. **(a)** Yes   **(b)** No   **(c)** Yes   **(d)** Yes
31. $(0,-4)$, $(2,-2)$, $(4,0)$, $(1,-3)$

33. $(0,3)$, $(2,2)$, $(6,0)$, $(-2,4)$ **35.** $(0,-8)$, $(1,-4)$, $(2,0)$, $(1,-4)$ **37.** $(0,2)$, $\left(-1,\dfrac{8}{3}\right)$, $(3,0)$, $(6,-2)$

39. $\left(0,-\dfrac{7}{4}\right)$, $(1,-1)$, $\left(\dfrac{7}{3},0\right)$, $\left(3,\dfrac{1}{2}\right)$

41.

x	y
0	0
-1	-3
-2	-6
2	6

43.

x	y
0	-3
1	-1
-2	-7
$\dfrac{1}{2}$	-2

45.

x	y
0	7
$\dfrac{7}{3}$	0
-1	10
$\dfrac{1}{3}$	6

47.

x	y
0	2
4	5
-4	-1
-1	$\dfrac{5}{4}$

49.

x	y
0	$-\dfrac{9}{5}$
3	0
-2	-3
$\dfrac{4}{3}$	-1

Exercises 8.2, Page 267

1.

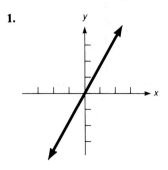

3.

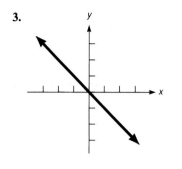

5.

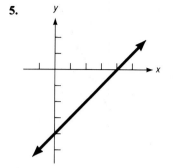

7.

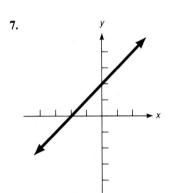

9.

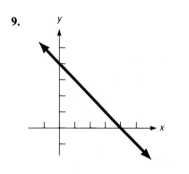

11.

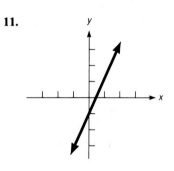

13.

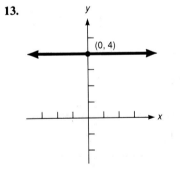

(0, 4)

15.

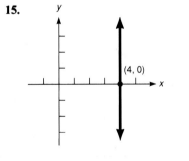

(4, 0)

17.

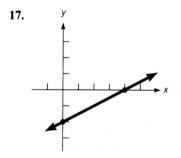

19.

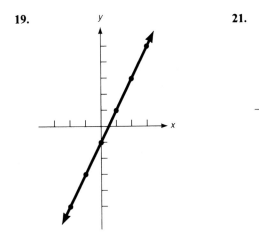

21.

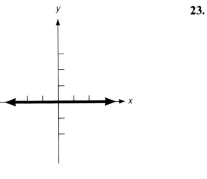

23.

25.

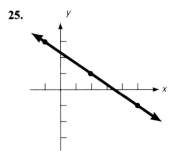

27.

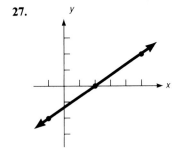

29.

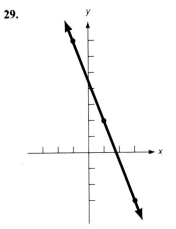

31.

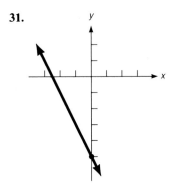

33.

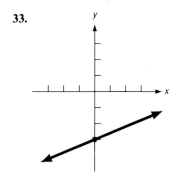

35.

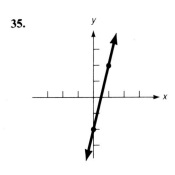

37.

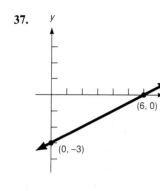

(6, 0)

(0, −3)

39.

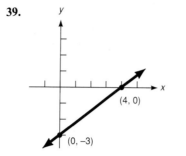

(4, 0)

(0, −3)

41.

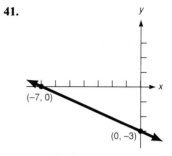

(−7, 0)

(0, −3)

43.

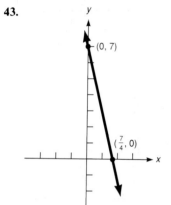

(0, 7)

$\left(\frac{7}{4}, 0\right)$

45.

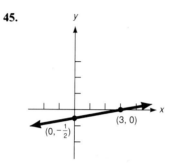

(3, 0)

$\left(0, -\frac{1}{2}\right)$

47.

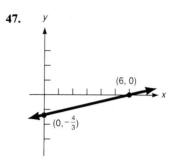

(6, 0)

$\left(0, -\frac{4}{3}\right)$

49.

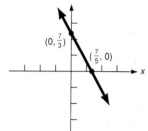

$\left(0, \frac{7}{3}\right)$

$\left(\frac{7}{5}, 0\right)$

Exercises 8.3, Page 276

1. $m = -5$ **3.** $m = \dfrac{8}{7}$ **5.** $m = \dfrac{1}{8}$ **7.** $m = \dfrac{3}{10}$ **9.** $m = \dfrac{5}{4}$

11. $y = \dfrac{2}{3}x$ **13.** $y = -\dfrac{3}{4}x - 3$ **15.** $y = -\dfrac{5}{3}x + 3$

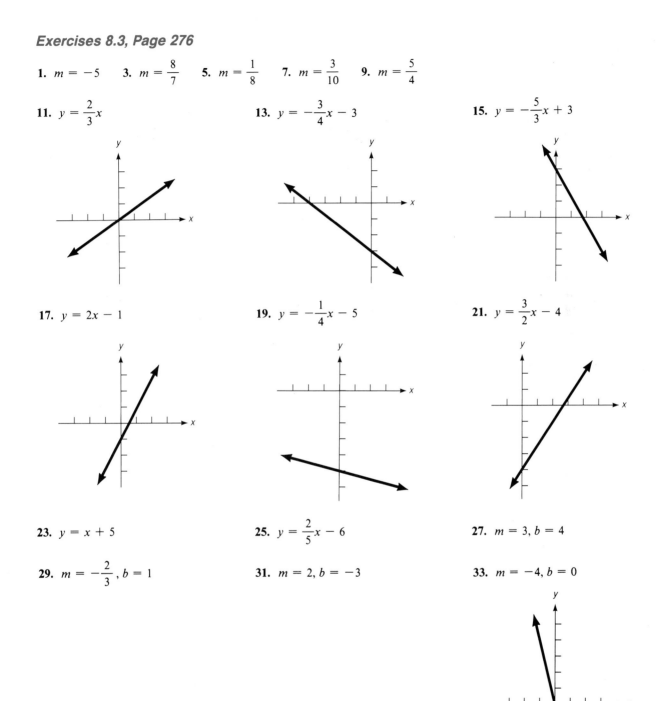

17. $y = 2x - 1$ **19.** $y = -\dfrac{1}{4}x - 5$ **21.** $y = \dfrac{3}{2}x - 4$

23. $y = x + 5$ **25.** $y = \dfrac{2}{5}x - 6$ **27.** $m = 3, b = 4$

29. $m = -\dfrac{2}{3}, b = 1$ **31.** $m = 2, b = -3$ **33.** $m = -4, b = 0$

35. $m = -\dfrac{1}{4}$, $b = -2$ **37.** $m = \dfrac{2}{5}$, $b = -2$ **39.** $m = -\dfrac{7}{2}$, $b = 2$

41. $m = -\dfrac{3}{8}$, $b = -2$ **43.** $m = -\dfrac{2}{3}$, $b = \dfrac{8}{3}$ **45.** $m = \dfrac{5}{2}$, $b = -\dfrac{3}{2}$

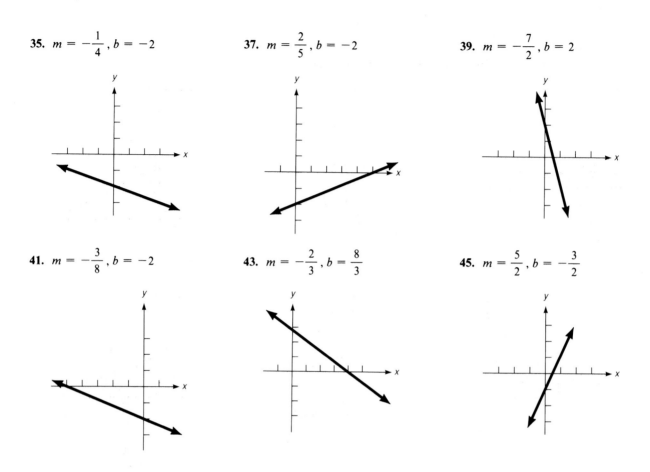

Exercises 8.4, Page 281

1. $y = 2x + 3$ or $-2x + y = 3$ **3.** $y = -\dfrac{2}{5}x + \dfrac{13}{5}$ or $2x + 5y = 13$ **5.** $y = \dfrac{3}{4}x + \dfrac{17}{4}$ or $-3x + 4y = 17$

7. $y = -1$ or $y + 1 = 0$ **9.** $x = 2$ or $x - 2 = 0$ **11.** $y = \dfrac{3}{2}x - \dfrac{5}{6}$ or $9x - 6y = 5$

13. $y = -\dfrac{1}{3}x + \dfrac{7}{3}$ or $x + 3y = 7$ **15.** $y = -\dfrac{2}{3}x + 6$ or $2x + 3y = 18$ **17.** $y = 2$ or $y - 2 = 0$

19. $y = \dfrac{1}{3}x + \dfrac{11}{3}$ or $x - 3y = -11$ **21.** $x = -2$ or $x + 2 = 0$ **23.** $y = -\dfrac{3}{10}x + \dfrac{4}{5}$ or $3x + 10y = 8$

25. $y = 5$ **27.** $x = -1$ **29.** $3x - y = 0$ **31.** $y = 1$ **33.** $2x - y = 2$ **35.** $c = 0.15x + 30$

37. $s = 0.09x + 200$ **39.** $P = -\dfrac{1}{200}x + 12$

Exercises 8.5, Page 286

1. (a) No (b) No (c) Yes (d) No **3.** (a) Yes (b) No (c) Yes (d) No

5. Consistent **7.** Inconsistent **9.** Consistent

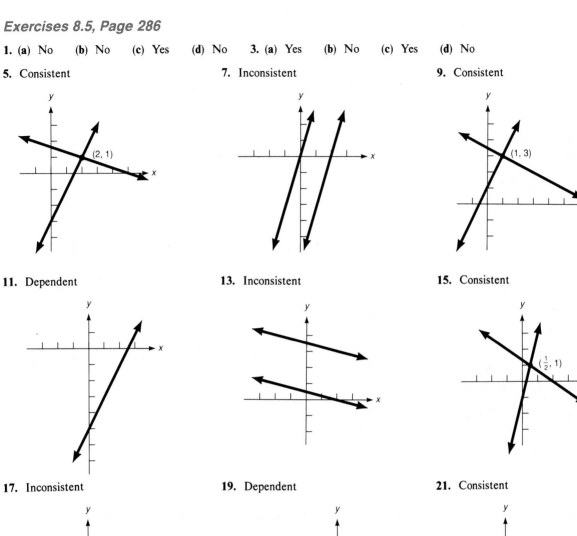

11. Dependent **13.** Inconsistent **15.** Consistent

17. Inconsistent **19.** Dependent **21.** Consistent

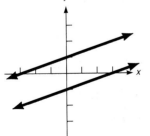

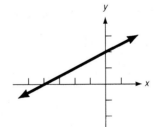

23. Consistent

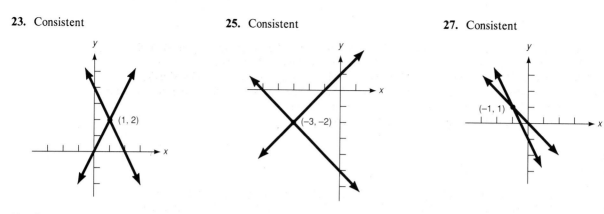

(1, 2)

25. Consistent

(−3, −2)

27. Consistent

(−1, 1)

29. Consistent

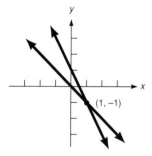

(1, −1)

Exercises 8.6, Page 292

1. (2,4) **3.** (1,−2) **5.** (−6,−2) **7.** (4,1) **9.** No solution, inconsistent **11.** (3,2) **13.** Dependent

15. (2,2) **17.** $\left(\frac{1}{2}, -4\right)$ **19.** $\left(2, \frac{5}{2}\right)$ **21.** $\left(-2, \frac{1}{2}\right)$ **23.** (2,−10) **25.** (−5,2) **27.** (4,−1)

29. $\left(\frac{13}{5}, -\frac{39}{5}\right)$ **31.** (10,4) **33.** (9,−3) **35.** (6,−4) **37.** 26, 14 **39.** $5500 @ 6%; $3500 @ 10%

41. $7400 @ 5.5%, $2600 @ 6% **43.** 11 m by 18 m **45.** 40 m by 80 m

Exercises 8.7, Page 298

1. (3,−1) **3.** $\left(1, -\frac{3}{2}\right)$ **5.** Inconsistent **7.** (1,−5) **9.** Dependent **11.** (4,3) **13.** (−3,−1)

15. (2,−1) **17.** (−2,−3) **19.** (5,−6) **21.** (10,2) **23.** (7,0) **25.** (5,6) **27.** (−3,7)

29. (−4,−2) **31.** (6,4) **33.** (4,7) **35.** $y = 5x - 7$ **37.** $y = \frac{1}{9}x + \frac{13}{9}$ **39.** $y = \frac{11}{4}x - \frac{49}{4}$

41. 20 lb—20%; 30 lb—70% **43.** 20 oz—30%; 30 oz—20% **45.** 10 mph; 2 mph

47. 2 hr @ 56 mph; $1\frac{1}{2}$ hr @ 52 mph **49.** 50 mph **51.** (3,−1) **53.** (1.2,1.6) **55.** (4.50,0.84)

Exercises 8.8, Page 301

1. 20 nickels, 10 dimes   **3.** Curt—4 years; Jill—12 years   **5.** 11 pairs @ $28; 5 pairs @ $35
7. 18 jackets @ $120; 22 jackets @ $95   **9.** 140 lb of newspapers; 40 lb of cans
11. 450 lb @ 35%; 1350 lb @ 15%   **13.** $14, $27   **15.** 7 model X radios, 10 model Y radios
17. Method I: 12 chairs; Method II: 16 chairs   **19.** 12 brand X sound systems; 0 brand Y sound systems
21. $8   **23.** (a) $C = 400 + 0.15x; R = 0.95x$ (b) 500 mousetraps
25. (a) $C = 520 + 0.40x; R = 0.80x$ (b) 1300 hot dogs

Chapter 8 Review, Page 304

1. $(-3,1), (-2,0), (-1,1), (1,1), (3,3)$ **2.** $(-3,1), (-1,3), (-1,-1), (1,1), (4,3), (4,-2)$
3. $(-2,-1), (0,0), (2,1), (2,2), (3,3), (4,4)$ **4.** $(-3,-1), (-1,-1), (0,-1), (1,1), (2,1), (3,1)$
5. $(-5,4), (-1,1), \left(0,-\dfrac{9}{2}\right), (1,3), (4,0)$ **6.**

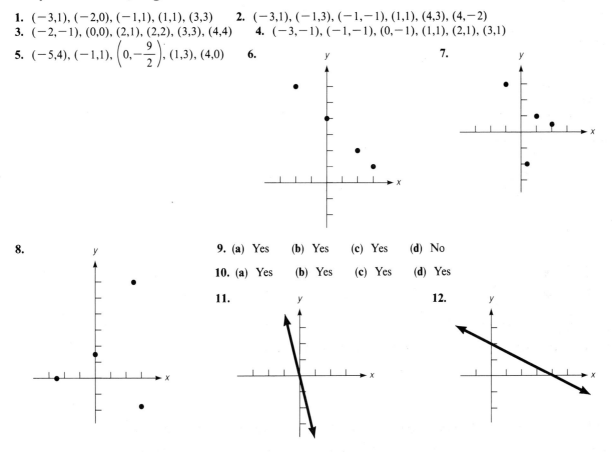

7.

8.

9. (a) Yes   (b) Yes   (c) Yes   (d) No
10. (a) Yes   (b) Yes   (c) Yes   (d) Yes

11.

12.

13.

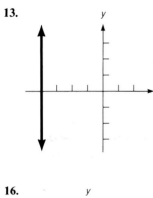

14.

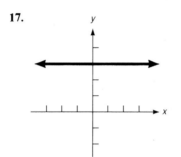

15.

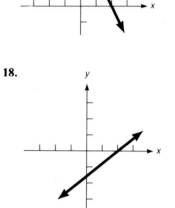

16.

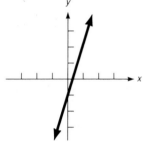

17.

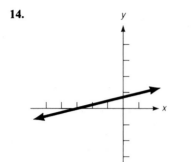

18.

19. $m = 2, b = 3$

20. $m = -\dfrac{2}{5}, b = 2$

21. Slope undefined, no *y*-intercept

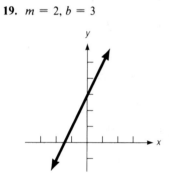

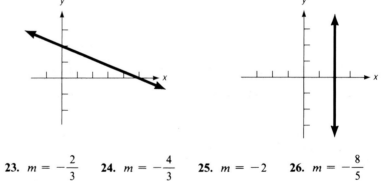

22. $m = \dfrac{4}{3}, b = -\dfrac{7}{3}$

23. $m = -\dfrac{2}{3}$ **24.** $m = -\dfrac{4}{3}$ **25.** $m = -2$ **26.** $m = -\dfrac{8}{5}$

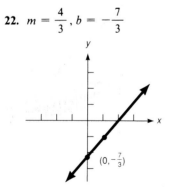

$\left(0, -\dfrac{7}{3}\right)$

27. $y = -\dfrac{5}{2}x + \dfrac{29}{2}$ or $5x + 2y = 29$ **28.** $y = 2$ **29.** $y = -1$

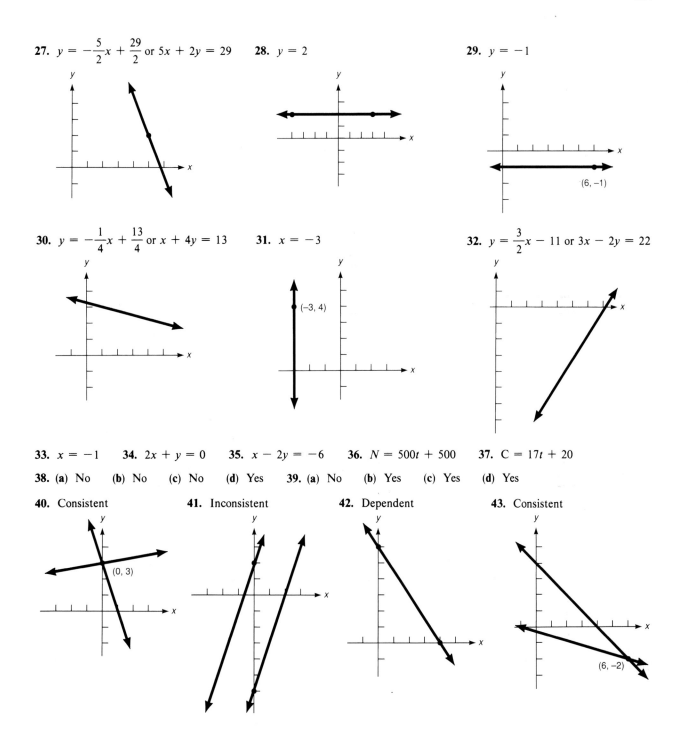

30. $y = -\dfrac{1}{4}x + \dfrac{13}{4}$ or $x + 4y = 13$ **31.** $x = -3$ **32.** $y = \dfrac{3}{2}x - 11$ or $3x - 2y = 22$

33. $x = -1$ **34.** $2x + y = 0$ **35.** $x - 2y = -6$ **36.** $N = 500t + 500$ **37.** $C = 17t + 20$

38. (a) No   (b) No   (c) No   (d) Yes   **39.** (a) No   (b) Yes   (c) Yes   (d) Yes

40. Consistent   **41.** Inconsistent   **42.** Dependent   **43.** Consistent

44. $(-6,2)$ **45.** Inconsistent **46.** Dependent **47.** $(2,-4)$ **48.** $(0,0)$

49. $\left(\dfrac{1}{3},-1\right)$ **50.** $(2,3)$ **51.** $(-1,-6)$ **52.** Inconsistent **53.** Dependent **54.** $(-2,-1)$ **55.** $(1,1)$

56. $(1,2)$ **57.** $\left(\dfrac{54}{11},-\dfrac{5}{11}\right)$ **58.** Dependent **59.** $\left(-\dfrac{1}{3},\dfrac{1}{2}\right)$ **60.** $\left(-\dfrac{21}{10},-\dfrac{23}{10}\right)$ **61.** $\left(\dfrac{4}{5},\dfrac{9}{5}\right)$

62. $y = -7x + 20$ **63.** $y = -\dfrac{4}{3}x + \dfrac{7}{3}$ **64.** 45 mph; 40 mph **65.** 440 lb of 70%; 200 lb of 54%

66. 30 members **67.** Boat—14 mph; current—2 mph **68.** 11 m by 14 m
69. $C = 10x + 2400; R = 22x; x = 200$ gloves **70.** 4 of Model A; 18 of Model B

Chapter 8 Test, Page 308

1. (a) No **(b)** Yes **(c)** Yes **(d)** No **2.** $(-4,2), (-2,-1), (0,-3), (1,2), (3,4), (5,0)$

3. **4.** **5.** **6.** $m = -\dfrac{7}{4}$

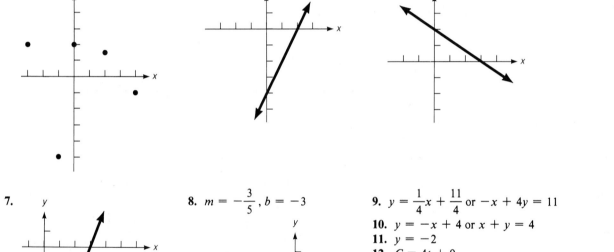

7. **8.** $m = -\dfrac{3}{5}, b = -3$ **9.** $y = \dfrac{1}{4}x + \dfrac{11}{4}$ or $-x + 4y = 11$

10. $y = -x + 4$ or $x + y = 4$
11. $y = -2$
12. $C = 4t + 9$
13. (a) No **(b)** No **(c)** Yes **(d)** No

14. Consistent

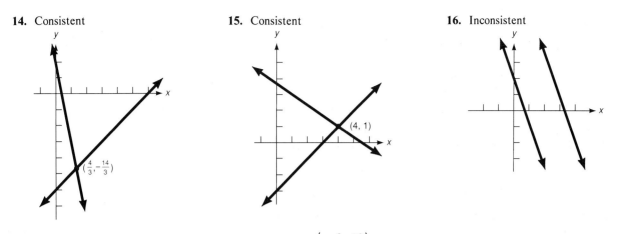

15. Consistent

16. Inconsistent

17. $(-8, -20)$ **18.** $(-2, 6)$ **19.** Dependent   **20.** $\left(-\dfrac{3}{14}, \dfrac{13}{7}\right)$ **21.** $(-3, 5)$ **22.** Inconsistent

23. $(3, -6)$ **24.** Pencils—\$0.08; pens—\$0.79   **25.** 13 in. by 17 in.

Cumulative Review (8), Page 309

1. x^3 **2.** y^6 **3.** $(2x - 1)(x + 8)$ **4.** $(4x - 3)(3x - 4)$ **5.** $\dfrac{3x + 1}{x + 1}$ **6.** $\dfrac{3x^2 + 15x + 50}{2(x - 5)^2(x + 5)}$

7. $x = -8$ **8.** $x = -6; x = \dfrac{3}{2}$ **9.** $x = -\dfrac{3}{2}$ **10.** $t = \dfrac{v - k}{g}$

11.   **12.** $m = -\dfrac{1}{2}, b = 2$ **13.** $x - 3y = -5$

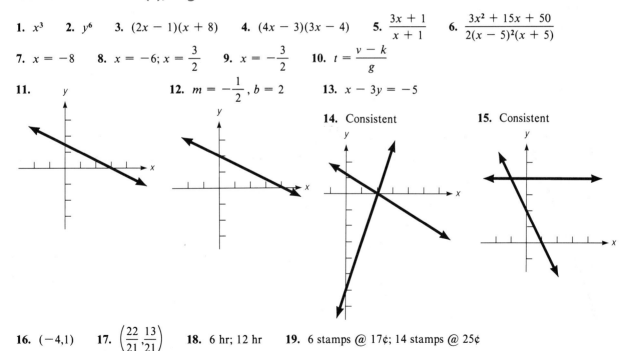

14. Consistent

15. Consistent

16. $(-4, 1)$ **17.** $\left(\dfrac{22}{21}, \dfrac{13}{21}\right)$ **18.** 6 hr; 12 hr   **19.** 6 stamps @ 17¢; 14 stamps @ 25¢

20. Boat—10 mph; stream—2 mph

CHAPTER 9

Exercises 9.1, Page 319

1. Rational   **3.** Irrational   **5.** Rational   **7.** Irrational   **9.** Rational   **11.** Rational   **13.** Irrational

15. Irrational   **17.** Rational   **19.** Rational   **21.** 6   **23.** $\dfrac{1}{2}$ **25.** -2 **27.** 1   **29.** $2\sqrt{2}$

31. $-2\sqrt{14}$ **33.** $\dfrac{\sqrt{5}}{3}$ **35.** $2\sqrt{3}$ **37.** $12\sqrt{2}$ **39.** $-6\sqrt{2}$ **41.** $-5\sqrt{5}$ **43.** $-4\sqrt[3]{2}$

45. $3\sqrt[3]{5}$ **47.** $-2\sqrt[3]{9}$ **49.** $\dfrac{-\sqrt{11}}{8}$ **51.** $\dfrac{2\sqrt{7}}{5}$ **53.** $\dfrac{4\sqrt{2}}{9}$ **55.** $\dfrac{\sqrt[3]{3}}{2}$ **57.** $\dfrac{5\sqrt[3]{3}}{2}$

59. $\dfrac{1-\sqrt{3}}{2}$ **61.** $\dfrac{1+\sqrt{6}}{2}$ **63.** $\dfrac{4+\sqrt{5}}{5}$ **65.** $\dfrac{5-3\sqrt{3}}{2}$ **67.** $\dfrac{8-\sqrt{15}}{6}$ **69.** $\dfrac{3-2\sqrt{3}}{7}$

71. 8.062   **73.** 4.610   **75.** 3.009   **77.** 3.359   **79.** 3.053

Exercises 9.2, Page 322

1. $8\sqrt{2}$ **3.** $7\sqrt{5}$ **5.** $-3\sqrt{10}$ **7.** $13\sqrt[3]{3}$ **9.** $-\sqrt{11}$ **11.** $3\sqrt{a}$ **13.** $7\sqrt{x}$ **15.** $4\sqrt{2}+3\sqrt{3}$

17. $8\sqrt{b}-4\sqrt{a}$ **19.** $13\sqrt[3]{x}-2\sqrt[3]{y}$ **21.** $5\sqrt{3}$ **23.** 0   **25.** $17\sqrt[3]{2}$ **27.** $8\sqrt{2}-6\sqrt{3}$ **29.** $\sqrt{5}-6$

31. $\sqrt{3}-4\sqrt{2}$ **33.** $5\sqrt[3]{2}-8\sqrt[3]{3}$ **35.** $4\sqrt{2x}$ **37.** $2y\sqrt{2y}$ **39.** $-4x\sqrt{3xy}$ **41.** $3\sqrt{2}-8$ **43.** 18

45. $-8\sqrt{3}$ **47.** $\sqrt{6}+12$ **49.** $\sqrt{xy}+2y$ **51.** $13-2\sqrt{2}$ **53.** $8-7\sqrt{6}$ **55.** $x-9$ **57.** -5

59. $x+2\sqrt{x}-15$

Exercises 9.3, Page 325

1. $\dfrac{5\sqrt{2}}{2}$ **3.** $-\dfrac{3\sqrt{7}}{7}$ **5.** $2\sqrt{3}$ **7.** 3   **9.** 3   **11.** $\dfrac{1}{3}$ **13.** $\dfrac{2\sqrt{3}}{3}$ **15.** $\dfrac{3\sqrt{2}}{2}$ **17.** $\dfrac{\sqrt{x}}{x}$ **19.** $\dfrac{\sqrt{2xy}}{y}$

21. $\dfrac{\sqrt{2y}}{y}$ **23.** $\dfrac{3\sqrt{7}}{5}$ **25.** $\dfrac{-\sqrt{2y}}{5}$ **27.** $\sqrt{6}+2$ **29.** $\dfrac{\sqrt{5}+3}{-4}$ **31.** $\dfrac{-6(5+3\sqrt{2})}{7}$ **33.** $\dfrac{\sqrt{3}(\sqrt{2}-5)}{23}$

35. $\dfrac{\sqrt{7}(1+3\sqrt{5})}{-44}$ **37.** $\dfrac{\sqrt{3}+\sqrt{5}}{-2}$ **39.** $5(\sqrt{2}-\sqrt{3})$ **41.** $\dfrac{4(\sqrt{x}-1)}{x-1}$ **43.** $\dfrac{5(6-\sqrt{y})}{36-y}$

45. $\dfrac{8(2\sqrt{x}-3)}{4x-9}$ **47.** $\dfrac{2y\sqrt{5}+2\sqrt{3y}}{5y-3}$ **49.** $\dfrac{3(\sqrt{x}+\sqrt{y})}{x-y}$ **51.** $\dfrac{x(\sqrt{x}-2\sqrt{y})}{x-4y}$ **53.** $\dfrac{5+3\sqrt{3}}{-1}$

55. $\dfrac{11-5\sqrt{5}}{-4}=\dfrac{5\sqrt{5}-11}{4}$ **57.** $\dfrac{x+2\sqrt{x}+1}{x-1}$ **59.** $\dfrac{x\sqrt{3}+2\sqrt{3x}-y\sqrt{x}-2y}{3x-y^2}$ or $\dfrac{(\sqrt{x}+2)(\sqrt{3x}-y)}{3x-y^2}$

Exercises 9.4, Page 329

1. $x=33$ **3.** $x=5$ **5.** No solution   **7.** $x=9$ **9.** $x=-3$ **11.** No solution   **13.** $x=6$

15. $x=-1$ **17.** $x=-22$ **19.** $x=40$ **21.** $x=2$ **23.** $x=7$ **25.** $x=4$ **27.** $x=-2$

29. $x=3, x=2$ **31.** $x=2$ **33.** $x=12$ **35.** $x=-2$ **37.** $x=6$ **39.** $x=-\dfrac{1}{4}$

Chapter 9 Review, Page 331

1. Rational   **2.** Irrational   **3.** Rational   **4.** Irrational   **5.** Rational   **6.** Irrational   **7.** 13   **8.** 14

9. $\dfrac{3}{4}$ **10.** $\dfrac{15}{7}$ **11.** $-4\sqrt{3}$ **12.** $3\sqrt{6}$ **13.** $2\sqrt[3]{5}$ **14.** $-5\sqrt[3]{2}$ **15.** $\dfrac{\sqrt{7}}{14}$ **16.** $-\dfrac{\sqrt{5}}{11}$ **17.** $-\dfrac{\sqrt{5}}{4}$

18. $\dfrac{5\sqrt{3}}{8}$ **19.** $\dfrac{\sqrt[3]{3}}{5}$ **20.** $\dfrac{3\sqrt[3]{5}}{4}$ **21.** $4 + \sqrt{3}$ **22.** $\dfrac{5 - \sqrt{3}}{2}$ **23.** $\dfrac{6 + \sqrt{5}}{10}$ **24.** $\dfrac{3 + \sqrt{2}}{2}$

25. $\dfrac{\sqrt{3} - \sqrt{5}}{2}$ **26.** $\sqrt{3} + \sqrt{2}$ **27.** $16\sqrt{2}$ **28.** $5\sqrt{5}$ **29.** $2\sqrt[3]{2}$ **30.** $\sqrt{2x}$ **31.** $\sqrt{3}$

32. $9\sqrt{3} - 2\sqrt{2}$ **33.** $17 - 2\sqrt{14}$ **34.** $\sqrt[3]{3} + 5\sqrt[3]{2}$ **35.** $-5\sqrt{3y}$ **36.** 12   **37.** $20\sqrt{6}$ **38.** $8\sqrt{6}$

39. $7\sqrt{6}$ **40.** -22 **41.** $17 + 8\sqrt{2}$ **42.** $12 + 15\sqrt{3}$ **43.** $x - 2$

44. $32 - 2\sqrt{3}$ **45.** $2x - 13\sqrt{x} - 24$ **46.** $3\sqrt{3}$ **47.** $\dfrac{\sqrt{15}}{10}$ **48.** $\dfrac{\sqrt{15}}{9}$ **49.** $\dfrac{\sqrt{10}}{5}$ **50.** $\dfrac{2\sqrt{7x}}{x}$

51. $-\dfrac{3\sqrt{2y}}{2y}$ **52.** $\dfrac{2\sqrt{5y}}{5y}$ **53.** $-\dfrac{\sqrt{3} + 5}{11}$ **54.** $\dfrac{3 - \sqrt{2}}{7}$ **55.** $-2(\sqrt{6} + 3)$ **56.** $-\dfrac{\sqrt{5} + \sqrt{13}}{2}$

57. $\dfrac{\sqrt{3}(\sqrt{x} + 7)}{x - 49}$ **58.** $\dfrac{2(\sqrt{5} - \sqrt{2})}{3}$ **59.** $3(\sqrt{7} - \sqrt{3})$ **60.** $\dfrac{y - 4\sqrt{y} + 3}{y - 9}$ **61.** $x = 1$ **62.** $x = 7$

63. $x = 29$ **64.** $x = 38$ **65.** $x = 2$ **66.** $x = 4$ **67.** $x = 4$ **68.** $x = 10$ **69.** $x = 2$ **70.** $x = \dfrac{2}{3}$

Chapter 9 Test, Page 332

1. Irrational   **2.** Rational   **3.** $5\sqrt{5}$ **4.** 12   **5.** $2\sqrt[3]{9}$ **6.** $-\dfrac{1}{2}$ **7.** $-\dfrac{\sqrt{6}}{3}$ **8.** $\dfrac{2\sqrt[3]{6}}{5}$ **9.** $\dfrac{\sqrt{6} + 3}{6}$

10. $\dfrac{3 - \sqrt{2}}{2}$ **11.** $13\sqrt{2}$ **12.** $22\sqrt{3}$ **13.** $-5\sqrt{2x}$ **14.** $\sqrt[3]{5} + 3\sqrt[3]{4}$ **15.** 75   **16.** $3\sqrt{5} + 5\sqrt{3}$

17. 14   **18.** $28 - \sqrt{2}$ **19.** $\dfrac{\sqrt{2}}{2}$ **20.** $\dfrac{2\sqrt{3}}{15}$ **21.** $\sqrt{3} + 1$ **22.** $\sqrt{6} - 2$ **23.** $x = 34$ **24.** $x = 2$

25. No solution

Cumulative Review (9), Page 332

1. $x^2 + 8x + 16 = (x + 4)^2$ **2.** $x^2 - 14x + 49 = (x - 7)^2$ **3.** $\dfrac{2(x + 4)}{x - 3)}$ **4.** $\dfrac{x^2 - 4x + 5}{(x + 6)(x - 2)}$

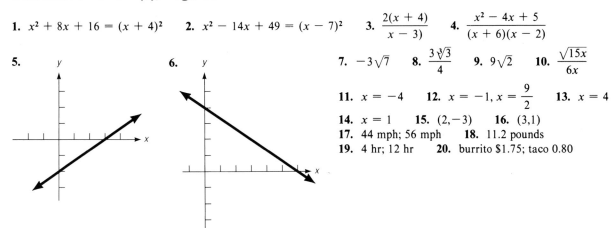

5.

6.

7. $-3\sqrt{7}$ **8.** $\dfrac{3\sqrt[3]{3}}{4}$ **9.** $9\sqrt{2}$ **10.** $\dfrac{\sqrt{15x}}{6x}$

11. $x = -4$ **12.** $x = -1, x = \dfrac{9}{2}$ **13.** $x = 4$

14. $x = 1$ **15.** $(2, -3)$ **16.** $(3, 1)$

17. 44 mph; 56 mph   **18.** 11.2 pounds

19. 4 hr; 12 hr   **20.** burrito $1.75; taco 0.80

CHAPTER 10

Exercises 10.1, Page 337

1. $x = \pm 11$ **3.** $x = \pm 6$ **5.** $x = \pm\sqrt{35}$ **7.** $x = \pm\sqrt{62}$ **9.** $x = \pm 3\sqrt{5}$ **11.** $x = \pm 3\sqrt{2}$

13. $x = \pm\dfrac{2}{3}$ **15.** $x = -1; x = 3$ **17.** No solutions   **19.** $x = -\dfrac{3}{2}; x = -\dfrac{1}{2}$ **21.** $x = \dfrac{7}{3}; x = \dfrac{11}{3}$

23. $x = 6 \pm 3\sqrt{2}$ **25.** $x = 7 \pm 2\sqrt{3}$ **27.** $x = \dfrac{-4 \pm 3\sqrt{3}}{3}$ **29.** $x = \dfrac{2 \pm 3\sqrt{7}}{5}$ **31.** Is a right triangle

33. Is a right triangle   **35.** $c = 15$ **37.** $b = 6\sqrt{3}$ **39.** 4 ft; 8 ft   **41.** $3\sqrt{2}$ cm   **43.** $x = \pm 25.44$

45. $x = \pm 5.25$ **47.** $x = \pm 1.70$ **49.** $x = \pm 4.13$

Exercises 10.2, Page 342

1. $x^2 + 12x + \underline{36} = (x + 6)^2$ **3.** $2x^2 - 16x + \underline{32} = 2(x - 4)^2$ **5.** $x^2 - 3x + \underline{\dfrac{9}{4}} = \left(x - \dfrac{3}{2}\right)^2$

7. $x^2 + x + \underline{\dfrac{1}{4}} = \left(x + \dfrac{1}{2}\right)^2$ **9.** $2x^2 + 4x + \underline{2} = 2(x + 1)^2$ **11.** $x = -7, x = 1$ **13.** $x = 9, x = -5$

15. $x = 8, x = -5$ **17.** $x = -\dfrac{4}{3}, x = 1$ **19.** $x = -\dfrac{1}{2}, x = \dfrac{3}{2}$ **21.** $x = -3 \pm \sqrt{6}$

23. $x = -1 \pm \sqrt{6}$ **25.** $x = -\dfrac{1}{3}, x = 1$ **27.** $x = -\dfrac{1}{2} \pm \dfrac{\sqrt{13}}{2} = \dfrac{-1 \pm \sqrt{13}}{2}$

29. $x = -\dfrac{3}{4} \pm \dfrac{\sqrt{17}}{4} = \dfrac{-3 \pm \sqrt{17}}{4}$ **31.** $x = \dfrac{9}{2} \pm \dfrac{\sqrt{73}}{2} = \dfrac{9 \pm \sqrt{73}}{2}$ **33.** $x = -\dfrac{7}{2} \pm \dfrac{\sqrt{105}}{2} = \dfrac{-7 \pm \sqrt{105}}{2}$

35. $x = 13, x = -2$ **37.** $x = -2, x = -\dfrac{1}{2}$ **39.** $x = 1$ **41.** $x = -\dfrac{5}{2} \pm \dfrac{\sqrt{33}}{2} = \dfrac{-5 \pm \sqrt{33}}{2}$

43. $x = \dfrac{2}{3} + \dfrac{\sqrt{10}}{6}$ or $x = \dfrac{2}{3} + \sqrt{\dfrac{5}{18}}$ **45.** $x = -\dfrac{7}{4} \pm \dfrac{\sqrt{17}}{4} = \dfrac{-7 \pm \sqrt{17}}{4}$ **47.** $x = \dfrac{1}{4} \pm \dfrac{\sqrt{13}}{4} = \dfrac{1 \pm \sqrt{13}}{4}$

49. $x = -\dfrac{5}{3}, x = -1$

Exercises 10.3, Page 347

1. $a = 1, b = -3, c = -2$ **3.** $a = 2, b = -1, c = 6$ **5.** $a = 7, b = -4, c = -3$

7. $a = 3, b = -9, c = -4$ **9.** $a = 2, b = 5, c = -3$ **11.** $a = 2, b = 3, c = -3$ **13.** $x = -2 \pm \sqrt{5}$

15. $x = -1, x = 4$ **17.** $x = -\dfrac{1}{2}, x = 1$ **19.** $x = -1, x = \dfrac{2}{5}$ **21.** $x = \dfrac{3 \pm \sqrt{33}}{12}$ **23.** $x = \pm\sqrt{7}$

25. $x = 0, x = -1$ **27.** $x = 0, x = -\dfrac{4}{3}$ **29.** $x = \dfrac{-5 \pm \sqrt{65}}{4}$ **31.** $x = \dfrac{-6 \pm \sqrt{42}}{3}$

33. $x = -3, x = -\dfrac{1}{2}$ **35.** $x = \dfrac{1 \pm \sqrt{2}}{3}$ **37.** $x = \dfrac{1 \pm \sqrt{3}}{2}$ **39.** $x = \dfrac{-2 \pm \sqrt{2}}{3}$ **41.** $x = \dfrac{7 \pm \sqrt{149}}{10}$

43. $x = \dfrac{3}{2}$ **45.** $x = -2, x = \dfrac{5}{3}$ **47.** $x = \dfrac{2 \pm \sqrt{3}}{3}$ **49.** $x = -\dfrac{1}{6}, x = 2$

Exercises 10.4, Page 351

1. 11   **3.** 7 m by 9 m   **5.** 11 m by 18 m   **7.** 20 ft by 30 ft   **9.** 10, 11   **11.** 4   **13.** 5 cm
15. 5 m by 12 m   **17.** $6\sqrt{2}$ cm   **19.** 30 members   **21.** 50 mph   **23.** 3 mph   **25.** 20 members
27. 13 in. square   **29.** 40 min

Chapter 10 Review, Page 354

1. $x = \pm 7$ **2.** $x = \pm 5$ **3.** $x = \pm\dfrac{1}{2}$ **4.** $x = \pm 3\sqrt{2}$ **5.** $x = 0, x = -4$ **6.** $x = -2, x = 4$

7. $x = 5 \pm \sqrt{7}$ **8.** $x = 4 \pm 2\sqrt{2}$ **9.** Is a right triangle   **10.** Is a right triangle   **11.** 11 ft   **12.** $2\sqrt{41}$ cm

13. $x^2 - 4x + 4 = (x - 2)^2$ **14.** $3x^2 + 18x + 27 = 3(x + 3)^2$ **15.** $x^2 - \dfrac{2}{3}x + \underline{\dfrac{1}{9}} = \left(x - \dfrac{1}{3}\right)^2$

16. $x^2 - x + \underline{\dfrac{1}{4}} = \left(x - \dfrac{1}{2}\right)^2$ **17.** $x^2 + 9x + \underline{\dfrac{81}{4}} = \left(x + \dfrac{9}{2}\right)^2$ **18.** $5x^2 - 10x + \underline{5} = 5(x - 1)^2$

19. $x = -1, x = 7$ **20.** $x = -2 \pm \sqrt{2}$ **21.** $x = -1 \pm \sqrt{5}$ **22.** $x = -2 \pm \dfrac{3\sqrt{2}}{2} = \dfrac{-4 \pm 3\sqrt{2}}{2}$

23. $x = -\dfrac{1}{2} \pm \dfrac{\sqrt{7}}{2} = \dfrac{-1 \pm \sqrt{7}}{2}$ **24.** $x = \dfrac{1}{3} \pm \dfrac{\sqrt{7}}{3} = \dfrac{1 \pm \sqrt{7}}{3}$ **25.** $x = -8, x = 2$ **26.** $x = -12, x = 4$

27. $x = -4, x = \dfrac{1}{2}$ **28.** $x = 0, x = \dfrac{6}{5}$ **29.** $x = 1 \pm \dfrac{\sqrt{10}}{5} = \dfrac{5 \pm \sqrt{10}}{5}$

30. $x = -\dfrac{1}{4} \pm \dfrac{\sqrt{13}}{4} = \dfrac{-1 \pm \sqrt{13}}{4}$ **31.** $x = -1 \pm \dfrac{2\sqrt{6}}{3} = \dfrac{-3 \pm 2\sqrt{6}}{3}$ **32.** $x = 0, x = -3$

33. $a = 2, b = 3, c = -2; x = -2, x = \dfrac{1}{2}$ **34.** $a = 3, b = 1, c = -4; x = -\dfrac{4}{3}, x = 1$

35. $a = 2, b = -2, c = -1; x = \dfrac{1 \pm \sqrt{3}}{2}$ **36.** $a = 2, b = 5, c = -6; x = \dfrac{-5 \pm \sqrt{73}}{4}$

37. $a = 5, b = 1, c = -1; x = \dfrac{-1 \pm \sqrt{21}}{10}$ **38.** $a = \dfrac{3}{4}, b = -2, c = \dfrac{1}{8}; x = \dfrac{8 \pm \sqrt{58}}{6}$ **39.** $x = \pm 6$

40. $x = 2 \pm \sqrt{17}$ **41.** $x = 3 \pm \sqrt{5}$ **42.** $x = -1, x = \dfrac{7}{5}$ **43.** $x = -5, x = 12$ **44.** $x = \dfrac{3 \pm \sqrt{73}}{8}$

45. $x = -1, x = 4$ **46.** $x = \dfrac{2 \pm \sqrt{10}}{3}$ **47.** 12, 14   **48.** 25 cm, 11 cm   **49.** 12 m, 9 m   **50.** 6 ft

51. 6 mph   **52.** 12 hr, 6 hr   **53.** $12, 25 shirts   **54.** 10 in. by 30 in.

Chapter 10 Test, Page 355

1. $x = -2, x = 12$ **2.** $x = \pm 2\sqrt{3}$ **3.** $b = 4\sqrt{6}$ **4.** $x^2 - 24x + \underline{144} = (x - 12)^2$

5. $x^2 + 9x + \underline{\dfrac{81}{4}} = \left(x + \dfrac{9}{2}\right)^2$ **6.** $3x^2 + 9x + \underline{\dfrac{27}{4}} = 3\left(x + \dfrac{3}{2}\right)^2$ **7.** $x = 1, x = 2$

8. $x = 0, x = -\dfrac{12}{5}$ **9.** $x = 2, x = -\dfrac{5}{3}$ **10.** $x = \dfrac{3}{2}, x = 4$ **11.** $x = -4, x = -2$ **12.** $x = 1 \pm \sqrt{6}$

13. $x = -\dfrac{3}{4} \pm \dfrac{\sqrt{33}}{4} = \dfrac{-3 \pm \sqrt{33}}{4}$ **14.** $x = \dfrac{-1 \pm \sqrt{13}}{6}$ **15.** $x = \dfrac{-4 \pm \sqrt{10}}{3}$ **16.** $x = \dfrac{2 \pm \sqrt{10}}{2}$

17. $x = 1 \pm \sqrt{5}$ **18.** $x = -3 \pm 2\sqrt{3}$ **19.** $x = -3 \pm \sqrt{11}$ **20.** $x = \dfrac{1}{3}, x = 2$ **21.** $x = \dfrac{3 \pm \sqrt{17}}{4}$

22. $x = -\dfrac{3}{5}, x = 1$ **23.** $x = 18\sqrt{2}$ in. **24.** 4 mph **25.** 20 in. by 23 in.

Cumulative Review (10), Page 356

1. $9x^4 y^2$ **2.** $-4x^6$ **3.** $x \le \dfrac{17}{4}$ **4.** $x \le -5$ **5.** $\dfrac{x + 4}{(x - 1)(x + 5)}$

6. $\dfrac{x^2 + 3x + 5}{(x + 3)^2}$ **7.** **8.** **9.** $3\sqrt{3} + 6\sqrt{2}$ **10.** $2 - 7\sqrt{3}$

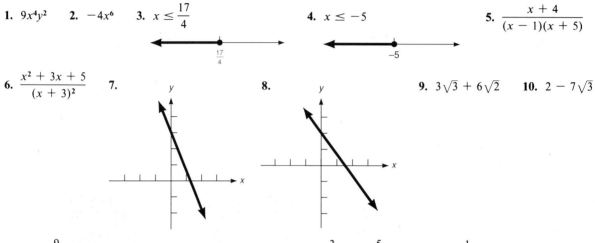

11. $x = \dfrac{9}{2}$ **12.** $x = 1, x = 12$ **13.** $x = 1$ **14.** $x = \dfrac{3}{2}, x = \dfrac{5}{3}$ **15.** $x = \dfrac{1}{2}, x = 2$ **16.** $(-2, -3)$

17. 4, 13 **18.** 3600 women, 2800 men **19.** $1500 @ 5.5%; $1200 @ 7.2% **20.** 10 mph, 15 mph

CHAPTER 11

Exercises 11.1, Page 361

1. $\sqrt{16} = 4$ **3.** $\sqrt[3]{-8} = -2$ **5.** $\sqrt[4]{81} = 3$ **7.** $\sqrt[5]{-32} = -2$ **9.** $\sqrt{0.0004} = 0.02$ **11.** $\dfrac{2}{3}$

13. $\dfrac{2}{5}$ **15.** $\dfrac{1}{6}$ **17.** $\dfrac{1}{2}$ **19.** $\dfrac{1}{3}$ **21.** 8 **23.** 9 **25.** 243 **27.** $\dfrac{1}{8}$ **29.** $\dfrac{1}{4}$ **31.** $x^{2/3}$

33. $5^{-1} = \dfrac{1}{5}$ **35.** $x^{3/5}$ **37.** $x^{1/2}$ **39.** $x^{2/5}$ **41.** $2^1 = 2$ **43.** $a^{7/6}$ **45.** $x^{5/8}$ **47.** $x^{-1/10} = \dfrac{1}{\sqrt[10]{x}}$

49. $6^{3/2}$ **51.** $x^{5/4}$ **53.** x^2 **55.** $196x^{7/6}$ **57.** $6x^2$ **59.** $\dfrac{2}{3}x$

Exercises 11.2, Page 366

1. (a) 4   **(b)** $(-3,4)$ **3. (a)** 3   **(b)** $\left(\dfrac{11}{2},-3\right)$ **5. (a)** $\dfrac{7}{2}$ **(b)** $\left(\dfrac{5}{4},1\right)$ **7. (a)** $\sqrt{2}$ **(b)** $\left(\dfrac{5}{2},\dfrac{1}{2}\right)$

9. (a) $\sqrt{13}$ **(b)** $\left(0,\dfrac{7}{2}\right)$ **11. (a)** 13   **(b)** $\left(-\dfrac{1}{2},-1\right)$ **13. (a)** $\dfrac{5}{7}$ **(b)** $\left(\dfrac{3}{14},\dfrac{2}{7}\right)$ **15. (a)** 5   **(b)** $\left(\dfrac{11}{2},3\right)$

17. (a) 13   **(b)** $\left(-4,\dfrac{1}{2}\right)$ **19. (a)** 10   **(b)** $(0,-2)$ **21. (a)** 5   **(b)** $\left(2,-\dfrac{3}{2}\right)$ **23. (a)** 2   **(b)** $\left(-\dfrac{1}{5},\dfrac{2}{7}\right)$

25. $|AB|^2 = 45$
$|BC|^2 = 20$
$|AC|^2 = 65$
yes, since $65 = 45 + 20$

27. $|AB| = \sqrt{80}$
$|BC| = \sqrt{32}$
$|AC| = \sqrt{80}$
yes, since $|AB| = |AC|$

29. $|AB| = 4$
$|BC| = 4$
$|AC| = 4$

31. $|AC| = \sqrt{61}$
$|BD| = \sqrt{61}$

33. $|AB| = \sqrt{80}$
$|BC| = 5$
$|AC| = 5$
$P = 10 + \sqrt{80}$

35. $|AB| = \sqrt{41}$
$|BC| = \sqrt{10}$
$|AC| = \sqrt{65}$
$P = \sqrt{41} + \sqrt{10} + \sqrt{65}$

37. $x = 5, y = 13$

Exercises 11.3, Page 371

1.   **3.**   **5.**   **7.**

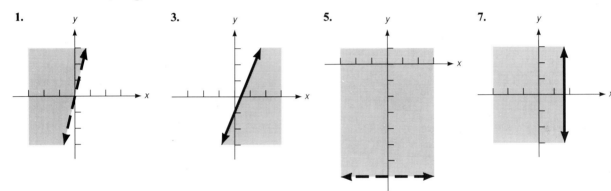

9.   **11.**   **13.**   **15.**

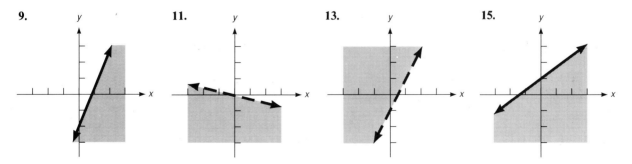

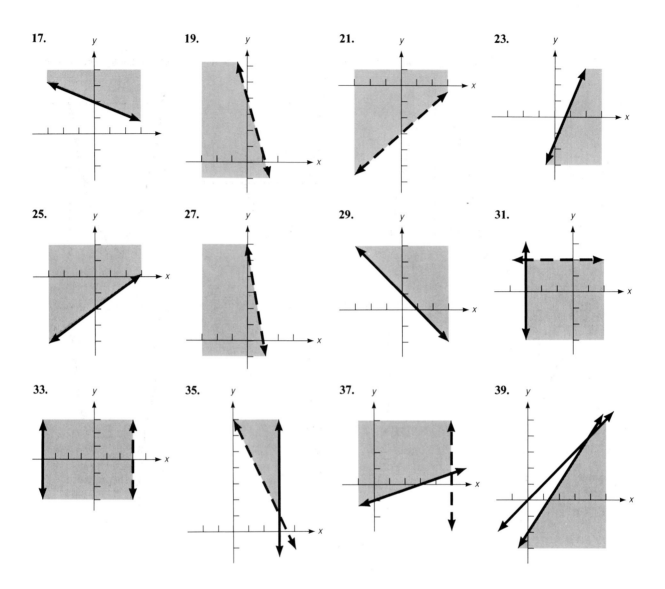

Exercises 11.4, Page 376

1. Domain = $\{0, 3, -2, 1, 2\}$; Range = $\{0, 1, 4\}$; is a function   **3.** Domain = $\{0, -1, 2, 3\}$; Range = $\{2, 1, 4, 5, -1\}$;
is not a function   **5.** Domain = $\{0, -1, 3, 2, 1\}$; Range = $\{1, 3, 4, -5, 2\}$; is a function
7. Domain = $\{0, 1, -1, \sqrt{3}, -2\}$; Range = $\{-1, 0, -2, 2, -3\}$; is a function   **9.** Domain = $\{-3, 1, 0, \sqrt{2}, 3\}$;
Range = $\{4\}$; is a function   **11.** Domain = $\{1, 2, 4, 0\}$; Range = $\{\sqrt{3}, 7, -1, 0, \sqrt{2}\}$; is not a function
13. Is a function   **15.** Is a function   **17.** Is not a function   **19.** Is not a function   **21.** Is a function
23. Is not a function   **25. (a)** 1   **(b)** 4   **(c)** 5   **(d)** 6   **(e)** 7   **27. (a)** 0   **(b)** 1   **(c)** 2   **(d)** 3   **(e)** −1
29. (a) −5   **(b)** 3   **(c)** 1   **(d)** −7   **(e)** −2   **31. (a)** −5   **(b)** 4   **(c)** −1   **(d)** 4   **(e)** −4
33. (a) 2   **(b)** 0   **(c)** 6   **(d)** 0   **(e)** 2   **35. (a)** 6   **(b)** 6   **(c)** 6   **(d)** 6
37. (a) 0   **(b)** 2   **(c)** $\sqrt{3}$ **(d)** $\dfrac{3}{2}$ **39. (a)** 2   **(b)** 3   **(c)** $\sqrt{2}$ **(d)** $\dfrac{7}{3}$

Exercises 11.5, Page 384

1. (a) (0,4) **(b)** $x = 0$ **3. (a)** (0,8) **(b)** $x = 0$ **5. (a)** (1,−4) **(b)** $x = 1$ **7. (a)** (−3,−9) **(b)** $x = −3$

9. (a) (−2,6) **(b)** $x = −2$ **11. (a)** $\left(\dfrac{5}{2}, -\dfrac{19}{2}\right)$ **(b)** $x = \dfrac{5}{2}$ **13. (a)** $\left(\dfrac{3}{2}, -\dfrac{13}{4}\right)$ **(b)** $x = \dfrac{3}{2}$

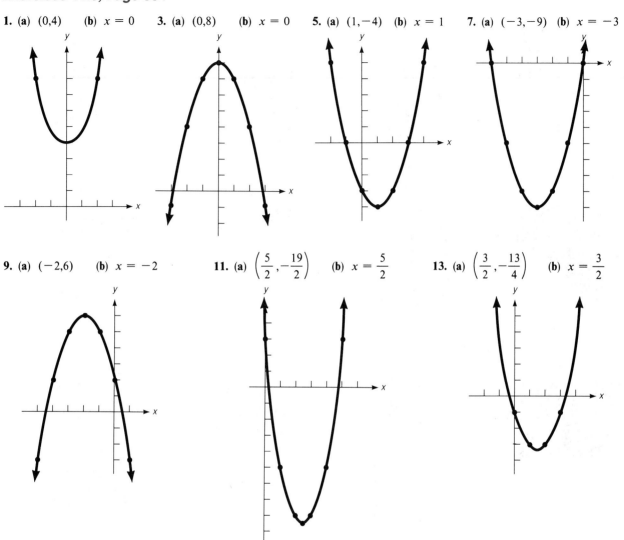

15. (a) $\left(\dfrac{1}{2}, -\dfrac{11}{4}\right)$ **(b)** $x = \dfrac{1}{2}$ **17. (a)** $\left(-\dfrac{7}{4}, -\dfrac{81}{8}\right)$ **(b)** $x = -\dfrac{7}{4}$ **19. (a)** $\left(-\dfrac{5}{6}, -\dfrac{1}{12}\right)$ **(b)** $x = -\dfrac{5}{6}$

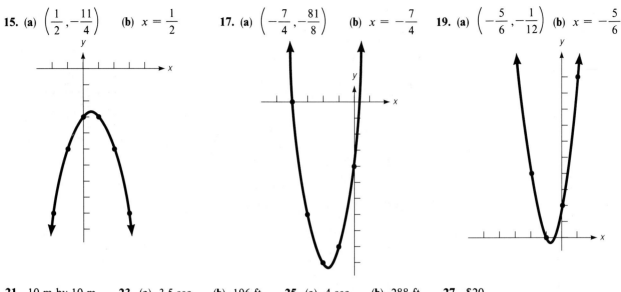

21. 10 m by 10 m   **23. (a)** 3.5 sec   **(b)** 196 ft   **25. (a)** 4 sec   **(b)** 288 ft   **27.** $20

29. (a) $20   **(b)** $800

Chapter 11 Review, Page 386

1. $\sqrt{9} = 3$ **2.** $\sqrt[3]{-0.027} = -0.3$ **3.** $\left(\sqrt[4]{\dfrac{16}{81}}\right)^3 = \dfrac{8}{27}$ **4.** $(\sqrt{36})^{-3} = \dfrac{1}{216}$ **5.** $x^{3/5}$ **6.** $x^{3/2}$ **7.** $y^{4/3}$

8. $y^{1/4}$ **9.** $a^{-1/6} = \dfrac{1}{a^{1/6}}$ **10.** a **11.** $x^{3/10}$ **12.** $x^{7/6}$ **13.** $30x^{5/6}$ **14.** $18x^{19/12}$ **15.** $4x^2$

16. $\dfrac{1}{2}x^{-1/6}$ **17. (a)** $2\sqrt{17}$ **(b)** $(0,6)$ **18. (a)** $2\sqrt{5}$ **(b)** $(4,3)$ **19. (a)** $\sqrt{10}$ **(b)** $\left(-\dfrac{9}{2}, -\dfrac{1}{2}\right)$

20. (a) $2\sqrt{2}$ **(b)** $\left(-\dfrac{1}{2},3\right)$ **21. (a)** 2   **(b)** $\left(-\dfrac{1}{2},4\right)$ **22. (a)** 1   **(b)** $\left(\dfrac{5}{6},2\right)$

23. (a) $\sqrt{10}$ **(b)** $\left(-\dfrac{7}{2}, -\dfrac{1}{2}\right)$ **24.** $\sqrt{61}, \left(-\dfrac{7}{2},5\right)$

25.   **26.**   **27.**

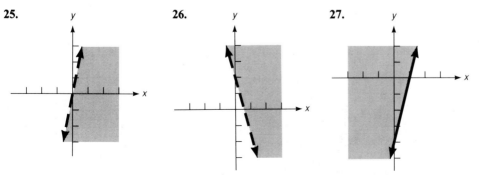

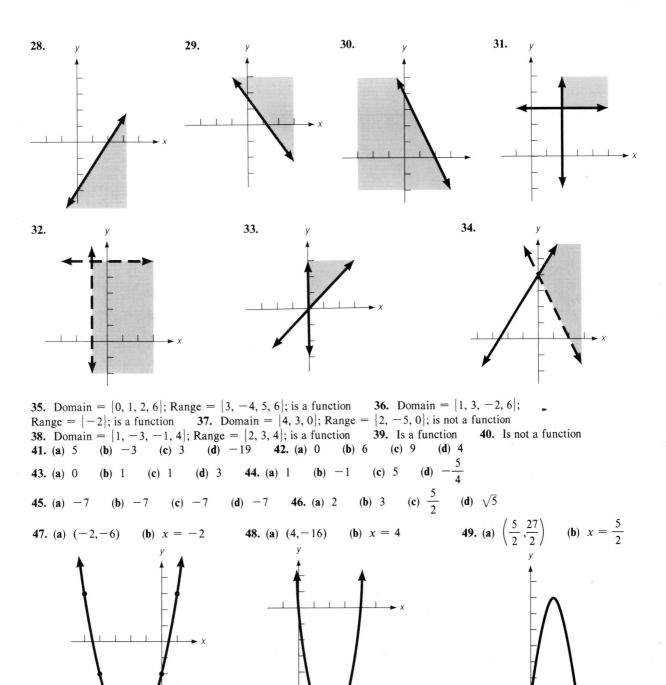

28. **29.** **30.** **31.**

32. **33.** **34.**

35. Domain $= \{0, 1, 2, 6\}$; Range $= \{3, -4, 5, 6\}$; is a function   **36.** Domain $= \{1, 3, -2, 6\}$;
Range $= \{-2\}$; is a function   **37.** Domain $= \{4, 3, 0\}$; Range $= \{2, -5, 0\}$; is not a function
38. Domain $= \{1, -3, -1, 4\}$; Range $= \{2, 3, 4\}$; is a function   **39.** Is a function   **40.** Is not a function
41. **(a)** 5   **(b)** -3 **(c)** 3   **(d)** -19 **42.** **(a)** 0   **(b)** 6   **(c)** 9   **(d)** 4

43. **(a)** 0   **(b)** 1   **(c)** 1   **(d)** 3   **44.** **(a)** 1   **(b)** -1 **(c)** 5   **(d)** $-\dfrac{5}{4}$

45. **(a)** -7 **(b)** -7 **(c)** -7 **(d)** -7 **46.** **(a)** 2   **(b)** 3   **(c)** $\dfrac{5}{2}$ **(d)** $\sqrt{5}$

47. **(a)** $(-2, -6)$ **(b)** $x = -2$ **48.** **(a)** $(4, -16)$ **(b)** $x = 4$ **49.** **(a)** $\left(\dfrac{5}{2}, \dfrac{27}{2}\right)$ **(b)** $x = \dfrac{5}{2}$

50. (a) $(-3,11)$ **(b)** $x = -3$ **51. (a)** $\left(\dfrac{3}{2}, \dfrac{19}{4}\right)$ **(b)** $x = \dfrac{3}{2}$ **52. (a)** $\left(\dfrac{5}{4}, -\dfrac{17}{8}\right)$ **(b)** $x = \dfrac{5}{4}$

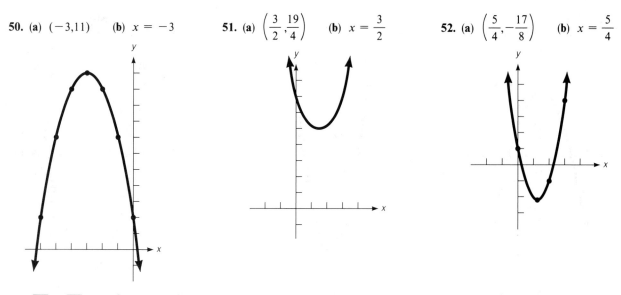

53. $\overline{AB} = \overline{BC} = 3\sqrt{2}$ **54.** $\overline{AC} = \overline{BC} = 2\sqrt{10}$ **55.** $7 + \sqrt{41}$ **56.** $5 + \sqrt{34} + \sqrt{5}$ **57.** 324 ft

58. $3200   **59.** 35¢   **60.** 256 cm²

Chapter 11 Test, Page 388

1. 8   **2.** $\dfrac{27}{8}$ **3.** $a^{1/4}$ **4.** $y^{11/10}$ **5.** $36x^3$ **6.** $4x^2$ **7. (a)** $8\sqrt{2}$ **(b)** $(-1,3)$

8. (a) $\sqrt{157}$ **(b)** $\left(-1, -\dfrac{1}{2}\right)$

9.   **10.**   **11.**   **12.**

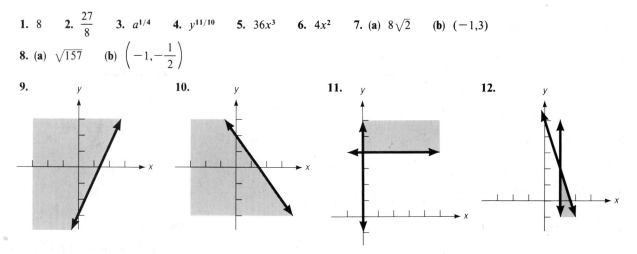

13. Domain $= \{1, -2, 3\}$; Range $= \{7, 5, 3, 1\}$; is not a function

14. Domain $= \left\{3, 4, 1, \dfrac{2}{3}\right\}$; Range $= \left\{\dfrac{1}{2}, 1, 4, 6\right\}$; is a function   **15.** Is a function

16. (a) 7   **(b)** 5   **(c)** 11   **(d)** 4   **17. (a)** 11   **(b)** 11   **(c)** 11   **(d)** 11

18. (a) -2 **(b)** 0   **(c)** 0   **(d)** $-\dfrac{5}{4}$ **19. (a)** 1   **(b)** 4   **(c)** 0   **(d)** $\sqrt{7}$

20. (a) $(0,4)$ **(b)** $x = 0$ **21. (a)** $(-1,-5)$ **(b)** $x = -1$

20. (a) (0,4)   **(b)** $x = 0$ **21. (a)** $(-1,-5)$ **(b)** $x = -1$ **22. (a)** $\left(-\dfrac{5}{2}, -\dfrac{29}{4}\right)$ **(b)** $x = -\dfrac{5}{2}$

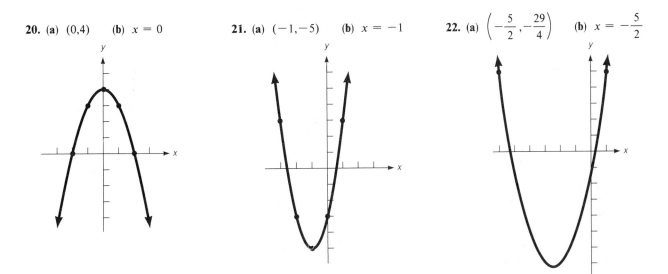

23. $18 + 2\sqrt{17}$ **24.** 6 seconds   **25.** 19 ft by 19 ft

Cumulative Review (11), Page 389

1. $x - 3$ **2.** $8x + 3$ **3.** $x = 11$ **4.** $x = -24$ **5.** $x = 5; x = 3$ **6.** $x = -9, x = 4$ **7.** $x = \dfrac{5}{3}$

8. $x = 0, x = 4$ **9.** $x = 7, x = 1$ **10.** $x = -3 \pm \sqrt{6}$ **11.** $x = -2 \pm \sqrt{6}$ **12.** $x = \dfrac{3 \pm \sqrt{41}}{4}$

13. $x = 7$ **14.** $x = -3; x = 5$ **15.** $x \le 3$ **16.** $x \ge -13$ **17.** $\left(\dfrac{32}{13}, \dfrac{17}{13}\right)$

18. $(1, -5)$ **19.** $b = \dfrac{p - a}{2}$ **20.** $a = \dfrac{2A}{h} - b$ or $a = \dfrac{2A - hb}{h}$ **21.** $2x^2$ **22.** $8x^4$ **23.** $\dfrac{n^4}{m^2}$ **24.** $\dfrac{49}{9}x^2 y^4$

25. $(x - 6)^2$ **26.** $(5x + 7)(5x - 7)$ **27.** $(3x + 1)(x + 3)$ **28.** $3(x + 6)(x - 4)$

29. $(5x + 2)(3x - 4)$ **30.** $(x - 8)(x^2 + 4)$ **31.** $7x^2 - 14$ **32.** $-3x^2 - 4x - 1$ **33.** $\dfrac{x}{x + 2}$

34. $\dfrac{3(x - 1)}{2x(x + 1)}$ **35.** $\dfrac{-x^2 + 6x + 5}{(x + 5)(x - 5)}$ **36.** $\dfrac{5x^2 + 2x + 8}{(x - 4)(x + 1)(x + 2)}$ **37.** $3\sqrt{7}$ **38.** $9\sqrt{3}$ **39.** $5\sqrt[3]{2}$

40. $\dfrac{5\sqrt{3}}{6}$ **41.**   **42.**   **43.**

44.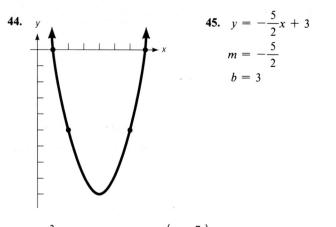

45. $y = -\dfrac{5}{2}x + 3$

$m = -\dfrac{5}{2}$

$b = 3$

46. (a) $\dfrac{3}{8}$ **(b)** $\sqrt{73}$ **(c)** $\left(-1, \dfrac{7}{2}\right)$ **(d)** $3x - 8y = -31$ or $y = \dfrac{3}{8}x - \dfrac{31}{8}$ **47.** 48 mph; 56 mph

48. $1300 @ 8%; $2100 @ 5%   **49.** $4\dfrac{4}{9}$ hr   **50.** 2 mph   **51.** 20 liters   **52.** 8 in by 15 in.

53. 4 pairs @ $3.00, 3 pairs @ $2.50   **54.** 5 cups of X, 2 cups of Y

POWERS, ROOTS, AND PRIME FACTORIZATIONS

No.	Square	Square Root	Cube	Cube Root	Prime Factorization
1	1	1.0000	1	1.0000	
2	4	1.4142	8	1.2599	prime
3	9	1.7321	27	1.4423	prime
4	16	2.0000	64	1.5874	2 · 2
5	25	2.2361	125	1.7100	prime
6	36	2.4495	216	1.8171	2 · 3
7	49	2.6458	343	1.9129	prime
8	64	2.8284	512	2.0000	2 · 2 · 2
9	81	3.0000	729	2.0801	3 · 3
10	100	3.1623	1000	2.1544	2 · 5
11	121	3.3166	1331	2.2240	prime
12	144	3.4641	1728	2.2894	2 · 2 · 3
13	169	3.6056	2197	2.3513	prime
14	196	3.7417	2744	2.4101	2 · 7
15	225	3.8730	3375	2.4662	3 · 5
16	256	4.0000	4096	2.5198	2 · 2 · 2 · 2
17	289	4.1231	4913	2.5713	prime
18	324	4.2426	5832	2.6207	2 · 3 · 3
19	361	4.3589	6859	2.6684	prime
20	400	4.4721	8000	2.7144	2 · 2 · 5
21	441	4.5826	9261	2.7589	3 · 7
22	484	4.6904	10,648	2.8020	2 · 11
23	529	4.7958	12,167	2.8439	prime
24	576	4.8990	13,824	2.8845	2 · 2 · 2 · 3
25	625	5.0000	15,625	2.9240	5 · 5
26	676	5.0990	17,576	2.9625	2 · 13
27	729	5.1962	19,683	3.0000	3 · 3 · 3
28	784	5.2915	21,952	3.0366	2 · 2 · 7
29	841	5.3852	24,389	3.0723	prime
30	900	5.4772	27,000	3.1072	2 · 3 · 5
31	961	5.5678	29,791	3.1414	prime
32	1024	5.6569	32,768	3.1748	2 · 2 · 2 · 2 · 2
33	1089	5.7446	35,937	3.2075	3 · 11
34	1156	5.8310	39,304	3.2396	2 · 17
35	1225	5.9161	42,875	3.2711	5 · 7
36	1296	6.0000	46,656	3.3019	2 · 2 · 3 · 3
37	1369	6.0828	50,653	3.3322	prime
38	1444	6.1644	54,872	3.3620	2 · 19
39	1521	6.2450	59,319	3.3912	3 · 13
40	1600	6.3246	64,000	3.4200	2 · 2 · 2 · 5
41	1681	6.4031	68,921	3.4482	prime
42	1764	6.4807	74,088	3.4760	2 · 3 · 7
43	1849	6.5574	79,507	3.5034	prime
44	1936	6.6333	85,184	3.5303	2 · 2 · 11
45	2025	6.7082	91,125	3.5569	3 · 3 · 5
46	2116	6.7823	97,336	3.5830	2 · 23
47	2209	6.8557	103,823	3.6088	prime
48	2304	6.9282	110,592	3.6342	2 · 2 · 2 · 2 · 3
49	2401	7.0000	117,649	3.6593	7 · 7
50	2500	7.0711	125,000	3.6840	2 · 5 · 5